Statistical and Machine-Learning Data Mining

Techniques for Better Predictive Modeling and Analysis of Big Data

Third Edition

Statistical and Machine-Learning Data Mining

Techniques for Better Predictive Modeling and Analysis of Big Data

Third Edition

Bruce Ratner

CRC Press is an imprint of the
Taylor & Francis Group, an **informa** business
A CHAPMAN & HALL BOOK

CRC Press
Taylor & Francis Group
6000 Broken Sound Parkway NW, Suite 300
Boca Raton, FL 33487-2742

First issued in paperback 2020

ISBN 13: 978-0-367-57360-7 (pbk)
ISBN 13: 978-1-4987-9760-3 (hbk)

Library of Congress Cataloging-in-Publication Data

Names: Ratner, Bruce, author.
Title: Statistical and machine-learning data mining / Bruce Ratner.
Description: Third Edition. | Boca Raton, FL : CRC Press, 2017. | Revised edition of the author's Statistical and machine-learning data mining, c2003.
Identifiers: LCCN 2016048787 | ISBN 9781498797603 (978-1-4987-9760-3)
Subjects: LCSH: Database marketing--Statistical methods. | Data mining--Statistical methods. | Big data--Statistical methods.
Classification: LCC HF5415.126 .R38 2017 | DDC 658.8/72--dc23 LC record available at https://lccn.loc.gov/2016048787

Visit the Taylor & Francis Web site at
http://www.taylorandfrancis.com

and the CRC Press Web site at
http://www.crcpress.com

This book is dedicated to:

My father, Isaac—always encouraging, my role model who taught me by doing, not saying, other than to think positive.

My mother, Leah—always nurturing, my friend who taught me to love love and hate hate.

My daughter, Amanda—always for me, my crowning and most significant result.

Contents

Preface to Third Edition

Predictive analytics of big data has maintained a steady presence in the four years since the publication of the second edition. My decision to write this third edition is not a result of the success (units) of the second edition but is due to the countless positive feedback (personal correspondence from the readership) I have received. And, importantly, I have the need to share my work on problems that do not have widely accepted, reliable, or known solutions. As in the previous editions, John Tukey's tenets, necessary to advance statistics, flexibility, practicality, innovation, and universality, are the touchstones of each chapter's new analytic and modeling methodology.

My main objectives in preparing the third edition are to:

1. Extend the content of the core material by including strategies and methods for problems, which I have observed on the top of *statistics on the table* [1] by reviewing predictive analytics conference proceedings and statistical modeling workshop outlines.
2. Reedit current chapters for improved writing and tighter endings.
3. Provide the statistical subroutines used in the proposed methods of analysis and modeling. I use Base SAS© and STAT/SAS. The subroutines are also available for downloading from my website: http://www.geniq.net/articles.html#section9. The code is easy to convert for users who prefer other languages.

I have added 13 new chapters that are inserted between the chapters of the second edition to yield the greatest flow of continuity of material. I outline the new chapters briefly here.

The first new chapter, Chapter 2, follows Chapter 1 (Introduction). The chapter is entitled *Science Dealing with Data: Statistics and Data Science.* If one were not looking, then it would appear that someone hit the delete key on statistics and statisticians and replaced them by science and data scientists. I investigate whether the recently minted term data science implies statistics is a subset of a more developed and expanded domain or if data science a buzzed-up cloaking of the current state of statistics.

Chapter 8, *Market Share Estimation: Data Mining for an Exceptional Case,* follows the Chapter 7 about principal component analysis (PCA). In this chapter, a market share estimation model, unique in that it does not fit the usual survey-based market share scenario, uses PCA as the foundation for estimating market share for a real exceptional case study. I provide the SAS subroutines used in building the market share model for the exceptional case study.

Chapter 11, *Predicting Share of Wallet without Survey Data,* follows the chapter on logistic regression. The everyday approach for predicting share of wallet (SOW) is with survey data. This approach is always met with reluctance because survey work is time-consuming, expensive, and yields unreliable data. I provide a two-step method for predicting SOW without data, by defining a quasi-SOW and using simulation for estimating total dollars spent. The second step uses fractional logistic regression to predict SOW_q. Fractional response regression cleverly uses ordinary logistic regression for dependent variables that assume proportions or rates. I present a case study in detail as well as provide SAS subroutines that readers should find valuable for their toolkits.

Chapter 19, *Market Segmentation Based on Time-Series Data Using Latent Class Analysis,* follows the chapter on market segmentation via logistic regression. In this chapter, I propose the model-based clustering method of latent class analysis (LCA). The innovative strategy of this segmentation is in my use of time-series data. The times-series LCA model is radically distinctive and not equal, and it will prove to be a template for treating times-series data in cross-sectional datasets and the application of LCA instead of the popular data-based heuristic k-means. I provide SAS subroutines so that data miners can perform similar segmentations as presented, along with a unique way of incorporating time-series data in an otherwise cross-sectional dataset.

Chapter 20, *Market Segmentation: An Easy Way to Understand the Segments,* is well-placed after the LCA-based market segmentation. The literature is replete with clustering methodologies, of which any one can serve for conducting a market segmentation. In contrast, the literature is virtually sparse in the area of how to interpret the segmentation results. This chapter provides an easy way to understand the discovered customer segments. I illustrate the new method with an admittedly simple example that will not belie the power of the approach. I provide SAS subroutines for conducting the proposed technique so data miners can add this worthy statistical technique to their toolkits.

Chapter 21, *The Statistical Regression Model: An Easy Way to Understand the Model,* is an extension of the method of understanding a market segmentation presented in Chapter 20. Its purpose is to provide an easy way to understand the statistical regression model, that is, ordinary least squares and logistic regression (LR) models. I illustrate the proposed method with an LR model. The illustration brings out the power of the method, in that it imparts supplementary information, making up for a deficiency in the ever-relied-upon regression coefficient for understanding a statistical regression model. I provide the SAS subroutines, which serve as a valued addition to any bag of statistical methods.

Chapter 23, *Model Building with Big Complete and Incomplete Data,* follows the chapter that uses CHAD as a method for imputation. This chapter overhears missing data warning the statistician, "You can't win unless you learn how to accept me." Traditional data-based methods (complete case analysis), predating big data, are known to be problematic with virtually all datasets. These methods now open a greater concern as to their unknown ineffectiveness on big data. I propose a two-stage approach, in which modeling of response on complete-case data precedes modeling of response on incomplete-case data via PCA. The two models can be used separately or combined, depending on the goals of the task. I provide SAS subroutines for the proposed method, which should become a utile technique for the statistical model builder.

Chapter 24, *Art, Science, Numbers, and Poetry,* is a high-order blend of artwork, science, numbers, and poetry, all inspired by the Egyptian pyramids, da Vinci, and Einstein. Love it or hate it, this chapter makes you think.

Chapter 27, *Decile Analysis: Perspective and Performance,* complements the preceding chapter on assessment of marketing models. Marketers use decile analysis to assess predictive incremental gains of their response models over responses obtained by chance. I define two new metrics, response model decile-analysis precision and chance model decile precision, which allow marketers to make a more insightful assessment as to the incremental gain of a response model over the chance model. I provide the SAS subroutines for constructing the two new metrics and the proposed procedure, which will be a trusty tool for marketing statisticians.

Chapter 28, *Net T-C Lift Model: Assessing the Net Effects of Test and Control Campaigns,* extends the practice of assessing response models to the proper use of a control group (found in the literature under names such as the uplift or net lift model) instead of the chance model as discussed in Chapter 27. There is large literature, albeit confusing and conflicting, on the methodologies of net lift modeling. I propose another approach, the *net T-C lift* model, to moderate the incompatible literature on this topic by offering a simple, straightforward, reliable model that is easy to implement and understand. I provide the SAS subroutines for the Net T-C Lift Model to enable statisticians to conduct net lift modeling without purchasing proprietary software.

Chapter 34, *Opening the Dataset: A Twelve-Step Program for Dataholics,* has valuable content for statisticians as they embark on the first step of any journey with data. Set in prose, I provide a light reading on the expectant steps of what to do when cracking open the dataset. Enjoy. I provide SAS subroutines of the twelve-step program in case the reader wants to take a nip.

Chapter 43, *Text Mining: Primer, Illustration, and TXTDM Software,* has three objectives: First, to serve as a primer, readable, brief though detailed, about what text mining encompasses, and how to conduct basic text mining; second, to illustrate text mining with a small body of text, yet interesting in its content; and third, to make text mining available to interested readers, by providing my SAS subroutines, named *TXTDM.*

Chapter 44, *Some of My Favorite Statistical Subroutines,* includes specific subroutines referenced throughout the book and generic subroutines for some second-edition chapters for which I no longer have the data. Lastly, I provide some of my favorite statistical subroutines, helpful in almost all analyses.

If there are any corrections post-production of the text, I will post them to the errata link http://www.geniq.net/articles.html#section9.

Reference

1. Stigler, S. M., *Statistics on the Table,* Harvard University Press, Cambridge, MA, 2002.

Preface of Second Edition

This book is unique. It is the only book, to date, that distinguishes between statistical data mining and machine-learning data mining. I was an orthodox statistician until I resolved my struggles with the weaknesses of statistics within the big data setting of today. Now, as a reform statistician who is free of the statistical rigors of yesterday, with many degrees of freedom to exercise, I have composed by intellectual might the original and practical statistical data mining techniques in the first part of the book. The GenIQ Model, a machine-learning alternative to statistical regression, led to the creative and useful machine-learning data mining techniques in the remaining part of the book.

This book is a compilation of essays that offer detailed background, discussion, and illustration of specific methods for solving the most commonly experienced problems in predictive modeling and analysis of big data. The common theme among these essays is to address each methodology and assign its application to a specific type of problem. To better ground the reader, I spend considerable time discussing the basic methodologies of predictive modeling and analysis. While this type of overview has been attempted before, my approach offers a truly nitty-gritty, step-by-step approach that both tyros and experts in the field can enjoy playing with. The job of the data analyst is overwhelmingly to predict and explain the result of the target variable, such as RESPONSE or PROFIT. Within that task, the target variable is either a binary variable (RESPONSE is one such example) or a continuous variable (of which PROFIT is a good example). The scope of this book is purposely limited, with one exception, to dependency models, for which the target variable is often referred to as the "left-hand" side of an equation, and the variables that predict and/or explain the target variable is the "right-hand" side. This is in contrast to interdependency models that have no left- or right-hand side. I devote a chapter to one type of interdependency model, which is tied into a dependency model. Because interdependency models comprise a minimal proportion of the data analyst's workload, I humbly suggest that the focus of this book will prove utilitarian.

Therefore, these essays have been organized in the following fashion. Chapter 1 reveals the two most influential factors in my professional life: John W. Tukey and the personal computer (PC). The PC has changed everything in the world of statistics. The PC can effortlessly produce precise calculations and eliminate the computational burden associated with statistics. One need only provide the right questions. Unfortunately, the confluence of the PC and the world of statistics has turned generalists with minimal statistical backgrounds into quasi-statisticians and affords them a false sense of confidence.

In 1962, in his influential article, "The Future of Data Analysis" [1], John Tukey predicted a movement to unlock the rigidities that characterize statistics. It was not until the publication of *Exploratory Data Analysis* [2] in 1977 that Tukey led statistics away from the rigors that defined it into a new area, known as EDA (from the first initials of the title of his seminal work). At its core, EDA, known presently as data mining or formally as statistical data mining, is an unending effort of numerical, counting, and graphical detective work.

To provide a springboard to more esoteric methodologies, Chapter 2 covers the correlation coefficient. While reviewing the correlation coefficient, I bring to light several issues unfamiliar to many, as well as introduce two useful methods for variable assessment. Building on the concept of smooth scatterplot presented in Chapter 2, I introduce in Chapter 3 the smoother scatterplot based on CHAID (chi-squared automatic interaction detection). The new method has the potential of exposing a more reliable depiction of the unmasked relationship for paired-variable assessment than that of the smoothed scatterplot.

In Chapter 4, I show the importance of straight data for the simplicity and desirability it brings for good model building. In Chapter 5, I introduce the method of symmetrizing ranked data and add it to the paradigm of simplicity and desirability presented in Chapter 4.

Principal component analysis, the popular data reduction technique invented in 1901, is repositioned in Chapter 6 as a data mining method for many-variable assessment. In Chapter 7, I readdress the correlation coefficient. I discuss the effects the distributions of the two variables under consideration have on the correlation coefficient interval. Consequently, I provide a procedure for calculating an adjusted correlation coefficient.

In Chapter 8, I deal with logistic regression, a classification technique familiar to everyone, yet in this book, one that serves as the underlying rationale for a case study in building a response model for an investment product. In doing so, I introduce a variety of new data mining techniques. The continuous side of this target variable is covered in Chapter 9. On the heels of discussing the workhorses of statistical regression in Chapters 8 and 9, I resurface the scope of literature on the weaknesses of variable selection methods, and I enliven a notable solution for specifying a well-defined regression model in Chapter 10 anew. Chapter 11 focuses on the interpretation of the logistic regression model with the use of CHAID as a data mining tool. Chapter 12 refocuses on the regression coefficient and offers common misinterpretations of the coefficient that point to its weaknesses. Extending the concept of the coefficient, I introduce the average correlation coefficient in Chapter 13 to provide a quantitative criterion for assessing competing predictive models and the importance of the predictor variables.

In Chapter 14, I demonstrate how to increase the predictive power of a model beyond that provided by its variable components. This is accomplished by creating an interaction variable, which is the product of two or more component variables. To test the significance of the interaction variable, I make what I feel to be a compelling case for a rather unconventional use of CHAID. Creative use of well-known techniques is further carried out in Chapter 15, where I solve the problem of market segment classification modeling using not only logistic regression but also CHAID. In Chapter 16, CHAID is yet again utilized in a somewhat unconventional manner—as a method for filling in missing values in one's data. To bring an interesting real-life problem into the picture, I wrote Chapter 17 to describe profiling techniques for the marketer who wants a method for identifying his or her best customers. The benefits of the predictive profiling approach is demonstrated and expanded to a discussion of look-alike profiling.

I take a detour in Chapter 18 to discuss how marketers assess the accuracy of a model. Three concepts of model assessment are discussed: the traditional decile analysis, as well as two additional concepts, precision and separability. In Chapter 19, continuing in this mode, I point to the weaknesses in the way the decile analysis is used and offer a new approach known as the bootstrap for measuring the efficiency of marketing models.

The purpose of Chapter 20 is to introduce the principal features of a bootstrap validation method for the ever-popular logistic regression model. Chapter 21 offers a pair of graphics or visual displays that have value beyond the commonly used exploratory phase of analysis. In this chapter, I demonstrate the hitherto untapped potential for visual displays to describe the functionality of the final model once it has been implemented for prediction.

I close the statistical data mining part of the book with Chapter 22, in which I offer a data-mining alternative measure, the predictive contribution coefficient, to the standardized coefficient.

With the discussions just described behind us, we are ready to venture to new ground. In Chapter 1, I elaborated on the concept of machine-learning data mining and defined it as PC learning without the EDA/statistics component. In Chapter 23, I use a metrical modelogue, "To Fit or Not to Fit Data to a Model," to introduce the machine-learning method of GenIQ and its favorable data mining offshoots.

In Chapter 24, I maintain that the machine-learning paradigm, which lets the data define the model, is especially effective with big data. Consequently, I present an exemplar illustration of genetic logistic regression outperforming statistical logistic regression, whose paradigm, in contrast, is to fit the data to a predefined model. In Chapter 25, I introduce and illustrate brightly, perhaps, the quintessential data mining concept: data reuse. Data reuse is appending new variables, which are found when building a GenIQ Model, to the original dataset. The benefit of data reuse is apparent: The original dataset is enhanced with the addition of new, predictive-full GenIQ data-mined variables.

In Chapters 26–28, I address everyday statistics problems with solutions stemming from the data mining features of the GenIQ Model. In statistics, an outlier is an observation whose position falls outside the overall pattern of the data. Outliers are problematic: Statistical regression models are quite sensitive to outliers, which render an estimated regression model with questionable predictions. The common remedy for handling outliers is "determine and discard" them. In Chapter 26, I present an alternative method of moderating outliers instead of discarding them. In Chapter 27, I introduce a new solution to the old problem of overfitting. I illustrate how the GenIQ Model identifies a structural source (complexity) of overfitting, and subsequently instructs for deletion of the individuals who contribute to the complexity, from the dataset under consideration. Chapter 28 revisits the examples (the importance of straight data) discussed in Chapters 4 and 9, in which I posited the solutions without explanation as the material needed to understand the solution was not introduced at that point. At this point, the background required has been covered. Thus, for completeness, I detail the posited solutions in this chapter.

GenIQ is now presented in Chapter 29 as such a nonstatistical machine-learning model. Moreover, in Chapter 30, GenIQ serves as an effective method for finding the best possible subset of variables for a model. Because GenIQ has no coefficients—and coefficients furnish the key to prediction—Chapter 31 presents a method for calculating a quasi-regression coefficient, thereby providing a reliable, assumption-free alternative to the regression coefficient. Such an alternative provides a frame of reference for evaluating and using coefficient-free models, thus allowing the data analyst a comfort level for exploring new ideas, such as GenIQ.

References

1. Tukey, J.W., The future of data analysis, *Annals of Mathematical Statistics*, 33, 1–67, 1962.
2. Tukey, J.W., *Exploratory Data Analysis*, Addison-Wesley, Reading, MA, 1977.

Acknowledgments

This book, like all books—except the Bible—was written with the assistance of others. First and foremost, I acknowledge Hashem who has kept me alive, sustained me, and brought me to this season.

I am grateful to David Grubbs, my editor, who contacted me about outdoing myself by writing this book. I am indebted to the staff of the CRC Press/Taylor & Francis Group for their excellent work: Sherry Thomas, Editorial Assistant and Project Coordinator; Todd Perry, Project Editor; Victoria Jones, Copy Editor; Viswanath Prasanna, Senior Project Manager; Shanmuga Vadivu, Proofreader; Celia McCoy, Indexer, and Elise Weinger and Kevin Craig, Cover Designers.

Author

Bruce Ratner, PhD, The Significant Statistician™, is president and founder of DM STAT-1 Consulting, the boutique firm for statistical modeling and analysis, data mining, and machine-learning. Bruce specializes in all standard statistical techniques, as well as recognized and innovative machine-learning methods, such as the patented GenIQ Model. Bruce achieves the clients' goals across varied industries: direct and database marketing, banking, insurance, finance, retail, telecommunications, healthcare, pharmaceutical, publication and circulation, mass and direct advertising, catalog marketing, e-commerce, Web-mining, B2B, risk management, and nonprofit fundraising.

Bruce's par excellence the expertise is apparent as he is the author of the best-selling book, *Statistical Modeling and Analysis for Database Marketing: Effective Techniques for Mining Big Data*. Bruce ensures the optimal solution methodology for his clients' marketing problems with rapid startup and timely delivery of project results. Bruce executes the highest level of statistical practice for his clients' projects. He is an often-invited speaker at public industry events, such as the SAS Data Mining Conference, and private seminars at the request of *Fortune* magazine's top 100 companies.

Bruce has his footprint in the predictive analytics community as a frequent speaker at industry conferences and as the instructor of the advanced statistics course sponsored by the Direct Marketing Association for more than a decade. He is the author of more than 100 peer-reviewed articles on statistical and machine-learning procedures and software tools. He is a coauthor of the popular textbook, *The New Direct Marketing*, and is on the editorial board of the *Journal of Database Marketing and Customer Strategy*.

Bruce is also active in the online data mining industry. He is a frequent contributor to *KDnuggets Publications*, the top resource for the data mining community. His articles on statistical and machine-learning methodologies draw a huge monthly following. Other online venues in which Bruce participates are the networking sites LinkedIn and ResearchGate, in which his postings on statistical and machine-learning procedures for big data have sparked countless rich discussions. Also, he is the author of his own *DM STAT-1 Newsletter* on the web.

Bruce holds a doctorate in mathematics and statistics, with a concentration in multivariate statistics and response model simulation. His research interests include developing hybrid modeling techniques, which combine traditional statistics and machine-learning methods. He holds a patent for a unique application in solving the two-group classification problem with genetic programming.

1

Introduction

Whatever you are able to do with your might, do it.

—Kohelet 9:10

1.1 The Personal Computer and Statistics

The personal computer (PC) has changed everything—for both better and worse—in the world of statistics. The PC can effortlessly produce precise calculations and eliminate the computational burden associated with statistics. One needs only to provide the right information. With minimal knowledge of statistics, the user points to the location of the input data, selects the desired statistical procedure, and directs the placement of the output. Thus, tasks such as testing, analyzing, and tabulating raw data into summary measures as well as many other statistical criteria are fairly rote. The PC has advanced statistical thinking in the decision-making process as evidenced by visual displays such as bar charts and line graphs, animated three-dimensional rotating plots, and interactive marketing models found in management presentations. The PC also facilitates support documentation, which includes the calculations for measures such as mean profit across market segments from a marketing database; statistical output is copied from the statistical software and then pasted into the presentation application. Interpreting the output and drawing conclusions still require human intervention.

Unfortunately, the confluence of the PC and the world of statistics has turned generalists with minimal statistical backgrounds into quasi-statisticians and affords them a false sense of confidence because they can now produce statistical output. For instance, calculating the mean profit is standard fare in business. However, the mean provides a "typical value" only when the distribution of the data is symmetric. In marketing databases, the distribution of profit commonly has a positive skewness.[*] Thus, the mean profit is not a reliable summary measure.[†] The quasi-statistician would doubtlessly not know to check this supposition, thus rendering the interpretation of the mean profit as floccinaucinihilipilification.[‡]

Another example of how the PC fosters a "quick-and-dirty"[§] approach to statistical analysis is in the use of the ubiquitous correlation coefficient (second in popularity to the mean

* *Positive skewed* or *right skewed* means the distribution has a long tail in the positive direction.

† For moderately skewed distributions, the mode or median should be considered and assessed for a reliably typical value.

‡ Floccinaucinihilipilification (FLOK-si-NO-si-NY-HIL-i-PIL-i-fi-KAY-shuhn), noun: estimating something as worthless.

§ The literal translation of this expression clearly supports my claim that the PC is sometimes not a good thing for statistics. I supplant the former with "thorough and clean."

as a summary measure), which measures the association between two variables. There is an assumption (that the underlying relationship between the two variables is linear or a straight line) to be met for the proper interpretation of the correlation coefficient. Rare is the quasi-statistician who is aware of the assumption. Meanwhile, well-trained statisticians often do not check this assumption, a habit developed by the uncritical use of statistics with the PC.

The PC with its unprecedented computational strength has also empowered professional statisticians to perform proper analytical due diligence; for example, the natural seven-step cycle of statistical analysis would not be practical [1]. The PC and the analytical cycle comprise the perfect pairing as long as the information obtained starts at Step 1 and continues straight through Step 7, without a break in the cycle. Unfortunately, statisticians are human and succumb to taking shortcuts in the path through the seven-step cycle. They ignore the cycle and focus solely on the sixth step. A careful statistical endeavor requires performance of all the steps in the seven-step cycle.* The seven-step sequence is as follows:

1. *Definition of the problem*—Determining the best way to tackle the problem is not always obvious. Management objectives are often expressed qualitatively, in which case the selection of the outcome or target (dependent) variable is subjectively biased. When the objectives are clearly stated, the appropriate dependent variable is often not available, in which case a surrogate must be used.
2. *Determining technique*—The technique first selected is often the one with which the data analyst is most comfortable; it is not necessarily the best technique for solving the problem.
3. *Use of competing techniques*—Applying alternative techniques increases the odds that a thorough analysis is conducted.
4. *Rough comparisons of efficacy*—Comparing variability of results across techniques can suggest additional techniques or the deletion of alternative techniques.
5. *Comparison in terms of a precise (and thereby inadequate) criterion*—An explicit criterion is difficult to define. Therefore, precise surrogates are often used.
6. *Optimization in terms of a precise and inadequate criterion*—An explicit criterion is difficult to define. Therefore, precise surrogates are often used.
7. *Comparison in terms of several optimization criteria*—This constitutes the final step in determining the best solution.

The founding fathers of classical statistics—Karl Pearson and Sir Ronald Fisher—would have delighted in the PC's ability to free them from time-consuming empirical validations of their concepts. Pearson, whose contributions include regression analysis, the correlation coefficient, the standard deviation (a term he coined in 1893), and the chi-square test of statistical significance (to name but a few), would have likely developed even more concepts with the free time afforded by the PC. One can further speculate that the functionality of the PC would have allowed Fisher's methods (e.g., maximum likelihood estimation, hypothesis testing, and analysis of variance) to have immediate and practical applications.

The PC took the classical statistics of Pearson and Fisher from their theoretical blackboards into the practical classrooms and boardrooms. In the 1970s, statisticians were starting to

* The seven steps are attributed to Tukey. The annotations are my attributions.

acknowledge that their methodologies had the potential for wider applications. However, they knew an accessible computing device was required to perform their on-demand statistical analyses with an acceptable accuracy and within a reasonable turnaround time. Because the statistical techniques, developed for a small data setting consisting of one or two handfuls of variables and up to hundreds of records, the hand tabulation of data was computationally demanding and almost insurmountable. Accordingly, conducting the statistical techniques on large data (big data were not born until the late 2000s) was virtually out of the question. With the inception of the microprocessor in the mid-1970s, statisticians now had their computing device, the PC, to perform statistical analyses on large data with excellent accuracy and turnaround time. The desktop PCs replaced handheld calculators in the classroom and boardrooms. From the 1990s to the present, the PC has offered statisticians advantages that were imponderable decades earlier.

1.2 Statistics and Data Analysis

As early as 1957, Roy believed that classical statistical analysis was likely to be supplanted by assumption-free, nonparametric approaches that were more realistic and meaningful [2]. It was an onerous task to understand the robustness of the classical (parametric) techniques to violations of the restrictive and unrealistic assumptions underlying their use. In practical applications, the primary assumption of "a random sample from a multivariate normal population" is virtually untenable. The effects of violating this assumption and additional model-specific assumptions (e.g., linearity between predictor and dependent variables, constant variance among errors, and uncorrelated errors) are hard to determine with any exactitude. It is difficult to encourage the use of statistical techniques, given that their limitations are not fully understood.

In 1962, in his influential article, "The Future of Data Analysis," John Tukey expressed concern that the field of statistics was not advancing [1]. He felt there was too much focus on the mathematics of statistics and not enough on the analysis of data; he predicted a movement to unlock the rigidities that characterize the discipline. In an act of statistical heresy, Tukey took the first step toward revolutionizing statistics by referring to himself not as a statistician but as a data analyst. However, it was not until the publication of his seminal masterpiece, *Exploratory Data Analysis*, in 1977, that Tukey led the discipline away from the rigors of statistical inference into a new area known as EDA (the initialism from the title of the unquestionable masterpiece) [3]. For his part, Tukey tried to advance EDA as a separate and distinct discipline from statistics—an idea that never took hold. EDA offered a fresh, assumption-free, nonparametric approach to problem-solving in which the data guide the analysis and utilize self-educating techniques, such as iteratively testing and modifying the analysis as the evaluation of feedback, thereby improving the final analysis for reliable results.

Tukey's words best describe the essence of EDA:

> Exploratory data analysis is detective work—numerical detective work—or counting detective work—or graphical detective work. … [It is] about looking at data to see what it seems to say. It concentrates on simple arithmetic and easy-to-draw pictures. It regards whatever appearances we have recognized as partial descriptions, and tries to look beneath them for new insights. [3, p. 1]

EDA includes the following characteristics:

1. *Flexibility*—Techniques with greater flexibility to delve into the data
2. *Practicality*—Advice for procedures of analyzing data
3. *Innovation*—Techniques for interpreting results
4. *Universality*—Use all statistics that apply to analyzing data
5. *Simplicity*—Above all, the belief that simplicity is the golden rule

On a personal note, when I learned that Tukey preferred to be called a data analyst, I felt both validated and liberated because many of my analyses fell outside the realm of the classical statistical framework. Also, I had virtually eliminated the mathematical machinery, such as the calculus of maximum likelihood. In homage to Tukey, I use the terms *data analyst* and *statistician* interchangeably throughout this book.

1.3 EDA

Tukey's book is more than a collection of new and creative rules and operations; it defines EDA as a discipline, which holds that data analysts only fail if they fail to try many things. It further espouses the belief that data analysts are especially successful if their detective work forces them to notice the unexpected. In other words, the philosophy of EDA is a trinity of *attitude* and *flexibility* to do whatever it takes to refine the analysis and *sharp-sightedness* to observe the unexpected when it does appear. EDA is thus a self-propagating theory; each data analyst adds his or her contribution, thereby contributing to the discipline, as I hope to accomplish with this book.

The sharp-sightedness of EDA warrants more attention because it is an important feature of the EDA approach. The data analyst should be a keen observer of indicators that are capable of being dealt with successfully and should use them to paint an analytical picture of the data. In addition to the ever-ready visual graphical displays as *indicators* of what the data reveal, there are numerical indicators, such as counts, percentages, averages, and the other classical descriptive statistics (e.g., standard deviation, minimum, maximum, and missing values). The data analyst's personal judgment and interpretation of indicators are not considered as bad things because the goal is to draw informal inferences rather than those statistically significant inferences that are the hallmark of statistical formality.

In addition to visual and numerical indicators, there are the *indirect messages* in the data that force the data analyst to take notice, prompting responses such as "the data look like ..." or "they appear to be ...." Indirect messages may be vague, but their importance is to help the data analyst draw informal inferences. Thus, indicators do not include any of the hard statistical apparatus, such as confidence limits, significance tests, or standard errors.

With EDA, a new trend in statistics was born. Tukey and Mosteller quickly followed up in 1977 with the second distinctive and stylish EDA book, *Data Analysis and Regression*, commonly referred to as EDA II. EDA II recasts the basics of classical inferential procedures of data analysis and regression into an assumption-free, nonparametric approach guided by "(a) a sequence of philosophical attitudes ... for effective data analysis and (b) a flow of useful and adaptable techniques that make it possible to put these attitudes to work" [4, p. vii].

In 1983, Hoaglin, Mosteller, and Tukey succeeded in advancing EDA with *Understanding Robust and Exploratory Data Analysis,* which provides an understanding of how badly the classical methods behave when their restrictive assumptions do not hold and offers alternative robust and exploratory methods to broaden the effectiveness of statistical analysis [5]. It includes a collection of methods to cope with data in an informal way, guiding the identification of data structures relatively quickly and easily and trading off optimization of the objective for the stability of results.

In 1991, Hoaglin, Mosteller, and Tukey continued their fruitful EDA efforts with *Fundamentals of Exploratory Analysis of Variance* [6]. They refashioned the basics of the analysis of variance with the classical statistical apparatus (e.g., degrees of freedom, F-ratios, and p-values). The recasting was a host of numerical and graphical displays, which often give insight into the structure of the data, such as size effects, patterns, and interaction and behavior of residuals.

EDA set off a burst of activity in the visual portrayal of data. In 1983, *Graphical Methods for Data Analysis* (Chambers et al.) presented new and old methods—some of which require a computer, while others only paper and pencil—but all are powerful data analytical tools to learn more about data structure [7]. In 1986, du Toit, Steyn, and Stumpf came out with *Graphical Exploratory Data Analysis,* providing a comprehensive, yet simple presentation of the topic [8]. Jacoby, with *Statistical Graphics for Visualizing Univariate and Bivariate Data* (1997) and *Statistical Graphics for Visualizing Multivariate Data* (1998), carried out his objective to obtain pictorial representations of quantitative information by elucidating histograms, one-dimensional and enhanced scatterplots, and nonparametric smoothing [9,10]. Also, Jacoby successfully transferred graphical displays of multivariate data to a single sheet of paper, a two-dimensional space.

1.4 The EDA Paradigm

EDA presents a major paradigm shift in the model-building process. With the mantra, "Let your data be your guide," EDA offers a view that is a complete reversal of the classical principles that govern the usual steps of the model-building process. EDA declares the model must always follow the data, not the other way around, as in the classical approach.

In the classical approach, the problem is stated and formulated regarding an outcome variable, Y. It assumes that the *true* model explaining all the variations in Y is known. Specifically, the structures of the predictor variables, $X_{i,}$ affecting Y are known and present in the model. For example, if Age affects Y, but the log of Age reflects the true relationship with Y, then the log of Age is present in the model. Once the model is specified, the model-specific analysis of the data provides the results regarding numerical values associated with the structures or estimates of the coefficients of the true predictor variables. Then, the modeling process concludes with the interpretation of the model. Interpretation includes: declaring whether X_i is an important predictor; if X_i is important, assessing how X_i affects the prediction of Y; and ranking X_i in order of predictive importance.

Of course, the data analyst never knows the true model. So, familiarity with the content domain of the problem is used to put forth explicitly the true *surrogate* model, which yields good predictions of Y. According to Box, "All models are wrong, but some are useful" [11]. In this case, the model selected provides serviceable predictions of Y. Regardless of

```
Problem ==> Model ===>  Data     ===> Analysis ===> Results/interpretation (classical)
Problem <==> Data <===> Analysis <===> Model   ===> Results/interpretation (EDA)

        Attitude, flexibility, and sharp-sightedness (EDA trinity)
```

FIGURE 1.1
EDA paradigm.

the model used, the assumption of knowing the truth about Y sets the statistical logic in motion to cause likely bias in the analysis, results, and interpretation.

In the EDA approach, the only assumption is having some prior experience with a content domain of the problem. Attitude, flexibility, and sharp-sightedness are the forces behind the data analysts, who assess the problem and let the data direct the course of the analysis, which then suggests the structures in the model. If the model passes the validity check, then it is considered final and ready for results and interpretation. If the validity check fails, the data analysts, with these forces still behind their back, revisit the analysis and data until new structures yield a sound and validated model. Then, the sought-after final results and interpretation are made (see Figure 1.1). Without exposure to assumption violations, the EDA paradigm offers a degree of confidence that its prescribed exploratory efforts are not biased, at least in the manner of the classical approach. Of course, no analysis is bias-free because all analysts admit their bias into the equation.

1.5 EDA Weaknesses

With all its strengths and determination, EDA as originally developed had two minor weaknesses that could have hindered its wide acceptance and great success. One is of a subjective or psychological nature, and the other is a misconceived notion. Data analysts know that failure to look into a multitude of possibilities can result in a flawed analysis and that they can thus find themselves in a competitive struggle against the data itself. Thus, EDA can foster a sense of insecurity within data analysts that their work is never complete. The PC can assist data analysts in being thorough with their analytical due diligence but bears no responsibility for the arrogance EDA engenders.

The belief that EDA, originally developed for the small data setting, does not work as well with large samples is a misconception. Indeed, some of the graphical methods, such as the stem-and-leaf plots, and some of the numerical and counting methods, such as folding and binning, do break down with large samples. However, the majority of EDA methodology is unaffected by data size. The manner by which the methods are carried out and the reliability of the results remain in place. In fact, some of the most powerful EDA techniques scale up nicely but do require the PC to perform the serious number crunching of *big data** [12]. For example, techniques such as the ladder of powers, reexpressing,† and smoothing are valuable tools for large sample or big data applications.

* Authors Weiss and Indurkhya and I use the general concept of "big" data. However, we stress different characteristics of the concept.

† Tukey, via his groundbreaking EDA book, put the concept of "reexpression" in the forefront of EDA data mining tools; yet, he never provided any definition. I assume he believed that the term is self-explanatory. Tukey's first mention of reexpression is in a question on page 61 of his work: "What is the single most needed form of re-expression?" I, for one, would like a definition of reexpression, and I provide one further in the book.

1.6 Small and Big Data

I would like to clarify the general concept of small and big data. Size, like beauty, is in the mind of the data analyst. In the past, small data fit the conceptual structure of classical statistics. Small always referred to the sample size, not the number of variables, which were always a handful. Depending on the method used by the data analyst, small was seldom less than 5 individuals; sometimes between 5 and 20; frequently between 30 and 50 or between 50 and 100; and rarely between 100 and 200. In contrast to today's big data, defined by complex tabular displays of rows (observations or individuals) and columns (variables or features), small data, defined by the simple tabular display, fit on a few sheets of paper.

In addition to the compact area they occupy, small data are neat and tidy. They are clean, in that they contain no improbable or impossible values, except for those due to primal data collection error. They do not include the statistical outliers and influential points or the EDA far-out and outside points. They are in the "ready-to-run" condition required by classical statistical methods.

Regarding big data, there are two views. One view is that of classical statistics, which considers big as simply *not small*. Theoretically, big is the sample size after which asymptotic properties of the method "kick in" for valid results. The other view is that of contemporary statistics, which considers big with regard to lifting (mathematical calculation) the observations and learning from the variables. Big depends on who is analyzing the data. That is, data are big if the data analyst feels they are big. Regardless of on which side the data analyst is working, EDA scales up for both rows and columns of the data table.

1.6.1 Data Size Characteristics

There are three distinguishable characteristics of data size: condition, location, and population. *Condition* refers to the state of readiness of the data for analysis. Data that require minimal time and cost to clean, before conducting reliable analysis, are said to be well-conditioned. Data that involve a substantial amount of time and cost are said to be ill-conditioned. Small data are typically clean and therefore well-conditioned.

Big data are an outgrowth of today's digital environment, which generates data flowing continuously from all directions at unprecedented speed and volume. They are considered "dirty" mainly because of the merging of multiple sources. The merging process is inherently time-intensive because multiple passes of the sources must be made to get a sense of how the combined sources fit together. Because of the iterative nature of the process, the logic of matching individual records across sources is at first fuzzy, and then fine-tuned to some level of soundness. The resultant data more times than not consist of unexplainable, seemingly random, nonsensical values. Thus, big data are usually ill-conditioned.

Location refers to where the data reside. Unlike the rectangular sheets for small data, big data reside in databases consisting of multidimensional tabular tables. The link among the data tables can be hierarchical (rank- or level-dependent) or sequential (time- or event-dependent). The merging of multiple data sources, each consisting of many rows and columns, produces data of even greater numbers of rows and columns, clearly suggesting bigness.

Population refers to the group of individuals having qualities or characteristics in common and related to the study under consideration. Small data ideally represent a random sample of a known population that is not expected to encounter changes in its composition in the short term. The data are collected to answer a specific problem, permitting straightforward answers from a given problem-specific method. In contrast, big data often represent multiple, nonrandom samples of unknown populations, shifting in composition within the short term. As such, big data are "secondary" in nature. Data originally collected for a specific purpose that are used for purposes other than the original intent are referred to as secondary data. Big data are available from the hydra of marketing information for use on any post hoc problem and may not have a straightforward solution.

It is interesting to note that Tukey never talked specifically about big data per se. However, he did predict that the cost of computing, in both time and dollars, would be cheap, which arguably suggests that he knew big data were coming. Regarding the cost, clearly, today's PCs bear this out.

1.6.2 Data Size: Personal Observation of One

The data size discussion raises the following question: "How large should a sample be?" Sample size can be anywhere from folds of 10,000 up to 100,000. In my experience as a statistical modeler and data mining consultant for more than 15 years and as a statistics instructor who analyzes deceivingly simple cross tabulations with the basic statistical methods as my data mining tools, I have observed that the less-experienced and less-trained data analyst uses sample sizes that are unnecessarily large. I see analyses and models that use samples too large by factors ranging from 20 to 50. Although the PC can perform the heavy calculations, the extra time and cost in getting the larger data out of the data warehouse and then processing and thinking about them are almost never justified. Of course, the only way a data analyst learns that extra big data are a waste of resources is by performing small versus big data comparisons, a step I recommend.

1.7 Data Mining Paradigm

The term *data mining* emerged from the database marketing community sometime between the late 1970s and early 1980s. Statisticians did not understand the excitement and activity caused by this new technique because the discovery of patterns and relationships (structure) in the data was not new to them. They had known about data mining for a long time, albeit under various names, such as data fishing, snooping, dredging, and—most disparaging—ransacking the data. Because any discovery process inherently exploits the data, producing spurious findings, statisticians did not view data mining in a positive light.

To state one of the numerous paraphrases of Maslow's hammer,* "If you have a hammer in hand, you tend eventually to start seeing nails." The statistical version of this maxim is, "Looking for structure typically results in finding structure." All data have spurious structures, formed by the "forces" that make things come together, such as chance. The bigger the data, the greater are the odds that spurious structures abound. Thus, an expectation of

* Abraham Maslow brought a fresh perspective to the world of psychology with his concept of "humanism," which he referred to as the "third force" of psychology—after Pavlov's "behaviorism" and Freud's "psychoanalysis." Maslow's hammer is frequently used without anybody seemingly knowing about the originator of this unique pithy statement, expressing a rule of conduct. Maslow's Jewish parents migrated from Russia to the United States to escape from harsh conditions and sociopolitical turmoil. He was born in Brooklyn, New York, in April 1908 and died from a heart attack in June 1970.

data mining is that it produces structures, both real and spurious, without any distinction between them.

Today, statisticians accept data mining only if it embodies the EDA paradigm. They define data mining as any process that finds unexpected structures in data and that uses the EDA framework to ensure that the process explores the data, not exploits it (see Figure 1.1). Note the word *unexpected*. It suggests the process is exploratory, not confirmatory, in discovering unexpected structures. By finding what one expects to find, there is no longer uncertainty regarding the existence of the structure.

Statisticians are mindful of the inherent nature of data mining and try to make adjustments to minimize the number of spurious structures identified. In classical statistical analysis, statisticians have explicitly modified most analyses that search for interesting structures, such as adjusting the overall alpha level/type I error rate or inflating the degrees of freedom [13,14]. In data mining, the statistician has no explicit analytical adjustments available, only the implicit adjustments affected by using the EDA paradigm itself. The steps discussed next outline the data mining/EDA paradigm. As expected from EDA, the steps are defined by *soft* rules.

Suppose the objective is to find a structure to help make good predictions of response to a future mail campaign. The following represents the steps required.

- *Obtain* the database that has similar mailings to the future mail campaign.
- *Draw* a sample from the database. Size can be several folds of 10,000 up to 100,000.
- *Perform* many exploratory passes of the sample. Carry out all desired calculations to determine interesting or noticeable structures.
- *Stop* the calculations used for finding the noticeable structure.
- *Count* the number of noticeable structures that emerge. The structures are not necessarily the results and should not be declared significant findings.
- *Seek* out indicators, visual and numerical, and the indirect messages.
- *React or respond* to all indicators and indirect messages.
- *Ask* questions. Does each structure make sense by itself? Do any of the structures form natural groups? Do the groups make sense? Is there consistency among the structures within a group?
- *Try* more techniques. Repeat the many exploratory passes with several fresh samples drawn from the database. Check for consistency across the multiple passes. If results do not behave in a similar way, there may be no structure to predict response to a future mailing because chance may have infected your data. If results behave similarly, then assess the variability of each structure and each group.
- *Choose* the most stable structures and groups of structures for predicting response to a future mailing.

1.8 Statistics and Machine Learning

Coined by Samuel in 1959, the term *machine learning* (ML) was given to the field of study that assigns computers the ability to learn without being explicitly programmed [15]. In other words, ML investigates ways in which the computer can acquire knowledge directly from data and thus learn to solve problems. It would not be long before ML would influence the statistical community.

In 1963, Morgan and Sonquist led a rebellion against the restrictive assumptions of classical statistics [16]. They developed the automatic interaction detection (AID) regression tree, a methodology without assumptions. AID is a computer-intensive technique that finds or learns multidimensional patterns and relationships in data and serves as an assumption-free, nonparametric alternative to regression prediction and classification analysis. Many statisticians believe that AID marked the beginning of an ML approach to solving statistical problems. There have been many improvements and extensions of AID: theta AID (THAID), multivariate AID (MAID), chi-squared AID (CHAID), and classification and regression trees (CART), which are now quite reliable and accessible data mining tools. CHAID and CART have emerged as the most popular today.

I consider AID and its offspring as quasi-ML methods. They are computer-intensive techniques that need the PC, a necessary condition for an ML method. However, they are not true ML methods because they use explicitly statistical criteria (e.g., theta, chi-squared and the F-test) for the learning. The PC enables a genuine ML method to learn via mimicking the way humans think. Thus, I must use the term quasi. Perhaps a more appropriate and suggestive term for AID-type procedures and other statistical problems using the PC is statistical ML.

Independent from the work of Morgan and Sonquist, ML researchers had been developing algorithms to automate the induction process, which provided another alternative to regression analysis. In 1979, Quinlan used the well-known concept learning system developed by Hunt, Marin, and Stone to implement one of the first intelligent systems—ID3—which was succeeded by C4.5 and C5.0 [17,18]. These algorithms are also considered data mining tools but have not successfully crossed over to the statistical community.

The interface of statistics and ML began in earnest in the 1980s. ML researchers became familiar with the three classical problems facing statisticians: regression (predicting a continuous outcome variable), classification (predicting a categorical outcome variable), and clustering (dividing a population of individuals into k subpopulations such that individuals within a group are as similar as possible, and the individuals among the groups are as dissimilar as possible). They started using their machinery (algorithms and the PC) for a nonstatistical, assumption-free nonparametric approach to the three problem areas. At the same time, statisticians began harnessing the power of the desktop PC to influence the classical problems they know so well, thus relieving themselves from the starchy parametric road.

The ML community has many specialty groups working on data mining: neural networks, support vector machines, fuzzy logic, genetic algorithms and programming, information retrieval, knowledge acquisition, text processing, inductive logic programming, expert systems, and dynamic programming. All areas have the same objective in mind but accomplish it with their tools and techniques. Unfortunately, the statistics community and the ML subgroups have no real exchanges of ideas or best practices. They create distinctions of no distinction.

1.9 Statistical Data Mining

In the spirit of EDA, it is incumbent on data analysts to try something new and retry something old. They can benefit not only from the computational power of the PC in doing the heavy lifting of big data but also from the ML ability of the PC in uncovering structure nestled in big data. In the spirit of trying something old, statistics still has a lot to offer.

Thus, today's data mining is about three conceptual components:

1. *Statistics with emphasis on EDA proper:* This includes using the descriptive and non inferential parts of classical statistical machinery as indicators. The parts include the sum of squares, degrees of freedom, F-ratios, chi-squared values, and p-values but exclude inferential conclusions.
2. *Big data:* Big data are given special mention because of today's digital environment. However, because small data are a component of big data, they are not excluded.
3. *Machine learning:* The PC is the learning machine, the *essential processing unit*, having the ability to learn without being explicitly programmed and the intelligence to find structure in the data. Moreover, the PC is essential for big data because it can always do what it is explicitly programmed to do.

The three concepts define the data mining mnemonic: Data Mining = Statistics + Big Data + Machine Learning and Lifting. Thus, *data mining* is all of statistics and EDA for big and small data with the power of the PC for lifting data and learning the structures within the data. Explicitly referring to big and small data implies the process works equally well on both.

Again, in the spirit of EDA, it is prudent to parse the mnemonic equation. Lifting and learning require two different aspects of the data table. Lifting focuses on the rows of the data table and uses the capacity of the PC regarding million instructions per second (MIPS), the speed by which program codes explicitly execute. Calculating the average income of one million individuals is an example of PC lifting.

Learning focuses on the columns of the data table and the ability of the PC to find the structure within the columns without being explicitly programmed. Learning is more demanding of the PC than lifting in the same way that learning from books is always more demanding than merely lifting the books. An example of PC learning is identifying structure, such as the square root of $(a^2 + b^2)$.

When there are indicators that the population is not homogeneous (i.e., there are subpopulations or clusters), the PC has to learn the rows and their relationships to each other to identify the row structures. Thus, when necessary, such as lifting and learning of the rows along with learning within the columns, the PC must work exceptionally hard but can yield extraordinary results.

Based on the preceding presentation, *statistical data mining* is the EDA/statistics component with PC lifting. Later in this book, I elaborate on *machine-learning data mining*, which I define as PC learning without the EDA/statistics component.

References

1. Tukey, J.W., The future of data analysis, *Annals of Mathematical Statistics*, 33, 1–67, 1962.
2. Roy, S.N., *Some Aspects of Multivariate Analysis*, Wiley, New York, 1957.
3. Tukey, J.W., *Exploratory Data Analysis*, Addison-Wesley, Reading, MA, 1977.
4. Mosteller, F., and Tukey, J.W., *Data Analysis and Regression*, Addison-Wesley, Reading, MA, 1977.
5. Hoaglin, D.C., Mosteller, F., and Tukey, J.W., *Understanding Robust and Exploratory Data Analysis*, Wiley, New York, 1983.

6. Hoaglin, D.C., Mosteller, F., and Tukey, J.W., *Fundamentals of Exploratory Analysis of Variance*, Wiley, New York, 1991.
7. Chambers, M.J., Cleveland, W.S., Kleiner, B., and Tukey, P.A., *Graphical Methods for Data Analysis*, Wadsworth & Brooks/Cole, Pacific Grove, CA, 1983.
8. du Toit, S.H.C., Steyn, A.G.W., and Stumpf, R.H., *Graphical Exploratory Data Analysis*, Springer-Verlag, New York, 1986.
9. Jacoby, W.G., *Statistical Graphics for Visualizing Univariate and Bivariate Data*, Sage, Thousand Oaks, CA, 1997.
10. Jacoby, W.G., *Statistical Graphics for Visualizing Multivariate Data*, Sage, Thousand Oaks, CA, 1998.
11. Box, G.E.P., Science and statistics, *Journal of the American Statistical Association*, 71, 791–799, 1976.
12. Weiss, S.M., and Indurkhya, N., *Predictive Data Mining*, Morgan Kaufman, San Francisco, CA, 1998.
13. Dun, O.J., Multiple comparison among means, *Journal of the American Statistical Association*, 54, 52–64, 1961.
14. Ye, J., On measuring and correcting the effects of data mining and model selection, *Journal of the American Statistical Association*, 93, 120–131, 1998.
15. Samuel, A., Some studies in machine learning using the game of checkers, in Feigenbaum, E., and Feldman, J., Eds., *Computers and Thought*, McGraw-Hill, New York, 14–36, 1963.
16. Morgan, J.N., and Sonquist, J.A., Problems in the analysis of survey data, and a proposal, *Journal of the American Statistical Association*, 58, 415–435, 1963.
17. Hunt, E., Marin, J., and Stone, P., *Experiments in Induction*, Academic Press, New York, 1966.
18. Quinlan, J.R., Discovering rules by induction from large collections of examples, In Mite, D., Ed., *Expert Systems in the Micro Electronic Age*, Edinburgh University Press, Edinburgh, UK, 143–159, 1979.

2

Science Dealing with Data: Statistics and Data Science

2.1 Introduction

Data science and data scientists have recently come into prominence, if not become a craze. Some say the pairing is reflective of substantial changes in the domain of statistics and statisticians, respectively. This chapter investigates whether the recently minted term data science implies statistics is a subset of a more developed and expanded domain or is a buzzed-up cloaking of the current state of statistics. Also, I probe to understand whether a data scientist is a super-statistician, whose new appellation signifies a larger skill set than that of the current statistician or, trivially, is the term data scientist a reimaging of the professional with a pretentious catchword of little exact meaning.

2.2 Background

Statistik (German word for statistics) was popularized and perhaps coined by German political scientist Gottfried Achenwall in 1749 [1]. By 1770, statistik had taken on the meaning *"science dealing with data* of the condition of a state or community" [2]. A considered opinion can declare that Achenwall's phrase is equivalent to an early stage meaning of statistics.

I am inclined to infer that Achenwall coined the originative term *data science.* As I reflect on science as a content-specific branch of knowledge requiring theory and methods, Achenwall's data science is indeed the definition of the equivalent branch of statistics. Expatiating on Achenwall's phrase, I define data science/statistics as the practice of

1. Collecting data
2. Analyzing data within a problem domain
3. Interpreting findings with graphics
4. Drawing conclusions

The delimitated 1770s' definition is extraordinary in light of today's popularized data science. To modernize the seminal definition, I incorporate the effects of the Internet.

The Internet accounts for big data, which include not only numbers but also text, voice, images, and so on. Big data account for the necessity of the computer. Also, big data account for the birth and continuing upping of the speed of high-performance statistical programs. So, I put forth a *modernized* definition of data science/statistics as a four-step process, which consists of the following:

1. Collecting data—originally small, today big—including numbers (traditionally structured), text (unstructured), spatial, voice, images, and so on.
2. Analyzing for accurate inferences and modeling while lessening uncertainty in a problem domain; tasks involving big data use computer-intensive statistical computations.
3. Interpreting findings (e.g., insights, patterns) with graphics and their derivative visualization methods, which continually improve to put *k*-dimensionality output on a two-dimensional flat surface.
4. Drawing conclusions.

The thesis of this chapter is that the newly minted data science and current statistics are identical. Accordingly, I undertake the task of gathering data—spanning two centuries from the modernized 1770s' data science/statistics to today's data science definition—from earlier texts, events, and prominent figures to make an honest comparison of the terms.

As to why I enter upon this undertaking, the answer is perhaps as old as Achenwall's data science itself. Comparisons are matches between "A and B" for the purpose of determining whether or not one is better than the other. In a nonsocial setting, comparisons could mean, for example, monetary gain for a drug manufacturer, relief for pain sufferers, or the saving of lives. In a social setting, as is the case here, comparison focuses on the drive of an individual to gain accurate self-evaluation [3]. I want to know whether the term *data scientist* implies having a larger skill set than mine or, trivially, whether I am behind the curve in reimaging myself with a pretentious catchword of little exact meaning.

To this end, I journey into the literature by browsing for mentions (e.g., citations, references, remarks, acknowledgments, credits, and statements) of the term data science. I remove sources with unsupported callouts of the term. For example, I exclude conference proceedings with data science in the title even though the programs' abstracts contain, for example, misstatements or no mention of the term. Such instances reflect a marketing ploy to increase attendance at conferences or workshops by taking advantage of the latest buzzwords. Ironically, these hot topics are counterproductive because they quickly become nonsense through endless repetition.

Before I discuss my journey of the mentions in the literature, I briefly consider what fads and trends are, and how they get started. A fad is a "seemingly irrational imitative behavior" [4], a spontaneous combustion of an awareness of a person, place, or thing, as well as an idea, concept, thought, or opinion. A fad disappears as quickly as it appears. As to who sets the spark, occasionally it is known. For example, in the 1960s, the rock-and-roll group The Beatles were the spark, which gave rise to the fad of the mop-top haircut. The Cabbage Patch doll of the 1980s was one of the most popular toy fads in history, without anyone knowing how it caught on. Sometimes the small fiery particle is "without obvious external stimuli" [4]. In contrast, a trend is a fad with staying power. For either fad or trend, the questions remain: Who sets it in motion, and who popularizes it?

The purpose of this chapter is to go beyond the expedition of Achenwall's 1770 data science to my acceptance of the unthought-of phrase, which I detail in the four-step process of the modernized definition of data science/statistics to ensure the meaning of the now-popular 2016 term data science. Specifically, I seek to determine whether data science and statistics are one and the same thing. Along the way, I search for who started it and who popularized it.

2.3 The Statistics and Data Science Comparison

As a well-schooled and highly practiced statistician, I use the lens of statistics—the four-step process of the modernized Achenwall's phrase—to examine the data science mentions collected. The four steps serve as touchstones for assessing whether a given mention in the data science space is describing a domain similar to statistics or extends statistics into the newly minted data science. I exercise a critical appraisal of a given mention against the touchstones but do not undertake a literal step-by-step comparison. The unstructured nature of citations, remarks, observations, quotations, and acknowledgments make a literal comparison impossible.

The research effort for finding mentions was unexpectedly tedious in that my go-to resource (Google) was exceptionally unproductive in locating good searches. Unexpectedly, the first pages from googling "data science" resulted in four advertisements for data scientist training and those seeking data science jobs: (1) IBM Data Science Summit, (2) Data Science Jobs from Indeed.com, (3) Data Science—12 Weeks, and (4) Data Science—Job Searching from Hired.com. I reviewed the "training" links for a definition or an explanation of data science. I found nothing suitable for the study at hand.

Interestingly, on a subsequent page, I found a Microsoft link to a posting of a data science curriculum, which included: querying data, introductions to R and Python, machine learning, and more topics about programming—essentially, tasks related to the information technology (IT) professional. The next few pages included more offerings of data science training and links to IT content. In sum, my Google search was not fruitful. Google produced a disproportionate number of advertisements for data science training and blogs of data science narratives (not definitions), which were turgid incoherent presentations.

My relevant collection of mentions starts in earnest with a useful reference, which came from the website Kaggle, after which one mention begot another. The final collection is small. I stopped searching when the mentions retrieved added no new content, facts, or knowledge. (I elaborate on my stopping rule in the discussion section.) I proceed with the assembled mentions chronologically. I conduct a text-by-text comparative investigation to determine whether data science is identical or at least similar to statistics. The result of my comparative investigation is in the next section.

2.3.1 Statistics versus Data Science

In 1960, Danish computer science pioneer Peter Naur used the term "data science" as a substitute for computer science, which he disliked.* Naur suggested the term "computer science" be renamed "datalogy" or "data science."

* Peter Naur (1928–2016) was a Danish computer science pioneer and Turing Award winner.

This entry is interesting yet not relevant to the current investigation. Curiously, the entry is not far afield from statistics because today's statistics require computer-intensive computation. Naur's data science may have been the spontaneous spark of something coming to statistics. This investigation is not about the context of Naur's citation.

1. Within the 1971 International Federation for Information *Processing* (IFIP) Guide to Data Processing [5], two brief descriptions of data science can be found.
   a. Data science is the *science of dealing with data*, once they have been established, while the relation of data to what they represent is delegated to other fields and sciences.
   b. A basic principle of data science is this: The data representation must be chosen with due regard to the transformation to be achieved and the data processing tools available. This stresses the importance of concern for the characteristics of the data processing tools.

   I am of two minds as to these quotes regarding data science. Oddly, the first quote goes back to Achenwall's leading phrase with the identical wording "science of dealing with data." However, the second part of the description implies statisticians delegate into which other fields the data should go. Delegation is not what statisticians do. Thus, the first quote does not comport with statistics as understood by any person dealing with data nor with the four-step process of the rooted modernized definition of statistics. The second quote only emphasizes computer power, which is a part but not the essence of the four-step process of statistics. So, based on the combined contents of the two quotes, I conclude data science is not similar to statistics.

   Decision I: The IFIP Guide to Data Processing indicates data science is not similar to statistics.

   Running tally of data science is identical or similar to statistics: 0 out of 1.

2. In 1997, C. F. Jeff Wu, an academic statistician, gave a lecture entitled "Statistics = Data Science?" If Wu's lecture title is put forward as a null hypothesis (H0), then the proposition is: Does Wu's lecture provides evidence to reject the null hypothesis? If there is no sufficient evidence, then one concludes "Statistics = Data Science" is true. That is, data science is equal to statistics. In his lecture, Wu characterized statistical work as a trilogy of data collection, data modeling and analysis, and decision-making. He called upon statisticians to initiate the usage of the term "data science" and advocated that statistics be replaced by data science and statisticians renamed data scientists (http://www2.isye.gatech.edu/~jeffwu/presentations/datascience.pdf).

   Wu's trilogy of statistics is fairly close to the four-step process. (He failed to provide evidence to reject H0: Statistics = Data Science.) Thus, Wu's reference supports the statement that data science is identical to statistics. Wu is obviously a statistician, and his choice of lecture topic suggests he is up-to-date on the activities in the statistics community. Given Wu's attentiveness to the trending term, I am perplexed as to why Wu advocates renaming statistics and statistician.

   Decision II: Wu's 1997 lecture "Statistics = Data Science?" indicates data science is identical to statistics.

   Running tally of data science is identical or similar to statistics: 1 out of 2.

3. In 2001, William S. Cleveland, an academic statistician, introduced data science as an independent discipline, extending the field of statistics to incorporate "advances in computing with data" [6].

   Cleveland's data science is statistics. Cleveland's recognition of big data necessitating advances in computer power implies his acknowledgment of statistics, especially the functional part—Step 2 of the four-step process.

   Decision III: Cleveland's quotes clearly indicate data science is identical to statistics.

   Running tally of data science is identical or similar to statistics: 2 out of 3.

4. In 2003, Columbia University began publishing *The Journal of Data Science* (http://www.jstage.jst.go.jp/browse/dsj/_vols), which provides a platform for all data workers to present their views and exchange ideas. The journal is largely devoted to the application of statistical methods and quantitative research.

   This citation, which indicates Columbia University's *The Journal of Data Science*, is a platform for data workers, and it provides excellent face validity that data science is about statistics but nothing that differentiates between data science and statistics. If the intent of the new journal is to demarcate data science and statistics, Columbia University ought to revisit the journal's mission statement. This citation does not indicate data science is similar to statistics.

   Decision IV: Columbia University's *The Journal of Data Science* platform unquestionably indicates data science is not similar to statistics.

   Running tally of data science is identical or similar to statistics: 2 out of 4.

5. In 2005, the National Science Board defined "data scientists as the information and computer scientists, database and software engineers and programmers, disciplinary experts, curators and expert annotators, librarians, archivists, and others, who are crucial to the successful management of a digital data collection" [7].

   This definition includes experts of varying disciplines, ranging from computer scientists to programmers to librarians and others, but oddly omits statisticians. Notwithstanding the omission of the focal element of data science, the definition pretermits any part of the four-step process.

   Decision V: The National Science Board's definition of data science, inferred from its definition of data scientists, is abundantly wanting and indicates data science is not similar to statistics.

   Running tally of data science is identical or similar to statistics: 2 out of 5.

6. In 2007, Fudan University in Shanghai, China, established the Research Center for Dataology and Data Science. In 2009, two of the center's researchers, Yangyong Zhu and Yun Xiong, both academic computer scientists, published "Introduction to Dataology and Data Science," in which they state "Dataology and Data Science takes data in cyberspace as its research object. It is a new science" [8].

   This reference, although limited in substance, sets apart the twins, dataology and data science, without providing definitions or context. Strangely, the reference considers data as the focal point in cyberspace, where the twins lie. Statistics, if implied to be either dataology or data science, is not the command station of cyberspace.

   Decision VI: Zhu and Xiong paint a picture that data science is in cyberspace, where there are no recorded sightings of statistics. The authors' data science does not indicate data science is similar to statistics.

   Running tally of data science is identical or similar to statistics: 2 out of 6.

7. In 2008, the Joint Information Systems Committee (JISC) published the final report of a study it commissioned to "examine and make recommendations on the role and career development of data scientists and the associated supply of specialist data curation skills to the research community." The study's final report defines "data scientists as people who work where the research is carried out—or, in the case of data centre personnel, in close collaboration with the creators of the data—and may be involved in creative enquiry and analysis, enabling others to work with digital data, and developments in database technology" (http://www.dcc.ac.uk/news/jisc-funding-opportunity-itt-skills-role-and-career-structure-datascientists-and-curators).

   JISC's citation provides its definition of data scientists as personnel, in part, who work closely with the creators of the data and whose work may involve creative analysis. There is no mention of statistics or the four-step process, except for the ubiquitous "creative analysis."

   Decision VII: The JISC definition of data science is void of any defining feature of statistics. JISC's data science is not similar to statistics.

   Running tally of data science is identical or similar to statistics: 2 out of 7.

8. In 2009, Michael Driscoll wrote in "The Three Sexy Skills of Data Geeks" "… with the Age of Data upon us, those who can model, mung, and visually communicate data—call us statisticians or data geeks—are a hot commodity" [9]. In 2010, Driscoll followed up with "The Seven Secrets of Successful Data Scientists" (http://medriscoll.com/post/4740326157/the-seven-secrets-of-successful-data-scientists). Driscoll holds a PhD in bioinformatics, a field whose curriculum greatly overlaps that of statistics. Thus, he is a first cousin of statisticians.

   Driscoll interchangeably uses the trio statisticians, data scientists, and data geeks as he defines them as those who model, mung, and effectively visualize data. Driscoll's PhD validates his understanding that data science and statistics are the same branches of knowledge.

   Decision VIII: Driscoll's trio clearly indicates data science is identical to statistics.

   Running tally of data science is identical or similar to statistics: 3 out of 8.

9. In 2009, Hal Varian, Google's chief economist (who holds a PhD in economics, a field whose curriculum significantly overlaps that of statistics), told the *McKinsey Quarterly*: "I keep saying the sexy job in the next ten years will be statisticians. People think I'm joking, but who would've guessed that computer engineers would've been the sexy job of the 1990s? The ability to take data—to be able to understand it, to process it, to extract value from it, to visualize it, to communicate it—that's going to be a hugely important skill in the next decades" (http://www.conversion-rate-experts.com/the-datarati/).

   Varian's quote indicates he has a very good understanding of the science of data, which requires statisticians and their statistics. I infer from Varian's mention of the sexy statistics job that he would likely use data science and statistics interchangeably.

   Decision IX: Varian's implied interchangeable use of data science and statistics indicates Varian's data science is identical to statistics.

   Running tally of data science is identical or similar to statistics: 4 out of 9.

10. In 2009, Nathan Yau wrote in "Rise of the Data Scientist": "As we've all read by now, Google's chief economist Hal Varian commented in January that the next

sexy job in the next ten years would be statisticians. Obviously, I whole-heartedly agree. Heck, I'd go a step further and say they're sexy now" (https://flowingdata.com/2009/06/04/rise-of-the-data-scientist/).

Yau's implied interchangeable use of data science and statistics clearly indicates he considers statisticians and data scientists are the same.

Decision X: Yau's data science is identical to statistics.

Running tally of data science is identical or similar to statistics: 5 out of 10.

11. In 2010, Kenneth Cukier, well-versed in data and digital products, wrote in *The Economist, Special Report* "Data, Data Everywhere," "... A new professional has emerged, the data scientist, who combines the skills of software programmer, statistician and storyteller/artist to extract the nuggets of gold hidden under mountains of data" (http://www.economist.com/node/15557443).

    Cukier's quote implies, in principle, that the data scientist is a statistician, notwithstanding his declaring the emergence of data scientist as a new profession. He adds the already recognized prerequisite of programming and whimsically adds a bit of art and storyteller. I push aside slightly Cukier's assertion that the data scientist's "primary" task is to find nuggets of information.

    Something about Cukier's quote was pulling me to find out about his academic background. I could not find even a torn, dog-eared page about his education from his many web pages. I do not know whether his educational background includes any level of statistical training. However, based on his high-volume publications on data, I consider Cukier a chronicler of statistics.

    Decision XI: Aside from his flourishes, Cukier's data science is identical to statistics.

    Running tally of data science is identical or similar to statistics: 6 out of 11.

12. In 2010, Mike Loukides, who is not a statistician but who does have a quantitative background with a BS in electrical engineering, wrote in "What Is Data Science?" (https://www.oreilly.com/ideas/what-is-data-science): "Data scientists combine entrepreneurship with patience, the willingness to build data products incrementally, the ability to explore, and the ability to iterate over a solution. They are inherently interdisciplinary. They can tackle all aspects of a problem, from initial data collection and data conditioning to drawing conclusions. They can think outside the box to come up with new ways to view the problem, or to work with very broadly defined problems: 'Here's a lot of data, what can you make from it?'"

    Loukides' long-winded citation is disappointing given his background is surely quantitative though not statistical. He comes off as a great salesperson for data science. I view Loukides' definition of data science as a word salad of things he knows about data science. My impression of Loukides is that he can talk about anything persuasively. Mike Loukides, vice president of content strategy for O'Reilly Media, Inc., edits many books on technical subjects. "Most recently, he's been fooling around with data and data analysis ..." (http://radar.oreilly.com/mikel). Someone fooling around with data and data analysis is not qualified to write about data science, as Loukides' citation demonstrates.

    Decision XII: Loukides' data science is not similar to statistics.

    Running tally of data science is identical or similar to statistics: 6 out of 12.

13. In 2013, *Forbes* ran an article by Gil Press entitled, "Data Science: What's the Half-Life of a Buzzword?" [10]. Press' article, which is a complication of interviews with

academics in the field of data analytics and journalists who cover business analytics, points out that although the use of the term "data science" has exploded in business environments, "there is more or less a consensus about the lack of consensus regarding a clear definition of data science." Gil Press, not a statistician but has an academic background in finance and marketing, concludes that data science is a buzzword without a clear definition and has simply replaced 'business analytics' in contexts such as graduate degree programs."

Press may know what business analytics is, but he definitely is uninformed about statistics, and any nexus between statistics and data science.

Decision XIII: Press's buzzy data science is not similar to statistics.

Running tally of data science is identical or similar to statistics: 6 out of 13.

14. In 2013, New York University (NYU) started a multi-million-dollar initiative to establish the country's leading Center for Data Science (CDS) training and research facilities. NYU states that "Data science overlaps traditionally strong disciplines at NYU such as mathematics, statistics, and computer science." "By combining aspects of statistics, computer science, applied mathematics, and visualization, data science can turn the vast amounts of data the digital age generates into new insights and new knowledge" (http://datascience.nyu.edu/what-is-data-science/).

    NYU's data science is explained in broad but accurate terms. CDS defines data science as encompassing the elements of the four-step process: the focal statistics, the implied computer science as the root computer-intensive statistical computations, and visualization. The inclusion of applied mathematics is the surrogate for the presumed fifth step of the process, the necessary theory of mathematical statistics, which serves as the foundation for the growth of statistics itself. Lastly, CDS's vision addresses the objective of data science, turning data into knowledge.

    Decision XIV: CDS's data science is identical to statistics.

    Running tally of data science is identical or similar to statistics: 7 out of 14.

15. In 2013, in the question-and-answer section of his keynote address at the Joint Statistical Meetings of the American Statistical Association, a noted applied statistician Nate Silver said, "I think data scientist is a sexed up term for a statistician. Statistics is a branch of science. Data scientist is slightly redundant in some way and people shouldn't berate the term statistician" [11].

    Silver's commentary on data science distinctly reflects the view of many statisticians, notwithstanding his view that the term berates statisticians.

    Decision XIV: Silver's quote plainly supports data science is identical to statistics.

    Running tally of data science is identical or similar to statistics: 8 out of 15.

16. In 2015, the *International Journal of Data Science and Analytics (IJDSA)* was launched by Springer to publish original work on data science and big data analytics. Its mission statement includes: *IJDSA* is the first scientific journal in data science and big data analytics. Its goals are to publish original, fundamental, and applied research outcomes in data and analytics theories, technologies, and applications, and to promote new scientific and technological approaches to strategic value creation in data-rich applications. It provides self-identification with the key words: artificial intelligence, bioinformatics, business information systems, database management, and information retrieval (http://www.springer.com/computer/database+management+%26+information+retrieval/journal/41060).

IJDSA's central point is being the first publication of original data science and big data. However, *IJDSA*'s mission points and key words do not acknowledge any content of the four-step process. There is no mention of the focal statistics, implied or otherwise.

Decision XVI: *IJDSA*'s data science is not similar to statistics.

Running tally of data science is identical or similar to statistics: 8 out of 16.

17. In 2016, on Kaggle, a self-described "world's largest community of data scientists compete to solve your most valuable problems content," data science is defined as "a newly emerging field dedicated to analyzing and manipulating data to derive insights and build data products. It combines skill sets ranging from computer science to mathematics to art" (Kaggle.com).

    Kaggle's definition does not acknowledge statistics proper, although it does include the perfunctory analysis data to derive insights, which are rhetorical statistics.

    Decision XVII: Kaggle's data science is not similar to statistics.

    Running tally of data science is identical or similar to statistics: 8 out of 17.

18. In 2016, KDnuggets (KDN), a self-proclaimed "official resource for data mining, analytics, big data, and data science" defined data science as "the extraction of knowledge from large volumes of data that aren't structured, which is a continuation of the field of data mining and predictive analytics, also known as knowledge discovery and data mining" (KDnuggets.com).

    KDN's definition limits data science to unstructured big data, which is part of data mining and predictive analytics. This definition is small in scope and misses the essence of the four-step process and central statistics.

    Decision XVIII: KDN's data science is not similar to statistics.

    Running tally of data science is identical or similar to statistics: 8 out of 18.

19. In 2016, on University of California (UC) Berkeley's website, there is a link, "What is Data Science?" in which, under the heading "A New Field Emerges, "the definition of data science is: "There is significant and growing demand for data-savvy professionals in businesses, public agencies, and nonprofits. The supply of professionals who can work effectively with data at scale is limited, and is reflected by rapidly rising salaries for data engineers, data scientists, statisticians, and data analysts" (https://datascience.berkeley.edu/about/what-is-data-science/27).

    UC Berkeley's definition includes an exhibit of being uninformed of the basic features of data science. Moreover, UC Berkeley includes data scientists and statisticians in its definition, which implies it considers that the two are different.

    Decision XIX: UC Berkeley's data science is not similar to statistics.

    Running tally of data science is identical or similar to statistics: 8 out of 19 (= 42.11%).

2.4 Discussion: Are Statistics and Data Science Different?

I perform the classical significance test to assess my hypothesis:

H0: Data Science and Statistics are identical ($p = p0$)

H1: Data Science and Statistics are not identical ($p \neq p0$)

I conduct the exact one-sample test for a proportion due to the small sample size.* Under the null hypothesis, $p0 = 50\%$. The formula for exact p-value is 2* PROBBNML(.50, 19, 8). The exact p-value is 0.6476, which is greater than alpha = 0.05. Thus, the null hypothesis is not rejected, and the decision is data science and statistics are one and the same.

The perfunctory use of the statistical test is not informative, so I do some data mining of the 42.11%. The numbers behind the observed 42.11% are:

1. All six statisticians and their academic cousins consider data science is statistics.
   a. C. F. Jeff Wu, William S. Cleveland, Michael Driscoll, Hal Varian, Nathan Yau, and Nate Silver
2. One chronicler of statistics considers data science is statistics.
   a. Kenneth Cukier
3. One academic institution considers data science is statistics.
   a. NYU's CDS
4. The remaining 11 items, which consider data science is not statistics, consist of:
   a. Seven commercial entities—the IFIP, National Science Board, Mike Loukides, Gil Press, *IJDSA*, Kaggle, and KDN.
   b. Four academic institutions—Columbia University, the Research Center for Dataology and Data Science, JISC, and UC Berkeley.

At this point, I clarify my previously cited stopping rule on the size of the collected mentions. I stopped searching when mentions retrieved provided no useful knowledge for the study. I observed that statisticians and one chronicler of statistics declared equivalence of statistics and data science. All other mentions with their varying meanings of data science contend data science and statistics are not similar. Continuing to collect additional mentions, I would not yield a significant change in the findings.

2.4.1 Analysis: Are Statistics and Data Science Different?

The first observation, all statisticians agree data science is statistics, is not surprising. Statisticians, who make up the core of the statistics community, should know what is happening inside and outside their community. Notwithstanding why trends start, the sole remark of Silver's—data science is a sexed up labeling of statistics—might be the answer as to the cloaking of statistics. Other than Silver's anger over berating statisticians by using the term data science, I did not find the unknown thread from where the term arose.

As for academic institutions that train statisticians, statements from them should indicate that data science and statistics are the same. However, four out of five academic institutions declare data science and statistics are dissimilar. NYU's CDS is the known holdout for this group of academia. I can only conclude that these four institutions of higher learning have taken the low road of marketing by going for the sexy hype of Silver to generate interest, step-up enrollment, and increase revenue.

The observation of greatest insight in this exercise is that all 11 mentions, which indicate the dissimilarity of statistics and data science, have one thing in common: virtually

* An exact test is used in situations when the sample size is small. Applying the well-known and virtually always used large sample tests render p-values that are not close approximations for the true p-values.

nothing. In general, the 11 mentions are data science narratives—not definitions, which are a hodgepodge of incongruous ideas. This finding suggests the changes stemming from the Internet's big data created a state of uncertainty about what are the changes on the essential statistics. There should be a new term to reflect the changes. I conclude either the changes are not significant enough or "the times they are *not* a changing," just not yet.

2.5 Summary

Statisticians think in a like manner. Statisticians know who they are. They will turn around to engage, not keep walking, if called data scientists. All others believe that the big data-related changes necessitate a redefining of statistics and statisticians, yet they have not even proposed a working definition that differentiates data science and data scientist form statistics and statisticians, respectively.

I wonder what John W. Tukey—an influential statistician who coined the word bit (contraction of binary digit), the father of exploratory data analysis (EDA), a great contributor to the statistics literature, and who considered himself simply as a data analyst—would think about the terms data scientist and data science.

2.6 Epilogue

"Statistics by Any Other Name Would Be as Sweet"
'Tis but the name that is the rival;
Thou art thyself, though not a Montague.
What's Montague? It is nor rows, nor columns
Nor the table itself, nor any other part
Belonging to statistics. O, be some other name!
Data science. What's in a name?
That which we call statistics,
By any other name would be as sweet.

Bruce "Shakespeare" Ratner

References

1. Vorbereitung zur Staatswissenschaft. The political constitution of the present principal European countries and peoples.
2. *The Columbia Electronic Encyclopedia*, 6th ed., Columbia University Press, NY, 2012.
3. Gruder, C.L., Determinants of social comparison choices, *Journal of Experimental Social Psychology*, 7(5), 473–489, 1971.

4. Bikhchandani, S., Hirshleifer, D., and Welch, I., A theory of fads, fashion, custom, and cultural change as informational cascade, *Journal of Political Economy*, 100, 992–1026, 1992.
5. Gould, I. H., (ed.), *IFIP Guide to Concepts and Terms in Data Processing*, North-Holland Publ. Co., Amsterdam, 1971.
6. Cleveland, W. S., Data science: An action plan for expanding the technical areas of the field of statistics, *International Statistical Review/Revue Internationale de Statistique*, 21–26, 2001.
7. *Long-Lived Digital Data Collections Enabling Research and Education in the 21st Century*, National Science Board, 2005, p. 19. http://www.nsf.gov/pubs/2005/nsb0540/nsb0540.pdf.
8. Zhu, Y. Y., and Xiong, Y., *Dataology and Data Science*, Fudan University Press, 2009.
9. Akerkar, R., and Sajja, P.S., *Intelligent Techniques for Data Science*, Springer, Switzerland, 2016. http://medriscoll.com/post/4740157098/the-three-sexy-skills-of-data-geeks.
10. Press, G., Data science: What's the half-life of a buzzword? Forbes Magazine, 2013.
11. Silver, N., *What I Need from Statisticians*, Statistics Views, 2013.

3

Two Basic Data Mining Methods for Variable Assessment

3.1 Introduction

Assessing the relationship between a predictor variable and a dependent variable is an essential task in the model-building process. If the identified relationship is tractable, then the predictor variable is reexpressed to reflect the uncovered relationship and consequently tested for inclusion into the model. The correlation coefficient is the key statistic, albeit often misused, in variable assessment methods. The linearity assumption of the correlation coefficient is frequently not subjected to testing in which case the utility of the coefficient is unknown. The purpose of this chapter is twofold: to present (1) the *smoothed scatterplot* as an easy and effective data mining method and (2) a *general association* nonparametric test for assessing the relationship between two variables. The intent of Point 1 is to embolden the data analyst to test the linearity assumption to ensure the proper use of the correlation coefficient. The intent of Point 2 is an effectual data mining method for assessing the indicative message of the smoothed scatterplot.

I review the correlation coefficient with a quick tutorial, which includes an illustration of the importance of testing the linearity assumption and outline the construction of the smoothed scatterplot, which serves as an easy method for testing the linearity assumption. Next, I introduce the *general association test* as a data mining method for assessing a general association between two variables.

3.2 Correlation Coefficient

The correlation coefficient, denoted by r, is a measure of the strength of the straight-line or linear relationship between two variables. The correlation coefficient takes on values ranging between +1 and –1. The following points are the accepted guidelines for interpreting the correlation coefficient:

1. 0 indicates no linear relationship.
2. +1 indicates a perfect positive linear relationship: As one variable increases in its values, the other variable also increases in its values via an exact linear rule.
3. –1 indicates a perfect negative linear relationship: As one variable increases in its values, the other variable decreases in its values via an exact linear rule.

4. Values between 0 and 0.3 (0 and –0.3) indicate a weak positive (negative) linear relationship via a shaky linear rule.
5. Values between 0.3 and 0.7 (–0.3 and –0.7) indicate a moderate positive (negative) linear relationship via a fuzzy-firm linear rule.
6. Values between 0.7 and 1.0 (–0.7 and –1.0) indicate a strong positive (negative) linear relationship via a firm linear rule.
7. The value of r^2 is the percent of the variation in one variable explained by the other variable or the percent of variation shared between the two variables.
8. Linearity assumption: The correlation coefficient requires the underlying relationship between the two variables under consideration to be linear. If the relationship is known to be linear, or the observed pattern between the two variables appears to be linear, then the correlation coefficient provides a reliable measure of the strength of the linear relationship. If the relationship is known to be nonlinear, or the observed pattern appears to be nonlinear, then the correlation coefficient is not useful or is at least questionable.

The calculation of the correlation coefficient for two variables X and Y is simple to understand. Let zX and zY be the standardized versions of X and Y, respectively. That is, zX and zY are both reexpressed to have means equal to zero, and standard deviations (std) equal to one. The reexpressions used to obtain the standardized scores are in Equations 3.1 and 3.2:

$$zX_i = \left[X_i - \text{mean}(X)\right]/\text{std}(X) \tag{3.1}$$

$$zY_i = \left[Y_i - \text{mean}(Y)\right]/\text{std}(Y) \tag{3.2}$$

The correlation coefficient is the mean product of the paired standardized scores (zX_i, zY_i) as expressed in Equation 3.3.

$$r_{X,Y} = \text{sum of}\left[zX_i * zY_i\right]/(n-1) \tag{3.3}$$

where n is the sample size.

For a simple illustration of the calculation of the correlation coefficient, consider the sample of five observations in Table 3.1. Columns zX and zY contain the standardized scores of X and Y, respectively. The last column is the product of the paired

TABLE 3.1

Calculation of Correlation Coefficient

obs	X	Y	zX	zY	zX*zY
1	12	77	–1.14	–0.96	1.11
2	15	98	–0.62	1.07	–0.66
3	17	75	–0.27	–1.16	0.32
4	23	93	0.76	0.58	0.44
5	26	92	1.28	0.48	0.62
Mean	18.6	87		sum	1.83
std	5.77	10.32			
n	5			r	0.46

standardized scores. The sum of these scores is 1.83. The mean of these scores (using the adjusted divisor n – 1, not n) is 0.46. Thus, $r_{X,Y} = 0.46$.

3.3 Scatterplots

The testing of the linearity assumption of the correlation coefficient uses the *scatterplot*, which is a mapping of the paired points (X_i, Y_i) in an X–Y graph. X_i and Y_i are typically assigned as the predictor and dependent variables, respectively; index i represents the observations from 1 to n, where n is the sample size. The scatterplot provides a visual display of a discoverable relation between two variables within a framework of a horizontal X-axis perpendicular to a vertical Y-axis graph (without a causality implication suggested by the designation of dependent and predictor variables). If the scatter of points in the scatterplot appears to overlay a straight line, then the assumption has been satisfied, and $r_{X,Y}$ provides a meaningful measure of the linear relationship between X and Y. If the scatter does not appear to overlay a straight line, then the assumption has not been satisfied, and the $r_{X,Y}$ value is at best questionable. Thus, when using the correlation coefficient to measure the strength of the linear relationship, it is advisable to construct the scatterplot to test the linearity assumption. Unfortunately, many data analysts do not construct the scatterplot, thus rendering any analysis based on the correlation coefficient as potentially invalid. The following illustration is presented to reinforce the importance of evaluating scatterplots.

Consider the four datasets with 11 observations in Table 3.2 [1]. There are four sets of (X, Y) points, with the same correlation coefficient value of 0.82. However, the X–Y relationships are distinct from one another, reflecting a different underlying structure, as depicted in the scatterplots in Figure 3.1.

The scatterplot for X1–Y1 (upper left) indicates a linear relationship. Thus, the $r_{X1,Y1}$ value of 0.82 correctly suggests a strong positive linear relationship between X1 and Y1. The scatterplot for X2–Y2 (upper right) reveals a curved relationship; $r_{X2,Y2} = 0.82$. The scatterplot for X3–Y3 (lower left) reveals a straight line except for the "outside" observation 3, data point (13, 12.74); $r_{X3,Y3} = 0.82$. The scatterplot for X4–Y4 (lower right) has a "shape of its

TABLE 3.2

Four Pairs of (X, Y) with the Same Correlation Coefficient ($r = 0.82$)

obs	X1	Y1	X2	Y2	X3	Y3	X4	Y4
1	10	8.04	10	9.14	10	7.46	8	6.58
2	8	6.95	8	8.14	8	6.77	8	5.76
3	13	7.58	13	8.74	13	12.74	8	7.71
4	9	8.81	9	8.77	9	7.11	8	8.84
5	11	8.33	11	9.26	11	7.81	8	8.47
6	14	9.96	14	8.1	14	8.84	8	7.04
7	6	7.24	6	6.13	6	6.08	8	5.25
8	4	4.26	4	3.1	4	5.39	19	12.5
9	12	10.84	12	9.13	12	8.15	8	5.56
10	7	4.82	7	7.26	7	6.42	8	7.91
11	5	5.68	5	4.74	5	5.73	8	6.89

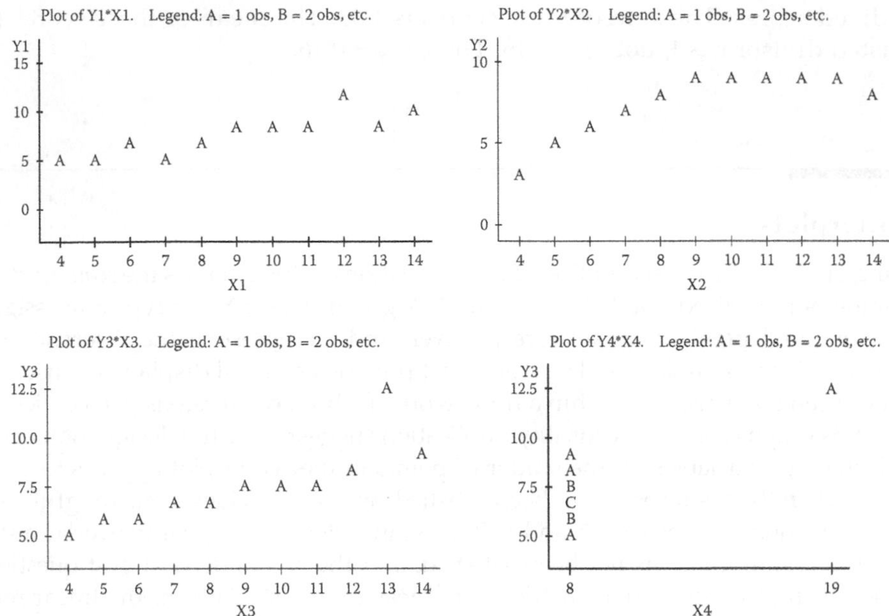

FIGURE 3.1
Four different datasets with the same correlation coefficient.

own," which is clearly not linear; $r_{X4,Y4}$= 0.82. Accordingly, the correlation coefficient value of 0.82 is not a meaningful measure for the last three X–Y relationships.

3.4 Data Mining

Data mining—the process of revealing unexpected relationships in data—is needed to unmask the underlying relationships in scatterplots filled with big data. Big data, so much a part of the information world, have rendered the scatterplot overloaded with data points or information. Paradoxically, scatterplots based on more information are less informative. With a quantitative target variable, the scatterplot typically becomes a cloud of points with sample-specific variation, called *rough*, which masks the underlying relationship. With a qualitative target variable, there is *discrete* rough, which masks the underlying relationship. In either case, removal of rough from the big data scatterplot reveals the underlying relationship. After presenting two examples that illustrate how scatterplots filled with more data can provide less information, I outline the construction of the *smoothed scatterplot*, a rough-free scatterplot, which reveals the underlying relationship in big data.

3.4.1 Example 3.1

Consider the quantitative target variable Toll Calls (TC) in dollars and the predictor variable Household Income (HI) in dollars from a sample of size 102,000. The calculated $r_{TC,HI}$ is 0.09. The TC–HI scatterplot in Figure 3.2 shows a cloud of points obscuring the underlying relationship within the data (assuming a relationship exists). This scatterplot is uninformative regarding an indication for the reliable use of the calculated $r_{TC,HI}$.

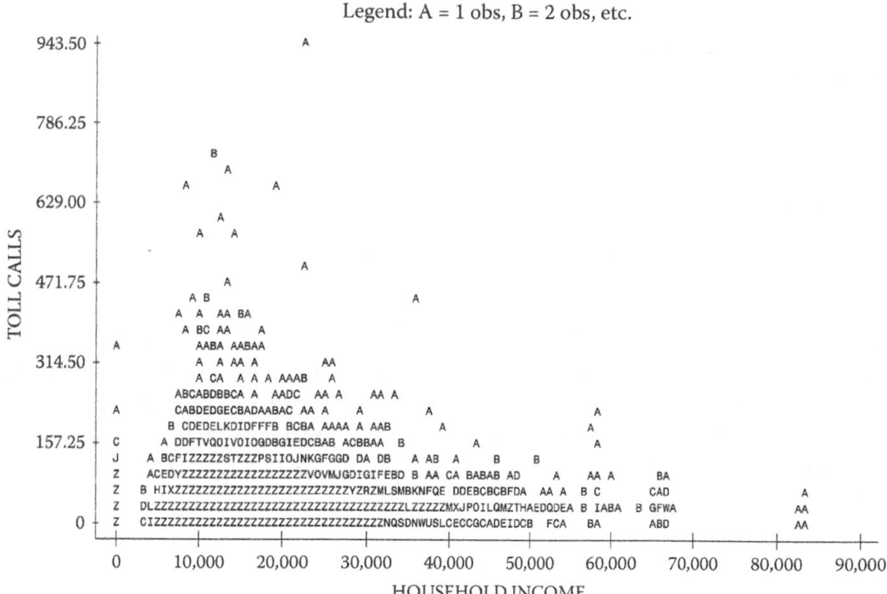

FIGURE 3.2
Scatterplot for TOLL CALLS and HOUSEHOLD INCOME.

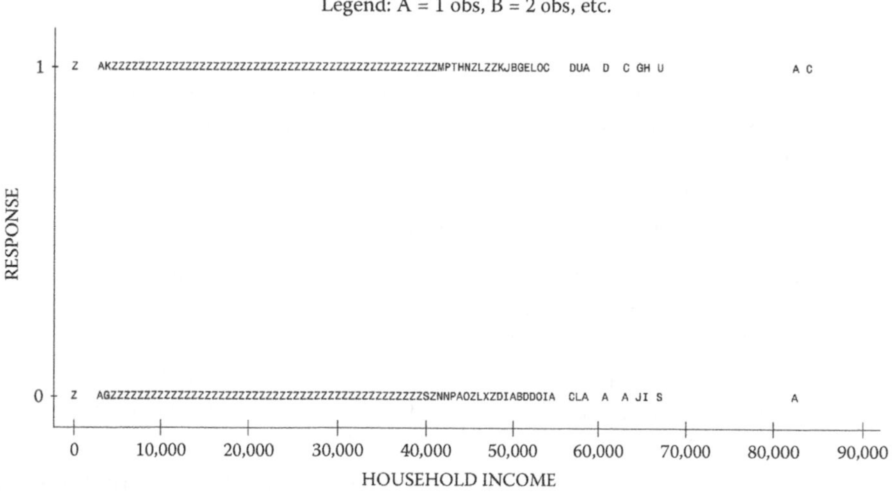

FIGURE 3.3
Scatterplot for RESPONSE and HOUSEHOLD INCOME.

3.4.2 Example 3.2

Consider the qualitative target variable Response (RS), which measures the response to a mailing, and the predictor variable HI from a sample of size of 102,000. RS assumes yes and no values, which are coded as 1 and 0, respectively. The calculated $r_{RS,HI}$ is 0.01. The RS–HI scatterplot in Figure 3.3 shows "train tracks" obscuring the underlying relationship within the data (assuming a relationship exists). The tracks appear because the target variable takes on only two values, 0 and 1. As in the first example, this scatterplot is uninformative regarding an indication for the reliable use of the calculated $r_{RS,HI}$.

3.5 Smoothed Scatterplot

The *smoothed scatterplot* is the desired visual display for revealing a rough-free relationship lying within big data. *Smoothing* is a method of removing the rough and retaining the predictable underlying relationship (the *smooth*) in data by averaging within *neighborhoods* of similar values. Smoothing an X–Y scatterplot involves taking the averages of both the target (dependent) variable Y and the continuous predictor (independent) variable X, within X-based neighborhoods [2]. The six-step procedure to construct a smoothed scatterplot follows:

1. Plot the (X_i, Y_i) data points on an X–Y graph.
2. For a continuous X variable, divide the X-axis into distinct and nonoverlapping neighborhoods (slices). A common approach to dividing the X-axis is creating 10 equal-sized slices (also known as *deciles*), whose aggregation equals the total sample [3–5]. Each slice accounts for 10% of the sample. For a categorical X variable, slicing per se cannot be performed. The categorical labels (levels) define *single-point* slices. Each single-point slice accounts for a percentage of the sample dependent on the distribution of the categorical levels in the sample.
3. Take the average of X within each slice. The average is either the mean or median. The average of X within each slice is known as a smooth X value, or smooth X. Notation for smooth X is sm_X.
4. Take the average of Y within each slice.
   a. For a continuous Y, the mean or median serves as the average.
   b. For a categorical Y that assumes only two levels, the levels are reassigned typically numeric values 0 and 1. Clearly, only the mean can be calculated. This coding yields Y proportions or Y rates.
   c. For a multinomial Y that assumes more than two levels, say k, clearly, the average cannot be calculated. (Level-specific proportions are calculable, but they do not fit into any developed procedure for the intended task.)
      i. The appropriate procedure, which involves generating all combinations of pairwise-level scatterplots, is cumbersome and rarely used.
      ii. The procedure is that most often used because of its ease of implementation and efficiency is discussed in Section 18.5.
   d. Notation for smooth Y is sm_Y.
5. Plot the smooth points (smooth Y, smooth X), constructing a *smooth scatterplot.*
6. Connect the smooth points, starting from the first left smooth point through the last right smooth point. The resultant *smooth trace* line reveals the underlying relationship between X and Y.

I provide subroutines for generating smoothplots in Chapter 44, Some of My Favorite Statistical Subroutines.

I return to Examples 1 and 2. The HI data, grouped into 10 equal-sized slices, consist of 10,200 observations. The averages (means) or smooth points for HI with both TC and RS within the slices (numbered from 0 to 9) are in Tables 3.3 and 3.4, respectively. The smooth points are plotted and connected.

The TC smooth trace line in Figure 3.4 clearly indicates a linear relationship. Thus, the $r_{TC, HI}$ value of 0.09 is a reliable measure of a weak positive linear relationship between TC and HI. Moreover, testing for the inclusion of HI (without any reexpression) in the TC model is possible and recommended. Note the small r value does not preclude testing HI for model inclusion. The discussion of this point is in Section 6.5.1.

The RS smooth trace line in Figure 3.5 indicates the relationship between RS and HI is not linear. Thus, the $r_{RS,HI}$ value of 0.01 is invalid. The nonlinearity raises the following question: Does the RS smooth trace line indicate a general association between RS and HI, implying a nonlinear relationship, or does the RS smooth trace line indicate a random scatter, implying no relationship between RS and HI? The answer lies with the graphical nonparametric general association test [5].

TABLE 3.3

Smooth Points: TOLL CALLS and HOUSEHOLD INCOME

Slice	Average TOLL CALLS	Average HOUSEHOLD INCOME
0	$31.98	$26,157
1	$27.95	$18,697
2	$26.94	$16,271
3	$25.47	$14,712
4	$25.04	$13,493
5	$25.30	$12,474
6	$24.43	$11,644
7	$24.84	$10,803
8	$23.79	$9,796
9	$22.86	$6,748

TABLE 3.4

Smooth Points: Response and HOUSEHOLD INCOME

Slice	Average Response (%)	Average HOUSEHOLD INCOME
0	2.8	$26,157
1	2.6	$18,697
2	2.6	$16,271
3	2.5	$14,712
4	2.3	$13,493
5	2.2	$12,474
6	2.2	$11,644
7	2.1	$10,803
8	2.1	$9,796
9	2.3	$6,748

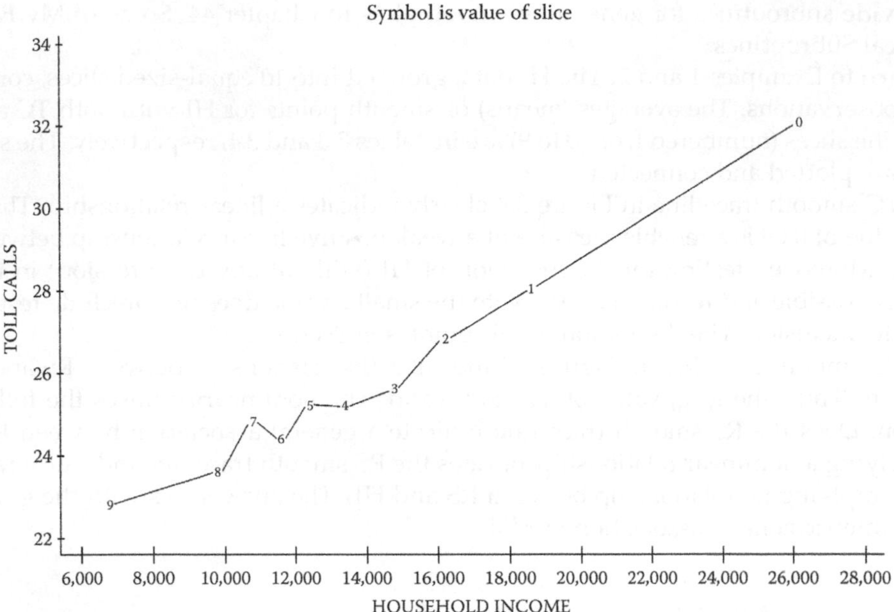

FIGURE 3.4
Smoothed scatterplot for TOLL CALLS and HOUSEHOLD INCOME.

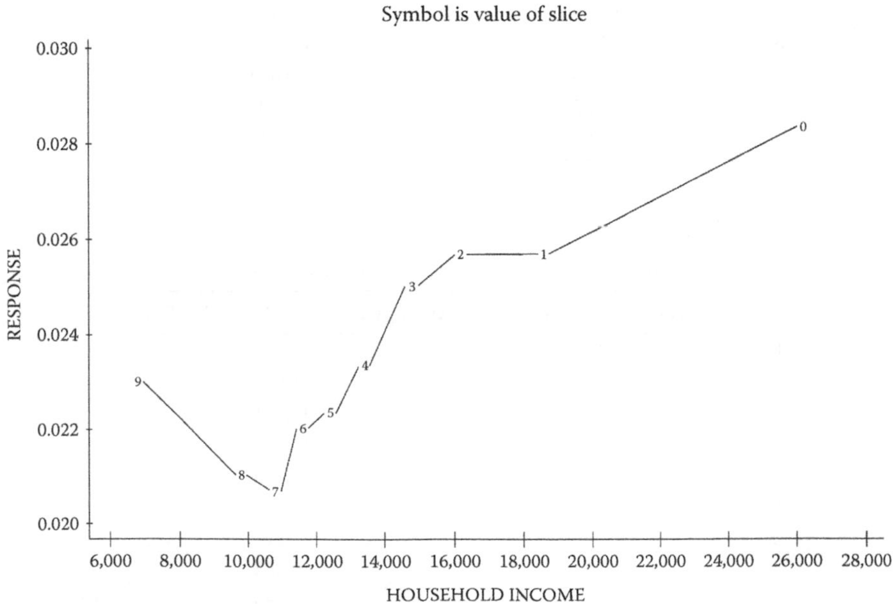

FIGURE 3.5
Smoothed scatterplot for RESPONSE and HOUSEHOLD INCOME.

3.6 General Association Test

Here is the general association test:

1. *Plot* the N smooth points in a scatterplot and draw a horizontal medial line that divides the N points into two equal-sized groups.
2. *Connect* the N smooth points starting from the first left-hand smooth point. N – 1 line segments result. Count the number m of line segments that cross the medial line.
3. *Test* for significance. The null hypothesis: There is no association between the two variables at hand. The alternative hypothesis: There is an association between the two variables.
4. *Consider* the test statistic TS is N – 1 – m.

Reject the null hypothesis if TS is greater than or equal to the cutoff score in Table 3.5. The conclusion is there is an association between the two variables. The smooth trace line indicates the "shape" or structure of the association.

Fail to reject the null hypothesis if TS is less than the cutoff score in Table 3.5. The conclusion is there is no association between the two variables.

TABLE 3.5

Cutoff Scores for General Association Test (95% and 99% Confidence Levels)

N	95%	99%
8–9	6	—
10–11	7	8
12–13	9	10
14–15	10	11
16–17	11	12
18–19	12	14
20–21	14	15
22–23	15	16
24–25	16	17
26–27	17	19
28–29	18	20
30–31	19	21
32–33	21	22
34–35	22	24
36–37	23	25
38–39	24	26
40–41	25	27
42–43	26	28
44–45	27	30
46–47	29	31
48–49	30	32
50–51	31	33

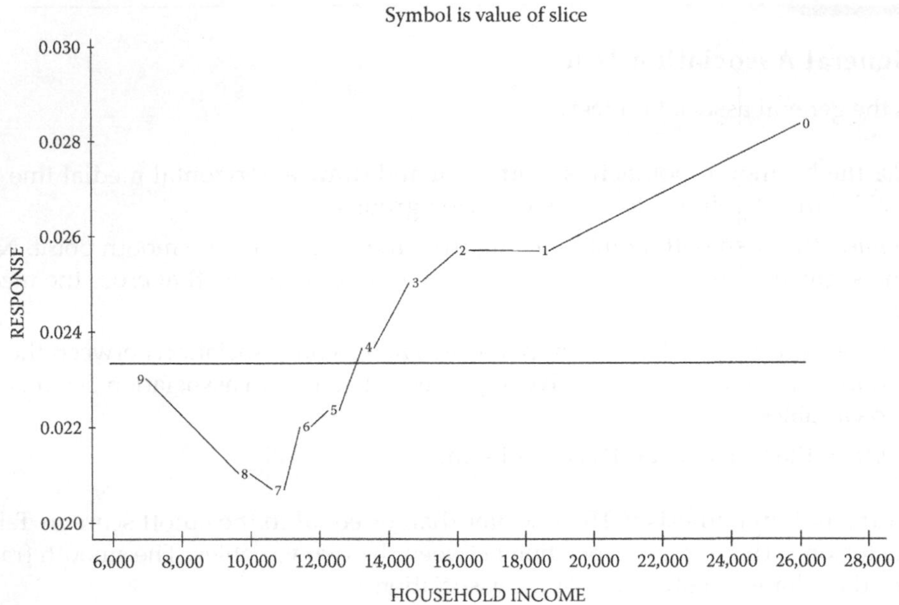

FIGURE 3.6
General association test for smoothed RS–HI scatterplot.

Returning to the smoothed scatterplot of RS and HI, I determine the following:

1. There are ten smooth points, N = 10.
2. The medial line divides the smooth points such that Points 5 to 9 are below the line and Points 0 to 4 are above.
3. The line segment formed by Points 4 and 5 in Figure 3.6 is the only segment that crosses the medial line. Accordingly, m = 1.
4. TS equals 8 (= 10 – 1 – 1), which is greater than or equal to the 95% and 99% confidence cutoff scores 7 and 8, respectively.

Thus, there is a 99% (and of course 95%) confidence level that there is an association between RS and HI. The RS smooth trace line in Figure 3.5 suggests the observed relationship between RS and HI appears to be polynomial to the third power. Accordingly, the linear (HI), quadratic (HI^2), and cubed (HI^3) Household Income terms should be tested for entering the Response model.

3.7 Summary

It should be clear that an analysis based on the uncritical use of the coefficient correlation is problematic. The strength of a relationship between two variables cannot simply be the calculated r value itself. The testing for the linearity assumption, which is made easy by the simple scatterplot or smoothed scatterplot, is necessary for a thorough-and-ready analysis.

If the observed relationship is linear, then the r value can be taken at face value for the strength of the relationship at hand. If the observed relationship is not linear, then the r value must be disregarded or used with extreme caution.

When a smoothed scatterplot for big data does not reveal a linear relationship, its scatter undergoes the proposed nonparametric method to test for randomness or a noticeable general association. If the former (randomness) is true, then it is concluded there is no association between the variables. If the latter (noticeable association) is true, then the predictor variable is reexpressed to reflect the observed relationship and therefore tested for inclusion in the model.

References

1. Anscombe, F.J., Graphs in statistical analysis, *American Statistician*, 27, 17–22, 1973.
2. Tukey, J.W., *Exploratory Data Analysis*, Addison-Wesley, Reading, MA, 1997.
3. Hardle, W., *Smoothing Techniques*, Springer-Verlag, New York, 1990.
4. Simonoff, J.S., *Smoothing Methods in Statistics*, Springer-Verlag, New York, 1996.
5. Quenouille, M.H., *Rapid Statistical Calculations*, Hafner, New York, 1959.

4

CHAID-Based Data Mining for Paired-Variable Assessment

4.1 Introduction

Building on the concepts of the scatterplot and the smoothed scatterplot as data mining methods presented in Chapter 3, I introduce a new data mining method: a *smoother scatterplot* based on chi-squared automatic interaction detection (CHAID). The new method has the potential of exposing a more reliable depiction of the unmasked relationship for paired-variable assessment than that of the scatterplot and the smoothed scatterplot. I use a new dataset to keep an edge on refocusing another illustration of the scatterplot and smoothed scatterplot. Then, I present a primer on CHAID, after which I bring the proposed subject to the attention of data miners, who are buried deep in data, to help exhume themselves along with the patterns and relationships within the data.

4.2 The Scatterplot

Big data are more than the bulk of today's information world. They contain valuable elements of the streaming current of digital data. Data analysts find themselves in troubled waters while pulling out the valuable elements. One impact of big data is on the basic analytical tool: the scatterplot has become overloaded with data points or with information. Paradoxically, the scatterplot based on more information is less informative. The scatterplot displays a cloud of data points, of which too many are due to sample variation, namely, the *rough* that masks a presuming existent underlying a relationship.* Removal of the rough from the cloudy scatterplot allows the *smooth,* hidden behind the cloud, to shine through and open the sought-after underlying relationship. I offer to view an exemplary scatterplot filled with too many data and then show the corresponding *smooth scatterplot,* a rough-free, smooth-full scatterplot, which reveals the sunny-side up of the inherent character of a paired-variable assessment.

* Scatterplots with the X or Y variables as qualitative dependent variables yield not a cloud, but either two or more parallel lines or a dot matrix corresponding to the number of categorical levels. In the latter situation, the scatterplot displays mostly the rough in the data.

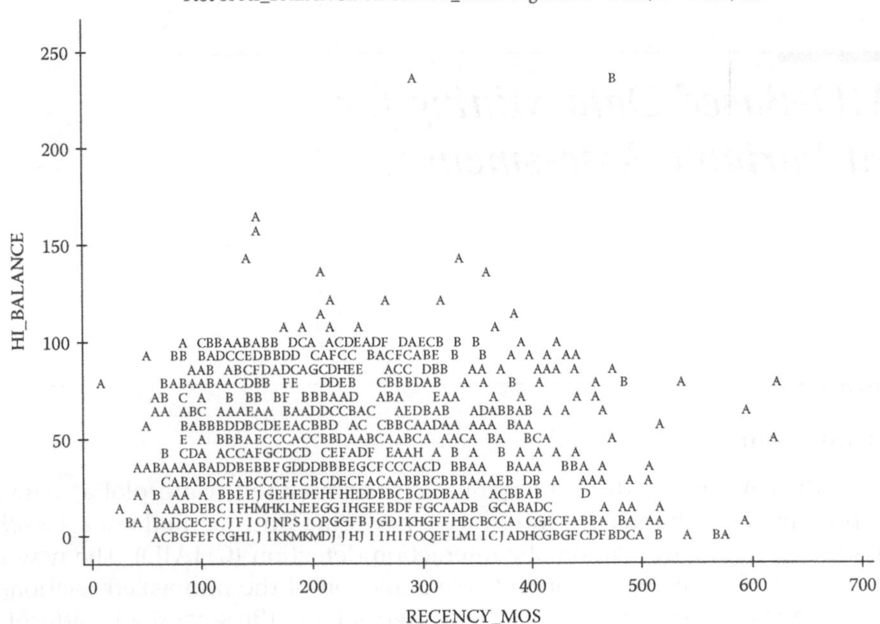

FIGURE 4.1
Scatterplot of HI_BALANCE and RECENCY_MOS.

4.2.1 An Exemplar Scatterplot

I data mine for an honest display of the relationship between two variables from a real study: HI_BALANCE (the highest balance attained among credit card transactions for an individual) and RECENCY_MOS (the number of months since last purchase). The first step in the data mining process is to produce the scatterplot of HI_BALANCE and RECENCY_MOS. It is clear the relationship between the two variables reflects an irregular and diffuse cloud of data in the scatterplot (Figure 4.1). To lift the cloudiness in the scatterplot (i.e., to remove the rough to expose the smooth in the data), I create the smooth scatterplot of the present scatterplot in the next section.

4.3 The Smooth Scatterplot

The smooth scatterplot is the appropriate visual display for uncovering the sought-after underlying relationship within an X–Y graph of raw big data. *Smoothing* is a method of averaging within *neighborhoods* of similar values—for removing the rough and retaining the smooth in data. Specifically, smoothing of a scatterplot involves taking the averages of both X and Y within X-based neighborhoods [1]. The six-step construction procedure for a smooth scatterplot is in Chapter 3.

The smooth scatterplot of HI_BALANCE and RECENCY_MOS is in Figure 4.2. The *SM_HI_BALANCE* and *SM_RECENCY_MOS* values per slice, which assumes numeric labels from 0 to 9, are in Table 4.1. The uncovered relationship is an S-shape trend. This relationship can be analyzed further using Tukey's bulging rule, discussed in Chapter 10.

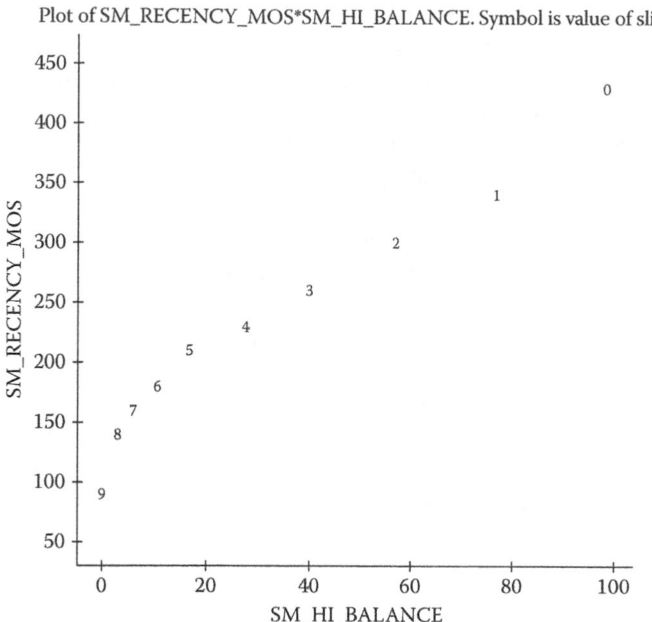

FIGURE 4.2
Smooth scatterplot of RECENCY_MOS and HI_BALANCE.

TABLE 4.1
Smooth Values for HI_BALANCE and RECENCY_MOS by Slice

Slice	SM_HI_BALANCE	SM_RECENCY_MOS
0	99.0092	430.630
1	77.6742	341.485
2	56.7908	296.130
3	39.9817	262.940
4	27.1600	230.605
5	17.2617	205.825
6	10.6875	181.320
7	5.6342	159.870
8	2.5075	135.375
9	0.7550	90.250

4.4 Primer on CHAID

Before discussing the CHAID-based *smoother* scatterplot, I provide a succinct primer on CHAID (the acronym for chi-squared automatic interaction detection, not detector). CHAID is a popular technique, especially among wannabe regression modelers with no significant statistical training because CHAID regression tree models are easy to build, understand, and implement. Also, CHAID's underpinnings are quite attractive: CHAID is an assumption-free method (i.e., there are no formal theoretical assumptions to meet), and CHAID is mighty powerful in its handling of a "big data" number of many predictor variables. In contrast, traditional regression models are assumption full, which makes them susceptible to risky results and inefficient in their managing of many predictor variables. Note, I use the following CHAID terms interchangeably: CHAID, CHAID tree, CHAID regression tree, CHAID regression tree model, and CHAID model.[*,†]

CHAID is a recursive technique that splits a population (Node 1) into nonoverlapping binary (two) subpopulations (nodes, bins, slices), defined by the "most important" predictor variable. Then, CHAID splits the first-level resultant nodes, defined by the next most important predictor variables, and continues to split second-level, third-level, …, and nth-level resultant nodes, until either stopping rules are satisfied or splitting criterion is not reached. For the splitting criterion,[‡] the variance of the dependent variable is minimized within each of the two resultant nodes and maximized between the two resultant nodes.

To clarify the recursive splitting process, after the first splitting of the population (virtually always[§]) produces resultant Nodes 2 and 3, further splitting is attempted on the resultant nodes. Nodes are split with respect to two conditions: (1) if the user-defined stopping rules (e.g., minimum-size node to split and maximum level of the tree) are not satisfied, and (2) if the splitting criterion bears resultant nodes with significantly different means for the Y variable. Assuming the conditions are in check, Node 2 splits into Nodes 4 and 5; Node 3 splits into Nodes 6 and 7. For each of Nodes 4–7, splitting is performed if either of the two conditions is true; otherwise, the splitting stops, and the CHAID tree is complete.

4.5 CHAID-Based Data Mining for a Smoother Scatterplot

I illustrate the CHAID model for making a *smoother scatterplot* with the HI_BALANCE and RECENCY_MOS variables of the previously mentioned study. The fundamental characteristic of the CHAID-based smoother scatterplot is: The CHAID model consists of *only one* predictor variable. I build a CHAID regression tree model, regressing HI_BALANCE on RECENCY_MOS. The stopping rules used are minimum-size node to split 10 and the maximum level of the tree 3. The CHAID regression tree model in Figure 4.3 reads as follows:

1. Node 1 mean for HI_BALANCE is 33.75 within the sample of size 2,000.
2. The splitting of Node 1 yields Nodes 2 and 3.
   a. Node 2 includes individuals whose RECENCY_MOS is ≤319.04836. Mean HI_BALANCE is 35.46 and node size is 1,628.
   b. Node 3 includes individuals whose RECENCY_MOS is >319.04836. Mean HI_BALANCE is 26.26 and node size is 372.
3. The splitting of Node 3 yields Nodes 10 and 11. The nodes are read similarly to Items 2a and 2b.

* There are many CHAID software packages in the marketplace. The best ones are based on the original automatic interaction detection (AID) algorithm.

† See *A Pithy History of CHAID and Its Offspring* on the author's website (http://www.geniq.net/res/Reference-Pithy-history-of-CHAID-and-Offspring.html).

‡ There are many splitting criterion metrics beyond the variance (e.g., gini, entropy, and misclassification cost).

§ There is no guarantee that the top node can be split, regardless of the number of predictor variables at hand.

4. The splitting of Node 11 yields Nodes 14 and 15. The nodes are read similarly to Items 2a and 2b.
5. The splitting criterion is not satisfied for eight nodes: 4–9, 12, and 13.

The *usual* interpretation of the instructive CHAID tree is moot as it is a *simple* CHAID model (i.e., it has one predictor variable). Explaining and predicting HI_BALANCE based on the singleton variable RECENCY_MOS need a fuller explanation and a more accurate prediction of HI_BALANCE, achieved with more than one predictor variable. However, the simple CHAID model is *not* abstractly academic as it is the vehicle for the proposed method.

The *unique* interpretation of the simple CHAID model is the core of the proposed CHAID-based data mining for a smoother scatterplot: CHAID-based smoothing. The end nodes of the CHAID model are *end-node slices,* which are numerous (in the hundreds) by user design. The end-node slices have quite a bit of accuracy because they are predicted (fitted) by a CHAID model *accounting for* the X-axis variable. The end-node slices by aggregation become 10 *CHAID slices.* The CHAID slices of the X-axis variable clearly produce *more accurate* (*smoother*) values (CHAID-based sm_X) than the smooth X values from the dummy slicing of the X-axis (sm_X from the smooth scatterplot). Ergo, the CHAID slices produce smoother Y values (CHAID-based sm_Y) than the smooth Y values from the dummy slicing of the X-axis (sm_Y from the smooth scatterplot). In sum, CHAID slices produce CHAID-based sm_X that is smoother than sm_X and CHAID-based sm_Y that is smoother than sm_Y.

Note, the instructive CHAID tree in Figure 4.3 does not have numerous end-node slices. For this section only, I ask the reader to assume the CHAID tree has numerous

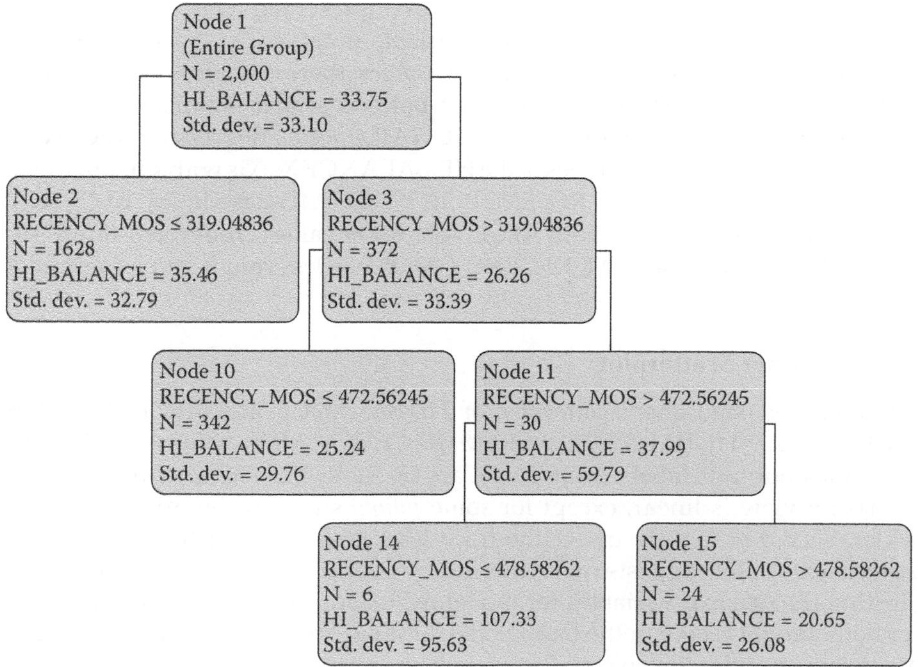

FIGURE 4.3
Illustration of CHAID tree of HI_BALANCE and RECENCY_MOS.

end nodes so I can bring forth the exposition of CHAID-based smoothing. The real study illustration of CHAID-based smoothing, in the next section, indeed has end nodes in the hundreds.

I continue with the CHAID tree in Figure 4.3 to assist the understanding of CHAID-based smoothing. The CHAID slices of the RECENCY_MOS X-axis produce smoother HI_BALANCE values, *CHAID-based SM_HI_BALANCE,* than the smooth HI_BALANCE values from the dummy slicing of the RECENCY_MOS X-axis. In other words, CHAID-based SM_HI_BALANCE has less rough and more smooth. The following sequential identities may serve as a mnemonic guide to the concepts of rough and smooth and to how CHAID-based smoothing works:

1. Data point/value = reliable value + *error* value, from theoretical statistics.
2. Data value = predicted/fitted value + residual value, from applied statistics.
   a. The residual is reduced.
   b. The fitted value is noticeable given a well-built model.
3. Data value = fitted value + residual value, a clearer restatement of Item 2.
4. Data = smooth + rough, from Tukey's exploratory data analysis (EDA) [1].
   a. The rough is reduced.
   b. The smooth is increased/more accurate, given a well-built model.
5. Data = smooth per slice + rough per slice, from the CHAID model.
   a. The rough per slice is reduced.
   b. The smooth per slice is observably accurate, given a well-built CHAID model.

To facilitate the discussion thus far, I have presented the dependent-predictor variable framework as the standard paired-variable notation (X_i, Y_i) suggests. However, when assessing the relationship between two variables, there is no dependent–independent variable framework, in which case the standard paired-variable notation is ($X1_i$, $X2_i$). Thus, analytical logic necessitates building a second CHAID model: Regressing RECENCY_MOS on HI_BALANCE yields the end nodes of the HI_BALANCE X-axis with smooth RECENCY_MOS values, *CHAID-based SM_RECENCY_MOS.* CHAID-based SM_RECENCY_MOS values are smoother than the smooth RECENCY_MOS values from the dummy slicing of HI_BALANCE. CHAID-based SM_RECENCY_MOS has less rough and more smooth.

4.5.1 The Smoother Scatterplot

The CHAID-based smoother scatterplot of HI_BALANCE and RECENCY_MOS is in Figure 4.4. The SM_HI_BALANCE and SM_RECENCY_MOS values per CHAID slice, which assumes numeric labels from 0 to 9, are in Table 4.2. The uncovered relationship in the majority view is linear, except for some *jumpy* slices in the middle (3–6), and for the stickler, Slice 0 is slightly under the trace line. Regardless of these last diagnostics, the *smoother* scatterplot suggests no further reexpressing of HI_BALANCE or RECENCY_MOS. Testing the original variables for model inclusion is the determinate of what shape either or both variables take in the final model. Lest one forgets, the data analyst compares the smoother scatterplot findings to the smooth scatterplot findings.

The full, hard-to-read CHAID HI_BALANCE model *accounting for* RECENCY_MOS is shown in Figure 4.5. The CHAID HI_BALANCE model has 181 end nodes stemming from

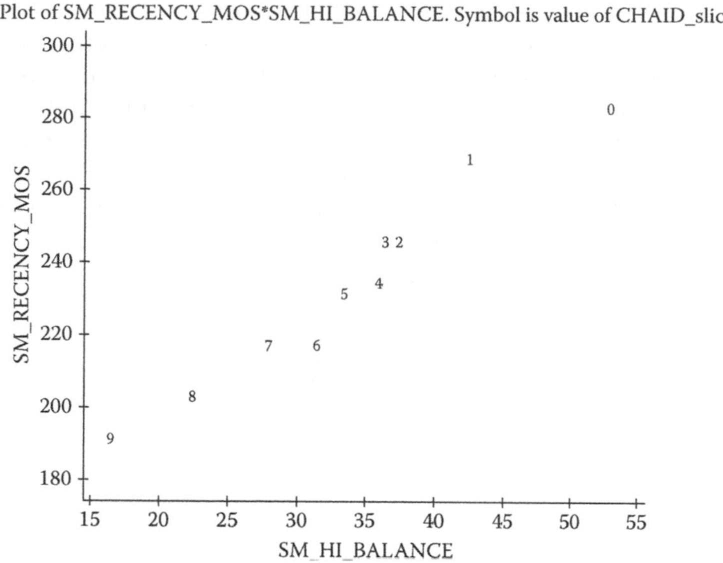

FIGURE 4.4
CHAID-based smoother scatterplot of RECENCY_MOS and HI_BALANCE.

TABLE 4.2

Smoother Values for HI_BALANCE and RECENCY_MOS by CHAID Slices

CHAID_slice	SM_HI_BALANCE	SM_RECENCY_MOS
0	52.9450	281.550
1	42.3392	268.055
2	37.4108	246.270
3	36.5675	245.545
4	36.1242	233.910
5	33.3158	232.155
6	31.5350	216.750
7	27.8492	215.955
8	22.6783	202.670
9	16.6967	191.570

FIGURE 4.5
CHAID regression tree: HI_BALANCE regressed on RECENCY_MOS.

minimum-size node to split 10 and the maximum tree level 10. A readable midsection of the tree with Node 1 (top and center) is shown in Figure 4.6. As this model's full-size output spans nine pages, I save a tree in the forest by capturing the hard-to-read tree as a JPEG image. The full, hard-to-read CHAID RECENCY_MOS model accounting for HI_BALANCE is shown in Figure 4.7. The CHAID RECENCY_MOS model has 121 end nodes stemming from minimum-size node to split 10 and the maximum tree level 14. A readable midsection of the tree with Node 1 (top and center) is shown in Figure 4.8. This model's full-size output spans four pages.

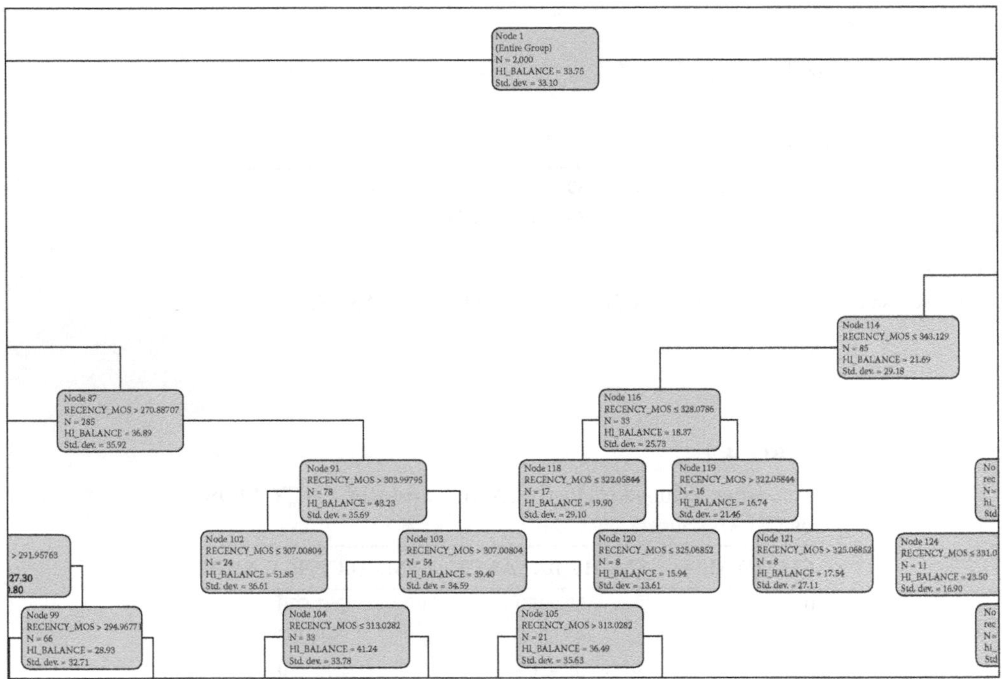

FIGURE 4.6
Midsection of CHAID regression tree: HI_BALANCE regressed on RECENCY_MOS.

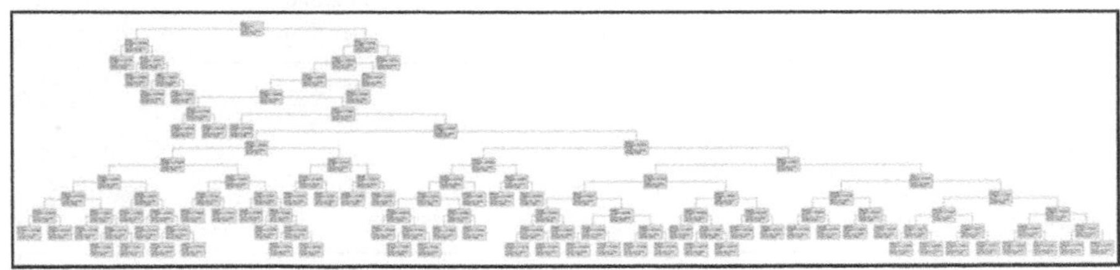

FIGURE 4.7
CHAID regression tree: RECENCY_MOS regressed on HI_BALANCE.

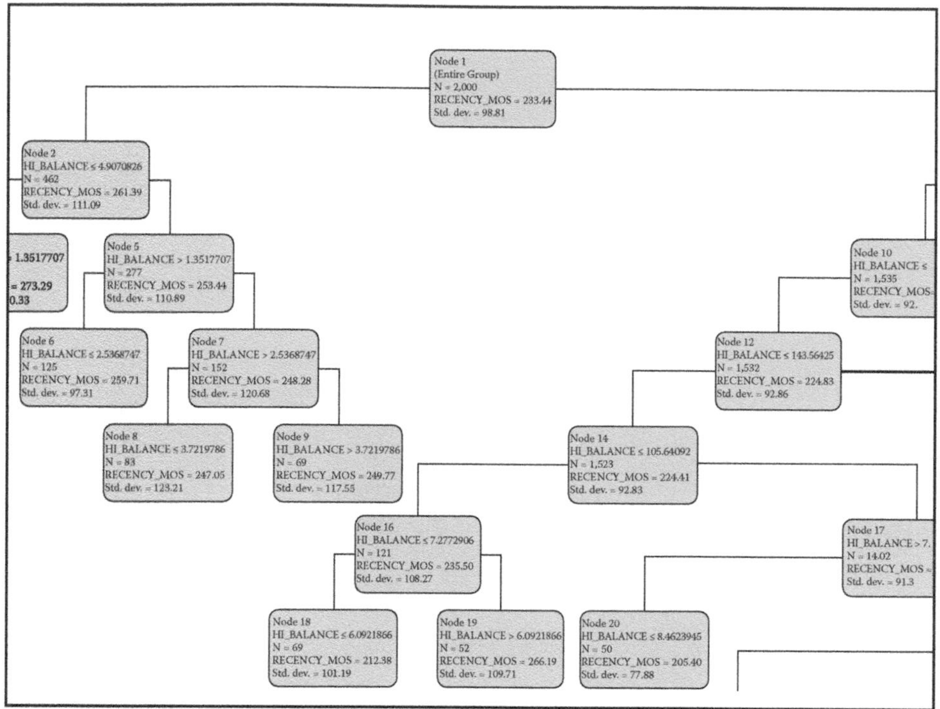

FIGURE 4.8
Midsection of CHAID regression tree: RECENCY_MOS regressed on HI_BALANCE.

4.6 Summary

The basic (raw data) scatterplot and smooth scatterplot are the current data mining methods for assessing the relationship between predictor and dependent variables, an essential task in the model-building process. Working with today's big data, containing valuable elements within, data analysts find themselves in troubled waters while pulling out the valuable elements. Big data have rendered the scatterplot overloaded with data points or information. Paradoxically, scatterplots based on more information are less informative. To data mine the information in the data-overloaded scatterplot, I review the smooth scatterplot for unmasking an underlying relationship as depicted in a raw data scatterplot. Then, I propose a CHAID-based method of data mining for paired-variable assessment, a new technique of obtaining a smoother scatterplot, which exposes a more reliable depiction of the unmasked relationship than that of the smooth scatterplot. The smooth scatterplot uses averages of raw data, and the smoother scatterplot uses averages of fitted values of CHAID end nodes. I illustrate the basic, the smooth, and smoother scatterplots using a real study.

Reference

1. Tukey, J.W., *Exploratory Data Analysis*, Addison-Wesley, Reading, MA, 1997.

5

The Importance of Straight Data: Simplicity and Desirability for Good Model-Building Practice

5.1 Introduction

The purpose of this chapter is to show the importance of straight data for the simplicity and desirability it brings for good model-building practice. I illustrate the content of the chapter title by giving details of what to do when an observed relationship between two variables depicted in a scatterplot is masking an acute underlying relationship. Data mining is employed to unmask and straighten the obtuse relationship. The correlation coefficient is used to quantify the strength of the exposed relationship, which possesses straight-line simplicity.

5.2 Straightness and Symmetry in Data

Today's data mechanics* (DMers), who consist of the entirety of statisticians, data analysts, data miners, knowledge discoverers, and the like, know that exploratory data analysis, better known as EDA, places special importance on straight data not in the least for the sake of simplicity itself. The paradigm of life is simplicity (at least for those of us who are older and wiser). In the physical world, Einstein uncovered one of the universe's ruling principles using only three letters: $E = mc^2$. In the visual world, however, simplicity is undervalued and overlooked. A smiley face is an unsophisticated, simple shape that nevertheless communicates effectively, clearly, and instantly. Why should DMers accept anything less than simplicity in their life's work? Numbers, as well, should communicate powerfully, unmistakably, and without much ado. Accordingly, DMers should seek two features that reflect simplicity: straightness and symmetry in data.

* Data mechanics (DMers) are a group of individuals skilled by the study of understanding data, stressing the throwing of light upon data by offering details or inferences previously unclear or only implicit. According to The Royal Society, founded in 1660, the first statistician was John Graunt, who believed that London's bills of mortality could provide information for more than just for gossip. He compiled and analyzed all bills from 1604 through 1661. Graunt's work was published as *Natural and Political Observations Made upon the Bills of Mortality*. Graunt became the first to put forth today's common statistics as: More boys are born than girls; women live longer than men; and, except in epidemics, the number of people dying each year is relatively constant. http://www.answers.com/topic/the-first-statistician.

There are five reasons why it is important to straighten data:

1. The straight-line (linear) relationship between two continuous variables X and Y *is simple as it gets*. As X increases (decreases) in its values, Y increases (decreases) in its values. In this case, X and Y are positively correlated. Or, as X increases (decreases) in its values, Y decreases (increases) in its values. In this case, X and Y are negatively correlated. As an example of this setting of simplicity (and of everlasting importance), Einstein's E and m have a perfect positive linear relationship.
2. With linear data, the data analyst without difficulty sees *what is going on within the data*. The class of linear data is the desirable element for good model-building practice.
3. Most marketing models, belonging to the class of innumerable varieties of the linear statistical model, *require linear relationships* between a dependent variable and (a) *each* predictor variable in a model and (b) *all* predictor variables considered jointly, regarding them as an array of predictor variables that have a multivariate normal distribution.
4. It is well known that *nonlinear models*, attributed with yielding good predictions with nonstraight data, in fact, *do better with straight data*.
5. I have not ignored the feature of symmetry. Not accidentally, there are theoretical reasons for *symmetry and straightness going hand in hand*. Straightening data often makes data symmetric and vice versa. Recall, symmetric data have values that are in correspondence in size and shape on opposite sides of a dividing line or middle value of the data. The iconic symmetric data profile is bell-shaped.

5.3 Data Mining Is a High Concept

Data mining is a high concept of three key elements: a) fast action in its development, b) glamor in its imagination for the unexpected, and c) mystique that feeds the curiosity of human thought. Conventional wisdom, in the DM space, has it that everyone knows *what data mining is* [1]. Everyone does it—that is what he or she says. I do not believe it. I know that everyone talks about it, but only a small, self-seeking group of data analysts genuinely does data mining. I make this bold and excessively self-confident assertion based on my consulting experience as a statistical modeler, data miner, and computer scientist for the many years that have been gifted to me.

5.4 The Correlation Coefficient

The term *correlation coefficient*, denoted by r, was coined by Karl Pearson in 1896. This statistic, over a century old, is still going strong. It is one of the most used statistics, second to the mean. The correlation coefficient weaknesses and warnings of misuse are well known. As a practiced consulting statistician and instructor of statistical modeling and data mining for continuing professional studies in the DM space,* I see too often that the weaknesses and misuses go unheeded perhaps because of their rare mention in practice. The correlation coefficient, whose values theoretically range within the left-closed right-closed interval [–1, +1], is restricted by the individual distributions of the two variables being correlated (see Chapter 9). The misuse of the correlation coefficient is the nontesting of the *linear assumption,* discussed in this section.

Assessing the relationship between dependent and predictor variables is an essential task in statistical linear and nonlinear regression model building. If the relationship is linear, then the modeler tests to determine whether the predictor variable has statistical importance to be included in the model. If the relationship is either nonlinear or indiscernible, then one or both of the two variables are reexpressed, that is, *data mined* to be voguish with terminology, to reshape the observed relationship into a data-mined linear relationship. As a result, the reexpressed variable(s) is(are) tested for inclusion into the model.

The everyday method of assessing a relationship between two variables—lest the data analyst forgets: *linear* relationships only—is the calculation of the correlation coefficient. The misuse of the correlation coefficient is due to *disregard for* the linearity assumption test, albeit simple to do. (I put forth an obvious reason, but still not acceptable, why the nontesting has a long shelf life later in the chapter.) I state the linear assumption, discuss the testing of the assumption, and provide how to interpret the observed correlation coefficient values.

The correlation coefficient requires that the underlying relationship between two variables is linear. If the observed pattern displayed in the scatterplot of two variables has an outward aspect of being linear, then the correlation coefficient provides a reliable measure of the *linear* strength of the relationship. If the observed pattern is either nonlinear or indiscernible, then the correlation coefficient is inutile or offers risky results. If the latter data condition exists, data mining efforts should be attempted to straighten the relationship. In the remote situation when the proposed data mining method is not successful, then *extra* data mining techniques, such as binning, should be explored. The latter techniques are outside the scope of this chapter. Sources for extra data mining are many [2–4].

If the relationship is deemed linear, then the *strength* of the relationship is quantified by an accompanying value of r. For convenience, I restate the accepted guidelines (from Chapter 3) for interpreting the correlation coefficient:

1. 0 indicates no linear relationship.
2. +1 indicates a perfect positive linear relationship: As one variable increases in its values, the other variable also increases in its values via an exact linear rule.
3. –1 indicates a perfect negative linear relationship: As one variable increases in its values, the other variable decreases in its values via an exact linear rule.
4. Values between 0 and 0.3 (0 and –0.3) indicate a weak positive (negative) linear relationship via a shaky linear rule.
5. Values between 0.3 and 0.7 (–0.3 and –0.7) indicate a moderate positive (negative) linear relationship via a fuzzy-firm linear rule.
6. Values between 0.7 and 1.0 (–0.7 and –1.0) indicate a strong positive (negative) linear relationship via a firm linear rule.

* DM space includes industry sectors such as direct and database marketing, banking, insurance, finance, retail, telecommunications, health care, pharmaceuticals, publication and circulation, mass and direct advertising, catalog marketing, e-commerce, Web-mining, business to business (B2B), human capital management, risk management, and the like.

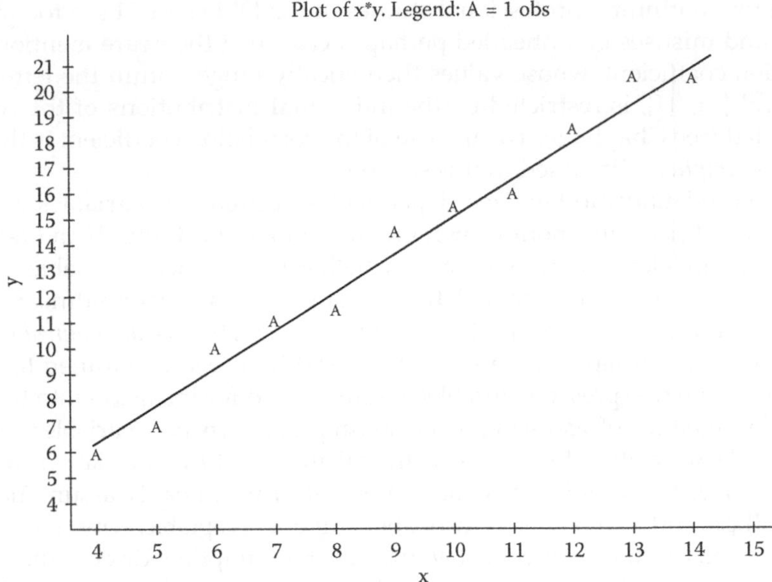

FIGURE 5.1
Sought-after scatterplot of paired variables (x, y).

I present the sought-after scatterplot of paired variables (x, y)—a cloud of data points that indicates a silver lining of a straight line. The correlation coefficient $r_{(x,y)}$, corresponding to this scatterplot, ensures that the value of r reliably reflects the strength of the linear relationship between x and y in Figure 5.1. The cloud of points in Figure 5.1 is not typical due to the small, eleven-observation dataset used in the illustration. However, the discussion still holds true, as if the presentation involves, say, 11,000 observations or greater. I take the freedom of writing to refer to the silver-lining scatterplot as a thin, wispy cirrus cloud.

5.5 Scatterplot of (xx3, yy3)

Consider the scatterplot of 11 data points of the third paired variables (x3, y3) in Table 5.1 regarding the Anscombe data. I construct the scatterplot of (x3, y3) in Figure 5.2, renaming (xx3, yy3) for a seemingly unnecessary inconvenience. (The reason for the renaming is explained later in the book.) Clearly, the relationship between xx3 and yy3 is problematic: It would be straightaway linear if not for the *far-out* point ID3 (13, 12.74). The scatterplot does not ocularly reflect a linear relationship. The nice large value of $r_{(xx3,\ yy3)} = 0.8163$ is meaningless and useless. I leave it to the reader to draw his or her underlying straight line.

5.6 Data Mining the Relationship of (xx3, yy3)

I data mine for the underlying structure of the paired variables (xx3, yy3) using a machine-learning approach to the discipline of evolutionary computation, specifically *genetic*

TABLE 5.1
Anscombe Data

ID	x1	y1	x2	y2	x3	y3	x4	y4
1	10	8.04	10	9.14	10	7.46	8	6.58
2	8	6.95	8	8.14	8	6.77	8	5.76
3	13	7.58	13	8.74	13	12.74	8	7.71
4	9	8.81	9	8.77	9	7.11	8	8.84
5	11	8.33	11	9.26	11	7.81	8	8.47
6	14	9.96	14	8.10	14	8.84	8	7.04
7	6	7.24	6	6.13	6	6.08	8	5.25
8	4	4.26	4	3.10	4	5.39	19	12.50
9	12	10.84	12	9.13	12	8.15	8	5.56
10	7	4.82	7	7.26	7	6.42	8	7.91
11	5	5.68	5	4.74	5	5.73	8	6.89

Source: Anscombe, F.J., *Am. Stat.*, 27, 17–21, 1973.

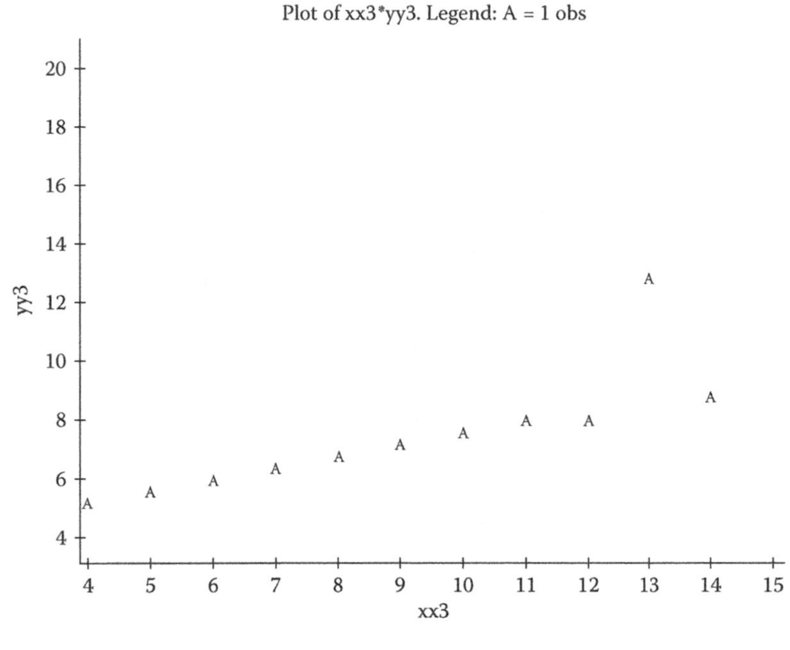

FIGURE 5.2
Scatterplot of (xx3, yy3).

programming (GP). The fruits of my data mining work yield the scatterplot in Figure 5.3. The data mining work is not an expenditure of time-consuming results or mental effort because the GP-based data mining (GP-DM) is a machine-learning adaptive intelligence process that is effective and efficient for straightening data. The data mining tool used is the GenIQ Model, which renames the data-mined variable with the prefix GenIQvar. Data-mined (xx3, yy3) is relabeled (xx3, GenIQvar(yy3)). (The GenIQ Model is formally introduced replete with eye-opening examples in Chapter 40.)

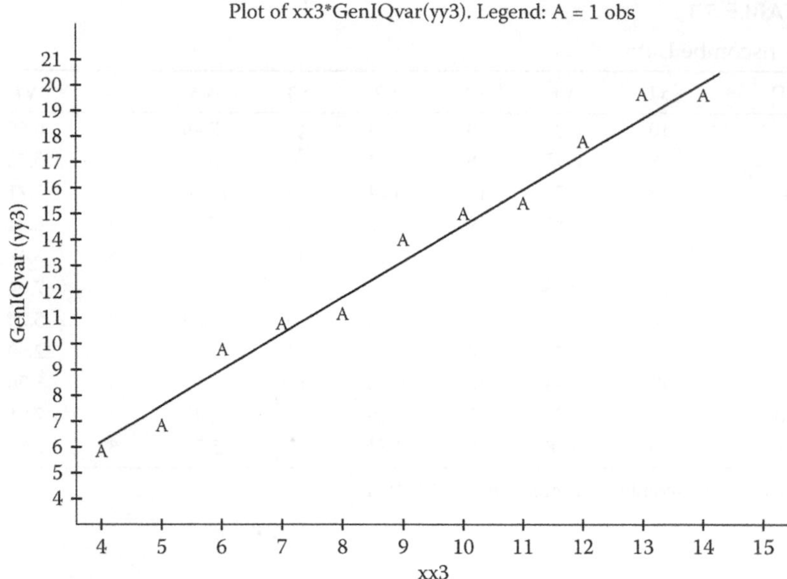

FIGURE 5.3
Scatterplot of (xx3, GenIQvar(yy3)).

TABLE 5.2
Reexpressed yy3, GenIQ(yy3), Descendingly Ranked by GenIQ(yy3)

xx3	yy3	GenIQ(yy3)
13	12.74	20.4919
14	8.84	20.4089
12	8.15	18.7426
11	7.81	15.7920
10	7.46	15.6735
9	7.11	14.3992
8	6.77	11.2546
7	6.42	10.8225
6	6.08	10.0031
5	5.73	6.7936
4	5.39	5.9607

The correlation coefficient $r_{(xx3, GenIQvar(yy3))} = 0.9895$ and the uncurtained underlying relationship in Figure 5.3 warrant a silver, if not gold, medal for straightness. The correlation coefficient is a reliable measure of the linear relationship between xx3 and GenIQvar(yy3). The almost-maximum value of $r_{(xx3, GenIQvar(yy3))}$ indicates an almost perfect linear relationship between the original xx3 and the data-mined GenIQvar(yy3). (Note: The scatterplot indicates GenIQvar_yy3 on the Y-axis and not the correct notation, GenIQvar(yy3), due to syntax restrictions of the graphics software.)

The values of the variables xx3, yy3, and GenIQvar(yy3) are in Table 5.2. The 11 data points are ordered based on the descending values of GenIQvar(yy3).

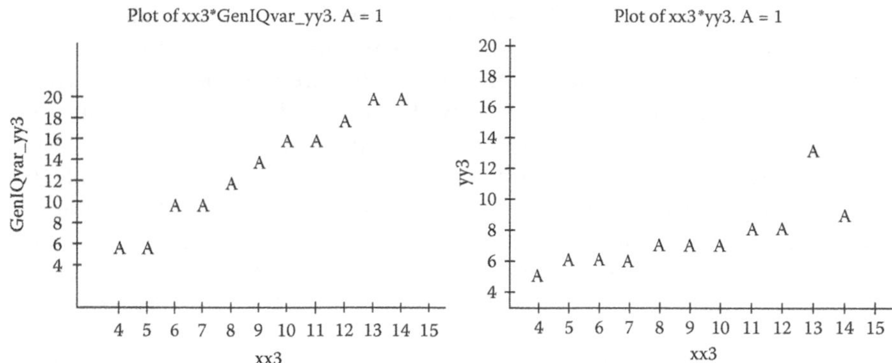

FIGURE 5.4
Side-by-side scatterplot.

5.6.1 Side-by-Side Scatterplot

A side-by-side scatterplot of all the goings-on is best pictured because it is worth 1,000 words. The side-by-side scatterplot speaks for itself of a data mining piece of work done well (see Figure 5.4).

5.7 What Is the GP-Based Data Mining Doing to the Data?

As evidenced by the detailed illustration, GP-DM maximizes the straight-line correlation between a genetically reexpressed dependent variable and a single predictor variable. When building a multiple (at least two predictor variables) regression model, GP-DM maximizes the straight-line correlations between a genetically reexpressed dependent variable with each predictor variable and the array of predictor variables considered jointly.

Although the GenIQ Model, as used here, is revisited in Chapter 39, it is not appropriate to leave without the definition of the reexpressed variable. The definition is in Equation 5.1:

$$\text{GenIQvar\_yy3} = \cos(2 * \text{xx3}) + (\text{xx3}/0.655) \tag{5.1}$$

The GenIQ Model did eminently good data mining that rendered excellent results: $r_{(\text{xx3, GenIQvar(yy3)})} = 0.9895$.

5.8 Straightening a Handful of Variables and a Baker's Dozen of Variables

A handful of variables (10 pairs of variables) can be dealt with presumably without difficulty. As for a baker's dozen of variables (78 pairs), they can be a *handful*. However, GP-DM, as outlined with its main features and points of functioning, *reduces* initially, effectively, and efficiently the 78 pairs to a practical number of variables. Examination of this reduction operation follows.

One requires clearness about a baker's dozen of variables. A data analyst cannot expect to straighten 78 pairs of variables. Many pairs of variables require many scatterplots. The generating of numerous scatterplots is the reason for abusing or ignoring the linearity assumption. GP-DM can. In fact, it does so with the speed of a Gatling gun. In practice, it is a common sight to work on a dataset of, say, 400 variables (79,800 pairs of variables). GP-DM, in its beginning step, deletes variables (single and paired variables) that are deemed to have no predictive power by dint of the probabilistic-selected biological operators of reproduction, mating, and mutation. During the second evolutionary step, GP-DM further decreases the number of variables to only handfuls. The remaining step of GP-DM is the full-fledged evolutionary process of GP proper, by which the data strengthening is carried out in earnest [5].

Ergo, GP-DM can handle efficiently virtually any number of variables as long as the computer can process all the variables of the original dataset initially. Illustrating GP-DM with an enlarged dataset of many, many pairs of variables is beyond the scope of this chapter. Moreover, demonstrating GP-DM with an expanded dataset would be spurious injustice and reckless wronging of an exposition demanding that which must go beyond two dimensions (2D) and even the 3D of the movie *Avatar.*

5.9 Summary

The objective of this chapter is to share my personal encounters: entering the mines of data, going deep to unearth acute underlying relationships, and rising from the inner workings of the data mine to show the importance of straight data for the simplicity and desirability it brings for good model-building practice. I discuss an illustration, simple but with an object lesson, of what to do when an observed relationship between two variables in a scatterplot is masking an acute underlying relationship. I propose the GP-DM method for directly unmasking and straightening the obtuse relationship. The correlation coefficient is used correctly because the exposed character of the exampled relationship has straight-line simplicity.

References

1. Ratner, B., Data mining: An ill-defined concept. 2009. http://www.geniq.net/res/data-mining-is-an-ill-defined-concept.html.
2. Han, J., and Kamber, M., *Data Mining: Concepts and Techniques,* Morgan Kaufmann, San Francisco, CA, 2001.
3. Bozdogan, H., ed., *Statistical Data Mining and Knowledge Discovery,* CRC Press, Boca Raton, FL, 2004.
4. Refaat, M., *Data Preparation for Data Mining Using SAS,* Morgan Kaufmann Series in Data Management Systems, Morgan Kaufmann, Maryland Heights, MO, 2006.
5. Ratner, B., *What is genetic programming?* 2007. http://www.geniq.net/Koza_GPs.html.

6

Symmetrizing Ranked Data: A Statistical Data Mining Method for Improving the Predictive Power of Data

6.1 Introduction

The purpose of this chapter is to introduce a new statistical data mining method, the *symmetrizing ranked data* method, and add it to the paradigm of simplicity and desirability for good model-building practice as presented in Chapter 5. The new method carries out the action of two basic statistical tools, symmetrizing and ranking variables, yielding new reexpressed variables with likely improved predictive power. I detail Steven's scales of measurement (nominal, ordinal, interval, and ratio). Then, I define an *approximate interval* scale that is an offspring of the new statistical data mining method. Next, I provide a quick review of the simplest of exploratory data analysis (EDA) elements: (1) the stem-and-leaf display and (2) the box-and-whiskers plot. Both methods are required for presenting the new method, which itself falls under EDA proper. Last, I illustrate the proposed method with two examples, which provide the data miner with a starting point for more applications of this useful statistical data mining tool.

6.2 Scales of Measurement

There are four scales of data measurement due to Steven's scales of measurement [1]:

1. Nominal data are classification *labels*, for example, color (red, white, and blue). There is no ordering of the data values. Clearly, arithmetic operations cannot be performed on nominal data. That is, one cannot add red + blue (= ?).
2. Ordinal data are *ordered* numeric labels in that higher/lower numbers represent higher/lower values on the scale. The intervals between the numbers are not necessarily equal.
   a. For example, consider the variables CLASS and AGE as per traveling on a cruise liner. I recode the CLASS labels (first, second, third, and crew) into the ordinal variable CLASS_, implying some measure of income. I recode the AGE labels (adult and child) into the ordinal variable AGE_, implying years old. Also, I create CLASS interaction variables with both AGE and GENDER, denoted by CLASS_AGE_

and CLASS_GENDER_, respectively. The definitions of the recoded variables are listed first, followed by the definitions of the interaction variables.

The three recoded individual variables are:

i. if GENDER = male then GENDER_ = 0
ii. if GENDER = female then GENDER_ = 1

i. if CLASS = first then CLASS_ = 4
ii. if CLASS = second then CLASS_ = 3
iii. if CLASS = third then CLASS_ = 2
iv. if CLASS = crew then CLASS_ = 1

i. if AGE = adult then AGE_ = 2
ii. if AGE = child then AGE_ = 1

The recoded CLASS interaction variables with both AGE and GENDER are:

i. if CLASS = second and AGE = child then CLASS_AGE_ = 8
ii. if CLASS = first and AGE = child then CLASS_AGE_ = 7
iii. if CLASS = first and AGE = adult then CLASS_AGE_ = 6
iv. if CLASS = second and AGE = adult then CLASS_AGE_ = 5
v. if CLASS = third and AGE = child then CLASS_AGE_ = 4
vi. if CLASS = third and AGE = adult then CLASS_AGE_ = 3
vii. if CLASS = crew and AGE = adult then CLASS_AGE_ = 2
viii. if CLASS = crew and AGE = child then CLASS_AGE_ = 1

i. if CLASS = first and GENDER = female then CLASS_GENDER_ = 8
ii. if CLASS = second and GENDER = female then CLASS_GENDER_ = 7
iii. if CLASS = crew and GENDER = female then CLASS_GENDER_ = 6
iv. if CLASS = third and GENDER = female then CLASS_GENDER_ = 5
v. if CLASS = first and GENDER = male then CLASS_GENDER_ = 4
vi. if CLASS = third and GENDER = male then CLASS_GENDER_ = 2
vii. if CLASS = crew and GENDER = male then CLASS_GENDER_ = 3
viii. if CLASS = second and GENDER = male then CLASS_GENDER_ = 1

b. One cannot assume the difference in income between CLASS_ = 4 and CLASS_ = 3 equals the difference in income between CLASS_ = 3 and CLASS_ = 2.
c. Arithmetic operations (e.g., subtraction) are not possible. With CLASS_ numeric labels, one cannot conclude 4 – 3 = 3 – 2.
d. Only the logical operators "less than" and "greater than" can be performed.
e. Another feature of an ordinal scale is that there is no "true" zero. True zero does not exist because the CLASS_ scale, which goes from 4 through 1, could have been recoded to go from 3 through 0.

3. Interval data are on a scale in which each position is equidistant from one another. Interval data allow for the distance between two pairs to be equivalent in some way.

a. Consider the HAPPINESS scale of 10 (= most happy) through 1 (= very sad). Four persons rate themselves on HAPPINESS:
   i. Persons A and B state 10 and 8, respectively.
   ii. Persons C and D state 5 and 3, respectively.
   iii. One can conclude that person-pair A and B (with a happiness difference of 2) represents the same difference in happiness as that of person-pair C and D (with a happiness difference of 2).
   iv. Interval scales *do not* have a true zero point. Therefore, it is not possible to make statements about how many times happier one score is than another.
      A. Interval data cannot be multiplied or divided. The common example of interval data is the Fahrenheit scale for temperature. Equal differences on this scale represent equal differences in temperature, but a temperature of 30° is not twice as warm as a temperature of 15°. That is, 30° – 20° = 20° – 10°, but 20°/10° is not equal to 2. That is, 20° is not twice as hot as 10°.

4. Ratio data are like interval data except they *have* true zero points. The common example is the Kelvin scale of temperature. This scale has an absolute zero. Thus, a temperature of 300 K is twice as high as a temperature of 150 K.
5. What is a true zero? Some scales of measurement have a true or natural zero.
   a. For example, WEIGHT has a natural zero at no weight. Thus, it makes sense to say that my dachshund Dappy weighing 26 pounds is twice as heavy as Dappy's sister dachshund Betsy weighing 13 pounds. WEIGHT is on a ratio scale.
   b. On the other hand, YEAR does not have a natural zero. The YEAR 0 is arbitrary, and it is not sensible to say that the year 2000 is twice as old as the year 1000. Thus, YEAR is measured on an interval scale.
6. Note: Some data analysts, unfortunately, make no distinction between interval or ratio data, calling them both continuous. Moreover, most data analysts put their heads in the sand to treat ordinal data, which assumes numerical values, as interval data. In both situations, this is not correct technically.

6.3 Stem-and-Leaf Display

The stem-and-leaf display is a graphical presentation of quantitative data to assist in visualizing the density and shape of a distribution. A salient feature of the display is that it retains the original data to at least two significant digits and puts the data in order. A basic stem-and-leaf display has two columns separated by a vertical line. The left column contains the *stems*, and the right column contains the *leaves*. Typically, the leaf contains the last digit of the number, and the stem contains all of the other digits. In the case of very large numbers, the data values, rounded to a fixed decimal place (e.g., the hundredths place), serve for the leaves. The remaining digits to the left of the rounded values serve as the stem. The stem-and-leaf display is also useful for highlighting outliers and finding the mode. The display is most useful for datasets of moderate EDA size (around 250 data points), after which the stem-and-leaf display becomes a histogram, rotated counterclockwise 90°, and the asterisk (*) represents the digits of the leaves.

6.4 Box-and-Whiskers Plot

The box-and-whiskers plot provides a detailed visual summary of various features of a distribution. The box stretches from the bottom horizontal line, the lower hinge, defined as the 25th percentile, to the top horizontal line, the upper hinge, defined as the 75th percentile. Adding the vertical lines on both ends of the hinges completes the box. The horizontal line within the box is the median. The "+" represents the mean.

The *H-spread* is the difference between the hinges, and a *step* is defined as 1.5 times the H-spread. *Inner fences* are one step beyond the hinges. *Outer fences* are two steps beyond the hinges. These fences are used to create the *whiskers*, which are vertical lines at either end of the box, extended up to the inner fences. An "o" indicates every value between the inner and outer fences. An "*" indicates a score beyond the outer fences. The stem-and-leaf display and box-and-whiskers plot for a symmetric distribution are shown in Figure 6.1. Note, I added the statistic skewness, which measures the lack of symmetry of a distribution. The skewness scale is at the interval level. Skewness = 0 means the distribution is symmetric.

If the skewness value is positive, then the distribution is said to be right-skewed or positive-skewed, which means the distribution has a long tail in the positive direction. Similarly, if the skewness value is negative, then the distribution is left-skewed or negative-skewed, which means the distribution has a long tail in the negative direction.

6.5 Illustration of the Symmetrizing Ranked Data Method

The best way of describing the proposed symmetrizing ranked data (SRD) method is by an exemplification. I illustrate the new statistical data mining method with two examples that provide the data miners with a starting point for their applications of the SRD method.

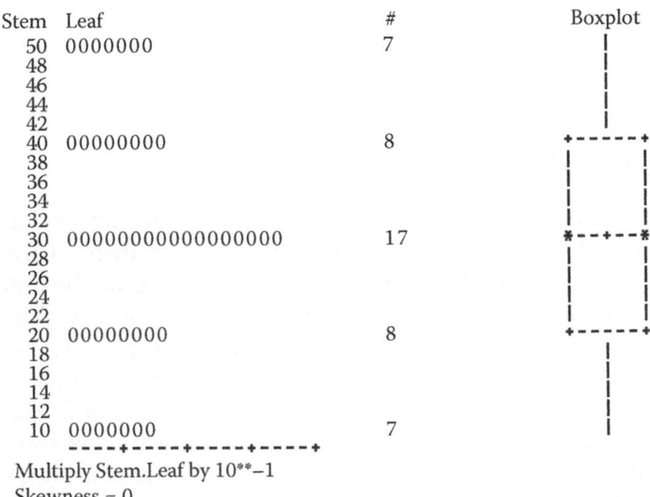

FIGURE 6.1
Stem-and-leaf display and box-and-whiskers plot of symmetric data.

6.5.1 Illustration 1

Consider the two variables from the real study presented in Chapter 4, HI_BALANCE (the highest balance attained among credit card transactions for an individual) and RECENCY_MOS (the number of months since last purchase). The SRD data mining process consists of the following two steps:

1. *Rank* the values of both HI_BALANCE and RECENCY_MOS and create the *rank score* variables rHI_BALANCE and rRECENCY_MOS, respectively. Use any method for handling tied rank scores.
2. *Symmetrize* the ranked variables, which have the same names as the rank scores variables, namely, rHI_BALANCE and rRECENCY_MOS.

The steps using the SAS© procedure RANK is as follows. The procedure creates the rank scores variables, and the option "normal = TUKEY" instructs for symmetrizing the rank scores variables. The input data is DTReg and the output data (i.e., the symmetrized-ranked data) is DTReg_NORMAL. The SAS program is:

```
PROC RANK data = DTReg_data_ normal = TUKEY out = DTReg_NORMAL;
var HI_BALANCE RECENCY_MOS;
ranks rHI_BALANCE rRECENCY_MOS;
run;
```

6.5.1.1 Discussion of Illustration 1

1. The stem-and-leaf displays and the box-and-whiskers plots for HI_BALANCE and rHI_BALANCE are shown in Figures 6.2 and 6.3, respectively. HI_BALANCE and rHI_BALANCE have skewness values of 1.0888 and 0.0098, respectively.

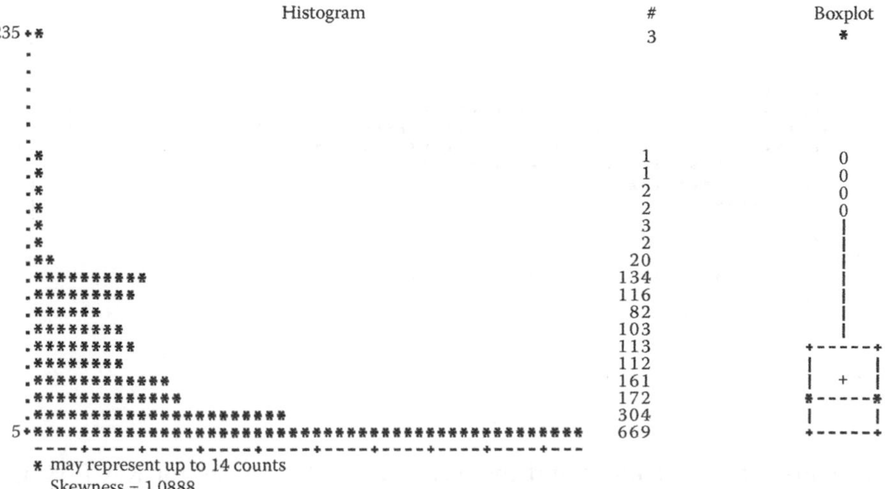

FIGURE 6.2
Histogram and boxplot of HI_BALANCE.

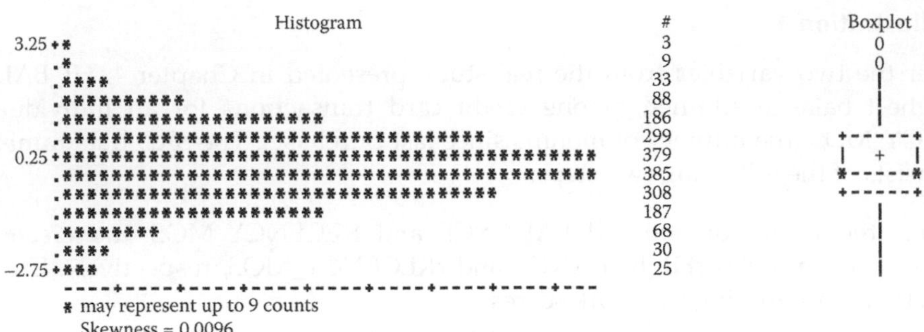

FIGURE 6.3
Histogram and boxplot for rHI_BALANCE.

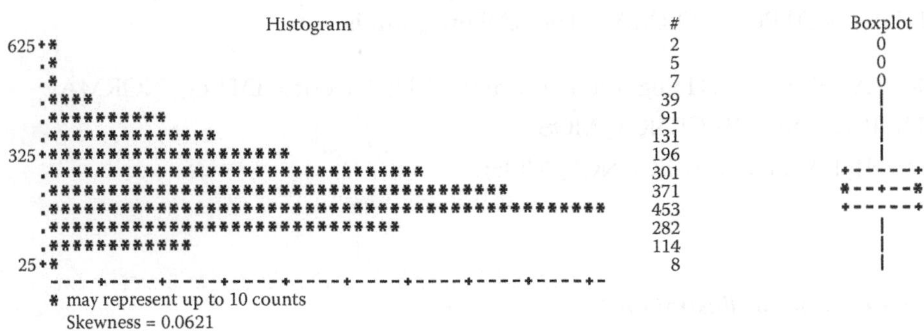

FIGURE 6.4
Histogram and boxplot for RECENCY_MOS.

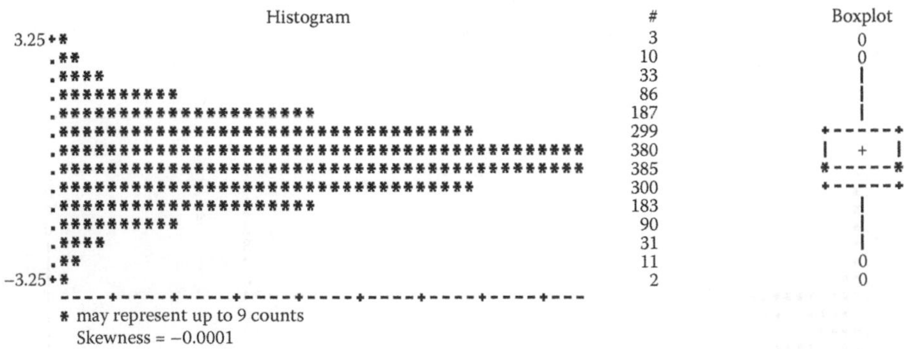

FIGURE 6.5
Histogram and boxplot for rRECENCY_MOS.

2. The stem-and-leaf displays and the box-and-whiskers plots for RECENCY_MOS and rRECENCY_MOS are shown in Figures 6.4 and 6.5, respectively. RECENCY_MOS and rRECENCY_MOS have skewness values of 0.0621 and −0.0001, respectively.

3. Note: The stem-and-leaf displays turn into histograms because the sample size is large, that is, 2,000. Regardless, the graphics provide the descriptive details about the shape of the distributions.

I *acknowledge hesitantly* throwing, with an overhand, my head in the sand, only to further the SRD method by treating ordinal data, which assumes recoded numeric values, as interval data.

Recall, symmetrizing data helps in straightening data. Accordingly, without generating scatterplots, the *reliable* correlation coefficients between the two pairs of variables, the raw variables HI_BALANCE and RECENCY_MOS, and the reexpressed variables, via the SRD method, rHI_BALANCE and rRECENCY_MOS, are −0.6412 and −0.10063, respectively (see Tables 6.1 and 6.2). Hence, the *SRD method has improved the strength* of the predictive relationship between the two raw variables by 56.9% (=abs(−0.10063) − abs(−0.06412))/abs(−0.06421)), where abs = the absolute value function, which ignores the negative sign. In sum, the paired variables (rHI_BALANCE, rRECENCY_MOS) have more predictive power than the originally paired variables and offer more potential in the model-building process.

6.5.2 Illustration 2

Since the fatal event of April 15, 1912, when the White Star liner *Titanic* hit an iceberg and went down in the North Atlantic, fascination with the disaster has never abated. In recent years, interest in the *Titanic* has increased dramatically because of the discovery of the wreck site by Dr. Robert Ballard in 1985. The century-old tragedy has become a national obsession. Any fresh morsel of information about the sinking is like a golden reef. I believe that the SRD method can satisfy the appetite of the *Titanic* aficionado. I build a *preliminary* Titanic *model* to identify survivors so when *Titanic II* sails it will know beforehand who will be most likely to survive an iceberg hitting with outcome odds of 2.0408e−12 to 1.* The *Titanic* model

TABLE 6.1

Correlation Coefficient between HI_BALANCE and RECENCY_MOS

Pearson Correlation Coefficients, *N* = 2,000 Prob > *r* under H0: Rho = 0

	HI_BALANCE	RECENCY_MOS
HI_BALANCE	1.00000	−0.06412
		0.0041
RECENCY_MOS	−0.06412	1.00000
	0.0041	

TABLE 6.2

Correlation Coefficient between rHI_BALANCE and rRECENCY_MOS

Pearson Correlation Coefficients, *N* = 2,000 Prob > *r* under H0: Rho = 0

	RHI_BALANCE	RECENCY_MOS
rRHI_BALANCE	1.00000	−0.10063
Rank for Variable rHI_BALANCE		<0.0001
rRECENCY_MOS	−0.10063	1.00000
Rank for Variable RECENCY_MOS	<0.0001	

* Source is unknown (actually, I lost it).

application, detailed in the remaining sections of the chapter, shows clearly the power of the SRD data mining technique, worthy of inclusion in every data miner's toolkit.

6.5.2.1 Titanic *Dataset*

There were 2,201* passengers and crew aboard the *Titanic*. Only 711 persons survived, resulting in a 32.2% survival rate. For all persons, their basic demographic variables are known: GENDER (female, male), CLASS (first, second, third, crew), and AGE (adult, child). The passengers fall into 14 patterns of GENDER–CLASS–AGE (Table 6.3). Also, Table 6.3 includes pattern, cell size (N), the number of survivors within each cell (S), and the survival rate (format is %).

Because there are only three variables and their scales are the least informative, nominal (GENDER), and ordinal (CLASS and AGE), building a *Titanic* model has been a challenge for the best of academics and practitioners [2–6]. The SRD method is an original and valuable data mining method that belongs to the *Titanic* modeling literature. In the following sections, I present the building of a *Titanic* model.

6.5.2.2 *Looking at the Recoded* Titanic *Ordinal Variables CLASS_, AGE_, GENDER_, CLASS_AGE_, and CLASS_GENDER_*

To see what the data look like, I generate stem-and-leaf displays and box-and-whisker plots for CLASS_, AGE_, and GENDER_ (Figures 6.6, 6.7, and 6.8, respectively). Also, I bring out the interaction variables CLASS_AGE_ and CLASS_GENDER_, which I created in Section 6.2. The graphics of CLASS_AGE_ and CLASS_GENDER_ are in Figures 6.9 and 6.10, respectively. Regarding the coding of ordinal values for the interaction variables, I use

TABLE 6.3
Titanic Dataset

Pattern	GENDER	CLASS	AGE	N	S	Survival Rate
1	Male	First	Adult	175	57	32.5
2	Male	First	Child	5	5	100.0
3	Male	Second	Adult	168	14	8.3
4	Male	Second	Child	11	11	100.0
5	Male	Third	Adult	462	75	16.2
6	Male	Third	Child	48	13	27.1
7	Male	Crew	Adult	862	192	22.3
8	Female	First	Adult	144	140	97.2
9	Female	First	Child	1	1	100.0
10	Female	Second	Adult	93	80	86.0
11	Female	Second	Child	13	13	100.0
12	Female	Third	Adult	165	76	46.1
13	Female	Third	Child	31	14	45.2
14	Female	Crew	Adult	23	20	87.0
			Total	2,201	711	32.2

* The actual number of passengers and survivors are in dispute. In my research, I most often see 2,201/711 passengers/survivors, but have also seen 2,208/712.

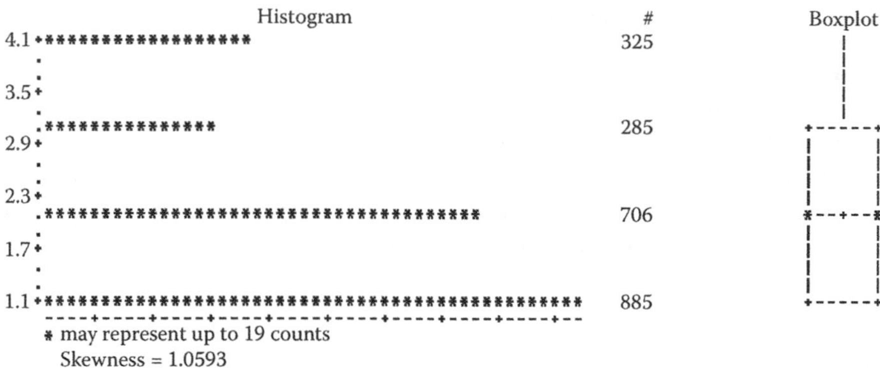

FIGURE 6.6
Histogram and boxplot of CLASS_.

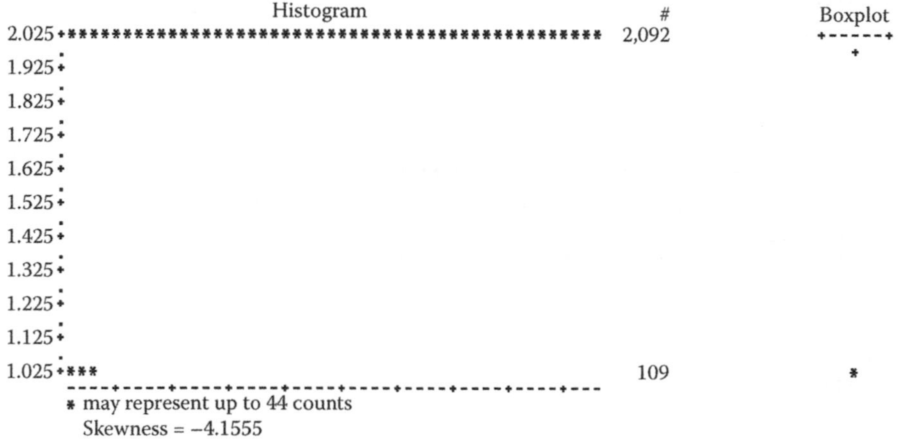

FIGURE 6.7
Histogram and boxplot for AGE_.

```
                     Histogram                               #        Boxplot
1.025+*************                                         470          *
     .
0.925+
     .
0.825+
     .
0.725+
     .
0.625+
     .
0.525+
     .
0.425+
     .
0.325+
     .
0.225+
     .
0.125+
     .
0.025+**********************************************       1731       +-----+
      ----+----+----+----+----+----+----+----+----+--
      * may represent up to 37 counts
        Skewness = 1.3989
```

FIGURE 6.8
Histogram and boxplot for GENDER_.

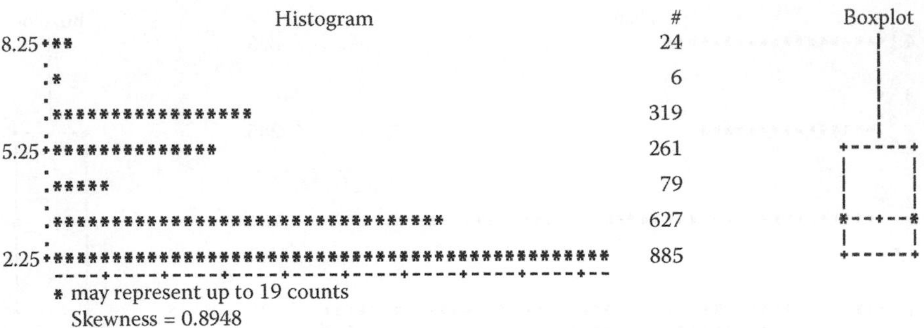

FIGURE 6.9
Histogram and boxplot for CLASS_AGE_.

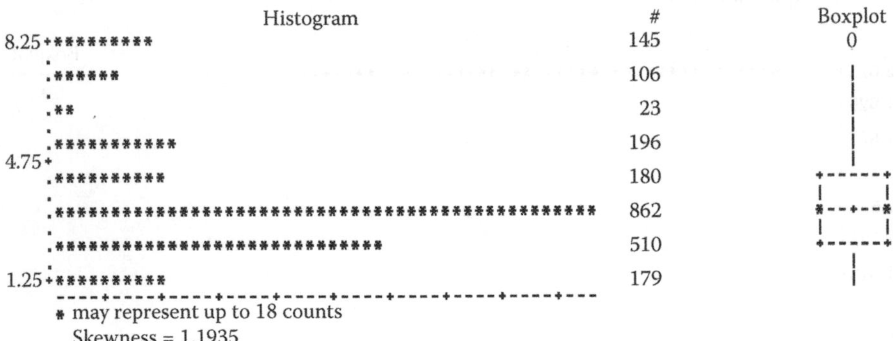

FIGURE 6.10
Histogram and boxplot for CLASS_GENDER_.

TABLE 6.4
Women and Children by SURVIVED

Row Pct Frequency	No	Yes	Total
Child	52	57	109
	47.71	52.29	
Female	109	316	425
	25.65	74.35	
Total	161	373	534

the commonly known refrain during a crisis, "Women and children, first." Survival rates for women and children are 74.35% and 52.29%, respectively (Table 6.4), which bear out the refrain.

6.5.2.3 *Looking at the Symmetrized-Ranked* Titanic *Ordinal Variables rCLASS_, rAGE_, rGENDER_, rCLASS_AGE_, and rCLASS_GENDER_*

The stem-and-leaf displays and box-and-whisker plots for the symmetrized-ranked variables rCLASS_, rAGE_, rGENDER_, rCLASS_AGE_, and rCLASS_GENDER_ are in Figures 6.11 through 6.15, respectively.

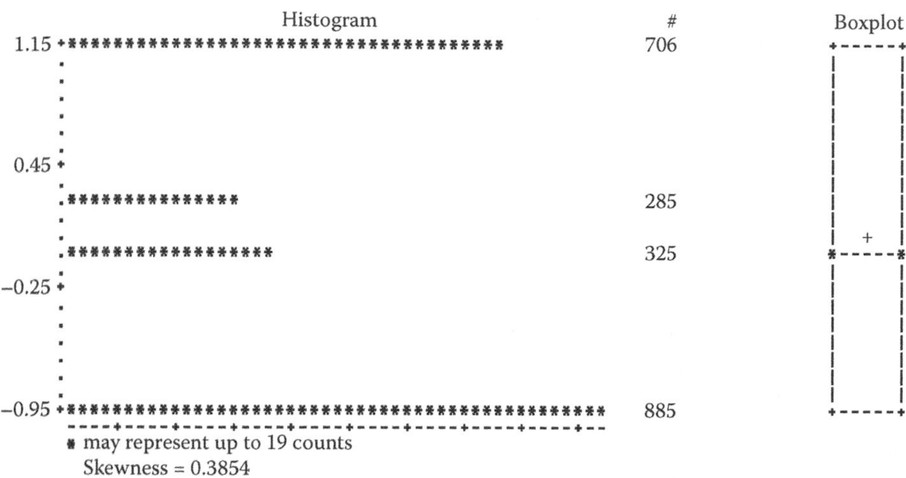

FIGURE 6.11
Histogram and boxplot for rCLASS_.

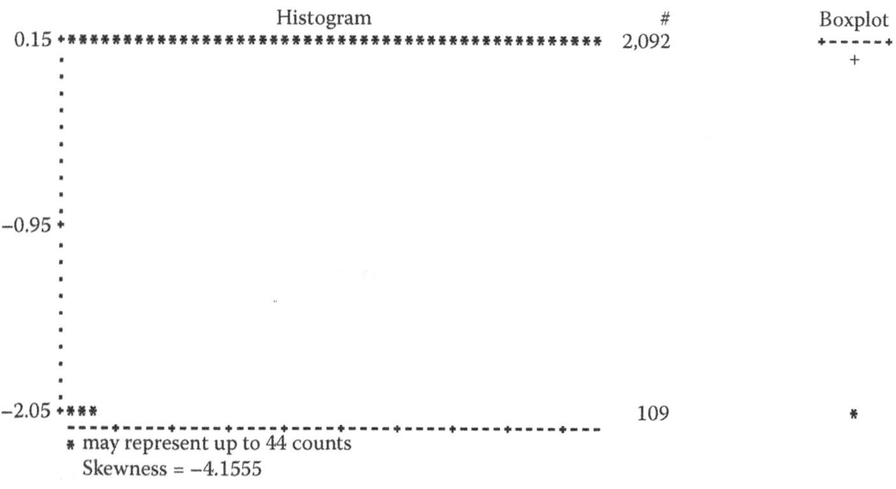

FIGURE 6.12
Histogram and boxplot for rAGE_.

The results of the SRD method are in Table 6.5, a comparison of skewness for the original and SRD variables. The CLASS_, CLASS_AGE_, and CLASS_GENDER_ variables have been reexpressed, rendering significant shifts in the corresponding skewed distributions to almost symmetric distributions: Skewness values decrease significantly in the direction of zero, although AGE_ and GENDER_ are moot variables with null (useless) graphics because the variables assume only two values, I include them as an object lesson.

6.5.2.4 *Building a Preliminary* Titanic *Model*

Reflecting on the definition of ordinal and interval variables, I know that the symmetrized-ranked variables are not ordinal variables. However, the scale property of the reexpressed variables rCLASS_, rCLASS_AGE_, and rCLASS_GENDER_ is not obvious. The variables

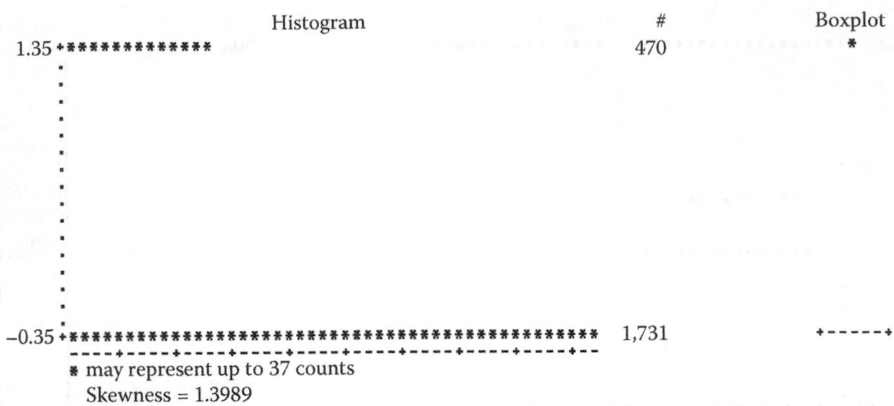

FIGURE 6.13
Histogram and boxplot for rGENDER_.

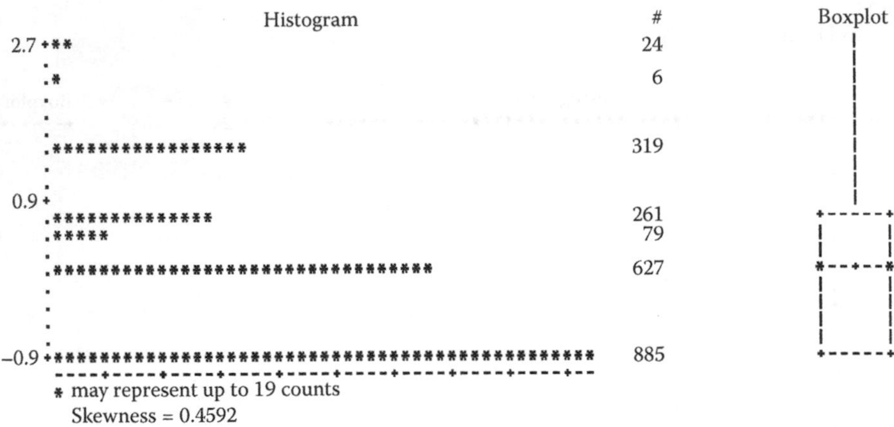

FIGURE 6.14
Histogram and boxplot for rCLASS_AGE_.

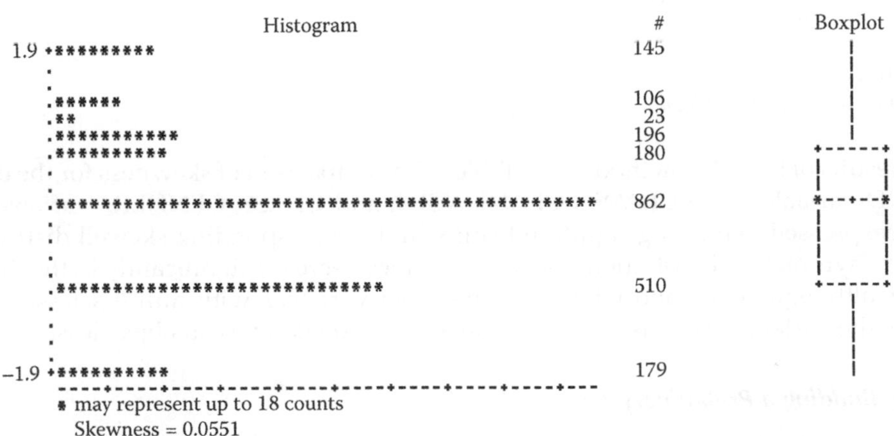

FIGURE 6.15
Histogram and boxplot for rCLASS_GENDER_.

TABLE 6.5

Comparison of Skewness Values for Original and Symmetrized-Ranked Variables

Variable	Skewness Value	Effect of the Symmetrized-Ranked Method (Yes, No, Null)
CLASS_	1.0593	Yes
rCLASS_	0.3854	
AGE_	–4.1555	Null. Binary variables render two parallel lines.
rAGE_	–4.1555	
GENDER_	1.3989	Null. Binary variables render two parallel lines.
rGENDER_	1.3989	
CLASS_AGE_	0.9848	Yes
rCLASS_AGE_	0.4592	
CLASS_GENDER_	1.1935	Yes
rCLASS_GENDER_	0.0551	

TABLE 6.6

Preliminary Titanic Model

The LOGISTIC Procedure: Analysis of Maximum Likelihood Estimates

Parameter	DF	Estimate	Standard Error	Wald Chi-Square	Pr > ChiSq
Intercept	1	–0.8690	0.0533	265.4989	<0.0001
rCLASS_AGE_	1	0.4037	0.0581	48.1935	<0.0001
rCLASS_GENDER_	1	1.0104	0.0635	252.9202	<0.0001

TABLE 6.7

Classification Table for the Preliminary Titanic Model

	Predicted Deceased	Predicted Survivors	Total
Actual deceased	1,199	291	1,490
Actual survivors	291	420	711
Total	1,490	711	2,201

are not on a ratio scale because a true zero value has no interpretation. Accordingly, I define the symmetrized-ranked variable as an *approximate interval* variable.

The preliminary *Titanic* model is a logistic regression model with the dependent variable SURVIVED, which assumes 1 = yes and 0 = no. The preliminary *Titanic* model, built by the SAS procedure LOGISTIC, and its definition, consisting of two interaction symmetrized-ranked variables rCLASS_AGE_ and rCLASS_GENDER_ are in Table 6.6.

The preliminary *Titanic* model result is that 59.1% (=420/711) survivors are classified correctly *among those predicted survivors*. See Table 6.7. This conditional survivor rate indicates a more accurate assessment of the predictiveness of a binary classification model—when there is a disproportionate large cell, such as the 1,199 passengers who are actual and predicted deceased (top row, first column). I present only the preliminary model because there is much work to be done, which includes testing the three-way interaction variables, before finalizing the Titanic model. Such work is beyond the scope of this chapter.

6.6 Summary

I introduce a new statistical data mining method, the SRD method, and add it to the paradigm of simplicity and desirability for good model-building practice. The method uses two basic statistical tools, symmetrizing and ranking variables, yielding new reexpressed variables with likely improved predictive power. First, I detail Steven's scales of measurement to provide a framework for the new symmetric reexpressed variables. Accordingly, I define the newly created SRD variables on an approximate interval scale. Then, I provide a quick review of the simplest of EDA elements, the stem-and-leaf display and the box-and-whiskers plot, as both are essential for presenting the underpinnings of the SRD method. Last, I illustrate the new method with two examples, which show the improved predictive power of the symmetric reexpressed variables over the raw variables. It is my intent that the examples are the starting point for model builders to apply the SRD method in their work.

References

1. Stevens, S.S., On the theory of scales of measurement, *American Association for the Advancement of Science*, 103(2684), 677–680, 1946.
2. Simonoff, J.S., The "Unusual Episode" and a second statistics course, *Journal of Statistics Education*, 5(1), 1997. http://www.amstat.org/publications/jse/v5n1/simonoff.html.
3. Friendly, M., Extending mosaic displays: Marginal, partial, and conditional views of categorical data, *Journal of Computational and Graphical Statistics*, 8, 373–395, 1999.
4. http://www.geniqmodel.com/res/TitanicGenIQModel.html, 1999.
5. Young, F.W., *Visualizing Categorical Data in ViSta*, University of North Carolina, 1993. http://forrest.psych.unc.edu/research/vista-frames/pdf/Categorical DataAnalysis.pdf.
6. SAS Institute, Introduction, 2009. http://support.sas.com/publishing/pubcat/chaps/56571.pdf.

7

Principal Component Analysis: A Statistical Data Mining Method for Many-Variable Assessment

7.1 Introduction

Principal component analysis (PCA), invented in 1901 by Karl Pearson[*†] as a classical data reduction technique, uncovers the interrelationships among many variables. The literature is sparse on PCA used as a reexpression[‡] technique. In this chapter, I reposition PCA as a data mining method. I illustrate PCA as a statistical data mining technique capable of serving in a *common* application with an expected solution and illustrate PCA in an *uncommon* application, yielding a reliable and robust solution. Also, I provide an original and valuable use of PCA in the construction of quasi-interaction variables, furthering the case of PCA as a powerful data mining method. I provide the SAS© subroutine used in the PCA construction of quasi-interaction variables.

7.2 EDA Reexpression Paradigm

Consider Table 7.1, the exploratory data analysis (EDA) reexpression paradigm, which indicates the objective of reexpression and method by the number of variables. In Section 5.2, "Straightness and Symmetry in Data," I discuss the relationship between the two concepts for one and two variables and provide a machine-learning data mining approach to straightening the data. The methods of the ladder of powers with the boxplot[§] and bulging rule, for one and two variables, are discussed in exacting details in Chapters 10 and 12. In Table 7.1, I put PCA in its proper place. PCA is used to retain variation among many variables by reexpressing the many original variables into a few new variables such that *most* of the variation among the many original variables is accounted for or retained by the few new uncorrelated variables. PCA, viewed as an EDA technique to identify structure, gives awareness that PCA is a valid (new) data mining tool.

* Pearson, K., On lines and planes of closest fit to systems of points in space, *Philosophical Magazine*, 2(6), 559–572, 1901.

† Harold Hotelling independently developed principal component analysis in 1933.

‡ Tukey coined the term *reexpression* without defining it. I need a definition of terms that I use. My definition of reexpression is changing the composition, structure, or scale of original variables by applying functions, such as arithmetic, mathematic, and truncation function, to produce reexpressed variables of the original variables. The objective is to bring out or uncover reexpressed variables that have more information than the original variables.

§ Boxplot is also known as box-and-whiskers plot.

TABLE 7.1

Objective of Reexpression and Method by Number of Variables

	Number of Variables		
	1	2	Many
Reexpression	Symmetrize	Straighten	Retain variation
Method	Ladder of powers with boxplot	Ladder of powers with bulging rule	PCA

7.3 What Is the Big Deal?

The use of many variables burdens the data miner with two costs:

1. Working with many variables takes more time and space. This task is self-evident—just ask a data miner.
2. Modeling dependent variable Y by including many predictor variables yields fitting many coefficients and renders the predicted Y with greater error variance if the fit to Y is adequate with few variables.

Thus, by reexpressing many predictor variables to a few new variables, the data miner saves time and space, and, importantly, reduces error variance of the predicted Y.

7.4 PCA Basics

PCA transforms a set of p variables, X1, X2, … , Xp into p linear combination variables PC1, PC2, … , PCp (PC for principal component) such that most of the information (variation) in the original set of variables can be represented in a smaller set of new variables, which are uncorrelated with each other. That is,

$$PC1 = a11*X1 + a12*X2 + ... + a1j*Xj + ... + a1p*Xp$$

$$PC2 = a21*X1 + a22*X2 + ... + a2j*Xj + ... + a2p*Xp$$

$$\vdots$$

$$PCi = ai1*X1 + ai2*X2 + ... + aij*Xj + ... + aip*Xp$$

$$\vdots$$

$$PCp = ap1*X1 + ap2*X2 + ... + apj*Xj + ... + app*Xp$$

where the aij's are constants called PC coefficients.

Note, for ease of presentation, the Xs are assumed to be standardized. Also, the PCs and the aij's have many algebraic and interpretive properties.

7.5 Exemplary Detailed Illustration

Consider the correlation matrix of four census EDUCATION variables (X_1, X_2, X_3, X_4) and the corresponding PCA output in Tables 7.2 and 7.3, respectively.

7.5.1 Discussion

1. Because there are four variables, it is possible to extract four PCs from the correlation matrix.
2. The basic statistics of PCA are
   a. The four variances: Latent Roots (LR1, LR2, LR3, LR4), which are ordered by size
   b. The associated weight (i.e., coefficient) vectors: Latent Vectors (a1, a2, a3, a4)
3. The total variance in the system or dataset is four—the sum of the variances of the four (standardized) variables.
4. Each Latent Vector contains four elements, one corresponding to each variable. For a1 there are

$$\left[-0.5514, -0.4041, 0.4844, 0.5457\right]$$

which are the four coefficients associated with the first and largest PC whose variance is 2.6620.

TABLE 7.2

Correlation Matrix of X1, X2, X3, and X4

	X1	X2	X3	X4
% less than high school (X1)	1.000	0.2689	–0.7532	–0.8116
% graduated high school (X2)		1.000	0.3823	–0.6200
% some college (X3)			1.000	0.4311
% college or more (X4)				1.000

TABLE 7.3

Latent Roots (Variance) and Latent Vectors (Coefficients) of Correlation Matrix

	Latent Vector			
Variable	**a1**	**a2**	**a3**	**a4**
X1	–0.5514	–0.4222	0.2912	0.6578
X2	–0.4042	0.7779	–0.3595	0.3196
X3	0.4844	0.4120	0.6766	0.3710
X4	0.5457	0.2162	–0.5727	0.5721
LRi	2.6620	0.8238	0.5141	0.0000
Prop. var %	66.55	20.59	12.85	0.0000
Cum. var %	66.55	87.14	100	100

5. The first PC is the linear combination.

$$PC1 = -0.5514*X1 - 0.4042*X2 + 0.4844*X3 + 0.5457*X4$$

6. PC1 explains (100 * 2.6620/4) = 66.55% of the total variance of the four variables.
7. The second PC is the linear combination

$$PC1 = -0.4222*X1 + 0.7779*X2 + 0.4120*X3 - 0.2162*X4$$

   with next-to-largest variance as 1.202, and this explains (100 * 0.8238/4) = 20.59% of the total variance of the four variables.
8. Together, the first two PCs account for 66.55% + 20.59% or 87.14% of the variance in the four variables.
9. For the first PC, the first two coefficients are negative, and the last two are positive. Accordingly, the interpretation of PC1 is:
   a. It is a *contrast* between persons who at most graduated from high school and persons who at least attended college.
   b. *High* scores on PC1 correspond with zip codes where the percentage of persons who at least attended college is *greater* than the percentage of persons who at most graduated from high school.
   c. *Low* scores on PC1 correspond with zip codes where the percentage of persons who at least attended college is *less* than the percentage of persons who at most graduated from high school.

7.6 Algebraic Properties of PCA

Almost always, PCA is performed on a correlation matrix. PCA performed on a correlation matrix implies that PCA uses standardized variables whose means = 0 and variances = 1.

1. Each PCi,

$$PCi = ai1*X1 + ai2*X2 + \ldots + aij*Xj + \ldots + aip*Xp$$

   has a variance, also called a latent root or eigenvalue, such that
   a. Var(PC1) is maximum.
   b. Var(PC1) > Var(PC2) >…> Var(PCp)
      i. Equality can occur but it is rare.
   c. Mean(PCi) = 0.
2. All PCs are uncorrelated.
3. Associated with each latent root i is a latent vector

$$\left(ai1, ai2, \ldots, aij, \ldots, aip\right),$$

which together are the weights for the linear combination of the original variables forming PCi

$$PCi = ai1*X1 + ai2*X2 + \ldots + aij*Xj + \ldots + aip*Xp.$$

4. The sum of the variances of the PCs (i.e., the sum of the latent roots) is equal to the sum of the variances of the original variables. Because the variables are standardized, the following identity results: The sum of latent roots = p.
5. The proportion of variance in the original p variables that k PCs account for is

$$= \frac{\text{sum of the latent roots for the first k PCs}}{p}$$

6. The correlation between Xi and PCj equals

$$aij * \text{sqrt}\left[\text{Var}(PCj)\right].$$

   This correlation is called a *PC loading.*
7. The rule of thumb for identifying significant loadings is: If loading aij satisfies the following inequality, then the aij is significant.

$$aij > 0.5 / \text{sqrt}\left[\text{Var}(PCj)\right]$$

8. The sum of the squares of loading across all the PCs for an original variable indicates how much variance for that variable (communality) is due to the PCs.
9. Var(PC) = small (less than 0.001) implies high multicollinearity.
10. Var(PC) = 0 implies a perfect collinear relationship exists.

7.7 Uncommon Illustration

The objective of this uncommon illustration is to examine the procedures for considering a categorical predictor variable R_CD, which assumes 64 distinct values, defined by six binary elemental variables (X_1, X_2, X_3, X_4, X_5, X_6), for inclusion in a binary RESPONSE predictive model.

The classic approach is to create (63) dummy variables and test the complete set of dummy variables for inclusion in the model regardless of the number of dummy variables that are declared nonsignificant. This approach is problematic: Putting all the dummy variables in the model effectively adds noise or unreliability to the model because nonsignificant variables are known to be noisy. Intuitively, a large set of inseparable dummy variables poses a difficulty in model building in that they quickly "fill up" the model, not allowing room for other variables.

An alternative approach is to break up the complete set of dummy variables. Even if the dummy variables are not considered as a set and regardless of the variable selection method used, too many dummy variables* are still included spuriously in a model. As with the classical approach, this tactic yields too many dummy variables that fill up the model, making it difficult for other candidate predictor variables to enter the model.

There are two additional alternative approaches for testing a categorical variable for inclusion in a model. One is *smoothing a categorical variable for model inclusion,* which is illustrated in a case study in Chapter 10. (At this point, I have not provided background for discussing the method of smoothing a categorical.) The second is the proposed PCA data mining procedure, which is effective, reliable, and easy to carry out. I present the procedure, in which the performance of PCA is on the six elemental variables X_1, X_2, X_3, X_4, X_5, and X_6, in the next section.

7.7.1 PCA of R_CD Elements (X_1, X_2, X_3, X_4, X_5, X_6)

The output of the PCA of R_CD six elemental variables is shown in Table 7.4.

7.7.2 Discussion of the PCA of R_CD Elements

1. Six elements of R_CD produce six PCs. PCA factoid: k original variables always produce k PCs.
2. The first two components, R1 and R2, account for 80.642% of the total variation. The first component R1 accounts for 50.634% of the total variation.
3. R1 is a contrast of X3 and X6 with X2, X4, and X5. The data-mined contrast is the fruit of the PCA data mining method.

TABLE 7.4

PCA of the Six Elements of R_CD (X1, X2, X3, X4, X5, X6)

	Eigenvalues of the Correlation Matrix			
	Eigenvalue	Difference	Proportion	Cumulative
R1	3.03807	1.23759	0.506345	0.50634
R2	1.80048	0.89416	0.300079	0.80642
R3	0.90632	0.71641	0.151053	0.95748
R4	0.18991	0.14428	0.031652	0.98913
R5	0.04563	0.02604	0.007605	0.99673
R6	0.01959	–	0.003265	1.00000

	Eigenvectors					
	R1	R2	R3	R4	R5	R6
X1	0.038567	–0.700382	0.304779	–0.251930	0.592915	0.008353
X2	0.473495	0.177061	0.445763	–0.636097	–0.323745	0.190570
X3	–0.216239	0.674166	0.102083	–0.234035	0.658377	0.009403
X4	0.553323	0.084856	0.010638	0.494018	0.260060	0.612237
X5	0.556382	0.112421	0.125057	0.231257	0.141395	–0.767261
X6	–0.334408	0.061446	0.825973	0.423799	–0.150190	0.001190

* Typically, dummy variables that reflect 100% and 0% response rates are based on a very small number of individuals.

4. R3 is a weighted average of all six positive elements. PCA factoid: A *weighted average* component, also known as a generalized component, is often produced. R3, a generalized component, is also one of the fruits of the PCA data mining method, as the generalized component is often used instead of any one or all of the many original variables.
5. Consider Table 7.5. The variables are ranked from most to least correlated to RESPONSE based on the absolute value of the correlation coefficient.
   a. PCs R1, R3, R4, and R6 have larger correlation coefficients than the original X variables.
   b. PCA factoid: It is typical for some PCs to have correlation coefficients larger than the correlation coefficients of some of the original variables.
   c. In fact, this is the reason to perform a PCA.
   d. Only R1 and R3 are statistically significant with p-values less than 0.0001. The other variables have p-values between 0.015 and 0.7334.

I build a RESPONSE model with the candidate predictor variable set consisting of the six original and six PC variables. (Details of the model are not shown.) I could only substantiate a two-variable model that includes (not surprisingly) R1 and R3. Regarding the predictive power of the model:

1. The model identifies the top 10% of the most responsive individuals with a response rate 24% *greater* than chance (i.e., the average response rate of the data file).
2. The model identifies the bottom 10% of the least responsive individuals with a response rate 68% *less* than chance.
3. Thus, the model's predictive power index (top 10%/bottom 10%) = 1.8 (=124/68). This index value indicates the model has moderate predictive power given that I use only two PC variables. With additional variables in the candidate predictor variable set, I know building a stronger predictive model is possible. And, I believe the PCs R1 and R3 will be in the final model.

Lest one thinks that I forgot about the importance of straightness and symmetry mentioned in Chapter 5, I note that PCs are typically normally distributed. And, because straightness and symmetry go hand in hand, there is little concern about checking the straightness of R1 and R3.

TABLE 7.5

Correlation Coefficients: RESPONSE with the Original and Principal Component Variables Ranked by Absolute Coefficient Values

	Original and Principal Component Variables					
	R1	**R3**	**R4**	**R6**	**X4**	**X6**
RESPONSE with	0.10048**	0.08797**	0.01257*	–0.01109*	0.00973*	–0.00959*
	R2	**X5**	**R5**	**X3**	**X2**	**X1**
	0.00082*	0.00753*	0.00741*	0.00729*	0.00538*	0.00258*

**$p < 0.0001$; *$0.015 < p < 0.7334$.

7.8 PCA in the Construction of Quasi-Interaction Variables

I provide an original and valuable use of PCA in the construction of quasi-interaction variables. I use SAS for this task and provide the program (in Section 7.8.1) after the steps of the construction are detailed. Consider the dataset IN in Table 7.6. There are two categorical variables: GENDER (assume M for male, F for female, and blank for missing) and MARITAL (assume M for married, S for single, D for divorced, and blank for missing).

I recode the variables, replacing the blank with the letter *x*. Thus, GENDER_ and MARITAL_ are the recoded versions of GENDER and MARITAL, respectively (see Table 7.7).

Next, I use the SAS procedure TRANSREG to create the dummy variables for both GENDER_ and MARITAL_. For each value of the variables, there are corresponding dummy variables. For example, for GENDER_ = M, the dummy variable is GENDER_M. The reference dummy variable is for the missing value of x (see Table 7.8).

I perform a PCA with GENDER_ and MARITAL_ dummy variables. This produces five quasi-interaction variables: GENDER_x_MARITAL_pc1 to GENDER_x_MARITAL_pc5. The output of the PCA is in Table 7.9. I leave the reader to interpret the results. Notwithstanding the detailed findings, it is clear that PCA is a powerful data mining method.

TABLE 7.6

Dataset IN

ID	GENDER	MARITAL
1	M	S
2	M	M
3	M	
4		
5	F	S
6	F	M
7	F	
8		M
9		S
10	M	D

TABLE 7.7

Dataset IN with Necessary Recoding of Variables for Missing Values

ID	GENDER	GENDER_	MARITAL	MARITAL_
1	M	M	S	S
2	M	M	M	M
3	M	M		x
4		x		x
5	F	F	S	S
8	F	F	M	M
7	F	F		x
8		x	M	M
9		x	S	S
10	M	M	D	D

TABLE 7.8

Dataset IN with Dummy Variables for GENDER_ and MARITAL_ Using SAS Proc TRANSREG

ID	GENDER_	MARITAL_	GENDER_F	GENDER_M	MARITAL_D	MARITAL_M	MARITAL_S
1	M	S	0	1	0	0	1
2	M	M	0	1	0	1	0
3	M	x	0	1	0	0	0
4	x	x	0	0	0	0	0
5	F	S	1	0	0	0	1
6	F	x	1	0	0	0	0
7	F	M	1	0	0	1	0
8	x	M	0	0	0	1	0
9	x	S	0	0	0	0	1
10	M	D	0	1	1	0	0

TABLE 7.9

PCA with GENDER_ and MARITAL_ Dummy Variables Producing Quasi-GENDER_x_ MARITAL_ Interaction Variables

The PRINCOMP Procedure

Observations	10
Variables	5

Simple Statistics

	GENDER_F	FIGURE	MARITAL_D	MARITAL_M	MARITAL_S
Mean	0.3000000000	0.4000000000	0.1000000000	0.3000000000	0.3000000000
StD	0.4830458915	0.5163977795	0.3162277660	0.4830458915	0.4830458915

Correlation Matrix

	GENDER_F	GENDER_M	MARITAL_D	MARITAL_M	MARITAL_S
GENDER_F	1.0000	–0.5345	–0.2182	0.0476	0.0476
GENDER_M	–0.5345	1.0000	0.4082	–0.0891	0.0891
MARITAL_D	–0.2182	0.4082	1.0000	–0.2182	0.2182
MARITAL_M	0.0476	–0.0891	–0.2182	1.0000	–0.4286
MARITAL_S	0.0476	–0.0891	–0.2182	0.4286	1.0000

Eigenvalues of the Correlation Matrix

	Eigenvalue	Difference	Proportion	Cumulative
1	1.84840072	0.41982929	0.3697	0.3697
2	1.42857143	0.53896598	0.2857	0.6554
3	0.88960545	0.43723188	0.1779	0.8333
4	0.45237357	0.07132474	0.0905	0.9238
5	0.38104883		0.0762	1.0000

(Continued)

TABLE 7.9 (CONTINUED)

PCA with GENDER_ and MARITAL_ Dummy Variables Producing Quasi-GENDER_x_ MARITAL_ Interaction Variables

	Eigenvectors				
	GENDER_x_ MARITAL_ pc1	GENDER_x_ MARITAL_ pc2	GENDER_x_ MARITAL_ pc3	GENDER_x_ MARITAL_ pc4	GENDER_x_ MARITAL_ pc5
GENDER_F	–0.543563	0.000000	0.567380	0.551467	–0.280184
GENDER_M	0.623943	0.000000	–0.209190	0.597326	–0.458405
MARITAL_D	0.518445	0.000000	0.663472	0.097928	0.530499
MARITAL_M	–0.152394	0.707107	–0.311547	0.405892	0.463644
MARITAL_S	–0.152394	–0.707107	–0.311547	0.405892	0.463644

7.8.1 SAS Program for the PCA of the Quasi-Interaction Variable

```
data IN;
input ID 2.0 GENDER $1. MARITAL $1.;
cards;
01MS
02MM
03M
04
05FS
08FM
07F
08 M
09 S
10MD
;
run;

PROC PRINT noobs data=IN;
title2 ' Data IN ';
run;

data IN;
set IN;
GENDER_ = GENDER; if GENDER =' ' then GENDER_ ='x';
MARITAL_= MARITAL; if MARITAL=' ' then MARITAL_='x';
run;

PROC PRINT noobs;
var ID GENDER GENDER_ MARITAL MARITAL_;
title2 ' ';
title3 ' Data IN with necessary Recoding of Vars. for Missing Values ';
title4 ' GENDER now Recoded to GENDER_, MARITAL now Recoded to MARITAL_';
title5 ' Missing Values, replaced with letter x ';
run;
```

```
/* Using PROC TRANSREG to create Dummy Variables for GENDER_ */
PROC TRANSREG data=IN DESIGN;
model class (GENDER_ / ZERO='x');
output out = GENDER_ (drop = Intercept _NAME_ _TYPE_);
id ID;
run;

/* Appending GENDER_ Dummy Variables */
PROC SORT data=GENDER; by ID;
PROC SORT data=IN; by ID;
run;

data IN;
merge IN GENDER_;
by ID;
run;

/* Using PROC TRANSREG to create Dummy Variables for GENDER_ */
PROC TRANSREG data=IN DESIGN;
model class (MARITAL_ / ZERO='x');
output out=MARITAL_ (drop= Intercept _NAME_ _TYPE_);
id ID;
run;

/* Appending MARITAL_ Dummy Variables */
PROC SORT data=MARITAL_; by ID;
PROC SORT data=IN; by ID;
run;
data IN;
merge IN MARITAL_; by ID;
run;

PROC PRINT data=IN (drop= GENDER MARITAL) noobs;
title2' PROC TRANSREG to create Dummy Vars. for both GENDER_ and MARITAL_ ';
run;

/* Running PCA with GENDER_ and MARITAL_ Variables Together */
/* This PCA of a Quasi-GENDER_x_MARITAL Interaction */
PROC PRINCOMP data= IN n=4 outstat=coef out=IN_pcs
        prefix=GENDER_x_MARITAL_pc std;
var GENDER_F GENDER_M MARITAL_D MARITAL_M MARITAL_S;
title2 ' PCA with both GENDER_ and MARITAL_ Dummy Variables ';
title3 ' This is PCA of a Quasi-GENDER_x_MARITAL Interaction ';
run;

PROC PRINT data=IN_pcs noobs;
title2 ' Data appended with the PCs for Quasi-GENDER_x_MARITAL Interaction ';
title3 ' ';
run;
```

7.9 Summary

I reposition the classical data reduction technique of PCA as a reexpression method of EDA. Then, I relabel PCA as a voguish data mining method of today. I illustrate PCA as a statistical data mining technique capable of serving in common applications with expected solutions. Specifically, the illustration details PCA as an exemplary presentation of a set of census EDUCATION variables. And, I illustrate PCA in an uncommon application of finding a structural approach for preparing a categorical predictive variable for possible model inclusion. The results are compelling as they highlight the power of the PCA data mining tool. Also, I provide an original and valuable use of PCA in the construction of quasi-interaction variables, along with the SAS program, to perform this novel PCA application.

8

Market Share Estimation: Data Mining for an Exceptional Case

8.1 Introduction

Market share is an essential metric that companies use to measure the performance of a product in the marketplace. Specifically, market share quantifies a company's customer preference for a given product over their competitions' customer preferences for that product. Companies do not readily have the competitive data necessary to estimate market share. Consequently, to build market share models, companies conduct market research surveys to obtain data on their competitors. Survey data are expensive and time-consuming to collect and are fraught with unreliability. The purpose of this chapter is to present a unique market share estimation model, in that it does not belong to the usual family of survey-based market share models. The proposed method uses the application of principal component analysis to build a market share model on an exceptional case study. The intended approach offers rich utility because companies typically have similar data to that of the case study. I provide the SAS© subroutines used in building the market share model for the exceptional case study. The subroutines are also available for downloading from my website: http://www.geniq.net/articles.html#section9.

8.2 Background

Market share is an essential statistic that companies use to measure the performance of a product or service (hereafter product is to encompass services) in the marketplace. A simple definition of market share, for product j in geographical area k, for Company ABC is ABC's sales (revenue or units) in period t divided by the relevant market* total sales in time t. Although the concept of market share is not complicated, it does raise many scenarios for the product manager. For example, the project manager must know how to parse the market share estimate by taking into account not only the company's actions but also the actions of its competitors. Additionally, seasonality and general economic conditions always affect the performance of products in the marketplace.

* Relevant market is the market at large of which Company ABC is a part and consists of potential buyers for product j.

Another consideration is that there is no universal goal for market share. A perfect market share (equal or near 100%) is not necessarily a good thing. A market leader with a large market share may have to expand the market for its growth with commensurate expenditures it may not be prepared to spend. The product manager has to monitor market share to ensure it is in harmony with the company's financial health.

Although market share is a simple concept, it is inherently more involved than sales analysis for a product. Market share requires taking into account competitive factors [1]. The literature is replete with theoretical market share models that address factors such as advertising elasticities and the effects of changes in the levels of marketing mix variables, namely, price, product, promotion, and place—known as the four Ps [2–4]. Moreover, the choice of the which model to use, as there are many, is another decision point for achieving the best forecasting in terms of accuracy as well as parameter validity.

There are two popular market share models, Multinomial Logit (MNL) and Multiplicative Competitive Interaction (MCI). The specifications of these models are beyond the scope of this chapter. It suffices to say that these models, their variants, and divergent models built on underpinnings poles apart allow for analyzing and predicting market share for a brand or a firm [5–7]. MNL and MCI use brand attributes and the four Ps for their predictions.

It is clear that data are what drive a model, notwithstanding the appropriate selection of the model, whether the naïve benchmark first-order autoregressive model or the complex market share attraction model. Data sources include retail store-level scanner data, wholesale warehouse withdrawals, consumer surveys, and diary panel data. The challenging data issue is knowing the current business position among the competition. The data issue requires acquiring and analyzing the competitors' information. A monitoring system of the market for the current business, at an industry-aggregated market level, yields customized reports centering the current business with respect to its competition.

8.3 Data Mining for an Exceptional Case

Regarding the literature review on market share estimation models previously discussed in this chapter, there is no market share model for an exceptional case study of the proposed method. The published market share estimation approaches are not applicable. Complete company and competitor's data (the Achilles heel of market share estimation) are necessary input for the naïve autoregressive model to the advanced models. The focal exceptional case involves market share estimation for a firm that has a *soft* knowledge of its market share and only one piece of market share data, namely, promotion.

8.3.1 Exceptional Case: Infant Formula YUM

Infant formula manufacturers know the brand of formula mothers receive in the hospital is the brand the mothers are most likely to continue to use throughout the first year or longer.

Formula manufacturers also have sought to create partnerships and brand loyalty with hospitals by providing free formula for use in the hospitals. Handing out free formula increases the likelihood of brand loyalty of new mothers getting free formula for feeding during their hospital stay. At the time of newborn hospital discharge, the manufacturers give new moms "discharge packs" containing an array of formula coupons. Some formula manufacturers offer both "breastfeeding" and "formula feeding" bags. The breastfeeding bags also contain formula and formula coupons.

A major infant formula manufacturer (let us call them RAL) tracks usage of its infant formula (let us call it YUM) among new mothers who breastfeed and use formula or only use formula up to 12 months. These new mothers define RAL's overall sales market. However, RAL supplies YUM to pre-selected hospitals, which defines its relevant market, for these new mothers during their stay in the hospital. Note, RAL is not the sole infant formula manufacturer who offers their infant formula in the hospital to new moms.

RAL gives new moms discharge packs. Six weeks after discharge, RAL follows up with a promotion mailing of coupons for purchasing YUM at a discount. RAL conducts a limited panel-like study, in which new moms report, three months after leaving the hospital, which infant formula they are using.

RAL wants to estimate their moms' market share of YUM over other brands of formula, which are also offered at the hospital. There are only two pieces of market share data available for each new mom:

1. A binary variable indicating whether or not the mom is using YUM at month 3 after the hospital stay.
2. The promotion coupon the mom received. There is no tracking of the mom having used the promotion coupon.

RAL wants the influence of promotion adjusted for in the calculation of market share. For the model building, the candidate predictor variables include the typical demographic, socioeconomic, and geographic variables, along with attitude, preference, and lifestyle variables. The marketing-mix variables are not in the set of candidate predictor variables.

8.4 Building the RAL-YUM Market Share Model

The predominant issue of building the RAL Market Share Model (for estimating RAL market shares of new mothers using YUM infant formula 3 months after their infants' births) is—statistically controlling for—the influence of promotion. I present the proposed methodology in a pedagogical manner due to various statistical procedures used and assumptions made to yield reliable RAL-YUM market share estimates. I present the data mining process step-by-step in building the RAL-YUM Market Share Model.

Step #1 – Frequency of Promotion code by YUM_3mos

I define YUM_3mos to perform Step #1. YUM_3mos defined as:

YUM_3mos = 1 if new mom uses YUM for first three months

YUM_3mos = 0 if new mom does not use YUM for first three months.

TABLE 8.1
RAL Market Share (Mean_YUM_3mos)
by Frequency of Promotion (PROMO_Code)

PROMO_Code	SIZE	MEAN_YUM_3mos
10	229	0.64192
7	207	0.56039
16	38	0.55263
6	127	0.55118
1	9,290	0.53617
5	333	0.53153
14	2,394	0.53091
12	946	0.52537
15	438	0.51598
13	1,003	0.51346
4	266	0.51128
11	3,206	0.51029
2	557	0.50090
3	2,729	0.44485
9	2,488	0.43368
8	2,602	0.41929

The frequency of promotion (PROMO_Code) by the mean of YUM_3mos is shown in Table 8.1. The table is ranked descending by MEAN_YUM_3mos, RAL's rough estimates of YUM_3mos market shares at the promotion code level. The table also includes SIZE, the number of promotions.

The influence of the promotions is apparent as RAL knows its YUM_3mos market share is a soft 0.20. The frequency table points to the objective of the study, namely, eliminating the effects of the promotions to yield an honest estimate of RAL-YUM market share.

Step #2 – Dummify PROMO_Code

To create dummy variables for all PROMO_Codes, I run the subroutine in Appendix 8.A, Dummify PROMO_Code, which automatically generates the code as follows.

If PROMO_Code = 1 then
PROMO_Code1 = 1; otherwise
PROMO_Code1 = 0

If PROMO_Code = 2 then
PROMO_Code2 = 1; otherwise
PROMO_Code2 = 0

...

If PROMO_Code = 16 then
PROMO_Code16 = 1; otherwise
PROMO_Code16 = 0

There are 16 promotion dummy variables, capturing all (100%) promotion information. The set of dummy variables assists in eliminating, as well as statistical controlling for, promotion effects.

Step #3 – PCA of PROMO_Code

I perform a principal component analysis (PCA) of PROMO_Code, using SAS Proc PRINCOMP. In Chapter 7, I provide a thorough review of the PCA model. Here, I focus on the two primary PCA statistics, eigenvalues and eigenvectors, used in statistical controlling for promotion effects.

Sixteen promotion dummy variables, PROMO_Code1 – PROMO_Code16, yield 16 principal components (PCs). The 16 PCs capture all information within the promotion dummy variables in a unique way with great statistical utility:

1. PCs are reliable and stable variables by the theoretical nature of their construction.
2. PCs are continuous and thus more stable than the original dummy variables.
3. PCs are fundamental to eliminate unwanted effects of other variables, as will be demonstrated.

I run Proc PRINCOMP code, in Appendix 8.B, PCA of PROMO_Code Dummy Variables, and obtain the 16 eigenvalues, in Table 8.2. Of note, the first eight PCs explain more than half (55.84%) of the total variance of PROMO_Code.

The second set of statistics from running Proc PRINCOMP of the PROMO_Code dummy variables are the eigenvectors in Tables 8.3 and 8.4. The naming convention for the eigenvectors (PCs) uses the prefix PROMO_Code_pc.

Step #4 – Eliminating the Influence of Promotion from YUM_3mos

The dependent variable (DepVar)YUM_3mos has a full data capture of PROMO_Code. DepVar and PROMO_Code dummy variables are data-fused. Invoking a basic statistical axiom generates the desired DepVar free of promotion effect.

TABLE 8.2

Eigenvalues of PCA of PROMO_Code Dummy Variables

	Eigenvalue	Difference	Proportion	Cumulative
1	1.37944062	0.25053430	0.0862	0.0862
2	1.12890632	0.01818890	0.0706	0.1568
3	1.11071742	0.00611858	0.0694	0.2262
4	1.10459884	0.00514552	0.0690	0.2952
5	1.09945332	0.05052025	0.0687	0.3639
6	1.04893307	0.01137866	0.0656	0.4295
7	1.03755441	0.01328778	0.0648	0.4944
8	1.02426663	0.00618906	0.0640	0.5584
9	1.01807757	0.00438791	0.0636	0.6220
10	1.01368967	0.00290790	0.0634	0.6854
11	1.01078176	0.00172483	0.0632	0.7485
12	1.00905694	0.00104677	0.0631	0.8116
13	1.00801016	0.00300506	0.0630	0.8746
14	1.00500510	0.00349694	0.0628	0.9374
15	1.00150816	1.00150816	0.0626	1.0000
16	0.00000000		0.0000	1.0000

TABLE 8.3

Eigenvectors of PCA of PROMO_Code Dummy Variables

	PROMO_ Code_pc1	PROMO_ Code_pc2	PROMO_ Code_pc3	PROMO_ Code_pc4	PROMO_ Code_pc5	PROMO_ Code_pc6	PROMO_ Code_pc7	PROMO_ Code_pc8
PROMO_Code1	−0.849192	−0.028771	−0.009891	−0.009136	−0.007133	−0.052172	−0.006089	−0.025351
PROMO_Code2	0.070919	0.021381	0.009245	0.009300	0.007832	0.180588	0.036573	0.830011
PROMO_Code3	0.220472	0.337257	−0.794960	−0.210305	−0.103634	−0.180422	−0.018312	−0.066588
PROMO_Code4	0.047265	0.013313	0.005649	0.005637	0.004710	0.088477	0.014938	0.123381
PROMO_Code5	0.053319	0.015242	0.006493	0.006490	0.005432	0.106072	0.018443	0.168341
PROMO_Code6	0.032117	0.008787	0.003700	0.003680	0.003066	0.053726	0.008646	0.062142
PROMO_Code7	0.041398	0.011515	0.004870	0.004852	0.004049	0.073728	0.012175	0.093983
PROMO_Code8	0.210117	0.239859	0.544302	−0.647716	−0.175960	−0.193263	−0.019324	−0.069403
PROMO_Code9	0.201152	0.188749	0.211250	0.685419	−0.505953	−0.206907	−0.020365	−0.072213
PROMO_Code10	0.043659	0.012200	0.005166	0.005150	0.004300	0.079181	0.013179	0.104149
PROMO_Code11	0.263586	−0.873479	−0.084243	−0.063358	−0.042934	−0.146332	−0.015455	−0.058193
PROMO_Code12	0.097277	0.032731	0.014647	0.014961	0.012789	0.529880	0.756561	−0.274451
PROMO_Code13	0.100948	0.034596	0.015579	0.015958	0.013680	0.669420	−0.650236	−0.238421
PROMO_Code14	0.193973	0.159567	0.138593	0.248137	0.836577	−0.220100	−0.021338	−0.074767
PROMO_Code15	0.061951	0.018142	0.007780	0.007798	0.006545	0.137057	0.025261	0.294764
PROMO_Code16	0.017384	0.004673	0.001959	0.001945	0.001617	0.027280	0.004286	0.028983

TABLE 8.4

Eigenvectors of PCA of PROMO_Code Dummy Variables

	PROMO_ Code_pc9	PROMO_ Code_pc10	PROMO_ Code_pc11	PROMO_ Code_pc12	PROMO_ Code_pc13	PROMO_ Code_pc14	PROMO_ Code_pc15	PROMO_ Code_pc16
PROMO_Code1	−0.013803	−0.012062	−0.009018	−0.005753	−0.003471	−0.005006	−0.003322	0.523367
PROMO_Code2	−0.454554	−0.166006	−0.089908	−0.049363	−0.027474	−0.032449	−0.017824	0.156809
PROMO_Code3	−0.034310	−0.028906	−0.021114	−0.013291	−0.007956	−0.011220	−0.007260	0.332449
PROMO_Code4	0.120133	0.231979	0.827199	−0.433931	−0.124681	−0.072155	−0.028364	0.108962
PROMO_Code5	0.196784	0.845121	−0.405641	−0.131678	−0.061284	−0.053520	−0.024437	0.121761
PROMO_Code6	0.050156	0.065911	0.073449	0.065851	0.052605	0.982015	−0.051203	0.075486
PROMO_Code7	0.082894	0.127209	0.187927	0.281328	0.906925	−0.115035	−0.033922	0.096228
PROMO_Code8	−0.035597	−0.029905	−0.021805	−0.013712	−0.008203	−0.011550	−0.007460	0.325475
PROMO_Code9	−0.036868	−0.030883	−0.022479	−0.014122	−0.008443	−0.011868	−0.007653	0.319012
PROMO_Code10	0.094889	0.155759	0.273262	0.837002	−0.390122	−0.093015	−0.031503	0.101170
PROMO_Code11	−0.030383	−0.025816	−0.018956	−0.011968	−0.007177	−0.010171	−0.006617	0.356754
PROMO_Code12	−0.100486	−0.071543	−0.047708	−0.028620	−0.016668	−0.021872	−0.013152	0.202840
PROMO_Code13	−0.092159	−0.067036	−0.045167	−0.027235	−0.015907	−0.021022	−0.012726	0.208632
PROMO_Code14	−0.038011	−0.031757	−0.023078	−0.014485	−0.008656	−0.012150	−0.007822	0.313531
PROMO_Code15	0.834506	−0.380535	−0.142646	−0.070377	−0.037354	−0.040119	−0.020583	0.139368
PROMO_Code16	0.021907	0.026213	0.025827	0.020264	0.014197	0.037841	0.996230	0.041360

Statistical Axiom

1. Regressing Y on X yields the estimate of Y (est_Y). Let *Y_due_X* denote est_Y.
2. If Y is binary, then 1 – est_Y represents *Y without effects of* X. Let *Y_wo_Xeffect* denote 1 – est_Y.

I run the subroutine for the logistic regression YUM_3mos based on the PROMO_Code dummy variables in Appendix 8.C, Logistic Regression YUM_3mos on PROMO_Code Dummy Variables. It is a well-known statistical factoid that among a set of k dummy variables, only k – 1 dummy variables can enter a model. In this case, the decision of excluding PROMO_Code16 is easy because its variance (eigenvalue) is zero in Table 8.2 (last row, second column). Based on the statistical axiom, the resultant logistic regression produces YUM_3mos_due_PROMO_Code.

For ease of presentation, I rename:

YUM_3mos_due_PROMO_Code to *YUM3mos_due_PROMO*, and

YUM_3mos_wo_PROMO_Codeeffect to *YUM3mos_wo_PROMOeff.*

The estimated YUM3mos_due_PROMO Model in Table 8.5 appear to indicate the model is ill-defined because there are many large p-values (Pr > ChiSq). However, all variables must be included in the model, significant ones with small p-values, and nonsignificant ones with large p-values because the intent of the model is to have est_YUM_3mos capture all promotion information (i.e., 100% of promotion variation as indicated in Table 8.2, last or penultimate row, last column). Thus, the large p-valued variables do not affect the model designed utility. All promotion variation is by definition captured by all the PROMO dummy variables (except PROMO_Code16). Thus, the YUM3mos_due_PROMO model achieves its objective.

TABLE 8.5

Maximum Likelihood Estimates of YUM3mos_due_PROMO Model

Parameter	DF	Estimate	Standard Error	Wald Chi-Square	Pr > ChiSq
Intercept	1	0.00414	0.0123	0.1140	0.7356
PROMO_Code_pcl	1	−0.1064	0.0122	75.4452	<0.0001
PROMO_Code_pc2	1	−0.0700	0.0122	32.7221	<0.0001
PROMO_Code_pc3	1	−0.0120	0.0123	0.9585	0.3276
PROMO_Code_pc4	1	0.0334	0.0123	7.3475	0.0067
PROMO_Code_pc5	1	0.0969	0.0123	62.4866	<0.0001
PROMO_Code_pc6	1	0.0615	0.0122	25.2594	<0.0001
PROMO_Code_pc7	1	0.0127	0.0122	1.0829	0.2981
PROMO_Code_pc8	1	0.0186	0.0122	2.3240	0.1274
PROMO_Code_pc9	1	0.0208	0.0122	2.8962	0.0888
PROMO_Code_pc10	1	0.0249	0.0122	4.1500	0.0416
PROMO_Code_pc11	1	0.0190	0.0123	2.4121	0.1204
PROMO_Code_pc12	1	0.0493	0.0126	15.3224	<0.0001
PROMO_Code_pc13	1	−0.00171	0.0124	0.0191	0.8902
PROMO_Code_pc14	1	0.00656	0.0123	0.2857	0.5930
PROMO_Code_pc15	1	0.00497	0.0123	0.1638	0.6857

The YUM3mos_due_PROMO model is defined in Equation 8.1:

$$\begin{aligned} \text{YUM3mos\_due\_PROMO} = &+0.0041 \\ &-0.1064*\text{PROMO\_Code\_pc1} \\ &-0.0700*\text{PROMO\_Code\_pc2} \\ &-0.0120*\text{PROMO\_Code\_pc3} \\ &+0.0334*\text{PROMO\_Code\_pc4} \\ &+0.0969*\text{PROMO\_Code\_pc5} \\ &+0.0615*\text{PROMO\_Code\_pc6} \\ &+0.0127*\text{PROMO\_Code\_pc7} \\ &+0.0186*\text{PROMO\_Code\_pc8} \\ &+0.0208*\text{PROMO\_Code\_pc9} \\ &+0.0249*\text{PROMO\_Code\_pc10} \\ &+0.0190*\text{PROMO\_Code\_pc11} \\ &+0.0493*\text{PROMO\_Code\_pc12} \\ &-0.0017*\text{PROMO\_Code\_pc13} \\ &+0.0065*\text{PROMO\_Code\_pc14} \\ &+0.0049*\text{PROMO\_Code\_pc15} \end{aligned} \tag{8.1}$$

Next, I run the subroutine in Appendix 8.D, Creating YUM_3mos_wo_PROMO_CodeEff, to calculate 1 – est_YUM_3mos. Therefore, I obtain YUM3mos_wo_PROMOeff based on the statistical axiom.

Step #5 – Creating the Market-Share Dependent Variable

In accordance with the principles of the statistical axiom of Step 4, I create the *dependent variable* MARKET-SHARE in five logical steps:

1. YUM_3mos, by definition, is promotion-effected and binary.
2. Any derivative of YUM_3mos is, therefore, a probability.
3. Thus, YUM3mos_due_PROMO is a probability.
4. YUM3mos_wo_PROMOeff, by definition, is an honest promotion-free variable.
5. Thus, MARKET-SHARE = YUM3mos_wo_PROMOeff.

Step #6 – Building the Preliminary YUM_3mos MARKET-SHARE Model

1. As for the candidate predictor variables, there are more than 1,200 candidate predictor variables: the typical demographic, socioeconomic, and geographic variables along with attitude, preference, and lifestyle variables. The marketing-mix variables are not in the set of candidate predictor variables.

2. The approach of variable selection applies the concepts* of the GenIQ Model of Chapter 40 for newly constructed variables and the final set of the predictor variables. Of course, data miners can use their preferred variable selection method.

Three predictor variables, SOFT_PCLUS1, SOFT_PCLUS2, and SOFT_PCLUS4 (predictor variables not shown), define the YUM_3mos Market-Share Model. The predictor variables are PCA-based predictor variables with a genetic programming influence. Each predictor variable is a *soft* version of a PC. For example, SOFT_PCLUS1 is a modified PC using only the signs of the PC coefficients, not the PC coefficients themselves. So, SOFT_PCLUS1 is the weighted sum of the 20 candidates, where the weights are the signs (+1 or –1) of the PC coefficients. The PC construction of the predictor variables is not shown but consists of 20 candidate variables.

Data miners at first blush are bewildered by the construction and value of soft PCs. They can test this approach by correlating the original (hard) PC with the soft version to observe, in virtually all situations, that the correlation is greater than 0.95. The advantages of the soft PCs are they are easy to understand and implement. Moreover, soft PCs offer greater stability than their hard PC counterparts because soft PCs do not use any degrees of freedom. A hard PC uses three degrees of freedom for each variable in the PC. So, a hard PC defined by, say, 20 variables uses 60 degrees of freedom. The YUM_3mos Market-Share Model with three soft PCs would otherwise lose 180 degrees of freedom.

The YUM_3mos MARKET-SHARE model, which is a *preliminary* version (for reasons to be explained in Step #7), is defined in Equation 8.2:

$$\begin{aligned} \text{MARKET-SHARE\_est} = &+ 0.40127 \\ &+ 0.02380 * \text{SOFT\_PCLUS1} \\ &+ 0.14041 * \text{SOFT\_PCLUS2} \\ &+ 0.09826 * \text{SOFT\_PCLUS4} \end{aligned} \tag{8.2}$$

Step #7 – Final YUM_3mos MARKET-SHARE Model

I generate the frequency plot along with the corresponding boxplot of the MARKET-SHARE Model scores (MktSh_est) in Figure 8.1. MktSh_est is bell-shaped, peaked with biased location (cf. market share of 0.495 versus RAL's soft market share of 0.20), and squeezed scale (0.519 – 0.471, with outlier 0.387).

With the working assumption of RAL's soft market share of 0.20, I shape MktSh_est to center the market share at 0.20, renaming the reshaped MktSh_est to MktSh_est20 in Figure 8.2. The frequency and boxplot indicate MktSh_est20 peak (centered) at 0.20; the distribution of MktSh_est20 is still bell-shaped but with a slight skewness (= 0.3477) to the right; its range is 0.729 (= 0.72 – 0.00) and does not appear to have outliers. I discussed how I created MktSh_est20 below. The effectiveness of

* Not the GenIQ Model itself.

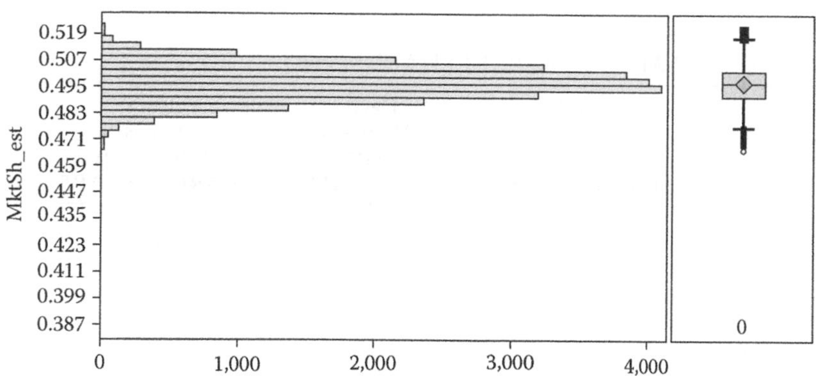

FIGURE 8.1
Frequency and boxplot of MARKET-SHARE Model scores (MktSh_est).

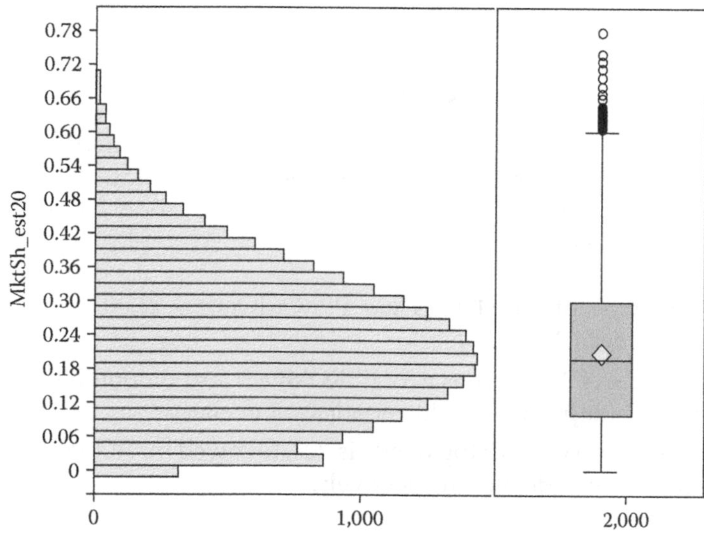

FIGURE 8.2
Frequency and boxplot of MARKET-SHARE Model centered at 0.20.

MktSh_est20 as a model for estimating YUM_3mos for new moms is discussed in the next section.

The statistical procedure to reshape MktSh_est into MktSh_est20 is described as follows. The subroutine in Appendix 8.E illustrates the transformation procedure.

1. SAS Proc RANK normalizes a variable by use of ranks.MktSh_est is bell-shaped, but Proc RANK is performed to ensure MktSh_est is as normally distributed as possible. (Note, bell-shaped is not necessarily normal-shaped.)
2. SAS Proc STANDARD "centers and spreads" a variable by subtracting a location value (minimum value: – 3.8138) and dividing by a scale value (maximum value: 3.9474/1.9961).
3. MktSh_est20 (adjusted to market share of 0.20) = (MktSh_est + 3.8138)/3.9474/1.9661.

TABLE 8.6
Decile Analysis of YUM_3mos MARKET-SHARE Model

Decile	Number of Moms	Total YUM3mos_wo_ PROMOeff MARKET-SHARE	Decile YUM3mos_wo_ PROMOeff MARKET-SHARE	Cum YUM3mos_wo_ PROMOeff MARKET-SHARE	Cum Lift
top	2,585	1,194.85	0.462	0.462	223
2	2,585	919.14	0.356	0.409	197
3	2,586	775.32	0.300	0.373	180
4	2,585	662.31	0.256	0.343	166
5	2,585	561.83	0.217	0.318	153
6	2,586	463.95	0.179	0.295	142
7	2,585	363.28	0.141	0.273	132
8	2,586	248.68	0.096	0.251	121
9	2,585	115.59	0.045	0.228	110
bottom	2,585	57.30	0.022	0.207	100
	25,853	5,362.24			

8.4.1 Decile Analysis of YUM_3mos MARKET-SHARE Model

I generate a bootstrapped decile analysis of the YUM_3mos MARKET-SHARE Model in Table 8.6. The decile analysis shows the YUM_3mos market share estimates centered at 0.207 (penultimate column, last row). The assessment of the model based on the usual statistics indicates the model is very good:

1. Top-to-bottom ratio of YUM3mos_wo_PPROMOeff is 21.00 (= 0.462/0.022; top decile to bottom decile).
2. The decile means of YUM3mos_wo_PPROMOeff are monotonically decreasing (no jumps up or bumps down in lower adjacent deciles).
3. The Cum Lifts are very good: top decile is 223, followed by 197, 180, and 166 for the second through fourth deciles, respectively.

8.4.2 Conclusion of YUM_3mos MARKET-SHARE Model

RAL management approved the final YUM_3mos MARKET-SHARE Model, which is at the individual-level of new moms. The model's value is hard to overstate in that only a new moms-level model can assist RAL in addressing the right decision on some key questions, such as:

1. How much of the market share can RAL conservatively take? What is the worth to the RAL organization financially and otherwise?
2. What kind of margins should RAL expect in this market? How do the margins align with the overall margins for the RAL organization?
3. Who are RAL's competitors in the market, and why do customers buy from them? What does RAL need to do to attract new moms?

As one can probably discern from these questions, it is important for RAL to understand not only what its organization brings to the market but the external forces that drive the market.

8.5 Summary

Digging into the literature for market share estimation models produces a wealth of theoretical models, of which some are useful for real business applications. For the project at hand, those models are of no use—even as a starting point from which I could make modifications for the current modeling project. Of all market share estimation models, in the literature and built by me, none accommodate for singular promotional datafused with the dependent variable. The only solution for building a YUM_3mos MARKET-SHARE Model is to invoke the salient features of exploratory data analysis characteristics:

1. Flexibility: I use data mining PCA because it is one among the best methods for extracting all the information from the data, especially when they are singular promotional datafused with the dependent variable.
2. Practicality: The model's performance is displayed in its tabular array, the decile analysis. A table of six columns and three statistics channel the functioning of the model and its accomplishment.
3. Innovation: Creating the market-share dependent variable is a demonstration of "Imagination is more important than knowledge."*
4. Universality: The proposed approach offers rich utility because companies typically have similar data to that of the exceptional case study.
5. Simplicity: All questions are simple once the revealed answers are known.†

Appendix 8.A Dummify PROMO_Code

```
PROC TRANSREG data= promo_ID DESIGN;
model class (PROMO_Code / ZERO='xx');
output out = PROMO_Code (drop = Intercept _NAME_ _TYPE_);
id ID;
run;

PROC SORT data= PROMO_Code; by ID;
PROC SORT data= RAL_data; by ID;
run;

data RAL_data1;
merge RAL_data PROMO_Code;
by ID;
run;
```

* Einstein said it. I just presented something to view.

† Like magic tricks.

Appendix 8.B PCA of PROMO_Code Dummy Variables

```
PROC PRINCOMP data= RAL_data1 n=16 outstat=coef out=RAL_data1_pcs
    prefix=PROMO_Code_pc std;
var
PROMO_Code1
PROMO_Code2
PROMO_Code3
PROMO_Code4
PROMO_Code5
PROMO_Code6
PROMO_Code7
PROMO_Code8
PROMO_Code9
PROMO_Code10
PROMO_Code11
PROMO_Code12
PROMO_Code13
PROMO_Code14
PROMO_Code15
PROMO_Code16
;
ods exclude cov corr SimpleStatistics;
run;
```

Appendix 8.C Logistic Regression YUM_3mos on PROMO_Code Dummy Variables

```
ods exclude ODDSRATIOS;
PROC LOGISTIC data=RAL_data1_pcs nosimple des outest=coef;
model YUM_3mos =
PROMO_Code_pc1-PROMO_Code_pc15;
run;
```

Appendix 8.D Creating YUM_3mos_wo_PROMO_CodeEff

```
PROC SCORE data=RAL_data1_pcs predict type=parms score=coef out=score;
var PROMO_Code_pc1-PROMO_Code_pc15;
run;

data score;
set score;
```

```
estimate=YUM_3mos2;
run;

data RAL_data1_wo_PromoEff;
set score;
prob_hat=exp(estimate)/(1+ exp(estimate));
YUM3mos_due_PROMO = prob_hat;
YUM3mos_wo_PROMOeff = 1- prob_hat;
run;
```

Appendix 8.E Normalizing a Variable to Lie Within [0, 1]

```
PROC RANK data=RAL_data1_wo_PromoEff normal=TUKEY
        out= X_RNORMAL ties=dense;
var YUM3mos_wo_PROMOeff;
ranks RX;
run;

PROC UNIVARIATE data=X_RNORMAL plot;
var RX;

PROC MEANS data = X_RNORMAL min max;
var RX;
run;

* Subtract min. value of RX, divide by max. value of RX;
data X_RNORMAL;
set X_RNORMAL;
RXX =(RX+3.8138025)/3.9474853/1.9661347;

PROC MEANS data = X_RNORMAL min max mean;
var RXX;
run;

* Center RXX at mean=0.20 fiddle with std values to yield 0<= RXX <=1;
PROC STANDARD data=X_RNORMAL mean=0.2 std=0.15
out=XRNORMALZ20;
var RXX;
run;

PROC UNIVARIATE data=X_RNORMALZ20 plot;
var RXX;
run;

title' MarketShare_est20=((RMarketShare+3.8138025)/ 3.9474853)/1.9661347 ';
data MKTShare_RNORMALZ20;
```

```
set X_RNORMALZ20;
MarketShare_est20=RXX;
run;
```

References

1. Cooper, L.S., and Masao Nakanishi, M., *Market Share Analysis: Evaluating Competitive Marketing Effectiveness*, Kluwer Academic Publishers, Boston, MA, 1988.
2. Ghosh, A., Neslin, S., and Shoemaker, R., A comparison of market share models and estimation procedures, *Journal of Marketing Research*, 21, 202–210, 1984.
3. Fraser, C., and Bradford, J.W., Competitive market structure analysis: Principal partitioning of revealed substitutabilities, *Journal of Consumer Research*, 10, 15–30, 1983.
4. Naert, P. A., and Weverbergh, M., On the prediction power of market share attraction models, *Journal of Marketing Research*, 18, 146–153, 1981.
5. Birch, K., Olsen, J. K., and Tjur, T., *Regression Models for Market-Shares*, Department of Finance Business School, Copenhagen, Denmark, 2005.
6. Fok, D., *Advanced Econometric Marketing Models*, Erasmus Research Institute of Management and Erasmus University Rotterdam, Rotterdam, 2003.
7. Basuroy, S., and Nguyen, D., Multinomial logit market share models: Equilibrium characteristics and strategic implications, *Management Science*, 44, 1396–1408, 1998.

9

The Correlation Coefficient: Its Values Range between Plus and Minus 1, or Do They?

9.1 Introduction

In 1896, the correlation coefficient was invented by Karl Pearson. This century-old statistic is still going strong today, second to the mean in the frequency of use. The correlation coefficient's weaknesses and warnings of misuse are well-known. Based on my consulting experience as a statistical modeler, data miner, and instructor of continuing professional studies in statistics for many years, I see too often that the weaknesses and warnings go unheeded. The weakness rarely mentioned is the correlation coefficient interval [–1, +1] is restricted by the distributions of the two variables under consideration. The purposes of this chapter are (1) to discuss the effects that the distributions of the two variables have on the correlation coefficient interval and (2) to provide a procedure for calculating an *adjusted correlation coefficient*, whose realized correlation coefficient interval is often shorter than the definitional correlation coefficient interval.

9.2 Basics of the Correlation Coefficient

The correlation coefficient, denoted by r, is a measure of the strength of the straight-line or linear relationship between two variables. The correlation coefficient—by definition—assumes any value in the interval between +1 and –1, including the end values ±1, namely, the closed interval denoted by [+1, –1].

The following points are the accepted guidelines for interpreting the correlation coefficient values:

1. 0 indicates no linear relationship.
2. +1 indicates a perfect positive linear relationship: As one variable increases in its values, the other variable also increases in its values via an exact linear rule.
3. –1 indicates a perfect negative linear relationship: As one variable increases in its values, the other variable decreases in its values via an exact linear rule.
4. Values between 0 and 0.3 (0 and –0.3) indicate a weak positive (negative) linear relationship via a shaky linear rule.

5. Values between 0.3 and 0.7 (–0.3 and –0.7) indicate a moderate positive (negative) linear relationship via a fuzzy-firm linear rule.
6. Values between 0.7 and 1.0 (–0.7 and –1.0) indicate a strong positive (negative) linear relationship via a firm linear rule.
7. The value of r squared, called the coefficient of determination, and denoted R-squared, is typically interpreted as the percent of the variation in one variable explained by the other variable or the percent of variation shared between the two variables. Here are good things to know about R-squared:
   a. r is the correlation between the observed and modeled (predicted) data values.
   b. R-squared can increase as the number of predictor variables in the model increases; R-squared cannot decrease as the number of predictor variables increases. Most modelers unwittingly think a model with a larger R-squared is better than a model with a smaller R-squared. As a result of this misunderstanding of R-squared, modelers have a tendency to include more (unnecessary) predictor variables in the model. Accordingly, an adjustment of R-squared was developed, appropriately called adjusted R-squared. The explanation of this statistic is the same as for R-squared, but it penalizes the R-squared when unnecessary variables are in the model.
   c. Specifically, the adjusted R-squared adjusts the R-squared for the sample size and the number of variables in the regression model. Therefore, the adjusted R-squared allows for an "apples-to-apples" comparison between models with different numbers of variables and different sample sizes. Unlike R-squared, adjusted R-squared does not necessarily increase as additional predictor variables enter the model.
   d. R-squared is a first-blush indicator of a good model. R-squared is often *misused** as the measure to assess which model produces better predictions. The root mean squared error (RMSE) is the measure for determining the better model. The smaller the RMSE value, the better the model is (viz., the more precise are the predictions). It is usually best to report the RMSE rather than the mean squared error (MSE) because the RMSE is measured in the *same units as the data,* rather than in squared units, and is representative of the size of a "typical" error. The RMSE is a valid indicator of relative model quality only if the model is *well-fitted* (i.e., if the model is neither overfitted nor underfitted).
8. Linearity assumption: The correlation coefficient requires that the underlying relationship between the two variables under consideration is linear. If the relationship is known to be linear, or the observed pattern between the two variables appears to be linear, then the correlation coefficient provides a reliable measure of the strength of the linear relationship. If the relationship is known to be nonlinear or the observed pattern appears to be nonlinear, then the correlation coefficient is not useful or at least is questionable.

I see too often the correlation coefficient in *misuse* due to *disregard for* the linearity assumption test, albeit this is simple to do.

* Misused, in part, as discussed earlier in Point b.

9.3 Calculation of the Correlation Coefficient

The calculation of the correlation coefficient for X and Y is simple to understand. Let zX and zY be the standardized versions of X and Y, respectively. That is, zX and zY are both reexpressed to have means equal to zero and standard deviations (std) equal to one. The reexpressions used to obtain the standardized scores and $r_{x,y}$ are in Equations 9.1, 9.2, and 9.3, respectively:

$$zX_j = [X_i - \text{mean } (X)]/\text{std}(X) \tag{9.1}$$

$$zY_j = [Y_i - \text{mean } (Y)]/\text{std}(Y) \tag{9.2}$$

The correlation coefficient is defined as the mean production of the paired standardized scores (zX_i, zY_i) as expressed in Equation 9.3.

$$r_{x,y} = \text{sum of } [zX_i{*}zY_i]/(n-1) \tag{9.3}$$

where n is the sample size.

For a simple illustration of the calculation, consider the sample of five observations in Table 9.1. Columns zX and zY contain the standardized scores of X and Y, respectively. The rightmost column is the product of the paired standardized scores. The sum of these scores is 1.83. The mean of these scores (using the adjusted divisor n – 1, not n) is 0.46. Thus, $r_{X,Y} = 0.46$.

For the sake of completeness, I provide the plot of the original data, Plot Y and X, in Figure 9.1. Unfortunately, the small sample size renders the plot visually unhelpful.

9.4 Rematching

As mentioned, the correlation coefficient by definition assumes values in the closed interval [+1, –1]. However, it is *not well-known* that the shapes (distributions) of the individual X and

TABLE 9.1

Calculation of the Correlation Coefficient

obs	X	Y	zX	zY	zX*zY
1	12	77	–1.14	–0.96	1.11
2	15	98	–0.62	1.07	–0.66
3	17	75	–0.27	–1.16	0.32
4	23	93	0.76	0.58	0.44
5	26	92	1.28	0.48	0.62
Mean	18.6	87.0		Sum	1.83
Std	5.77	10.32			
n	5			*r*	0.46

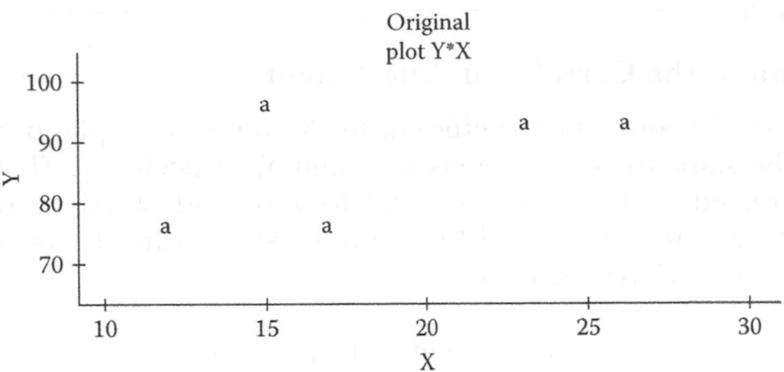

FIGURE 9.1
Original plot of Y with X.

Y data restrict the definitional correlation coefficient interval.* Specifically, the extent to which the shapes of the individual X and Y data are not the same, the length of the realized correlation coefficient interval is shorter than the definitional correlation coefficient interval. Clearly, a shorter realized correlation coefficient interval necessitates the calculation of the *adjusted correlation coefficient* (discussed in Section 9.5).

The process of *rematching* determines the length of the realized correlation coefficient interval. Rematching takes the original X–Y paired data to create new X–Y rematched-paired data such that the rematched-paired data produce the strongest positive and strongest negative relationships. The correlation coefficients of the strongest positive and strongest negative relationships yield the length of the realized correlation coefficient interval. The conceptual elements of the rematching procedure are:

1. The strongest positive relationship comes about with the pairings of the highest X value and the highest Y value; the second-highest X value and the second-highest Y value; and so on until the lowest X value and the lowest Y value.
2. The strongest negative relationship comes about with the pairings of the highest X value and the lowest Y value; the second-highest X value and the second-lowest Y value; and so on until the lowest X value and the highest Y value.

Continuing with the data in Table 9.1, I rematch the X–Y data in Table 9.2. The rematching produces

$$r_{X,Y}(\text{negative rematch}) = -0.99$$

and

$$r_{X,Y}(\text{positive rematch}) = +0.99$$

For the sake of completeness, I provide the rematched plots. Unfortunately, the small sample size renders the plots visually unhelpful. The plots of the negative and positive rematched data, Plot rnegY and rnegX and Plot rposY and rposX, are in Figures 9.2 and 9.3, respectively.

* My first sighting of the term restricted is in Tukey's EDA, 1977. However, I had known about it many years before: I guess from my days in graduate school. I cannot provide any pre-EDA reference.

TABLE 9.2
Rematched (X, Y) Data from Table 9.1

	Original (X, Y)		Positive Rematch		Negative Rematch	
obs	X	Y	X	Y	X	Y
1	12	77	26	98	26	75
2	15	98	23	93	23	77
3	17	75	17	92	17	92
4	23	93	15	77	15	93
5	26	92	12	75	12	98
r		0.46	+0.90		−0.99	

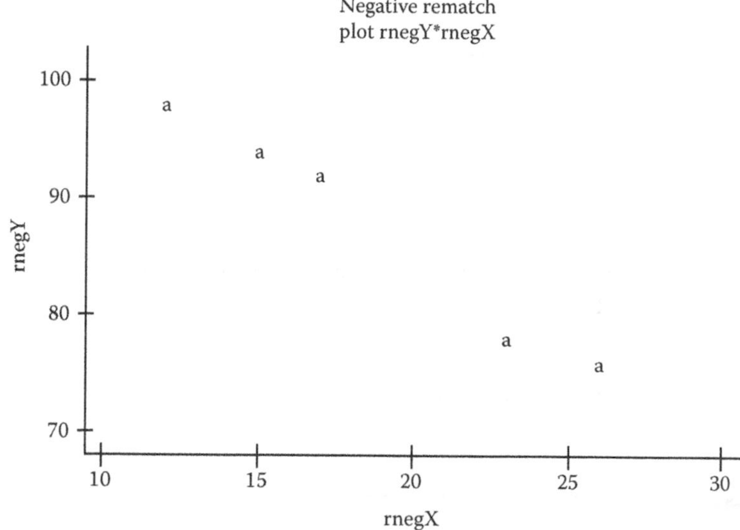

FIGURE 9.2
Negative rematch plot of Y with X.

Because there is an adjustment for R-squared, there is an adjustment for the correlation coefficient due to the shapes of the X and Y data. Thus, the adjusted realized correlation coefficient interval is [−0.99, +0.90]. The calculation of the adjusted correlation coefficient is discussed in the next section.

9.5 Calculation of the Adjusted Correlation Coefficient

The adjusted correlation coefficient is the quotient of the original correlation coefficient and the rematched correlation coefficient. The sign of the adjusted correlation coefficient is the sign of the original correlation coefficient. If the sign of the original r is *negative*, then the sign of the adjusted r is *negative*, even though the arithmetic of dividing two negative numbers yields a positive number.

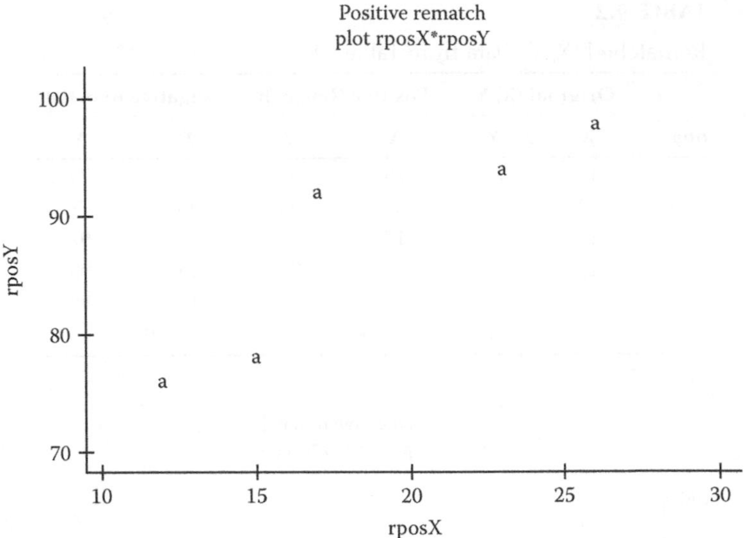

FIGURE 9.3
Positive rematch plot of Y with X.

For the current example, the adjusted correlation coefficient is shown in Equation 9.4. Thus, $r_{x,y}$(adjusted) = 0.51 (= 0.45/0.90), a 10.9% increase over the original correlation coefficient.

$$r_{X,Y}(\text{adjusted}) = r_{X,Y}(\text{orignal})/r_{X,Y}(\text{positive rematch}) \tag{9.4}$$

9.6 Implication of Rematching

The adjusted interval of the correlation coefficient is due to the observed shapes of the X and Y data. The shape of the data has the following effects:

1. Regardless of the shape of either variable, symmetric or otherwise, if one variable's shape is different from the other variable's shape, the correlation coefficient is restricted.
2. The rematch indicates the restriction.
3. It is not possible to obtain perfect correlation unless the variables have the same shape, symmetric or otherwise.
4. A condition that is necessary for a perfect correlation is that the shapes must be the same, but this does not guarantee a perfect correlation.

9.7 Summary

The everyday correlation coefficient is still going strong after its introduction more than 100 years ago. The statistic is well studied, and its weakness and warnings of misuse, unfortunately, at least from my experience, have not been heeded. The weakness rarely mentioned is that the correlation coefficient interval [–1, +1] is restricted by the distributions of the two variables under consideration. I discuss with a simple, yet compelling, illustration the effects that the distributions have on the correlation coefficient interval. I provide and illustrate a procedure for calculating an adjusted correlation coefficient whose realized correlation coefficient interval is often shorter than the definitional correlation coefficient interval.

10

Logistic Regression: The Workhorse of Response Modeling

10.1 Introduction

Logistic regression is a popular technique for classifying individuals into two mutually exclusive and exhaustive categories, for example, buyer–nonbuyer and responder–nonresponder. Logistic regression is the workhorse of *response* modeling as its results are considered the gold standard. Accordingly, it serves as the benchmark for assessing the superiority of new techniques, such as the machine-learning GenIQ Model. Also, it is used to determine the advantage of popular techniques, such as the chi-squared automatic interaction detection (CHAID) regression tree model. In a database marketing application, response to a prior solicitation is the binary dependent variable (defined by responder and nonresponder), and a logistic regression model (LRM) is built to classify an individual as either most likely or least likely to respond to a future solicitation.

To explain logistic regression, I first provide a brief overview of the technique and include a SAS© program for building and scoring an LRM. The program is a welcome addition to the toolkit of techniques used by model builders working on the two-group classification problem. Next, I provide a case study to demonstrate the building of a response model for an investment product solicitation. The case study presentation illustrates a host of statistical data mining techniques that include the following:

- Logit plotting
- Reexpressing variables with the ladder of powers and the bulging rule
- Measuring the straightness of data
- Assessing the importance of individual predictor variables
- Assessing the importance of a subset of predictor variables
- Comparing the importance between two subsets of predictor variables
- Assessing the relative importance of individual predictor variables
- Selecting the best subset of predictor variables
- Assessing goodness of model predictions
- Smoothing a categorical variable for model inclusion

The data mining techniques are basic skills that model builders, who are effectively acting like data miners, need to acquire. They are easy to understand, execute, and interpret. Model builders should master these techniques if they want to be captain of their data and master of their findings. At this point, in keeping within the context of this chapter, which emphasizes data mining, I use more often the term *data miner* than model builder. However, I do acknowledge that an astute model builder is a well-informed data miner.

10.2 Logistic Regression Model

Let Y be a binary dependent variable that assumes two outcomes or classes (typically labeled 0 and 1). The LRM classifies an individual into one of the classes based on the values of predictor (independent) variables $X_1, X_2, \ldots, X_n$ for that individual.

LRM estimates the logit of Y—a log of the odds of an individual belonging to Class 1. The definition of LRM is in Equation 10.1. The logit, which takes on values* between –7 and +7, is a virtually abstract measure for all but the experienced model builder. Fortunately, there is a simple transformation that takes the logit into the probability of an individual belonging to Class 1, Prob(Y = 1). The logit-probability transformation is in Equation 10.2.

$$\text{Logit Y} = b_0 + b_1{*}X_1 + b_2{*}X_2 + \ldots + b_n{*}X_n \tag{10.1}$$

$$\text{Prob}(Y = 1) = \exp(\text{Logit Y}) / (1+\exp(\text{Logit Y})) \tag{10.2}$$

Plugging in the values of the predictor variables for an individual in Equations 10.1 and 10.2 produces that individual's estimated (predicted) probability of belonging to Class 1. The b's are the logistic regression coefficients, determined by the calculus-based method of maximum likelihood. Note, unlike the other coefficients, b_0 (referred to as the intercept) has no corresponding predictor variable X_0.

As presented, the LRM is readily seen as the workhorse of response modeling as the yes–no response variable is an exemplary binary class variable. The illustration in the next section shows the rudiments of logistic regression response modeling.

10.2.1 Illustration

Consider Dataset A, which consists of 10 individuals and 3 variables, in Table 10.1. The binary variables are RESPONSE (Y), INCOME in thousands of dollars (X_1), and AGE in years (X_2). I perform a logistic analysis regressing response on INCOME and AGE using Dataset A.

The standard LRM output in Table 10.2 includes the logistic regression coefficients and other "columns" of information (a discussion of these is beyond the scope of this chapter). The "Parameter Estimates" column contains the coefficients for variables INCOME,

* The logit theoretically assumes values between plus and minus infinity. However, in practice, it rarely goes outside the range of plus and minus 7.

TABLE 10.1

Dataset A

RESPONSE (1 = yes, 0 = no)	INCOME ($000)	AGE (years)
1	96	22
1	86	33
1	64	55
1	60	47
1	26	27
0	98	48
0	62	23
0	54	48
0	38	24
0	26	42

TABLE 10.2

LRM Output

Variable	df	Parameter Estimate	Standard Error	Wald Chi-Square	Pr > Chi-Square
Intercept	1	−0.9367	2.5737	0.1325	0.7159
INCOME	1	0.0179	0.0265	0.4570	0.4990
AGE	1	−0.0042	0.0547	0.0059	0.9389

AGE, and the INTERCEPT. The INTERCEPT variable is a mathematical device; it is defined implicitly as X_0, which is always equal to one (i.e., intercept = X_0 = 1). The coefficient b_0 serves as a "start" value given to all individuals regardless of their specific values of predictor variables in the model.

The estimated LRM is defined by Equation 10.3:

$$\text{Logit of RESPONSE} = -0.9367 + 0.0179 * \text{INCOME} - 0.0042 * \text{AGE} \tag{10.3}$$

Do not forget that the LRM predicts the logit of RESPONSE not the probability of RESPONSE.

10.2.2 Scoring an LRM

The SAS program in Figure 10.1 produces the LRM built with Dataset A and scores an external Dataset B in Table 10.3. The SAS procedure LOGISTIC produces logistic regression coefficients and puts them in the "coeff" file, as indicated by the code "outest = coeff." The coeff files produced by SAS versions 6 and 8 (SAS 6, SAS 8) are in Tables 10.4 and 10.5, respectively. (The latest versions of SAS are 9.3 and 9.4. SAS currently supports versions 8 and 9 but not version 6.*) The procedure LOGISTIC, as presented here, holds true for SAS 9.

* There are still die-hard SAS 6 users, as an SAS technical support person informed me.

```
/****** Building the LRM on dataset A ************/
PROC LOGISTIC data = A nosimple des outest = coeff;
model RESPONSE =
INCOME AGE;
run;
/****** Scoring the LRM on dataset B ************/
PROC SCORE data = B predict type = parms score = coeff
out = B_scored;
var INCOME AGE;
run;
/******* Converting Logits into Probabilities ********/
                                SAS version 6
data B_scored;
set B_scored;
Prob_Resp = exp(Estimate)/(1 + exp(Estimate));
run;
                                SAS version 8
data B_scored;
set B_scored;
Prob_Resp = exp(RESPONSE)/(1 + exp(RESPONSE));
run;
```

FIGURE 10.1
SAS code for building and scoring an LRM.

TABLE 10.3
Dataset B

INCOME ($000)	AGE (years)
148	37
141	43
97	70
90	62
49	42

TABLE 10.4
Coeff File (SAS 6)

OBS	_LINK_	_TYPE_	_NAME_	INTERCEPT	INCOME	AGE	_LNLIKE_
1	LOGIT	PARMS	Estimate	–0.93671	0.017915	–0.0041991	–6.69218

TABLE 10.5
Coeff File (SAS 8)

OBS	_LINK_	_TYPE_	_STATUS_	_NAME_	INTERCEPT	INCOME	AGE	_LNLIKE_
1	LOGIT	PARMS	0 Converged	RESPONSE	–0.93671	0.017915	–0.0041991	–6.69218

TABLE 10.6

Dataset B_scored

INCOME ($000)	AGE (years)	Predicted Logit of Response: Estimate (SAS 6), Response (SAS 8)	Predicted Probability of Response: Prob_Resp
148	37	1.55930	0.82625
141	43	1.40870	0.80356
97	70	0.50708	0.62412
90	62	0.41527	0.60235
49	42	−0.23525	0.44146

Regardless, I present the SAS 6 program as it is instructive in that it highlights the difference between the uncomfortable-for-most logit and the always-desired probability. The coeff files differ in two ways:

1. An additional column in the SAS 8 coeff file is *_STATUS_*, which does not affect the scoring of the model.
2. The naming of the predicted logit is "Response" in SAS 8, which is indicated by _NAME_ = Response.

Although it is unexpected, the naming of the predicted logit in SAS 8 is the class variable used in the PROC LOGISTIC statement, as indicated by the code "model Response =." In this illustration, the predicted logit is called "Response," which is indicated by _NAME_ = Response, in Table 10.4. The SAS 8 naming convention is unfortunate as it may cause the model builder to think that the Response variable is a binary variable and not a logit.

The SAS procedure SCORE scores the five individuals in Dataset B using the LRM coefficients, as indicated by the code "score = coeff." Coeff serves to append the predicted logit variable (called Estimate when using SAS 6 and Response when using SAS 8), to the output file B_scored, as indicated by the code "out = B_scored," in Table 10.6. The probability of response (Prob_Resp) is easily obtained with the code at the end of the SAS program in Figure 10.1.

10.3 Case Study

By examining the following case study about building a response model for a solicitation for investment products, I illustrate a host of *data mining techniques*. To make the discussion of the techniques manageable, I use small data (a handful of variables, some of which take on few values, and a sample of size "petite grande") drawn from the original direct mail solicitation database. Reported results replicate the original findings obtained with slightly bigger data.

I allude to the issue of data size here in anticipation of data miners who subscribe to the idea that big data are better for analysis and modeling. Currently, there is a trend, especially in related statistical communities, such as computer science, knowledge discovery, and Web mining, to use extra big data. The trend is due to the false notion that extra big data are better than big data. A statistical factoid states if small data yield the true model, then a rebuilt true model with big or extra big data produces large prediction error variance. As model builders are never aware of the true model, they must be guided by the principle of simplicity. Therefore, it is wisest to build a model with the minimal amount of data that yields a good model. If the predictions are good, then the model is a good approximation of the true model. If predictions are not acceptable, then exploratory data analysis (EDA) procedures prescribe an increase in data size (by adding predictor variables and individuals) until the model produces good predictions. The data size with which the model produces good predictions is big enough. If the model build starts with extra big data, then unnecessary variables tend to creep into the model, thereby increasing the prediction error variance.

10.3.1 Candidate Predictor and Dependent Variables

Let TXN_ADD be the yes–no response dependent variable, which records the activity of existing customers who received a mailing intended to motivate them to purchase additional investment products. The yes–no response, which is coded 1–0, respectively, corresponds to customers who have/have not added at least one new product fund to their investment portfolio. The TXN_ADD response rate is 11.9%, which is typically large for a direct mail campaign, and is usual for solicitations intended to stimulate purchases among existing customers.

The five candidate predictor variables for predicting TXN_ADD whose values reflect measurement before the mailing are:

1. FD1_OPEN reflects the number of different types of accounts the customer has.
2. FD2_OPEN reflects the number of total accounts the customer has.
3. INVESTMENT reflects the customer's investment dollars in ordinal values: 1 = $25 to $499, 2 = $500 to $999, 3 = $1,000 to $2,999, 4 = $3,000 to $4,999, 5 = $5,000 to $9,999, and 6 = $10,000+.
4. MOS_OPEN reflects the number of months the account is opened in ordinal values: 1 = 0 to 6 months, 2 = 7 to 12 months, 3 = 13 to 18 months, 4 = 19 to 24 months, 5 = 25 to 36 months, and 6 = 37+ months.
5. FD_TYPE is the product type of the customer's most recent investment purchase: A, B, C, … , N.

10.4 Logits and Logit Plots

The LRM belongs to the family of linear models that advance the implied assumption that the underlying relationship between a given predictor variable and the logit is a linear or straight line. Bear in mind that, to model builders, the adjective *linear* refers to the explicit fact that the logit is the sum of weighted predictor variables, where the weights are the regression coefficients. In practice, however, the term refers to the implied assumption. Checking this assumption requires the *logit plot*. A logit plot is the plot of the binary dependent variable (hereafter, response variable) against the values of the predictor variable. Three steps are required to generate the logit plot:

1. Calculate the mean of the response variable corresponding to each value of the predictor variable. If the predictor variable takes on more than 10 distinct values, then use *typical* values, such as smooth decile values, as defined in Chapter 3.

2. Calculate the logit of response using the formula that converts the mean of response to logit of response: Logit = ln(mean/(1 – mean)), where ln is the natural logarithm.
3. Plot the logit-of-response values against the original distinct or the smooth decile values of the predictor variable.

A point worth noting: The logit plot is an aggregate-level, not individual-level, plot. The logit is an aggregate measure based on the mean of individual response values. Moreover, by using smooth decile values, the plot is further aggregated as each decile value represents 10% of the sample.

I provide the SAS subroutines for generating smooth logit and smooth probability plots in Chapter 43.

10.4.1 Logits for Case Study

For the case study, the response variable is TXN_ADD, and the logit of TXN_ADD is named LGT_TXN. For no particular reason, I start with candidate predictor variable FD1_OPEN, which takes on the distinct values 1, 2, and 3 in Table 10.7. Following the three-step construction for each FD1_OPEN value, I generate the LGT_TXN logit plot in Figure 10.2. I calculate the mean of TXN_ADD and use the mean-to-logit conversion formula. For example, for FD1_OPEN = 1, the mean of TXN_ADD is 0.07, and the logit LGT_TXN is –2.4 (= ln(0.07/(1 – 0.07)). Last, I plot the LGT_TXN logit values against the FD1_OPEN values.

The LGT_TXN logit plot for FD1_OPEN does not suggest an underlying straight-line relationship between LGT_TXN and FD1_OPEN. To use the LRM correctly, I need to

TABLE 10.7
FD1_OPEN

FD1_OPEN	Mean TXN_ADD	LGT_TXN
1	0.07	–2.4
2	0.18	–1.5
3	0.20	–1.4

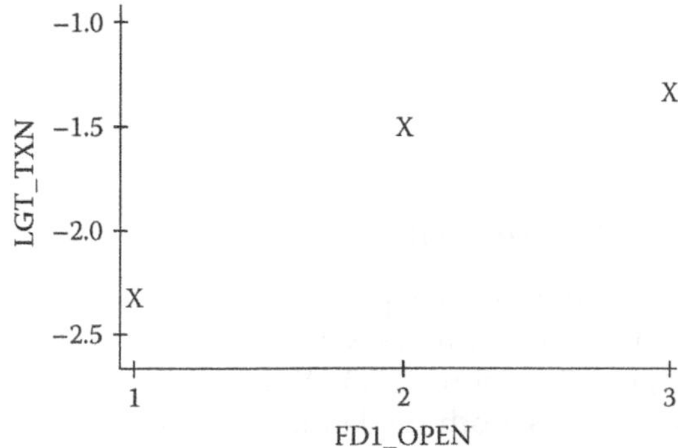

FIGURE 10.2
Logit plot for FD1_OPEN.

straighten the relationship. A very effective and simple technique for straightening data is reexpressing, which uses Tukey's ladder of powers and the bulging rule. Before presenting the details of the technique, it is worth discussing the importance of straight-line relationships or straight data.

10.5 The Importance of Straight Data

EDA places special importance on straight data, not in the least for the sake of simplicity itself. The paradigm of life is simplicity (at least for those of us who are older and wiser). In the physical world, Einstein uncovered one of the universe's ruling principles using only three letters: $E = mc^2$. In the visual world, however, simplicity is undervalued and overlooked. A smiley face is an unsophisticated, simple shape that nevertheless communicates effectively, clearly, and efficiently. Why should the data miner accept anything less than simplicity in his or her life's work? Numbers, as well, should communicate clearly, effectively, and immediately. In the data miner's world, there are two features that reflect simplicity: symmetry and straightness in the data. The data miner should insist that the numbers be symmetric and straight.

The straight-line relationship between two continuous variables X and Y is simple as it gets. As X increases (decreases) in its values, Y increases (decreases) in its values. In this case, X and Y are positively correlated. Or, as X increases (decreases) in its values, Y decreases (increases) in its values. In this case, X and Y are negatively correlated. As a further demonstration of its simplicity, Einstein's E and m have a perfect positively correlated straight-line relationship.

The second reason for the importance of straight data is that most response models require it as they belong to the class of innumerable varieties of the linear model. Moreover, nonlinear models, which pride themselves on making better predictions with nonstraight data, in fact, do better with straight data.

I have not ignored the feature of symmetry. Not accidentally, as there are theoretical reasons, symmetry and straightness go hand in hand. Straightening data often make data symmetric and vice versa. You may recall iconic symmetric data have the profile of the bell-shaped curve. However, symmetric data are defined as their data values have the same shape (identical on both sides) above and below the middle value of the entire data distribution.

10.6 Reexpressing for Straight Data

The ladder of powers is a method of reexpressing variables to straighten a bulging relationship between two continuous variables X and Y. Bulges in the data can be depicted as one of four shapes, as displayed in Figure 10.3. When the X–Y relationship has a bulge similar to any one of the four shapes, both the ladder of powers and the bulging rule, which guides the choice of "rung" in the ladder, are used to straighten out the bulge. Most data have bulges. However, when kinks or elbows characterize the data, then another approach is required, which is discussed further in the chapter.

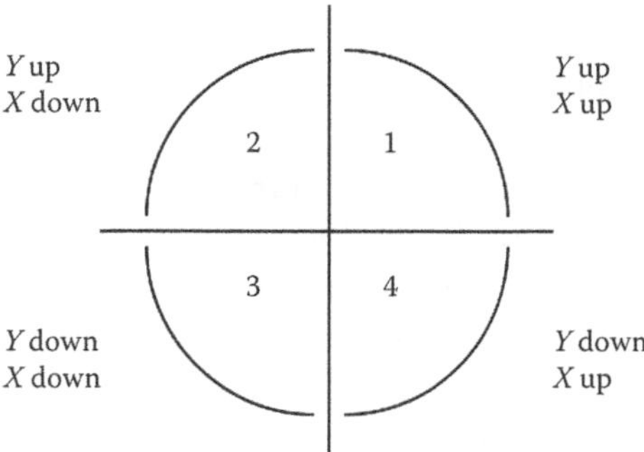

FIGURE 10.3
The bulging rule.

10.6.1 Ladder of Powers

Going up-ladder of powers means reexpressing a variable by raising it to a power p greater than 1. (Remember that a variable raised to the power of 1 is still that variable; $X^1 = X$ and $Y^1 = Y$). The most common p values used are 2 and 3. Sometimes values higher up-ladder and in-between values such as 1.33 are used. Accordingly, starting at p = 1, the data miner goes up-ladder, resulting in reexpressed variables, for X and Y, as follows:

$$\text{Starting at } X^1: X^2, X^3, X^4, X^5, \ldots$$

$$\text{Starting at } Y^1: Y^2, Y^3, Y^4, Y^5, \ldots$$

Some variables reexpressed via going up-ladder have special names. Corresponding to power values 2 and 3, they are called X squared and X cubed, respectively. Similarly, for the Y variables, they are called Y squared and Y cubed, respectively.

Going down-ladder of powers means reexpressing a variable by raising it to a power p that is less than 1. The most common p values are ½, 0, –½, and –1. Sometimes, values lower down-ladder and in-between values such as 0.33 are used. Also, for negative powers, the reexpressed variable now sports a negative sign (i.e., multiplied by –1). The reason for the multiplication is theoretical and beyond the scope of this chapter. Accordingly, starting at p = 1, the data miner goes down-ladder, resulting in reexpressed variables for X and Y, as follows:

$$\text{Starting at } X^1: X^{1/2}, X^0, -X^{-1/2}, -X^{-1}, \ldots$$

$$\text{Starting at } Y^1: Y^{1/2}, Y^0, -Y^{-1/2}, -Y^{-1}, \ldots$$

Some reexpressed variables via going down-ladder have special names. Corresponding to values ½, –½, and –1, they are called the square root of X, the negative reciprocal square root of X, and the negative reciprocal of X, respectively. Similarly, for the Y variables, they are called the square root of Y, the negative reciprocal square root of Y, and the negative reciprocal of Y, respectively. Reexpressions for p = 0 use log to base 10.* Thus, $X^0 = \log X$, and $Y^0 = \log Y$.

* Reexpression for p = 0 is not mathematically defined but is conveniently defined.

10.6.2 Bulging Rule

The bulging rule states the following:

1. If the data have a shape similar to that in the first quadrant, then the data miner tries reexpressing by going up-ladder for X, Y, or both.
2. If the data have a shape similar to that shown in the second quadrant, then the data miner tries reexpressing by going down-ladder for X or up-ladder for Y.
3. If the data have a shape similar to that in the third quadrant, then the data miner tries reexpressing by going down-ladder for X, Y, or both.
4. If the data have a shape similar to that in the fourth quadrant, then the data miner tries reexpressing by going up-ladder for X or down-ladder for Y.

Reexpressing is an important, yet fallible, part of EDA detective work. While it will typically result in straightening the data, it might result in a deterioration of information. Here is why: Reexpression (going down too far) has the potential to squeeze the data so much that its values become indistinguishable, resulting in a loss of information. Expansion (going up too far) can potentially pull apart the data so much that the new far-apart values lie within an artificial range, resulting in a spurious gain of information.

Thus, reexpressing requires a careful balance between straightness and soundness. Data miners can always go to the extremes of the ladder by exerting their will to obtain a little more straightness, but they must be mindful of a consequential loss of information. Sometimes, it is evident when one has gone too far up/down on the ladder: There is a power p, after which the relationship either does not improve noticeably or inexplicably bulges in the opposite direction due to a corruption of information. I recommend using discretion to avoid over-straightening and its potential deterioration of information. Also, I caution that extreme reexpressions are sometimes due to the extreme values of the original variables. Thus, always check the maximum and minimum values of the original variables to make sure they are reasonable before reexpressing the variables.

10.6.3 Measuring Straight Data

The correlation coefficient measures the strength of the straight-line or linear relationship between two variables X and Y discussed in detail in Chapter 3. However, there is an additional assumption to consider.

In Chapter 3, I refer to a "linear assumption," in that the underlying relationship between X and Y is linear. The second assumption is an implicit one: the (X, Y) data points are at the individual level. When analyzing the (X, Y) points at an aggregate level, such as in the logit plot and other plots presented in this chapter, the correlation coefficient based on "big" points tends to produce a "big" r value, which serves as a *gross* estimate of the individual-level r value. The aggregation of data diminishes the idiosyncrasies of the individual (X, Y) points, thereby increasing the resolution of the relationship, for which the r value also increases. Thus, the correlation coefficient on aggregated data serves as a gross indicator of the strength of the original X–Y relationship at hand. There is a drawback of aggregation: It often produces r values without noticeable differences because of loss of the power of the distinguishing individual-level information.

10.7 Straight Data for Case Study

Returning to the LGT_TXN logit plot for FD1_OPEN, whose bulging relationship is in need of straightening, I identify its bulge as the type in quadrant 2 in Figure 10.3. According to the bulging rule, I should try going up-ladder for LGT_TXN or down-ladder for FD1_OPEN. Applying the bulging rule to LGT_TXN is not logical because LGT_TXN is the explicit dependent variable as defined by the logistic regression framework. Reexpressing it would produce grossly illogical results. Thus, I do not go up-ladder for LGT_TXN.

To go down-ladder for FD1_OPEN, I use the powers ½, 0, –½, –1, and –2. Going down the ladder results in the square root of FD1_OPEN, labeled FD1_SQRT; the log to base 10 of FD1_OPEN, labeled FD1_LOG; the negative reciprocal root of FD1_OPEN, labeled FD1_RPRT; the negative reciprocal of FD1_OPEN, labeled FD1_RCP; and the negative reciprocal square of FD1_OPEN, labeled FD1_RSQ. The corresponding LGT_TXN logit plots for these reexpressed variables and the original FD1_OPEN (repeated here for convenience) are in Figure 10.4.

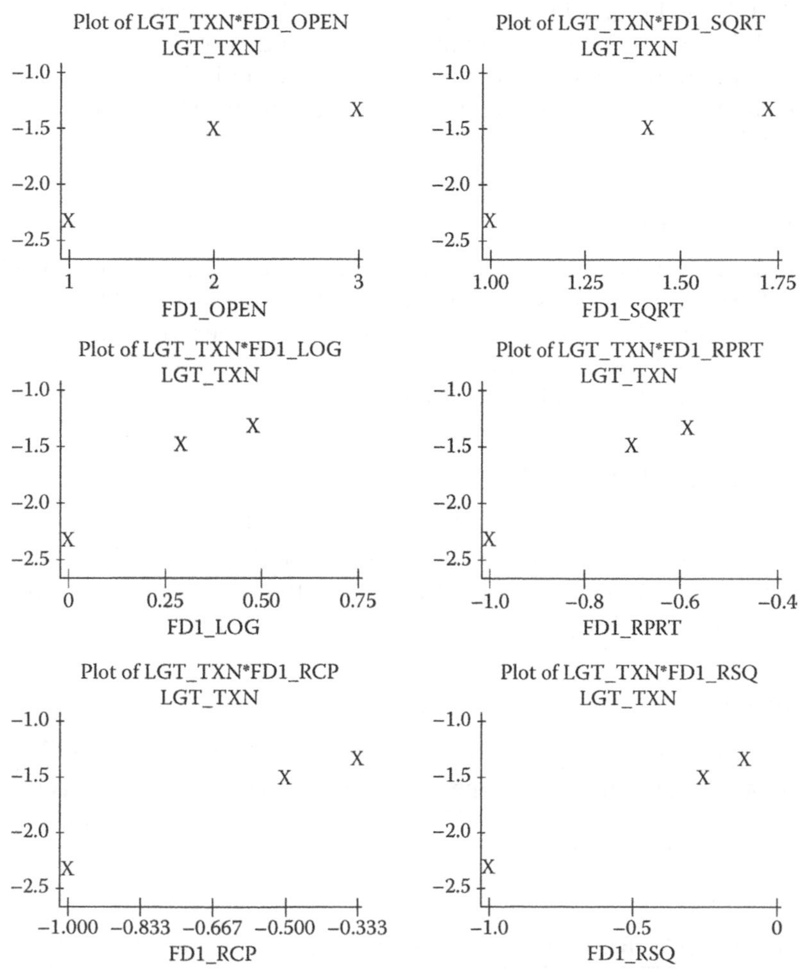

FIGURE 10.4
Logit plots for FD1_OPEN and its reexpressed variables.

Visually, it appears that reexpressed variables FD1_RSQ, FD1_RCP, and FD1_RPRT do an equal job of straightening the data. I could choose any of them but decide to do a little more detective work by looking at the numerical indicator—the correlation coefficient between LGT_TXN and the reexpressed variable—to support my choice of the best-reexpressed variable. The larger the correlation coefficient, the more effective the reexpressed variable is in straightening the data. Thus, the reexpressed variable with the largest correlation coefficient is declared the best-reexpressed variable, with exceptions guided by the data miner's experience with these visual and numerical indicators in the context of the problem domain.

The correlation coefficients for LGT_TXN with FD1_OPEN and with each reexpressed variable ranked in descending order are in Table 10.8. The correlation coefficients of the reexpressed variables represent *noticeable* improvements in straightening the data over the correlation coefficient for the original variable FD1_OPEN (r = 0.907). FD1_RSQ has the largest correlation coefficient (r = 0.998), but it is slightly greater than that for FD1_RCP (r = 0.988) and therefore not worthy of notice.

My choice of the best-reexpressed variable is FD1_RCP, which represents an 8.9% (= (0.988 – 0.907)/0.907) improvement in straightening the data over the original relationship with FD1_OPEN. I prefer FD1_RCP over FD1_RSQ and other extreme reexpressions down-ladder (defined by power p less than –2) because I do not want to select unwittingly a reexpression that might be too far down-ladder, resulting in loss of information. Thus, I go back one rung to power –1, hoping to get the right balance between straightness and minimal loss of information.

10.7.1 Reexpressing FD2_OPEN

Reexpressing for FD2_OPEN is virtually identical to the one presented for FD1_OPEN. Reexpressing FD2_OPEN is not surprising as FD1_OPEN and FD2_OPEN share a large amount of information. The correlation coefficient between the two variables is 0.97, meaning the two variables share 94.1% of their variation. Thus, I prefer FD2_RCP as the best-reexpressed variable for FD2_OPEN (see Table 10.9).

10.7.2 Reexpressing INVESTMENT

The relationship between the LGT_TXN and INVESTMENT, depicted in the plot in Figure 10.5, is somewhat straight with a negative slope and a slight bulge in the middle for INVESTMENT values 3, 4, and 5. I identify the bulge of the type in quadrant 3 in Figure 10.3. Thus, going down-ladder for powers ½, 0, –½, –1, and –2 results in the square root of

TABLE 10.8

Correlation Coefficients between LGT_TXN and Reexpressed FD1_OPEN

FD1_RSQ	FD1_RCP	FD1_RPRT	FD1_LOG	FD1_SQRT	FD1_OPEN
0.998	0.988	0.979	0.960	0.937	0.907

TABLE 10.9

Correlation Coefficients between LGT_TXN and Reexpressed FD2_OPEN

FD2_RSQ	FD2_RCP	FD2_RPRT	FD2_LOG	FD2_SQRT	FD2_OPEN
0.995	0.982	0.968	0.949	0.923	0.891

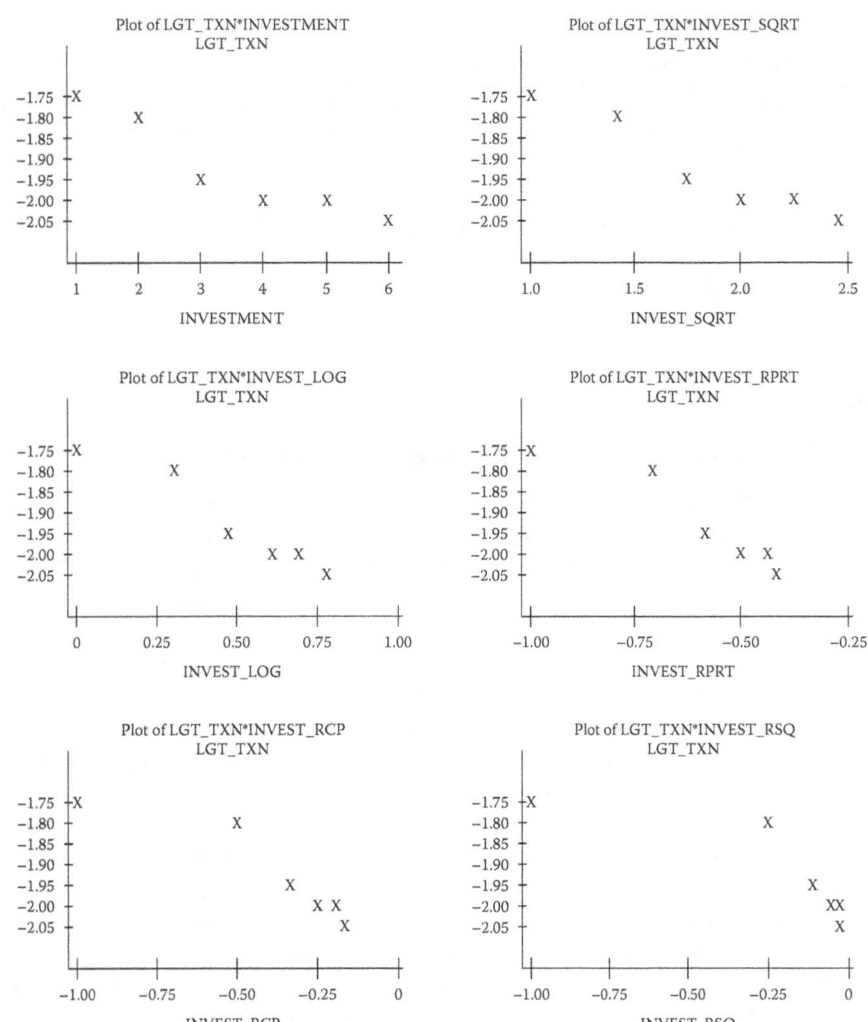

FIGURE 10.5
Logit plots for INVESTMENT and its reexpressed variables.

INVESTMENT, labeled INVEST_SQRT. The log to base 10 of INVESTMENT is labeled INVEST_LOG. The negative reciprocal root of INVESTMENT is labeled INVEST_RPRT. The negative reciprocal of INVESTMENT is labeled INVEST_RCP. And, the negative reciprocal square of INVESTMENT is labeled INVEST_RSQ. The corresponding LGT_TXN logit plots for these reexpressed variables and the original INVESTMENT are in Figure 10.5.

Visually, I like the straight-line produced by INVEST_SQRT. Quantitatively, INVEST_LOG has the largest correlation coefficient, which supports the statistical factoid that the log function is the appropriate reexpression for a variable in dollar units. The correlation coefficients for INVEST_LOG and INVEST_SQRT in Table 10.10 are −0.978 and −0.966, respectively. Admittedly, the correlation coefficients do not reflect a noticeable difference. My choice of the best reexpression for investment is INVEST_LOG because I prefer the statistical factoid to my visual choice. Only if a noticeable difference between correlation coefficients for INVEST_LOG and INVEST_SQRT existed would I sway from being guided

TABLE 10.10
Correlation Coefficients between LGT_TXN and Reexpressed INVESTMENT

INVEST_LOG	INVEST_SQRT	INVEST_RPRT	INVESTMENT	INVEST_RCP	INVEST_RSQ
–0.978	–0.966	–0.950	0.946	–0.917	–0.840

by the factoid. INVEST_LOG represents an improvement of 3.4% (= (0.978 – 0.946)/0.946; disregarding the negative sign) in straightening the data over the relationship with the original variable INVESTMENT (r = –0.946).

10.8 Techniques when the Bulging Rule Does Not Apply

I describe two plotting techniques for uncovering the correct reexpression when the bulging rule does not apply. After discussing the techniques, I return to the next variable for reexpression, MOS_OPEN. The relationship between LGT_TXN and MOS_OPEN is interesting and offers an excellent opportunity to illustrate the data mining flexibility of the EDA methodology.

It behooves the data miner to perform due diligence either to explain qualitatively or to account quantitatively for the relationship in a logit plot. Typically, the latter is easier than the former as the data miner is at best a scientist of data, not a psychologist of data. The data miner seeks to investigate the plotted relationship to uncover the correct representation or *structure* of the given predictor variable. Briefly, *structure* is an organization of variables and functions. In this context, variables are broadly defined as both raw variables (e.g., X_1, X_2, …, X_i, …) and numerical constants, which are variable-like in that they assume any single value k, that is, $X_i = k$. Functions include the arithmetic operators (addition, subtraction, multiplication, and division); comparison operators (e.g., equal to, not equal to, greater than); and logical operators (e.g., and, or, not, if-then-else). For example, $X_1 + X_2/X_1$ is a structure.

By definition, any raw variable X_i is considered a structure as it can be defined by $X_i = X_i + 0$, or $X_i = X_i*1$. A dummy variable (X_dum)—a variable that assumes two numerical values, typically 1 and 0, which indicate the presence and absence of a condition, respectively—is structure. For example, X_dum = 1 if X equals 6; X_dum = 0 if X does not equal 6. The condition is "equals 6."

10.8.1 Fitted Logit Plot

The *fitted logit* plot is a valuable visual aid in uncovering and confirming structure. The fitted logit plot is a plot of the *predicted* logit against a given structure. The steps required to construct the plot and its interpretation are as follows:

1. *Perform* a logistic regression analysis on the response variable with the given structure, obtaining the predicted logit of response, as outlined in Section 10.2.2.
2. *Identify* the values of the structure to use in the plot. Identify the distinct values of the given structure. If the structure has more than 10 values, identify its smooth decile values.
3. *Plot* the predicted (fitted) logit values against the identified values of the structure. Label the points by the identified values.

4. *Infer* that if the fitted logit plot reflects the shape in the original logit plot, the structure is the correct one. This further implies that the structure has some importance in predicting response. The extent to which the fitted logit plot is different from the original logit plot indicates the structure is a poor predictor of response.

10.8.2 Smooth Predicted-versus-Actual Plot

Another valuable plot for exposing the detail of the strength or weakness of a structure is the *smooth predicted-versus-actual* plot, defined as the plot of *mean* predicted response against *mean* actual response for values of a reference variable. The steps required to construct the plot and its interpretation are as follows:

1. Calculate the mean predicted response by averaging the individual predicted probabilities of response from the appropriate LRM for each value of the reference variable. Similarly, calculate the mean actual response by averaging the individual actual responses for each value of the reference variable.
2. The paired points (mean predicted response, mean actual response) are called *smooth points*.
3. Plot the smooth points and label them with the values of the reference variable. If the structure has more than 10 values, identify its smooth decile values.
4. Insert the 45° line in the plot. The line serves as a reference for visual assessment of the importance of a structure for predicting response thereby confirming that the structure under consideration is the correct one. Smooth points on the line imply that the mean predicted response and the mean actual response are equal, and there is great certainty that the structure is the correct one. The tighter the smooth points "hug" the 45° line, the greater the certainty of the structure. Conversely, the greater the scatter about the line, the lesser the certainty of the structure.

10.9 Reexpressing MOS_OPEN

The relationship between LGT_TXN and MOS_OPEN in Figure 10.6 is not straight in the full range of MOS_OPEN values from one to six but is straight between values one and five. The LGT_TXN logit plot for MOS_OPEN shows a check mark shape with vertex at MOS_OPEN = 5 as LGT_TXN jumps at MOS_OPEN = 6. Clearly, the bulging rule does not apply.

Accordingly, while uncovering the MOS_OPEN structure, I am looking for the organization of variables and functions that renders the ideal straight-line relationship between LGT_TXN and the MOS_OPEN structure. It will implicitly account for the jump in logit where LGT_TXN at MOS_OPEN = 6. After identifying the correct MOS_OPEN structure, I can include it in the TXN_ADD response model.

Exploring the structure of MOS_OPEN itself, I generate the LGT_TXN fitted logit plot (Figure 10.7), based on the logistic regression analysis on TXN_ADD with MOS_OPEN. The LRM, which provides the predicted logits, is defined in Equation 10.4:

$$\text{Logit(TXN\_ADD)} = -1.24 - 0.17 * \text{MOS\_OPEN} \qquad (10.4)$$

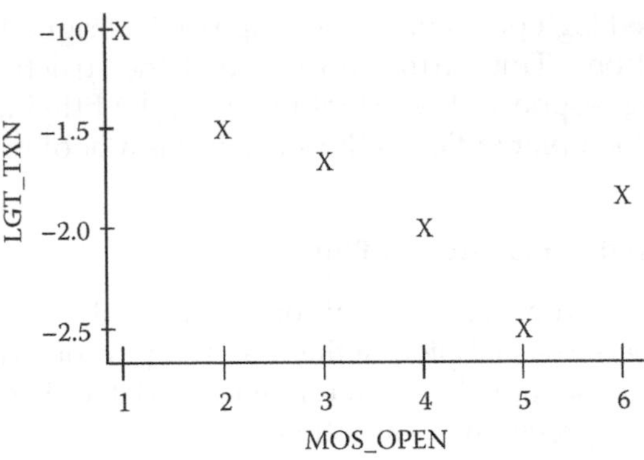

FIGURE 10.6
Logit plot of MOS_OPEN.

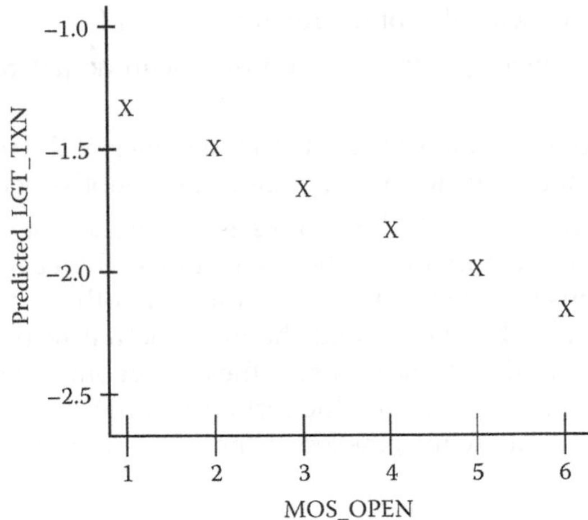

FIGURE 10.7
Fitted logit plot for MOS_OPEN.

Noticeably, MOS_OPEN has six distinct values. The fitted logit plot does not reflect the shape of the relationship in the original LGT_TXN logit plot in Figure 10.6. The predicted point at MOS_OPEN = 6 is way too low. The confirmation is MOS_OPEN alone is not the correct structure as it does not produce the shape in the original logit plot.

10.9.1 Plot of Smooth Predicted versus Actual for MOS_OPEN

I generate the TXN_ADD smooth predicted-versus-actual plot for MOS_OPEN (Figure 10.8), which depicts both the structure under consideration and the reference variable. The smooth predicted values are based on the LRM previously defined in Equation 10.4 and restated here in Equation 10.5 for convenience.

$$\text{Logit(TXN\_ADD)} = -1.24 - 0.17 * \text{MOS\_OPEN} \tag{10.5}$$

There are six smooth points, each labeled by the corresponding six values of MOS_OPEN. The scatter about the 45° line is wild, implying MOS_OPEN is not a good predictive

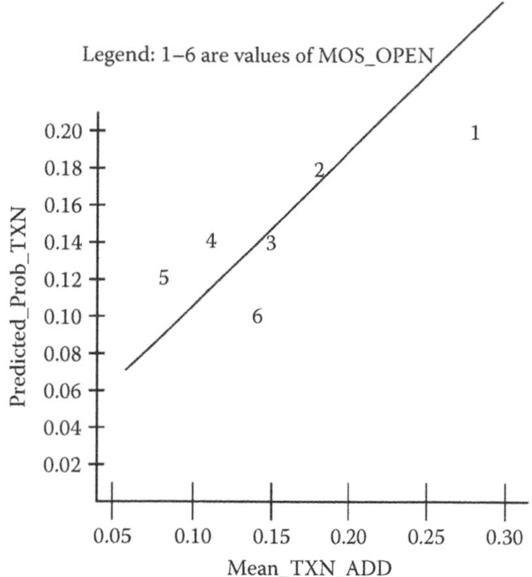

FIGURE 10.8
Plot of smooth predicted versus actual for MOS_OPEN.

structure, especially when MOS_OPEN equals 1, 5, 6, and 4, as their corresponding smooth points are not close to the 45° line. Point MOS_OPEN = 5 is understandable as it can be considered the springboard to jump into LGT_TXN at MOS_OPEN = 6. MOS_OPEN = 1, as the farthest point from the line, strikes me as inexplicable. MOS_OPEN = 4 may be within an acceptable distance from the line.

When MOS_OPEN is equal to 2 and 3, the prediction appears to be good as the corresponding smooth points are close to the line. Two good predictions of a possible six predictions result in a poor 33% accuracy rate. Thus, MOS_OPEN is not a good structure for predicting TXN_ADD. As before, the implication is that MOS_OPEN alone is not the correct structure to reflect the original relationship between LGT_TXN and MOS_OPEN in Figure 10.6. More detective work is needed.

It occurs to me that the major problem with MOS_OPEN is the jump point. To account explicitly for the jump, I create an MOS_OPEN dummy variable structure, defined as

$$\text{MOS\_DUM} = 1 \text{ if MOS\_OPEN} = 6;$$

$$\text{MOS\_DUM} = 0 \text{ if MOS\_OPEN not equal to } 6.$$

I generate a second LGT_TXN fitted logit plot in Figure 10.9, this time consisting of the predicted logits from regressing TXN_ADD on the structure consisting of MOS_OPEN and MOS_DUM. The LRM definition is in Equation 10.6:

$$\text{Logit(TXN\_ADD)} = -0.62 - 0.38 * \text{MOS\_OPEN} + 1.16 * \text{MOS\_DUM} \qquad (10.6)$$

This fitted plot accurately reflects the shape of the original relationship between TXN_ADD and MOS_OPEN in Figure 10.6. The implication is that MOS_OPEN and MOS_DUM

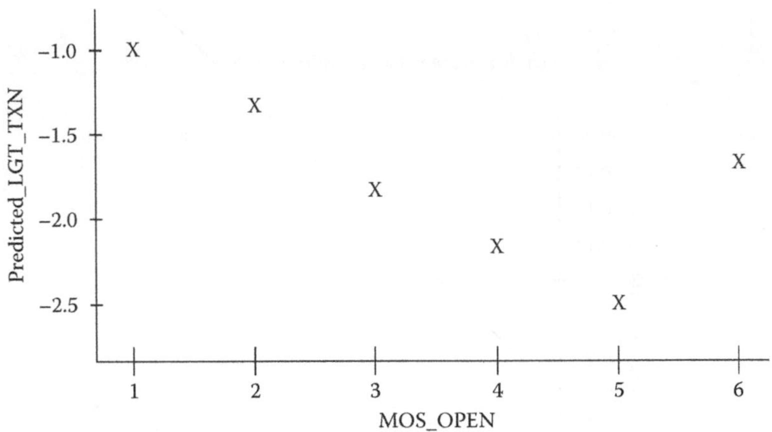

FIGURE 10.9
Fitted logit plot for MOS_OPEN and MOS_DUM.

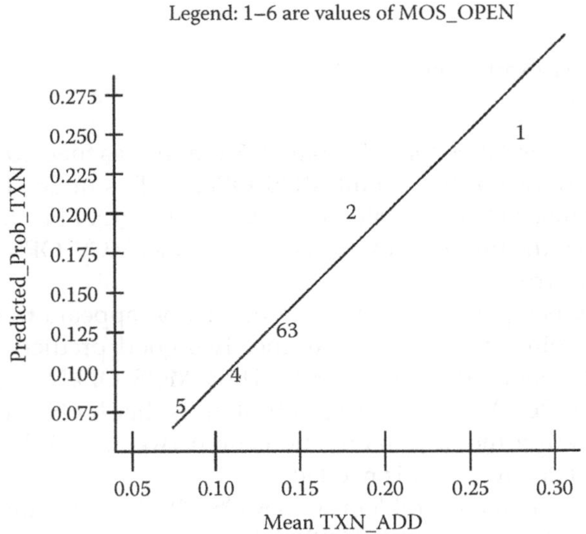

FIGURE 10.10
Plot of smooth predicted versus actual for MOS_OPEN and MOS_DUM.

make up the correct structure of the information carried in MOS_OPEN. The definition of the structure is the right side of the equation itself.

To complete my detective work, I create the second TXN_ADD smooth predicted-versus-actual plot in Figure 10.10, consisting of mean predicted logits of TXN_ADD against mean MOS_OPEN. The predicted logits come from the logistic regression equation 10.6, which includes the predictor variable pair MOS_OPEN and MOS_DUM. MOS_OPEN serves as the reference variable. The smooth points hug the 45° line nicely. The implication is that the MOS_OPEN structure defined by MOS_OPEN and MOS_DUM is again confirmed, and the two-piece structure is an important predictor of TXN_ADD.

10.10 Assessing the Importance of Variables

The classic approach for assessing the statistical significance of a variable considered for model inclusion is the well-known null hypothesis significance testing procedure based on the reduction in prediction error (actual response minus predicted response) associated with the variable in question. The statistical apparatus of the formal testing procedure for logistic regression analysis consists of the log-likelihood (LL) function, the G statistic, degrees of freedom (df), and the p-value. The procedure uses the apparatus within a theoretical framework with weighty and untenable assumptions. From a purist point of view, this could cast doubt on findings that have statistical significance. Even if findings of statistical significance are correctly accepted, they may not be of practical importance or have *noticeable* value to the study at hand. For the data miner with a pragmatic slant, the limitations and lack of scalability inherent in the classic system cannot be overlooked, especially within big data settings.

In contrast, the data mining approach uses the LL units, the G statistic, and degrees of freedom in an informal data-guided search for variables that suggest a *noticeable* reduction in prediction error. A point worth noting is that the informality of the data mining approach calls for a suitable change in terminology, from declaring a result as statistically significant to one worthy of notice or *noticeably important.*

Before I describe the data mining approach to variable assessment, I would like to comment on the objectivity of the classic approach as well as degrees of freedom. The classic approach is so ingrained in the analytic community that no viable alternative occurs to practitioners, especially an alternative based on an informal and sometimes highly individualized series of steps. Declaration of a variable as statistically significant appears to be purely objective as it is on sound probability theory and statistics machinery. However, the settings of the testing machinery defined by model builders could affect the results. The settings include the levels of rejecting a variable as significant when, in fact, it is not or accepting a variable as not significant when, in fact, it is. Determining the proper sample size is also a subjective setting as it depends on the amount budgeted for the study. Last, the model builder's experience sets the allowable deviation of violations of test assumptions. Therefore, by acknowledging the subjective nature of the classic approach, the model builder can be receptive to the alternative data mining approach, which is free of theoretical ostentation and mathematical elegance.

A word about degrees of freedom clarifies the discussion. The degrees of freedom, as typically described, is a generic measure of the number of independent pieces of information available for analysis. The measure often is subject to the mathematical adjustment "replace N with N–1" to ensure accurate results. The concept of degrees of freedom gives a deceptive impression of simplicity in counting the pieces of information. However, the principles used in counting are not easy for all but the mathematical statistician. To date, there is no generalized calculus for counting degrees of freedom. Fortunately, the counting already exists for many analytical routines. Therefore, the correct degrees of freedom are readily available; computer output automatically provides them, and there are lookup tables in older statistics textbooks. For the analyses in the following discussions, I provide degrees of freedom, eliminating the need for counting.

10.10.1 Computing the G Statistic

In data mining, the assessment of the importance of a subset of variables for predicting response involves the notion of a noticeable reduction in prediction error due to the subset of variables and the ratio of the G statistic to the degrees of freedom, G/df. The degrees of freedom is the number of variables in the subset. The G statistic is defined, in Equation 10.7, as the difference between two LL quantities, one corresponding to a model *without* the subset of variables and the other corresponding to a model *with* the subset of variables.

$$G = -2LL(\text{model without variables}) - -2LL(\text{model with variables}) \tag{10.7}$$

There are two points worth noting: first, the –2LL units replace the LL units, as a mathematical necessity; second, the term *subset* is used to imply there is always a large set of variables available from which the model builder considers the smaller subset, which can include a single variable.

In the following sections, I detail the decision rules in three scenarios for assessing the likelihood that the variables have some predictive power. In brief, the larger the average G value per degrees of freedom (G/df), the more important the variables are in predicting response.

10.10.2 Importance of a Single Variable

If X is the only variable considered for inclusion into the model, the definition of the G statistic is in Equation 10.8:

$$G = -2LL(\text{model with intercept only}) - -2LL(\text{model with X}) \tag{10.8}$$

The decision rule for declaring X an important variable in predicting response is as follows: If G/df* is greater than the standard G/df value 4, then X is an important predictor variable and should be considered for inclusion in the model. Note, the decision rule only indicates that the variable has *some* importance not how much importance. The decision rule implies that a variable X1 with a G/df value greater than variable X2's G/df value indicates that X1 has a greater likelihood of some importance than does X1. The decision rule does suggest X1 has greater importance than X2.

10.10.3 Importance of a Subset of Variables

When subset A consisting of k variables is the only subset considered for model inclusion, the definition of the G statistic is in Equation 10.9:

$$G = -2LL(\text{model with intercept}) - -2LL(\text{model with A(k)variables}) \tag{10.9}$$

The decision rule for declaring subset A important in predicting response is as follows: If G/k is greater than the standard G/df value 4, then subset A is an important subset of the predictor variable and is a candidate subset for inclusion in the model. As before, the decision rule only indicates that the subset has some importance not how much importance.

10.10.4 Comparing the Importance of Different Subsets of Variables

Let subsets A and B consist of k and p variables, respectively. The number of variables in each subset does not have to be equal. If they are equal, then all but one variable can be

* Obviously, G/df equals G for a single-predictor variable with df = 1.

the same in both subsets. The definitions of the G statistics for A and B are in Equations 10.10 and 10.11, respectively:

$$G(k) = -2LL(\text{model with intercept}) - -2LL(\text{model with "A" variables}) \quad (10.10)$$

$$G(p) = -2LL(\text{model with intercept}) - -2LL(\text{model with "B" variables}) \quad (10.11)$$

The decision rule for declaring which of the two subsets is more important (i.e., greater likelihood of having some predictive power) in predicting response is as follows:

1. If G(k)/k is greater than G(p)/p, then subset A is the more important predictor variable subset; otherwise, B is the more important subset.
2. If G(k)/k and G(p)/p are equal or have comparable values, then both subsets are to be regarded tentatively as of comparable importance. The model builder should consider additional indicators to assist in the decision about which subset is better.

It follows clearly from the decision rule that the "more important" subset defines the better model. Of course, this rule assumes that G(k)/k and G(p)/p are greater than the standard G/df value 4.

10.11 Important Variables for Case Study

The first step in variable assessment is to determine the baseline LL value for the data under study. The LRM for TXN_ADD without variables produces two essential bits of information in Table 10.11:

1. The baseline for this case study is –2LL equals 3606.488.
2. The LRM definition is in Equation 10.12:

$$\text{Logit}(TNX\_ADD = 1) = -1.9965 \quad (10.12)$$

TABLE 10.11

The LOGISTIC Procedure for TXN_ADD

Response Profile	
TXN_ADD	COUNT
1	589
0	4,337
–2LL = 3606.488	

Variable	Parameter Estimate	Standard Error	Wald Chi-Square	Pr > Chi-Square
Intercept	–1.9965	0.0439	2,067.050	0.0

LOGIT = 1.9965

ODDS = EXP (0.19965) = 0.1358

$$\text{PROB(TXN\_ADD = 1)} = \frac{\text{ODDS}}{1+\text{ODDS}} = \frac{0.1358}{1+0.1358} = 0.119$$

There are interesting bits of information in Table 10.11 that illustrate two useful statistical identities:

1. Exponentiation of both sides of Equation 10.12 produces odds of response equal to 0.1358. Recall, exponentiation is the mathematical operation of raising a quantity to a power. The exponentiation of a logit is the odds. Consequently, the exponentiation of –1.9965 is 0.1358. See Equations 10.13 through 10.15.

$$\text{Exp}(\text{Logit}(\text{TNX\_ADD} = 1)) = \text{Exp}(-1.9965) \tag{10.13}$$

$$\text{Odds}(\text{TNX\_ADD} = 1) = \text{Exp}(-1.9965) \tag{10.14}$$

$$\text{Odds}(\text{TNX\_ADD} = 1) = 0.1358 \tag{10.15}$$

2. The probability of (TNX_ADD = 1), hereafter the probability of RESPONSE, is easily obtained as the ratio of odds divided by 1 + odds. The implication is that the best estimate of RESPONSE—when no information or variables are available—is 11.9%, namely, the average response of the mailing.

10.11.1 Importance of the Predictor Variables

With the LL baseline value 3,606.488, I assess the importance of the five variables: MOS_OPEN and MOS_DUM, FD1_RCP, FD2_RCP, and INVEST_LOG. Starting with MOS_OPEN and MOS_DUM, as they must be together in the model, I perform a logistic regression analysis on TXN_ADD with MOS_OPEN and MOS_DUM. The output is in Table 10.12. From Equation 10.9, the G value is 107.022 (= 3,606.488 – 3,499.466). The degree of freedom is equal to the number of variables; df is 2. Accordingly, G/df equals 53.511, which is greater than the standard G/df value of 4. Thus, the pair MOS_OPEN and MOS_DUM are important predictor variables of TXN_ADD.

From Equation 10.8, the G/df value for each remaining variable in Table 10.12 is greater than 4. Thus, these five variables, each important predictors of TXN_ADD, comprise a starter subset for predicting TXN_ADD. I have not forgotten about FD_TYPE, discussed in Section 10.16.

I build a preliminary model by regressing TXN_ADD on the starter subset; the output is in Table 10.13. From Equation 10.9, the five-variable subset has a G/df value of 40.21 (= 201.031/5), which is greater than 4. Thus, this is an incipient subset of important variables for predicting TXN_ADD.

TABLE 10.12
G and df for Predictor Variables

Variable	–2LL	G	df	p
Intercept	3606.488			
MOS_OPEN + MOS_DUM	3499.466	107.023	2	0.0001
FD1_RCP	3511.510	94.978	1	0.0001
FD2_RCP	3503.993	102.495	1	0.0001
INV_LOG	3601.881	4.607	1	0.0001

TABLE 10.13

Preliminary Logistic Model for TXN_ADD with Starter Subset

	Intercept Only	Intercept and All Variables	All Variables	
–2LL	3606.488	3405.457	201.031 with 5 df ($p = 0.0001$)	
Variable	**Parameter Estimate**	**Standard Error**	**Wald Chi-Square**	**Pr > Chi-Square**
Intercept	0.9948	0.2462	16.3228	0.0001
FD2_RCP	3.6075	0.9679	13.8911	0.0002
MOS_OPEN	–0.3355	0.0383	76.8313	0.0001
MOS_DUM	0.9335	0.1332	49.0856	0.0001
INV_LOG	–0.7820	0.2291	11.6557	0.0006
FD1_RCP	–2.0269	0.9698	4.3686	0.0366

10.12 Relative Importance of the Variables

The "mystery" in building a statistical model is that the *true* subset of variables defining the *true* model is not known. The model builder can be most productive by seeking to find the *best* subset of variables that defines the final model as an intelli-guess of the true model. The final model reflects more of the model builder's effort given the data at hand than an estimate of the true model itself. The model builder's attention is toward the most noticeable, unavoidable collection of predictor variables, whose behavior is known to the extent the logit plots uncover their shapes and their relationships to response.

There is also magic in building a statistical model, in that the best subset of predictor variables consists of variables whose contributions to the predictions of the model are often unpredictable and unexplainable. Sometimes, the most important variable in the mix drops from the top, in that its contribution in the model is no longer as strong as it was without the presence of other variables. Other times, the least likely variable rises from the bottom, in that its contribution in the model is stronger than it was without the presence of other variables. In the best of times, the variables interact with each other such that their total effect on the predictions of the model is greater than the sum of their individual effects.

Unless the variables are uncorrelated with each other (the rarest of possibilities), it is impossible for the model builder to assess the unique contribution of a variable. In practice, the model builder can assess the *relative importance* of a variable, specifically, its importance in the presence of the other variables in the model. The Wald chi-square—as posted in logistic regression analysis output—serves as an indicator of the relative importance of a variable as well as for selecting the best subset. Discussion of Wald chi-square is in the next section.

10.12.1 Selecting the Best Subset

The decision rules for finding the best subset of important variables consists of the following steps:

1. *Select an initial subset of important variables.* Variables that are thought to be important are probably important. Let experience (the model builders and others) in the

problem domain be the rule. If there are many variables from which to choose, rank the variables based on the correlation coefficient r (between the response variable and each candidate predictor variable). One to two handfuls of the experience-based variables, the largest r-valued variables, and some small r-valued variables serve as the initial subset. Include the latter variables because small r values may falsely exclude important nonlinear variables. (Recall that the correlation coefficient is an indicator of linear relationship.) Categorical variables require special treatment as the correlation coefficient is not appropriate. (I illustrate with FD_TYPE how to include a categorical variable in a model in the last section.)

2. For the variables in the initial subset, *generate logit plots and straighten the variables as required.* The most noticeable handfuls of original and reexpressed variables comprise the starter subset.
3. *Perform the preliminary logistic regression analysis on the starter subset.* Delete one or two variables with Wald chi-square values less than the *Wald cutoff value of 4* from the model. Deletion of variables results in the first incipient subset of important variables.
4. *Perform another logistic regression analysis on the incipient subset.* Delete one or two variables with Wald chi-square values less than the Wald cutoff value of 4 from the model. The model builder can create an illusion of important variables appearing and disappearing with the deletion of different variables. The Wald chi-square values can exhibit bouncing above and below the Wald cutoff value as the variables undergo deletion. The bouncing effect is due to the correlation between the included variables and the deleted variables. A greater correlation implies greater bouncing (unreliability) of Wald chi-square values. Consequently, the greater the bouncing of Wald chi-square values implies the greater the uncertainty of declaring important variables.
5. *Repeat Step 4 until all retained predictor variables have comparable Wald chi-square values.* This step often results in different subsets as the model builder deletes judicially different pairings of variables.
6. *Declare the best subset by comparing the relative importance of the different subsets using the decision rule in Section 10.10.4.*

10.13 Best Subset of Variables for Case Study

I perform a logistic regression on TXN_ADD with the five-variable subset MOS_OPEN and MOS_DUM, FD1_RCP, FD2_RCP, and INVEST_LOG. The output is in Table 10.13. FD1_RCP has the smallest Wald chi-square value, 4.3686. FD2_RCP, which has a Wald chi-square of 13.8911, is highly correlated with FD1_RCP ($r_{FD1\_RCP,\,FD2\_RCP} = 0.97$), thus rendering their Wald chi-square values unreliable. However, without additional indicators for either variable, I accept their face values as an indirect message and delete FD1_RCP, the variable with the lesser value.

INVEST_LOG has the second smallest Wald chi-square value, 11.6557. With no apparent reason other than it just appears to have a less-relative importance given MOS_OPEN, MOS_DUM, FD1_RCP, and FD2_RCP in the model, I also delete INVEST_LOG from the model. Thus, the incipiently best subset consists of FD2_RCP, MOS_OPEN, and MOS_DUM.

TABLE 10.14

Logistic Model for TXN_ADD with Best Incipient Subset

	Intercept Only	Intercept and All Variables	All Variables	
–2LL	3606.488	3420.430	186.058 with 3 df ($p = 0.0001$)	
Variable	**Parameter Estimate**	**Standard Error**	**Wald Chi-Square**	**Pr > Chi-Square**
Intercept	0.5164	0.1935	7.1254	0.0076
FD2_RCP	1.4942	0.1652	81.8072	0.0001
MOS_OPEN	–0.3507	0.0379	85.7923	0.0001
MOS_DUM	0.9249	0.1329	48.4654	0.0001

I perform another logistic regression on TXN_ADD with the three-variable subset (FD2_RCP, MOS_OPEN, and MOS_DUM). The output is in Table 10.14. MOS_OPEN and FD2_RCP have comparable Wald chi-square values, 81.8072 and 85.7923, respectively, which are obviously greater than the Wald cutoff value 4. The Wald chi-square value for MOD_DUM is half of that of MOS_OPEN and is not comparable to the other values. However, MOS_DUM is staying in the model because it is empirically needed (recall Figures 10.9 and 10.10). I acknowledge that MOS_DUM and MOS_OPEN share information, which could be affecting the reliability of their Wald chi-square values. The actual amount of shared information is 42%, which indicates there is a minimal effect on the reliability of their Wald chi-square values.

I compare the importance of the current three-variable subset (FD2_RCP, MOS_OPEN, MOS_DUM) and the starter five-variable subset (MOS_OPEN, MOS_DUM, FD1_RCP, FD2_RCP, INVEST_LOG). The G/df values are 62.02 (= 186.058/3 from Table 10.14) and 40.21 (= 201.031/5 from Table 10.13) for the former and latter subsets, respectively. Based on the decision rule in Section 10.10.4, I declare the three-variable subset is better than the five-variable subset. Thus, I expect good predictions of TXN_ADD based on the three-variable model defined in Equation 10.16:

$$\begin{aligned} &\text{Predicted Logit\_TXN\_ADD} = \text{Predicted LGT\_TXN} \\ &= 0.5164 + 1.4942 * \text{FD2\_RCP} - 0.3507 * \text{MOS\_OPEN} + 0.9249 * \text{MOS\_DUM} \end{aligned} \quad (10.16)$$

10.14 Visual Indicators of Goodness of Model Predictions

In this section, I provide visual indicators of the quality of model predictions. The LRM itself is a variable as it is a sum of weighted variables with the logistic regression coefficients serving as the weights. As such, the logit model prediction (e.g., the predicted LGT_TXN) is a variable that has a mean, a variance, and all the other descriptive measures afforded any variable. Also, the logit model prediction can be graphically displayed as afforded any variable. Accordingly, I present three valuable plotting techniques, which reflect the EDA-prescribed graphic detective work, for assessing the goodness of model predictions.

10.14.1 Plot of Smooth Residual by Score Groups

The plot of *smooth residual by score groups* is the plot consisting of the mean residual against the mean predicted response by *score groups*, identified by the unique values created by preselected variables—typically the predictor variables in the model under consideration. For example, for the three-variable model, there are 18 score groups: three values of FD2_RCP multiplied by six values of MOS_OPEN. The two values of MOS_DUM are not unique as they are part of the values of MOS_OPEN.

The steps required to construct the plot of the smooth residual by score groups and its interpretation are as follows:

1. *Score* the data by appending the predicted logit as outlined in Section 10.2.2.
2. *Convert* the predicted logit to the predicted probability of response as outlined in Section 10.2.2.
3. *Calculate* the residual (error) for an individual: Residual = actual response minus predicted probability of response.
4. *Determine* the score groups by the unique values created by the preselected variables.
5. For each score group, *calculate* the mean (smooth) residual and mean (smooth) predicted response, producing a set of paired smooth points (smooth residual, smooth predicted response).
6. *Plot* the smooth points by score group.
7. *Draw* a straight line through mean residual = 0. This zero line serves as a reference line for determining whether a general trend exists in the scatter of smooth points. If the smooth residual plot looks like the ideal or *null* plot (i.e., has a random scatter about the zero line with about half of the points above the line and the remaining points below), then it is concluded that there is no general trend in the smooth residuals. Thus, the predictions aggregated at the score group level are considered good. The desired implication is that on average the predictions at the *individual level* are also good.
8. *Examine* the smooth residual plot for noticeable deviations from random scatter. Such examination is at best a subjective task as it is the unwitting nature of the model builder to find what is sought. To aid in an objective examination of the smooth residual plot, use the general association test discussed in Chapter 3 to determine whether the smooth residual plot is equivalent to the null plot.
9. When the smooth residual plot is declared null, *look for a local pattern*. It is not unusual for a small wave of smooth points to form a local pattern, which has no ripple effect to create a general trend in an otherwise null plot. A local pattern indicates a weakness or *weak spot* in the model in that there is a prediction bias for the score groups identified by the pattern.

10.14.1.1 Plot of the Smooth Residual by Score Groups for Case Study

I construct the plot of smooth residual by score groups to determine the quality of the predictions of the three-variable (FD2_RCP, MOS_OPEN, MOS_DUM) model. The smooth residual plot in Figure 10.11 is declared to be equivalent to the null plot based on the general association test discussed in Chapter 3. Thus, the overall quality of prediction is considered good. That is, on average, the predicted TXN_ADD is equal to the actual TXN_ADD.

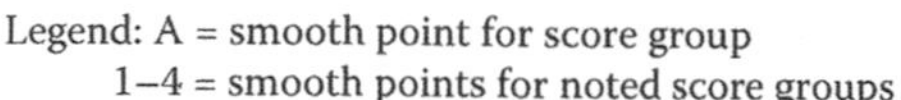

Legend: A = smooth point for score group
1–4 = smooth points for noted score groups

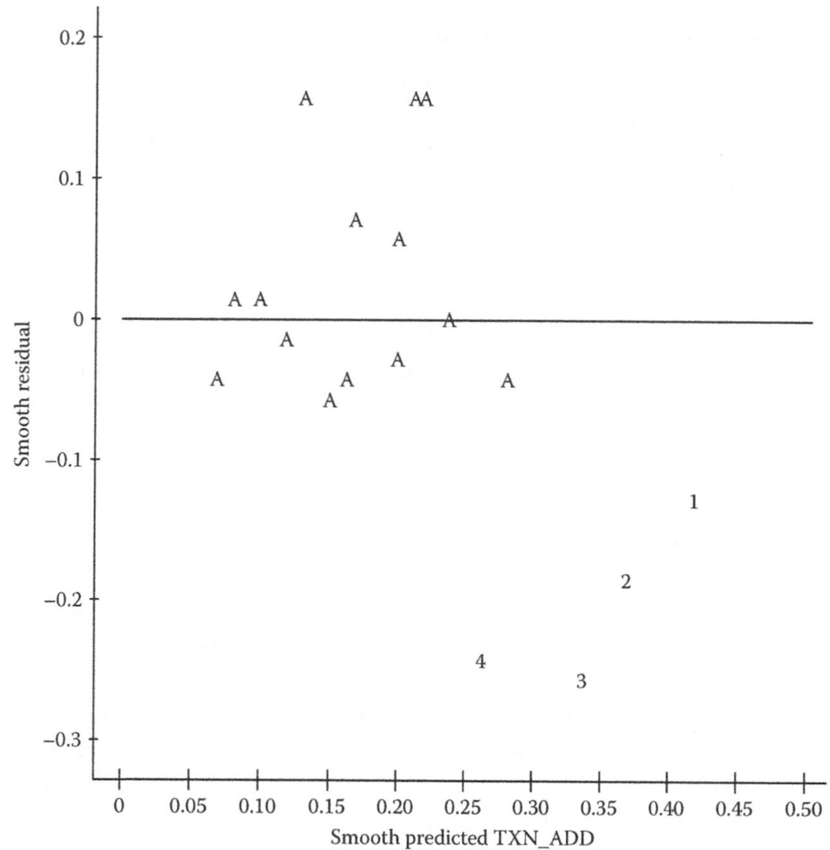

FIGURE 10.11
Plot of smooth residual by score group for three-variable (FD2_RCP, MOS_OPEN, and MOS_DUM) model.

Easily seen, but not easily understood (at this point in the analysis), is the local pattern defined by four score groups (labeled 1 through 4) in the lower right-hand side of the plot. The local pattern explicitly shows that the smooth residuals are noticeably negative. The local pattern indicates a weak spot in the model as its predictions for the individuals in the four score groups have, on average, a positive bias; that is, their predicted TXN_ADD tends to be larger than their actual TXN_ADD.

If the implementation of the model can afford "exception rules" for individuals in a weak spot, then model performance can be enhanced. For example, response models typically have a weak spot as prediction bias stems from limited information on new customers and outdated information on inactive customers. Thus, if model implementation on a solicitation database can include exception rules (e.g., new customers are always targeted—assigned to the top decile) and inactive customers placed in the middle deciles, then the overall quality of prediction is improved.

For use in further discussions, the descriptive statistics for the plot of the smooth residual by score groups/three-variable model are as follows: (1) For the smooth residuals, the minimum and maximum values and the range are −0.26, 0.16, and 0.42, respectively, and (2) the standard deviation of the smooth residuals is 0.124.

10.14.2 Plot of Smooth Actual versus Predicted by Decile Groups

The plot of *smooth actual versus predicted by decile groups* is the plot consisting of the mean actual response against the mean predicted response by *decile groups.* Decile groups are 10 equal-sized classes, based on the predicted response values from the LRM under consideration. Decile groupings are not an arbitrary partitioning of the data as most database models are implemented at the decile level and consequently are built and validated at the decile level.

The steps required to construct the plot of the smooth actual versus predicted by decile groups and its interpretation are as follows:

1. *Score* the data by appending the predicted logit as outlined in Section 10.2.2.
2. *Convert* the predicted logit to the predicted probability of response as outlined in Section 10.2.2.
3. *Determine* the decile groups. Rank in descending order the scored data by the predicted response values. Divide the scored-ranked data into 10 equal-sized classes. The first class has the largest mean predicted response, labeled "top"; the next class is labeled "2," and so on. The last class has the smallest mean predicted response, labeled "bottom."
4. For each decile group, *calculate* the mean (smooth) actual response and mean (smooth) predicted response, producing a set of 10 smooth points (smooth actual response, smooth predicted response).
5. *Plot* the smooth points by decile group, labeling the points by decile group.
6. *Draw* the 45° line on the plot. This line serves as a reference for assessing the quality of predictions at the decile group level. If the smooth points are either on or *hug the 45° line* in their proper order (top to bottom, or bottom to top), then predictions, on average, are considered *good.*
7. *Determine* the "tightness" of the hug of the smooth points about the 45° line. To aid in an objective examination of the smooth plot, use the correlation coefficient between the smooth actual and predicted response points. The correlation coefficient serves as an indicator of the amount of scatter about the 45° straight line. The larger the correlation coefficient is, the less scatter there will be and the better the overall quality of prediction.
8. As discussed in Section 10.6.3, the correlation coefficient based on big points tends to produce a big r value, which serves as a gross estimate of the individual-level r value. The correlation coefficient based on smooth actual and predicted response points is a gross measure of the individual-level predictions of the model. The correlation coefficient serves best as a *comparative indicator* in choosing the better model.

10.14.2.1 Plot of Smooth Actual versus Predicted by Decile Groups for Case Study

I construct a plot of the smooth actual versus predicted by decile groups based on Table 10.15 to determine the quality of the three-variable model predictions. The smooth plot in Figure 10.12 has a minimal scatter of the 10 smooth points about the 45° lines, with two noted exceptions. Decile groups 4 and 6 appear to be the farthest away from the line (regarding perpendicular distance). Decile groups 8, 9, and bot are on top of each

TABLE 10.15

Smooth Points by Deciles from Model Based on FD_RCP, MOS_OPEN, and MOS_DUM

	TXN_ADD		Predicted TXN_ADD		
Decile	***N***	**Mean**	**Mean**	**Min**	**Max**
top	492	0.069	0.061	0.061	0.061
2	493	0.047	0.061	0.061	0.061
3	493	0.037	0.061	0.061	0.061
4	492	0.089	0.080	0.061	0.085
5	493	0.116	0.094	0.085	0.104
6	493	0.085	0.104	0.104	0.104
7	492	0.142	0.118	0.104	0.121
8	493	0.156	0.156	0.121	0.196
9	493	0.185	0.198	0.196	0.209
bottom	492	0.270	0.263	0.209	0.418
Total	4,926	0.119	0.119	0.061	0.418

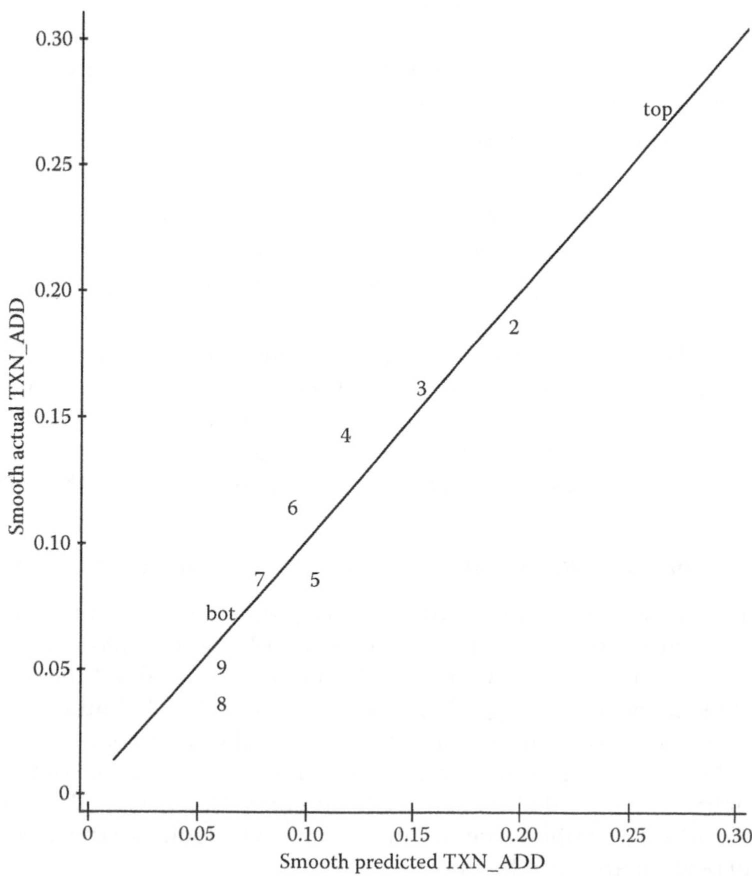

FIGURE 10.12
Plot of smooth actual versus predicted by decile group for three-variable (FD2_RCP, MOS_OPEN, and MOS_DUM) model.

other, which indicate the predictions are the same for these groups. The indication is that the model cannot discriminate among the least-responding individuals. But because the implementation of response models typically excludes the lower three or four decile groups, their spread about the 45° line and their (lack of) order are not as critical a feature in assessing the quality of the prediction. Thus, the overall quality of prediction is considered good.

The descriptive statistic for the plot of smooth actual versus smooth predicted by decile groups/three-variable model is the correlation coefficient between the smooth points $r_{\text{sm. actual, sm. predicted: decile group}}$ is 0.972.

10.14.3 Plot of Smooth Actual versus Predicted by Score Groups

The plot of *smooth actual versus predicted by score groups* is the plot consisting of the mean actual against the mean predicted response by the score groups. Its construction and interpretation are virtually identical to the plot for smooth actual versus predicted by decile groups. The painlessly obvious difference is that score groups replace decile groups, as in the discussion in Section 10.14.1 on the plot of the smooth residual by score groups.

I outline the steps compactly for the construction and interpretation of the plot of the smooth actual versus predicted by score groups:

1. *Score* the data by appending the predicted logit and convert the predicted logit to the predicted probability of response.
2. *Determine* the score groups and calculate their smooth values for actual response and predicted response.
3. *Plot* the smooth actual and predicted points by score group.
4. *Draw* a 45° line on the plot. If the smooth plot looks like the null plot, then it is concluded that the model predictions aggregated at the score group level are considered good.
5. *Use* the correlation coefficient between the smooth points to aid in an objective examination of the smooth plot. The correlation coefficient serves as an indicator of the amount of scatter about the 45° line. The larger the correlation coefficient is, the less scatter there is, and the better the overall quality of predictions is. The correlation coefficient serves best as a comparative measure in choosing the better model.

10.14.3.1 Plot of Smooth Actual versus Predicted by Score Groups for Case Study

I construct the plot of the smooth actual versus predicted by score groups based on Table 10.16 to determine the quality of the three-variable model predictions. The smooth plot in Figure 10.13 indicates the scatter of the 18 smooth points about the 45° line is good, except for the four points on the right-hand side of the line, labeled numbers 1 through 4. These points correspond to the four score groups, which became noticeable in the smooth residual plot in Figure 10.11. The indication is the same as that of the smooth residual plot: The overall quality of the prediction is considered good. However, if the implementation of the model can afford exception rules for individuals who look like the four score groups, then the model performance can be improved.

The profiling of the individuals in the score groups is immediate from Table 10.16. The original predictor variables, instead of the reexpressed versions, are used to make the interpretation of the profile easier. The sizes (20, 56, 28, and 19) of the four noticeable groups are

TABLE 10.16

Smooth Points by Score Groups from Model Based on FD2_RCP, MOS_OPEN, and MOS_DUM

		TXN_ADD		PROB_HAT
MOS_OPEN	FD2_OPEN	*N*	Mean	Mean
1	1	161	0.267	0.209
	2	56	0.268	0.359
	3	20	0.350	0.418
2	1	186	0.145	0.157
	2	60	0.267	0.282
	3	28	0.214	0.336
3	1	211	0.114	0.116
	2	62	0.274	0.217
	3	19	0.158	0.262
4	1	635	0.087	0.085
	2	141	0.191	0.163
	3	50	0.220	0.200
5	1	1,584	0.052	0.061
	2	293	0.167	0.121
	3	102	0.127	0.150
6	1	769	0.109	0.104
	2	393	0.186	0.196
	3	156	0.237	0.238
Total		4,926	0.119	0.119

quite small for groups 1 to 4, respectively, which may account for the undesirable spread about the 45° line. However, there are three other groups of a small size (60, 62, and 50) that do not have noticeable spread about the 45° line. So, perhaps group size is not the reason for the undesirable spread. Regardless of why the unwanted spread exists, the four noticeable groups indicate that the three-variable model reflects a small weak spot, a segment that accounts for only 2.5% (= (20 + 56 + 28 + 19)/4,926)) of the sample and, by extension, of the parent database population. Thus, implementation of the three-variable model is expected to yield good predictions, even if exception rules cannot be afforded to the weak-spot segment as its effects on model performance are hardly noticeable.

The descriptive profile of the weak-spot segment is as follows: Newly opened (less than 6 months) accounts of customers with two or three accounts; recently opened (between 6 months and 1 year) accounts of customers with three accounts; and older (between 1 and 1½ years) of customers with three accounts. The actual profile cells are

1. MOS_OPEN = 1 and FD2_OPEN = 3
2. MOS_OPEN = 1 and FD2_OPEN = 2
3. MOS_OPEN = 2 and FD2_OPEN = 3
4. MOS_OPEN = 3 and FD2_OPEN = 3

The descriptive statistic for the plot of smooth actual versus smooth predicted by score groups/three-variable model is the correlation coefficient between the smooth points $r_{\text{sm. actual, sm. predicted: score group}}$ is 0.848.

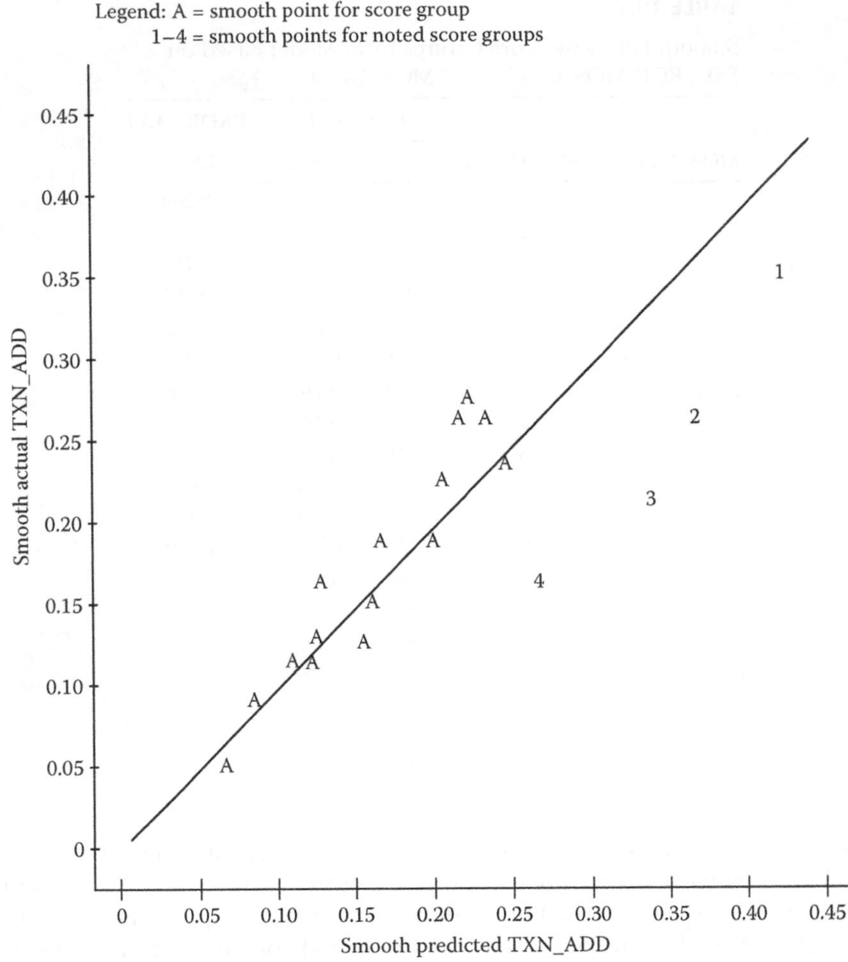

FIGURE 10.13
Plot of smooth actual versus predicted by score group for three-variable (FD2_RCP, MOS_OPEN, and MOS_DUM) model.

10.15 Evaluating the Data Mining Work

To appreciate the data mining analysis that produced the three-variable EDA model, I build a non-EDA model for comparison. I use the stepwise logistic regression variable selection process, which is a "good" choice *for* a non-EDA variable selection process, although the discussion of its weaknesses is in Chapter 13. The stepwise and other statistics-based variable selection procedures can be questionably classified as minimal data mining techniques as they only find the best subset of the original variables without generating potentially important variables. They do *not* generate structure in the search for the best subset of variables. Specifically, they do not create new variables such as reexpressed versions of the original variables or derivative variables such as dummy variables defined by the original variables. In contrast, the most productive data mining techniques generate structure from

TABLE 10.17

Best Non-EDA Model Criteria for Assessing Model Fit

	Intercept Only	Intercept and All Variables	All Variables	
−2LL	3606.488	3483.857	122.631 with 2 df (p = 0.0001)	
Variable	**Parameter Estimate**	**Standard Error**	**Wald Chi-Square**	**Pr > Chi-Square**
Intercept	−2.0825	0.1634	162.3490	0.0001
FD2_OPEN	0.6162	0.0615	100.5229	0.0001
MOS_OPEN	−0.1790	0.0299	35.8033	0.0001

the original variables and determine the best combination of those structures along with the original variables. More about variable selection is presented in Chapters 13 and 41.

I perform a stepwise logistic regression analysis on TXN_ADD with the original five variables. The analysis identifies the best non-EDA subset consisting of only two variables: FD2_OPEN and MOS_OPEN. The output is given in Table 10.17. The G/df value is 61.3 (= 122.631), which is comparable to the G/df value (62.02) of the three-variable (FD2_RCP, MOS_OPEN, MOS_DUM) EDA model. Based on the G/df indicator of Section 10.10.4, I cannot declare that the three-variable EDA model is better than the two-variable non-EDA model.

Could it be that all the EDA detective work was for naught—that the quick-and-dirty non-EDA model was the obvious one to build? The answer is no. Remember that an indicator is sometimes just an indicator that serves as a pointer to the next thing, such as moving on to the ladder of powers. Sometimes, it is an instrument for automatically making a decision based on visual impulses, such as determining if the relationship is straight enough or the scatter in a smooth residual plot is random. And sometimes, a lowly indicator does not have the force of its own to send a message until it is in the company of other indicators (e.g., smooth plots and their aggregate-level correlation coefficients).

I perform a simple comparative analysis of the descriptive statistics stemming from the EDA and non-EDA models to determine the better model. I need only to construct the three smooth plots—smooth residuals at the score group level, smooth actuals at the decile group level, and smooth actuals at the score group level—from which I obtain the descriptive statistics for the latter model. I already have the descriptive statistics for the former model.

10.15.1 Comparison of Plots of Smooth Residual by Score Groups: EDA versus Non-EDA Models

I construct the plot of the smooth residual by score groups in Figure 10.14 for the non-EDA model. The plot is not equivalent to the null plot based on the general association test. Thus, the overall quality of the predictions of the non-EDA model is not considered good. There is a local pattern of five smooth points in the lower right-hand corner below the zero line. The five smooth points, labeled I, II, III, IV, and V, indicate that the predictions for the individuals in the five score groups have, on average, a positive bias. That is, their predicted TXN_ADD tends to be larger than their actual TXN_ADD. There is a smooth point, labeled VI, at the top of the plot that indicates a group of individuals with a negative average bias. That is, their predicted TXN_ADD tends to be smaller than their actual TXN_ADD.

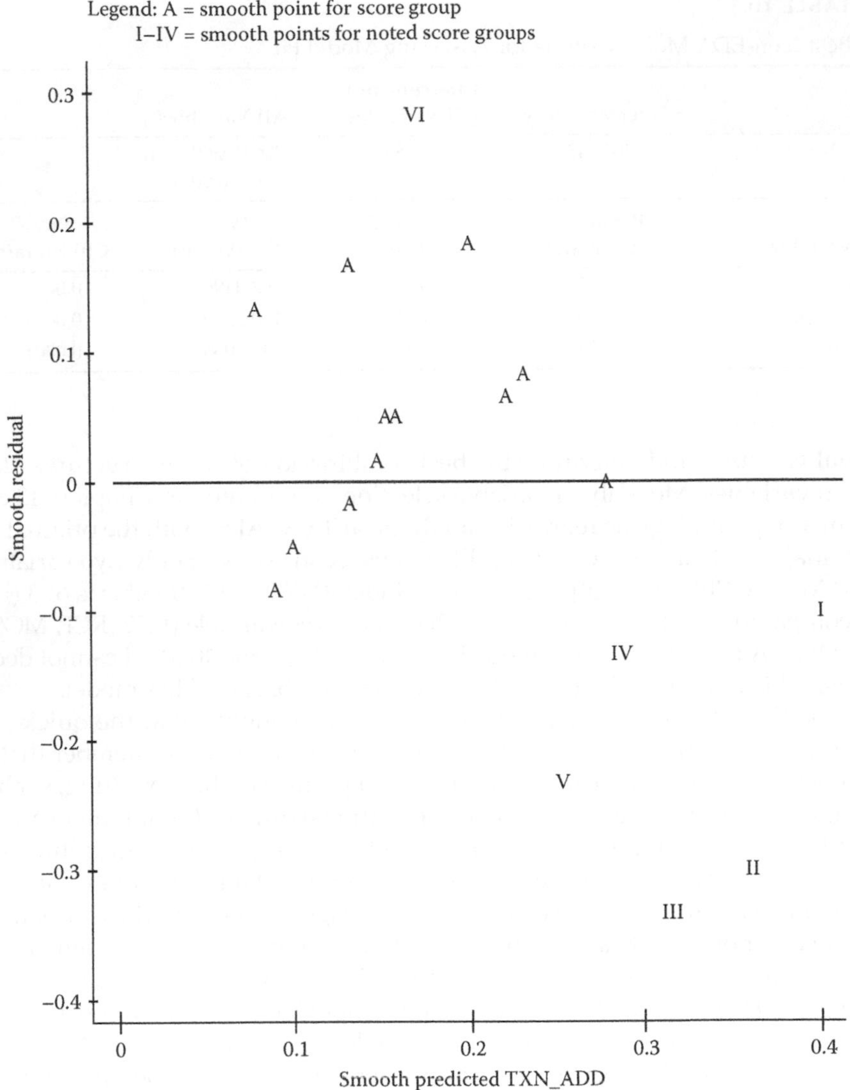

FIGURE 10.14
Smooth residual by score group plot for non-EDA (FD2_OPEN, MOS_OPEN) model.

The descriptive statistics for the plot of the smooth residual by score groups/non-EDA model are as follows: For the smooth residual, the minimum and maximum values and the range are −0.33, 0.29, and 0.62, respectively. The standard deviation of the smooth residuals is 0.167.

The comparison of the EDA and non-EDA smooth residuals indicates the EDA model produces smaller smooth residuals (prediction errors). The EDA smooth residual range is noticeably smaller than that of the non-EDA: 32.3% (= (0.62 – 0.42)/0.62) smaller. The EDA smooth residual standard deviation is noticeably smaller than that of the non-EDA: 25.7% (= (0.167 – 0.124)/0.167) smaller. The implication is that the EDA model has a better quality of prediction.

10.15.2 Comparison of the Plots of Smooth Actual versus Predicted by Decile Groups: EDA versus Non-EDA Models

I construct the plot of smooth actual versus smooth predicted by decile groups in Figure 10.15 for the non-EDA model. The plot clearly indicates a scatter that does not hug the 45° line well, as decile groups top and 2 are far from the line, and 8, 9, and bot are out of order, especially bot. The decile-based correlation coefficient between smooth points $r_{\text{sm. actual, sm. predicted: decile group}}$ is 0.759.

The comparison of the decile-based correlation coefficients of the EDA and non-EDA indicates that the EDA model produces a larger decile-based correlation coefficient and a tighter hug about the 45° line. The EDA correlation coefficient is noticeably larger than that of the non-EDA: 28.1% (= (0.972 – 0.759)/0.759) larger. The implication is that the EDA model has a better quality of prediction at the decile level.

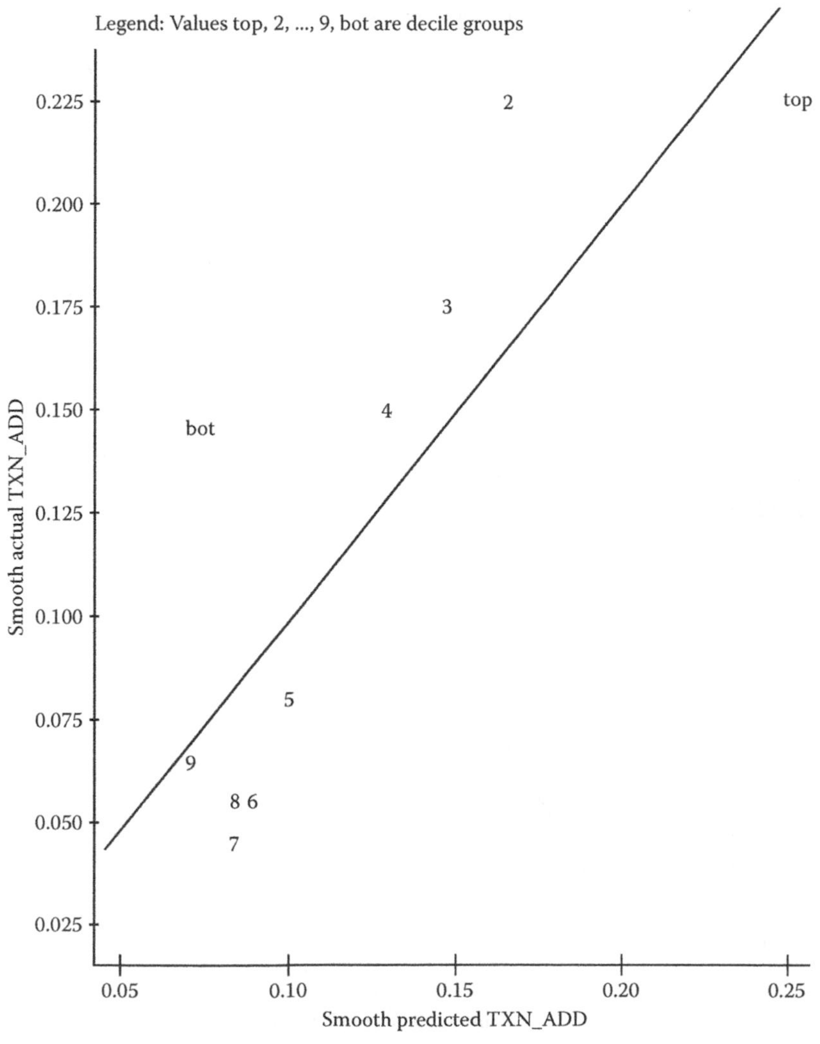

FIGURE 10.15
Smooth actual versus predicted by decile group plot for non-EDA (FD2_OPEN, MOS_OPEN) model.

10.15.3 Comparison of Plots of Smooth Actual versus Predicted by Score Groups: EDA versus Non-EDA Models

I construct the plot of smooth actual versus predicted by score groups in Figure 10.16 for the non-EDA model. The plot clearly indicates a scatter that does not hug the 45° line well, as score groups II, III, V, and VI are far from the line. The score-group-based correlation coefficient between smooth points $r_{\text{sm. actual, sm. predicted: score group}}$ is 0.635.

The comparison of the plots of the score-group-based correlation coefficient of the EDA and non-EDA smooth actual indicates that the EDA model produces a tighter hug about the 45° line. The EDA correlation coefficient is noticeably larger than that of the non-EDA: 33.5% (= (0.848 – 0.635)/0.635) larger. The implication is that the EDA model has a better quality of prediction at the score group level.

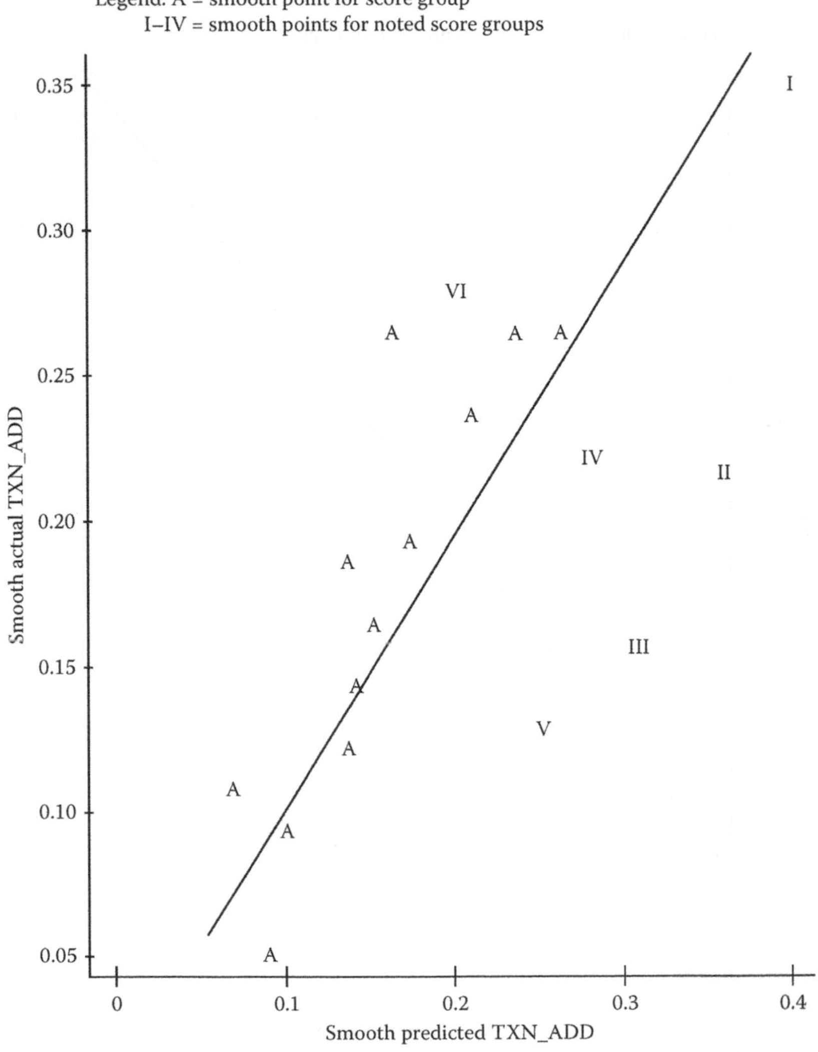

FIGURE 10.16
Smooth actual versus predicted by score group plot for non-EDA (FD2_OPEN, MOS_OPEN) model.

10.15.4 Summary of the Data Mining Work

From the comparative analysis, I have the following determinations:

1. The overall quality of the predictions of the EDA model is better than that of the non-EDA as the smooth residual plot of the former is null and that of the non-EDA model is not.
2. The prediction errors of the EDA model are smaller than those of the non-EDA as the smooth residuals of the former have less spread (smaller range and standard deviation). Also, the EDA model has better aggregate-level predictions than those of the non-EDA as the former model has less prediction bias (larger correlations between smooth actual and predicted values at decile and score group levels).
3. I conclude that the three-variable EDA model consisting of FD2_RCP, MOS_OPEN, and MOS_DUM is better than the two-variable non-EDA model consisting of FD2_OPEN and MOS_OPEN.

As the last effort to improve the EDA model, I consider the last candidate predictor variable FD_TYPE for data mining in the next section.

10.16 Smoothing a Categorical Variable

The classic approach to include a categorical variable into the modeling process involves *dummy variable coding*. A categorical variable with k classes of qualitative information is uniquely equivalent to a set of k − 1 quantitative dummy variables. The set of dummy variables replaces the categorical variable in the modeling process. The dummy variable assumes values of 1 or 0 for the presence or absence, respectively, of the class values. The class left out is called the *reference class*. The reference class is the baseline for comparing the other classes when interpreting the effects of dummy variables on the response variable. The classic approach instructs that the complete set of k − 1 dummy variables is included in the model regardless of the number of dummy variables that are declared nonsignificant. This approach is problematic when the number of classes is large, which is typically the case in big data applications. By chance alone, as the number of class values increases, the probability of one or more dummy variables declared nonsignificant increases. To put all the dummy variables in the model effectively adds noise or unreliability to the model as nonsignificant variables are known to be noisy. Intuitively, a large set of inseparable dummy variables poses difficulty in model building in that they quickly "fill up" the model, not allowing room for other variables.

The EDA approach of treating a categorical variable for model inclusion is a viable alternative to the classic approach as it explicitly addresses the problems associated with a large set of dummy variables. It reduces the number of classes by merging (smoothing or averaging) the classes with comparable values of the dependent variable under study, which for the application of response modeling is the response rate. The smoothed categorical variable, now with fewer classes, is less likely to add noise to the model and allows more room for other variables to get into the model.

There is an additional benefit offered by smoothing a categorical variable. The information captured by the smoothed categorical variable tends to be more reliable than that of the complete set of dummy variables. The reliability of information of the categorical variable is only as good as the aggregate reliability of information of the individual classes. Classes of small size tend to provide unreliable information. Consider the extreme situation of a class of size one. The estimated response rate for this class is either 100% or 0% because the sole individual either responds or does not respond, respectively. It is unlikely that the estimated response rate is the true response rate for this class. This class is considered to provide unreliable information regarding its true response rate. Thus, the reliability of information for the categorical variable itself decreases as the number of small class values increases. The smoothed categorical variable tends to have greater reliability than the set of dummy variables because it intrinsically has fewer classes and consequently has larger class sizes due to the merging process. The rule of thumb of EDA for small class size is that less than 200 is considered small.

CHAID is often the preferred EDA technique for smoothing a categorical variable. In essence, CHAID is an excellent EDA technique as it involves the three main elements of statistical detective work: numerical, counting, and graphical. CHAID forms new larger classes based on a numerical merging, or averaging, of response rates and counts the reduction in the number of classes as it determines the best set of merged classes. Last, the output of CHAID is conveniently presented in an easy to read and understand graphical display, a treelike box diagram with leaf boxes representing the merged classes.

The technical details of the merging process of CHAID are beyond the scope of this chapter. CHAID is discussed in detail in subsequent chapters, so here I briefly discuss and illustrate it with the smoothing of the last variable to be considered for predicting TXN_ADD response, namely, FD_TYPE.

10.16.1 Smoothing FD_TYPE with CHAID

Remember that FD_TYPE is a categorical variable that represents the product type of the customer's most recent investment purchase. It assumes 14 products (classes) coded A, B, C, …, N. The TXN_ADD response rate by FD_TYPE values are in Table 10.18.

TABLE 10.18
FD_TYPE

	TXN_ADD	
FD_TYPE	***N***	**MEAN**
A	267	0.251
B	2,828	0.066
C	250	0.156
D	219	0.128
E	368	0.261
F	42	0.262
G	45	0.244
H	225	0.138
I	255	0.122
J	57	0.193
K	94	0.202
L	126	0.222
M	19	0.421
N	131	0.160
Total	4,926	0.119

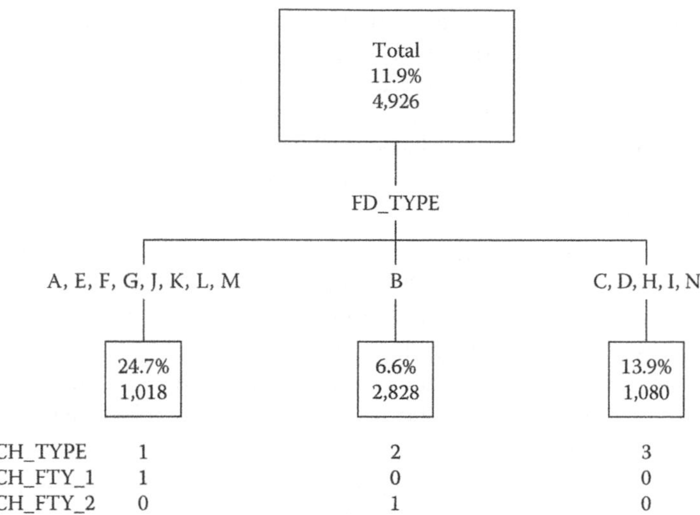

FIGURE 10.17
Double smoothing of FD_TYPE with CHAID.

There are seven small classes (F, G, J, K, L, M, and N) with sizes 42, 45, 57, 94, 126, 19, and 131, respectively. Their response rates—0.26, 0.24, 0.19, 0.20, 0.22, 0.42, and 0.16, respectively—can be considered potentially unreliable. Class B has the largest size, 2,828, with a surely reliable 0.06 response rate. The remaining six presumably reliable classes (A, C, D, E, and H) have sizes between 219 and 368.

The CHAID tree for FD_TYPE in Figure 10.17 is read and interpreted as follows:

1. The top box, the root of the tree, represents the sample of 4,926 with a response rate of 11.9%.
2. The CHAID technique smooths FD_TYPE by way of merging the original 14 classes into 3 merged (smoothed) classes, as displayed in the CHAID tree with three leaf boxes.
3. The leftmost leaf, which consists of the six of the small (N is excluded), unreliable classes and the two reliable classes A and E, represents a newly merged class with a reliable response rate of 24.7% based on a class size of 1,018. In this situation, the smoothing process increases the reliability of the small classes with two-step averaging. The first step combines all the small classes into a temporary class, which by itself produces a reliable average response rate of 22.7% based on a class size of 383. In the second step, which does not always occur in smoothing, the temporary class is further united with the already-reliable classes A and E because the latter classes have comparable response rates to the temporary class response rate. The *double-smoothed* newly merged class represents the average response rate of the seven small classes and classes A and E. When double smoothing does not occur, the temporary class is the final class.
4. The smoothing of a categorical variable clearly provides increased reliability. Consider class M with its unreliable estimated response rate of 42% based on class size 19. The smoothing process puts class M in the larger, more reliable leftmost leaf with a response rate of 24.7%. The implication is that class M now has a more

reliable estimate of response rate, namely, the response rate of its newly assigned class, 24.7%. Thus, the smoothing has effectively adjusted the originally estimated response rate of class M downward, from a positively biased 42% to a reliable 24.7%. In contrast, within the same smoothing process, the adjustment of class J is upward, from a negatively biased 19% to 24.7%. It is not surprising that the two reliable classes, A and E, remain noticeably unchanged, from 25% and 26% to 24.7%, respectively.

5. The middle leaf consists of only class B, defined by a large class size of 2,828 with a reliable response rate of 6.6%. Apparently, the low response rate of class B is not comparable to any class (original, temporary, or newly merged) response rate and does not warrant a merging. Thus, the originally estimated response rate of class B is unchanged after the smoothing process. The unchanged original estimate presents no concern over the reliability of class B because its class size is largest from the outset.
6. The rightmost leaf consists of large classes C, D, H, and I and the small class N for an average reliable response rate of 13.9% with a class size of 1,080. The smoothing process adjusts the response rate of class N downward, from 16% to a smooth 13.9%. The same adjustment occurs for class C. The remaining classes D, H, and I experience an upward adjustment.

I call the smoothed categorical variable CH_TYPE. Its three classes are labeled 1, 2, and 3, corresponding to the leaves from left to right, respectively (see the bottom of Figure 10.17). I also create two dummy variables for CH_TYPE:

1. CH_FTY_1 = 1 if FD_TYPE = A, E, F, G, J, K, L, or M; otherwise, CH_FTY_1 = 0;
2. CH_FTY_2 = 1 if FD_TYPE = B; otherwise, CH_FTY_2 = 0.
3. This dummy variable construction uses class CH_TYPE = 3 as the reference class. If an individual has values CH_FTY_1 = 0 and CH_FTY_2 = 0, then the individual has implicitly CH_TYPE = 3 and one of the original classes (C, D, H, I, or N).

10.16.2 Importance of CH_FTY_1 and CH_FTY_2

I assess the importance of the CHAID-based smoothed variable CH_TYPE by performing a logistic regression analysis on TXN_ADD with both CH_FTY_1 and CH_FTY_2 as the set dummy variable must be together in the model. The output is given in Table 10.19. The G/df value is 108.234 (= 216.468/2), which is greater than the standard G/df value of 4. Thus, CH_FTY_1 and CH_FTY_2 together are declared important predictor variables of TXN_ADD.

TABLE 10.19

G and df for CHAID-Smoothed FD_TYPE

Variable	−2LL	G	df	*p*
Intercept	3606.488			
CH_FTY_1 and CH_FTY_2	3390.021	216.468	2	0.0001

10.17 Additional Data Mining Work for Case Study

I try to improve the predictions of the three-variable (MOS_OPEN, MOS_DUM, and FD2_RCP) model with the inclusion of the smoothed variable CH_TYPE. I perform the LRM on TXN_ADD with MOS_OPEN, MOS_DUM, FD2_RCP, and CH_FTY_1 and CH_FTY_2. The output is in Table 10.20. The Wald chi-square value for FD2_RCP is less than 4. Thus, I delete FD2_RCP from the model and rerun the model with the remaining four variables.

The four-variable (MOS_OPEN, MOS_DUM, CH_FTY_1, and CH_FTY_2) model produces comparable Wald chi-square values for the four variables. The output is in Table 10.21. The G/df value equals 64.348 (= 257.395/4), which is slightly larger than the G/df (62.02) of the three-variable (MOS_OPEN, MOS_DUM, FD2_RCP) model. The G/df value is not a strong indication that the four-variable model has more predictive power than the three-variable model.

In Sections 10.17.1 through 10.17.4, I perform the comparative analysis, similar to the analysis of EDA versus non-EDA in Section 10.15, to determine whether the four-variable (4var-) EDA model is better than the three-variable (3var-) EDA model. I need the smooth plot descriptive statistics for the latter model as I already have the descriptive statistics for the former model.

TABLE 10.20

Logistic Model: EDA Model Variables plus CH_TYPE Variables

	Intercept Only	Intercept and All Variables	All Variables	
–2LL	3606.488	3347.932	258.556 with 5 df (p = 0.0001)	
Variable	**Parameter Estimate**	**Standard Error**	**Wald Chi-Square**	**Pr > Chi-Square**
Intercept	–0.7497	0.2464	9.253	0.0024
CH_FTY_1	0.6264	0.1175	28.4238	0.0001
CH_FTY_2	–0.6104	0.1376	19.6737	0.0001
FD2_RCP	0.2377	0.2212	1.1546	0.2826
MOS_OPEN	–0.2581	0.0398	42.0054	0.0001
MOS_DUM	0.7051	0.1365	26.6804	0.0001

TABLE 10.21

Logistic Model: Four-Variable EDA Model

	Intercept Only	Intercept and All Variables	All Variables	
–2LL	3606.488	3349.094	257.395 with 4 df (p = 0.0001)	
Variable	**Parameter Estimate**	**Standard Error**	**Wald Chi-Square**	**Pr > Chi-Square**
Intercept	–0.9446	0.1679	31.6436	0.0001
CH_FTY_1	0.6518	0.1152	32.0362	0.0001
CH_FTY_2	–0.6843	0.1185	33.3517	0.0001
MOS_OPEN	–0.2510	0.0393	40.8141	0.0001
MOS_DUM	0.7005	0.1364	26.3592	0.0001

10.17.1 Comparison of Plots of Smooth Residual by Score Group: 4var-EDA versus 3var-EDA Models

The plot of smooth residual by score group for the 4var-EDA model in Figure 10.18 is equivalent to the null plot based on the general association test. Thus, the overall quality of the predictions of the model is considered good. It is worthy of notice that there is a smooth far-out point, labeled FO, in the middle of the top of the plot. This smooth residual point corresponds to a score group consisting of 56 individuals (accounting for 1.1% of the data), indicating a weak spot.

The descriptive statistics for the smooth residual by score groups/4var-model plot are as follows: For the smooth residual, the minimum and maximum values and the range are −0.198, 0.560, and 0.758, respectively; the standard deviation of the smooth residual is 0.163.

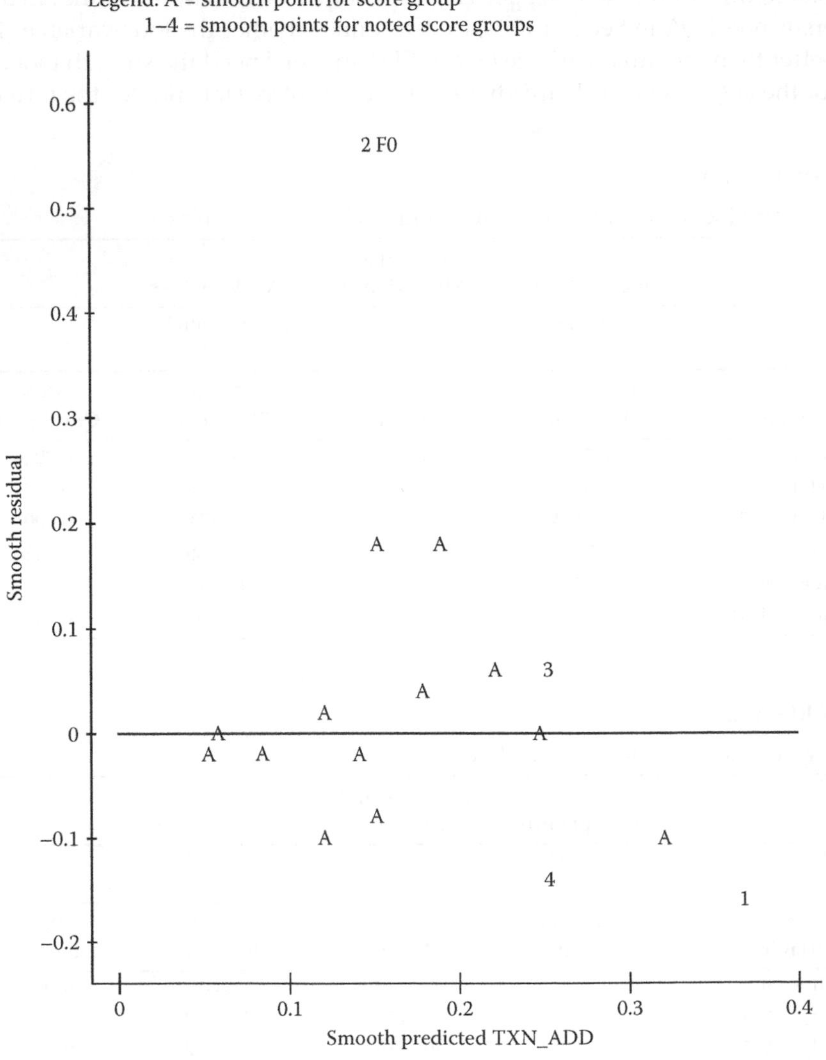

FIGURE 10.18
Smooth residual by score group plot for 4var-EDA (MOS_OPEN, MOS_DUM, CH_FTY_1, CH_FTY_2) model.

The descriptive statistics based on all the smooth points, *excluding* the FO smooth point, are worthy of notice because such statistics are known to be sensitive to far-out points, especially when they are smoothed and account for a very small percentage of the data. For the FO-adjusted smooth residual, the minimum and maximum values and the range are −0.198, 0.150, and 0.348, respectively; the standard deviation of the smooth residuals is 0.093.

The comparison of smooth residuals between the 3var-EDA and 4var-EDA models indicates the 3var-EDA model produces smaller smooth residuals than the 4var-EDA model. The 3var-EDA model smooth residual range is noticeably smaller than that of the latter model: 44.6% (= (0.758 − 0.42)/0.758) smaller. The 3var-EDA model smooth residual standard deviation is noticeably smaller than that of the 4var-EDA model: 23.9% (= (0.163 − 0.124/0.163) smaller. The implication is that the CHAID-based dummy variables carrying the information of FD_TYPE are not important enough to produce better predictions than that of the 3var-EDA model. In other words, the 3var-EDA model has a better quality of prediction.

However, if the implementation of the TXN_ADD model permits an exception rule for the FO score group/weak spot, the implication is the 4var-EDA FO-adjusted model has a better quality of predictions than the 4var-EDA model. The 4var-EDA FO-adjusted model produces smaller smooth residuals than the 4var-EDA model. The 4var-EDA FO-adjusted model smooth residual range is noticeably smaller than that of the 3var-EDA model: 17.1% (= (0.42 − 0.348)/0.42) smaller. The 4var-EDA FO-adjusted model smooth residual standard deviation is noticeably smaller than that of the 3var-EDA model: 25.0% (= (0.124 − 0.093)/0.124).

10.17.2 Comparison of the Plots of Smooth Actual versus Predicted by Decile Groups: 4var-EDA versus 3var-EDA Models

The plot of smooth actual versus smooth predicted by decile groups for the 4var-EDA model in Figure 10.19 indicates good hugging of scatter about the 45° lines, despite the two exceptions. First, there are two pairs of decile groups (6 and 7, and 8 and 9) where the decile groups in each pair are adjacent to each other. The two adjacent pairs indicate that the predictions are different for decile groups within each pair, which should have the same response rate. Second, the bot decile group is very close to the line but out of order and positioned between the two adjacent pairs. Because implementation of response models typically excludes the lower three or four decile groups, their spread about the 45° line and their (lack of) order is not as critical a feature in assessing the quality of predictions. Thus, overall, the plot is considered very good. The coefficient correlation between smooth actual and smooth predicted points $r_{\text{sm. actual, sm. predicted: decile group}}$ is 0.989.

The comparison of the decile-based correlation coefficient of the 3var-EDA and 4var-EDA models' smooth actual plots indicates the latter model produces a meagerly tighter hug about the 45° line. The 4var-EDA model correlation coefficient is hardly noticeably larger than that of the three-variable model: 1.76% (= (0.989 − 0.972)/0.972) smaller. The implication is that both models have equivalent quality of prediction at the decile level.

10.17.3 Comparison of Plots of Smooth Actual versus Predicted by Score Groups: 4var-EDA versus 3var-EDA Models

The score groups for the plot of the smooth actual versus predicted by score groups for the 4var-EDA model in Figure 10.20 are defined by the variables in the 3var-EDA model to make

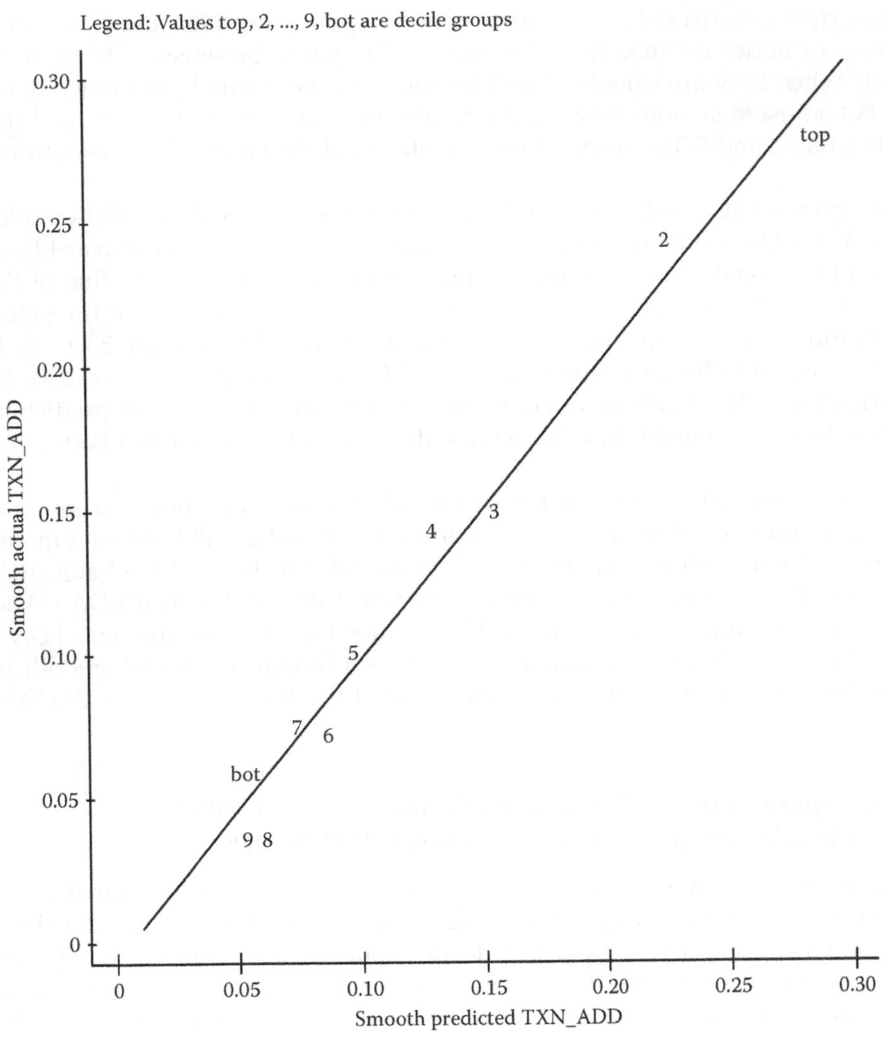

FIGURE 10.19
Smooth actual versus predicted by decile group plot for 4var-EDA (MOS_OPEN, MOS_DUM, CH_FTY_1, CH_FTY_2) model.

an uncomplicated comparison. The plot indicates a very nice hugging about the 45° line, except for one far-out smooth point, labeled FO, initially uncovered by the smooth residual plot in Figure 10.18. The score-group-based correlation coefficient between all smooth points $r_{\text{sm. actual, sm. predicted: score group}}$ is 0.784. The score-group-based correlation without the far-out score group FO, $r_{\text{sm. actual, sm. predicted: score group-FO}}$ is 0.915. The comparison of the score-group-based correlation coefficient of the 3var-smooth and 4var-smooth actual plots indicates that the 3var-EDA model produces a somewhat noticeably tighter hug about the 45° line. The 3var-EDA model score-group-based correlation coefficient is somewhat noticeably larger than that of the 4var-EDA model: 8.17% (= (0.848 – 0.784)/0.784) larger. The implication is the 3var-EDA model has a somewhat better quality of prediction at the score group level.

However, the comparison of the score group based correlation coefficient of the 3var-smooth and 4var-smooth actual plots without the FO score group produces a reverse implication.

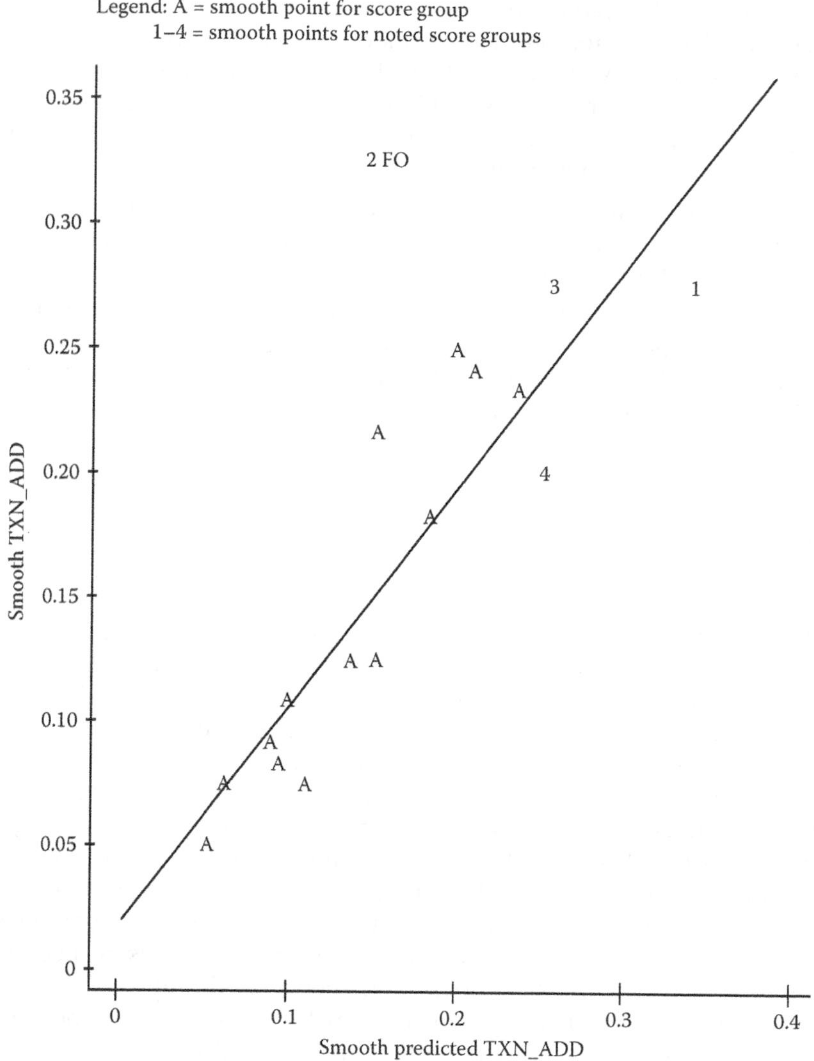

FIGURE 10.20
Smooth actual versus predicted by score group plot for 4var-EDA (MOS_OPEN, MOS_DUM, CH_FTY_1, CH_FTY_2) model.

The 4var-EDA model without the FO group score group based correlation coefficient is somewhat noticeably larger than that of the three-variable model: 7.85% (= (0.915 – 0.848)/0.848) larger. The implication is the 4var-EDA model without the FO score group has a somewhat better quality of prediction at the score group level than the 3var-EDA model.

10.17.4 Final Summary of the Additional Data Mining Work

The comparative analysis offers the following:

1. The overall quality of the 3var-EDA and 4var-EDA models is considered good as both models have a null smooth residual plot. Worthy of notice, there is a very small weak spot (FO score group accounting for 1.1% of the data) in the latter model.

2. The 3var-EDA model prediction errors are smaller than the prediction errors of the 4var-EDA model as the former model's smooth residuals have less spread (smaller range and standard deviation). The 3var-EDA model has equivalent, or somewhat better, aggregate-level predictions as it has equivalent or somewhat less prediction bias (equivalent/larger correlation between smooth actual and predicted values at the decile level/score group level).
3. If the model implementation can accommodate an exception rule for the FO weak spot, the indicators suggest that the 4var-EDA FO-adjusted model has less spread and somewhat better aggregate-level predictions.
4. In sum, I prefer the 3var-EDA model, consisting of MOS_OPEN, MOS_DUM, and FD2_RCP. If exception rules for the far-out score group can be effectively developed and reliably used, I prefer the 4var-EDA model, consisting of MOS_OPEN, MOS_DUM, CH_FTY_1, and CH_FTY_2.

10.18 Summary

I present the LRM as the workhorse of response modeling. As such, I demonstrate how it fulfills the desired analysis of a binary response variable in that it provides individual probabilities of response as well as yields a meaningful aggregation of individual-level probabilities into decile-level probabilities of response. The decile-level probabilities of response are often necessary for database implementation of response models. Moreover, I show the durability and usefulness of the 60-plus-year-old logistic regression analysis and modeling technique as it works well within the EDA/data mining paradigm of today.

I illustrate the rudiments of the LRM by discussing the SAS program for building and scoring an LRM. Working on a small dataset, I point out—and ultimately clarify—an often-vexing relationship between the actual and predicted response variables: The former assumes two nominal values, typically 1–0 for yes-no responses, respectively, yet the latter assumes logits, which are continuous values between –7 and +7.

Next, I present a case study that serves as the vehicle for introducing a host of data mining techniques, inspired by the EDA paradigm and tailored to logistic regression modeling. The techniques involve the concept of importance, not significance, of individual predictor variables and subsets of predictor variables as well as the indispensable use of smooth plots. The case study produces a final comparison of a data-guided EDA model and a non-EDA model. The EDA model is the preferred model with a better quality of predictions.

I provide the SAS subroutines for generating the essential smooth logit and smooth probability plots in Chapter 44.

11

Predicting Share of Wallet without Survey Data

11.1 Introduction

The share of wallet (SOW), the percentage of a customer's total spending that a company captures from the customer, is a key statistic used in implementing marketing strategies. Although simply defined, SOW calculations are problematically difficult because competitors' customer data are not readily available. The tried and true, and always used, approach for predicting SOW is with survey data. The purpose of this chapter is to present a two-step approach for predicting SOW without survey data. I illustrate the proposed method for the credit card industry who is a primary user of SOW. I present a case study in detail, as well as provide SAS© subroutines, that readers should find valuable for their toolkits. The subroutines are also available for downloading from my website: http://www.geniq.net/articles.html#section9.

11.2 Background

In today's world of big data, advancements in statistical methodologies, and gains in computer-intensive parallel processing, companies often find themselves not fully furnished with data about their customers. A company clearly has a complete history of its customers' transactional data but does not know about their customers' transactions with competitors. Depending upon its objectives, the company might be able to purchase external data (expensive and not always complete itself) or to conduct a survey (expensive, time-consuming, and biased) to fill the data gap.

SOW is a key measure used in implementing marketing strategies as well as in monitoring a company's vitals such as customer loyalty, trends in attrition, retention, and profit. A company develops a market campaign to stir any one of its primary spending categories. In developing the campaign, the company needs to know the SOW for the category of interest for each of its accounts.

I present the proposed method with a case study of a business credit card company. The choice of credit card study is quite appropriate because credit card companies routinely rely on SOW.

Say the category of interest is Supplies. Targeting to accounts with large Supplies-SOW values is done with the gentle voice. Optionally, if the company wants to harvest accounts with small Supplies-SOW values, then such accounts are targeted with a highly attractive message, embedded with alluring tokens such as discounts, additional bonus points, or deferred payment plans.

11.2.1 SOW Definition

The base definition of SOW is a customer's spending for a given product or service over a period divided by the customer's total spending over the period. Although simply defined, SOW calculations are problematically difficult because competitors' customer data are not readily available. A credit card company knows all details of its customers' purchases with its card but is at a significant disadvantage without information about its customers' purchase transactions made with its competitors' credit cards.

In practice, the popular approach for predicting SOW is with survey data. Survey data are the source for obtaining the purchases of the other credit cards—the competitor. The time and cost to conduct a survey are lengthy and expensive. More importantly, survey data are not known for their quality, that is, the reliability and validity of data [1,2]. I believe the literature has one article that uses a two-stage statistical model for predicting *wallet* (total dollars spent for a product or service) and SOW without survey data (Glady and Croux 2009). The model is restrictive, limited, and not ready for practical use. The authors conclude their wallet-SOW model is preliminary and call for additional research [3].

11.2.1.1 SOW_q Definition

The proposed quasi-SOW, labeled SOW_q, is defined as SOW for accounts that *appear* to have varying levels of total spending and use probabilistic weights to estimate a business account's total spending across all categories. I present thoroughly, including all important particulars, the calculating and modeling of SOW_q for a real case study of a business credit card company (the company name AMPECS is fictitious). Generalization of SOW_q to any industry sector, product, or service is straightforward.

11.2.1.2 SOW_q Likeliness Assumption

The underlying assumption for SOW_q for a company is:

1. A business account that shows large spending across all categories of expenditures for a given period has a large SOW_q because the account has *not likely* used other credit cards.
2. A business account with small spending across all categories of expenditures for a given period has a small SOW_q because the account *has likely* used other credit cards.
3. A business account that shows middling spending across all categories of expenditures for a given period has a middling SOW_q because the account has *somewhat likely* used other credit cards.

Because the likeliness assumption is the linchpin of the proposed SOW_q heuristic, I want a conservative approach to the SOW_q calculations to remove any upward bias inherent in the approach. So, I reduce the values of SOW_q by factors within SOW_q ranges. For example, one such rescaling is: if 0.20 ≤ SOW_q < 0.30, then SOW_q = SOW_q*0.20.

11.3 Illustration of Calculation of SOW_q

AMPECS, one of the nine major business credit card companies, is seeking a SOW model to affect a successful marketing campaign of its top-spending business accounts. AMPECS wants to use the industry metric SOW, but they do not want to use any resources for conducting a survey. I construct the proposed SOW_q metric, which serves as a surrogate for SOW with survey data. The tenableness of the SOW_q likeliness assumption and the offsetting of the competitors' estimated total spending by use of weighting AMPECS total spending yield reliable and actionable estimates of SOW and predictions of SOW.

AMPECS data for the full month (billing cycle) May 2016 consist of a sample (size 30,212) of business accounts' transactional data for the six industry-accepted business categories of expenditures: Services, Communications, Entertainment, Merchandise, Supplies, and Travel. The purchases (TRX) for the categories, for 10 randomly selected accounts, are displayed in Table 11.1. The category TRX fields are binary variables indicating whether an account made a purchase in a given category (yes = 1/no = 0). The last field, NUMBER of CATEGORIES USED, is a count variable showing the number of categories from which the account made purchases. See Table 11.1.

Among the 10 accounts, the number of categories used ranges from four to six. There are four accounts that use five and six categories and two accounts that use four categories. Of note, there are three accounts (#18065, #2060, and #20947) among the five-category accounts that have the same profile of category usage.

11.3.1 Query of Interest

It will be interesting to see, for accounts #18065, #2060, and #20947, whether their SOW_q values are equal or, if unequal, to what extent their SOW_q values differ.

11.3.2 DOLLARS and TOTAL DOLLARS

The DOLLAR fields corresponding to the categories of purchase, for the same 10 randomly selected accounts of Table 11.1, are in Table 11.2. Account #15047 spendings during AMPECS' May 2016 billing cycle are $26.74 for Services, $864.05 for Communications, and so on for the remaining four categories. The last field, AMPEC TOTAL DOLLARS, is the sum of the six category dollar fields. AMPECS TOTAL DOLLARS is the numerator of the SOW_q.

SOW_q, like any formulation of SOW, requires TOTAL DOLLARS. Before I detail the proposed estimation (simulation) procedure for TOTAL DOLLARS, I show TOTAL DOLLARS (WALLET*) fields corresponding to the DOLLARS fields, for the same 10 accounts in Table 11.1, in Table 11.3. The TOTAL DOLLARS per category are estimates of TOTAL DOLLARS spent *as if* using AMPECS and the competitors' credit cards.

* Wallet is also referred to as wallet size.

TABLE 11.1

Purchase (yes = 1, no = 0) per CATEGORY and NUMBER OF CATEGORIES USED

Acct	SERVICES TRX	COMMUNICATIONS TRX	ENTERTAINMENT TRX	MERCHANDISE TRX	SUPPLIES TRX	TRAVEL TRX	NUMBER OF CATEGORIES USED
15047	1	1	1	1	1	0	5
17855	1	1	1	1	1	1	6
18065	1	1	1	0	1	1	5
16322	1	1	1	1	1	1	6
9605	1	1	1	0	1	0	4
5965	1	1	1	1	1	1	6
7569	1	1	1	0	1	0	4
2060	1	1	1	0	1	1	5
20947	1	1	1	0	1	1	5
6137	1	1	1	1	1	1	6

TABLE 11.2

Dollars per CATEGORY and AMPECS TOTAL DOLLARS

Acct	SERVICES DOLLARS	COMMUNICATIONS DOLLARS	ENTERTAINMENT DOLLARS	MERCHANDISE DOLLARS	SUPPLIES DOLLARS	TRAVEL DOLLARS	AMPECS TOTAL DOLLARS
15047	$26.74	$864.05	$1,062.69	$722.37	$26.44	$0.00	$2,702.30
17855	$344.63	$803.60	$128.46	$398.68	$22.06	$1,178.40	$2,875.82
18065	$7.58	$2,916.61	$805.79	$0.00	$29.29	$1,782.52	$5,541.80
16322	$183.39	$1,030.67	$227.06	$21.82	$272.48	$2,676.98	$4,412.39
9605	$213.82	$1,672.97	$223.15	$0.00	$672.34	$0.00	$2,782.29
5965	$41.59	$440.51	$88.85	$871.31	$1,565.84	$1,569.43	$4,577.53
7569	$48.87	$1,811.69	$11.95	$0.00	$72.68	$0.00	$1,945.19
2060	$65.47	$380.54	$1,491.85	$0.00	$136.42	$3,358.40	$5,432.67
20947	$51.03	$3,152.64	$237.42	$0.00	$261.19	$1,263.52	$4,965.80
6137	$184.96	$4,689.42	$470.23	$638.19	$1,432.01	$1,291.37	$8,706.18

TABLE 11.3

Total Dollars (Wallet) per CATEGORY and TOTAL DOLLARS (WALLET)

Acct	TOTAL SERVICES DOLLARS (WALLET)	TOTAL COMMUNICATIONS DOLLARS (WALLET)	TOTAL ENTERTAINMENT DOLLARS (WALLET)	TOTAL MERCHANDISE DOLLARS (WALLET)	TOTAL SUPPLIES DOLLARS (WALLET)	TOTAL TRAVEL DOLLARS (WALLET)	TOTAL DOLLARS (WALLET)
15047	$26.74	$864.05	$1,062.69	$722.37	$79.32	$0.00	$2,755.18
17855	$689.27	$803.60	$128.46	$398.68	$88.22	$1,178.40	$3,286.62
18065	$7.58	$5,833.23	$805.79	$0.00	$58.58	$1,782.52	$8,487.71
16322	$550.17	$1,030.67	$227.06	$21.82	$817.43	$5,353.95	$8,001.10
9605	$213.82	$3,345.94	$223.15	$0.00	$1,344.69	$0.00	$5,127.60
5965	$166.38	$440.51	$88.85	$1,742.62	$3,131.67	$3,138.86	$8,708.89
7569	$146.60	$3,623.33	$11.95	$0.00	$145.36	$0.00	$3,927.29
2060	$65.47	$380.54	$4,475.54	$0.00	$136.42	$6,716.81	$11,774.77
20947	$51.03	$9,457.93	$237.42	$0.00	$522.37	$1,263.52	$11,532.28
6137	$369.92	$18,757.67	$470.23	$638.19	$5,728.06	$1,291.37	$27,255.44

The procedure for calculating the TOTAL DOLLARS fields is as follows:

1. For a given category, say, Service, the expected number of Service transactions (purchases) made by a business account, SERVICE_TRX, is simulated by treating SERVICE_TRX as a random binomial variable with
   a. p = probability of a purchase
   b. n = the number of days in the billing cycle
   c. p is set at 1/30 for Services, Communications, Entertainment, Merchandise, and Travel; for Supplies p is set at 3/30. These values are determined by assessing AMPECS' May 2016 data
   d. n is set at 30, obviously for all categories.
2. For a given category, say, Service, the expected total dollars made by a business account, TOTAL SERVICE DOLLARS, is equal to SERVICE_TRX * AMPECS SERVICE DOLLARS (from Table 11.2).
3. Steps #1 and #2 are repeated for each of the remaining categories, generating COMMUNICATIONS SERVICE DOLLARS, ENTERTAINMENT SERVICE DOLLARS, MERCHANDISE SERVICE DOLLARS, SUPPLIES SERVICE DOLLARS, and TRAVEL SERVICE DOLLARS.
4. TOTAL DOLLARS (wallet) = sum of the generated variables in Step #3.
5. AMPECS DOLLARS (wallet) = AMPECS TOTAL DOLLARS.
6. SOW_q = AMPECS DOLLARS (wallet)/TOTAL DOLLARS (wallet).

The subroutines for the six-step process are noted in Appendix 11.A. SOW_q values for the 10 accounts are shown in Table 11.4. Observing the relationship (for the 10 accounts) between TOTAL DOLLARS and SOW_q, one might conclude there is a negative correlation. As a point of fact, there is no such relationship for the sample of AMPECS' May 2016 billing cycle.

In the next section, I build the SOW_q Model. I display the frequency distribution of SOW_q to reveal the nature of SOW_q as a dependent variable in Table 11.5. A greater reveal

TABLE 11.4

SOW_q

Acct	AMPECS Dollars (Wallet)	Total Dollars (Wallet)	SOW_q
15047	$2,702.30	$2,755.18	0.9808
17855	$2,875.82	$3,286.62	0.8750
18065	$5,541.80	$8,487.71	0.6529
16322	$4,412.39	$8,001.10	0.5515
9605	$2,782.29	$5,127.60	0.5426
5965	$4,577.53	$8,708.89	0.5256
7569	$1,945.19	$3,927.29	0.4953
2060	$5,432.67	$11,774.77	0.4614
20947	$4,965.80	$11,532.28	0.4306
6137	$8,706.18	$27,255.44	0.3194

TABLE 11.5
SOW_q Frequency Distribution and Basic Statistics

SOW_q	Frequency	Percent	Cumulative Frequency	Cumulative Percent
1.0	1850	6.12	1850	6.12
0.9	629	2.08	2479	8.21
0.8	252	0.83	2731	9.04
0.7	916	3.03	3647	12.07
0.6	389	1.29	4036	13.36
0.5	1814	6.00	5850	19.36
0.4	3376	11.17	9226	30.54
0.3	6624	21.93	15850	52.46
0.2	6269	20.75	22119	73.21
0.1	7245	23.98	29364	97.19
0.0	848	2.81	30212	100.00

Minimum	Maximum	Mean	Median	Skewness
0.0400290	1.0000000	0.3122372	0.2526832	1.4926534

is the core statistics of any distribution for modeling or analysis at the bottom of Table 11.5. The SOW_q distribution is not diagnostically problematic: slight positive skewness (1.492), mean 0.312, and range 0.96 (= 1.000 – 0.040) indicate the data have substantial variation, which is a necessary condition for a fruitful model building effort.

11.4 Building the AMPECS SOW_q Model

SOW_q is a fractional continuous (dependent) variable, whose values are proportions or rates within the closed interval [0, 1]. Ordinary least-squares (OLS) regression seemingly is the appropriate method for modeling SOW_q. Similar to modeling a 0–1 dependent variable with OLS regression, which cannot guarantee the predicted values lie within the closed interval [0, 1], modeling SOW_q with OLS regression also cannot guarantee the predicted values lie within the closed interval [0, 1]. There is another conceptual issue for not using OLS regression to model SOW_q. The whole real line is the domain of the OLS dependent variable, which apparently SOW_q is not defined on. There are additional theoretical issues as to why OLS regression is not the appropriate method for a fractional dependent variable [4,5]. A possible approach to directly model SOW_q is to transform SOW_q into a logit, that is, logit (SOW_q) = log(SOW_q/(1– SOW_q)). This well-defined transformation, often used in other situations, is not suitable where there is a mass of observations at the end points of the closed interval, namely, 0 and 1.

Papke and Wooldridge made the first sound attempt, called fractional response regression (FRM) at modeling a fractional continuous dependent variable [6]. Since their 1996 seminal paper, there has been continuous work on FRM, refining it with theoretical niceties; comparative studies conclude there is no newer version significantly superior to the original FRM [7]. The literature on FRM includes references under various model names (e.g., fractional logistic regression, fractional logit model, and fractional regression).

I build the SOW_q Model, using binary logistic regression coded by Liu and Xin [7]. The procedure is comfortable to use because it uses the ordinary logistic regression; therefore, its output is familiar and easy to interpret. The process outline is as follows:

1. The original AMPECS (size = 30,212) data are doubly-stacked and called DATA2. Thus, the DATA2 has twice the number of observations.
2. Dependent variable Y is binary and appended to DATA2.
   a. Y assumes values 1 and 0 for the first and second halves of DATA2, respectively.
3. Observations with Y equal to 1 and 0 are assigned weight values SOW_q and 1 – SOW_q, respectively.
4. Logistic regression of Y on a set of predictor variables, with a weight statement, is performed.
5. SOW_q estimates are the familiar maximum likelihood estimates.
6. Interestingly, Liu and Xin do not mention the effects of the doubled sample size on the p-values. Primarily, I bootstrap DATA2 to a size of 30,000, shy of the original size, to remove noise in DATA2. Secondly, the p-values of data with the original size have face validity and are less unsettling.
7. The decile analysis of the SOW_q Model is the final determinant of model performance.

The subroutines for the seven-step process are noted in Appendix 11.B.

11.5 SOW_q Model Definition

I run the weighted logistic regression for SOW_q on four predictor variables. The weights are SOW_q and 1 – SOW_q for observations with Y equal to 1 and 0, respectively. The definitions of the predictor variables are:

1. BAL_TO_LIMIT is the balance to limit ratio as April 30, 2016, the month before the modeling period of May 2016.
2. PAY_AMOUNT_1 is the amount paid by the business account a month prior to the modeling period of May 2016.
3. PAY_AMOUNT_2 is the amount paid by the account 2 months prior to the modeling period of May 2016.
4. PAY_AMOUNT_3 is the amount paid by the account 3 months prior to the modeling period of May 2016.

The maximum likelihood estimates of the SOW_q Model, in Table 11.6, indicate the correct signs of the variables, along with statistical significance, as the p-values are quite small.

TABLE 11.6
Maximum Likelihood Estimates of SOW_q Model

Parameter	DF	Estimate	Standard Error	Wald Chi-Square	Pr > ChiSq
Intercept	1	−0.7014	0.0211	1102.4115	<0.0001
BAL_TO_LIMIT	1	4.564E−7	1.07E−7	18.2019	<0.0001
PAY_AMOUNT_1	1	3.488E−6	9.923E−7	12.3539	0.0004
PAY_AMOUNT_2	1	0.2455	0.0137	320.0763	<0.0001
PAY_AMOUNT_3	1	0.0539	0.0128	17.8548	<0.0001

The Logit of SOW_q uses the parameter estimates in its definition in Equation 11.1. The Logit provides the information to obtain Prob (SOW_q) in Equation 11.2.

$$\begin{aligned}\text{Logit}(Y = 1 \text{ with weight} = SOW\_q) = &- 0.7014 + 4.564E{-}7{*}BAL\_TO\_LIMIT \\ &+ 3.488E{-}6{*}PAY\_AMOUNT\_1 \\ &+ 0.2455{*}PAY\_AMOUNT\_2 \\ &+ 0.0539{*}PAY\_AMOUNT\_3\end{aligned} \tag{11.1}$$

$$\text{Prob}\ (Y = 1 \text{ with weight } SOW\_q) = \exp(\text{Logit})/(1 + \exp(\text{Logit})) \tag{11.2}$$

Before discussing the SOW_q Model results, I refer to the query in Section 11.3.1 where three illustrative accounts, #18065, #2060 and #20947, have the same profile of category usage.

The query of interest is whether these accounts would have equal SOW_q values or, if unequal, to what extent their SOW_q values would differ. Given the SOW_q Model is complete, the following results address this question:

1. For account #18065, predicted SOW_q = 0.26735.
2. For account #2060, predicted SOW_q = 0.44267.
3. For account #20947, predicted SOW_q = 0.25967.

The implication is not unexpected because similar transaction profiles do not necessarily imply similar dollars and total dollars spent, which both affect SOW_q.

11.5.1 SOW_q Model Results

The decile analysis in Table 11.7 readily displays the results and performance of the SOW_q Model.* The DECILE column, as a noncalculated vertical identifier, renders the five arithmetically derived columns, Columns #2 through #6. The business accounts are ranked from high to low based on the Logit (or Prob (SOW_q)). Ten equal-sized

* Chapter 26 details thoroughly the construction and interpretation of the decile analysis. Indeed, the reader can take a quick detour to Chapter 26 and then return to this section. Or, the reader can go through the model results presented here, and after reading Chapter 26, the reader can revisit this section.

TABLE 11.7

Decile Analysis of SOW_q Model

Decile	Number of Accounts	Number of Accounts w/Large SOW_q	MEAN SOW_q (%)	CUM SOW_q (%)	CUM LIFT (%)
top	2,929	1,595	54.5	54.5	176
2	2,887	1,159	40.1	47.4	153
3	3,079	1,014	32.9	42.4	137
4	3,044	869	28.5	38.8	125
5	2,945	751	25.5	36.2	117
6	2,931	715	24.4	34.3	110
7	3,090	764	24.7	32.8	106
8	3,030	902	29.8	32.5	105
9	2,993	795	26.6	31.8	103
bottom	3,072	738	24.0	31.0	100
	30,000	9,302			

groups or deciles are created based on the ranked file. The five columns implicatively labeled show:

1. The sample size is 30,000, and there are 9,302 accounts with Y = 1. Thus, the CUM SOW_q is 31.0, in Column #5, bottom decile.
2. In the fourth column, MEAN SOW_q (%) are the means at the decile level. The top decile mean (54.4%) and the bottom decile mean (24.0%) show a top-to-bottom ratio 2.27. This ratio value indicates that the model *significantly discriminates* among the business accounts.
3. The last column, CUM LIFT (%), presents the performance of the model. The top decile, CUM LIFT 176, means that the model identifies the top 10% business accounts, whose average top 10% SOW_q values are 1.76 times (76% greater than) the average SOW_q 31.0.
4. The CUM LIFT (%) for the top two deciles, 153, indicates the model identifies the top 20% (top and second deciles) of business accounts, whose mean SOW_q are 1.53 times (53% greater than) the CUM SOW_q 31.0.
5. For the remaining deciles, the interpretation of CUM LIFT(%) is similar.

In sum, the SOW_q Model has significant discriminatory power and identifies the best business accounts with large SOW_q values for effective target marketing campaigns.

11.6 Summary

SOW is a key statistic used in implementing marketing strategies. The everyday approach for predicting SOW uses survey data. This approach is always met with some reluctance because conducting a survey is time-consuming, expensive, and yields unreliable data.

I provide a two-step approach for predicting SOW without data. Step #1 defines a quasi-SOW, SOW_q. SOW_q uses simulation for estimating total dollars spent, effectively eliminating the need for survey data. Step #2 uses fractional logistic regression to predict SOW_q. I illustrate the two-step approach with a real case study. The derived SOW_q dependent variable is reliable because it has a well-defined distribution, producing an SOW_q Model with significant discriminatory power for identifying the best business accounts with large SOW_q values for effective target marketing campaigns. I provide SAS subroutines for the new method, which the readers should find valuable for their toolkits.

Appendix 11.A Six Steps

```
libname sq 'c://0-SOW_q';
option pageno=1;

data simulate_trx;
set  sq.AMPECS_data;
call streaminit(12345);
do i=1 to 30212;

x1=rand('binomial',(1/30), 30);
x2=rand('binomial',(1/30), 30);
x3=rand('binomial',(1/30), 30);
x4=rand('binomial',(1/30), 30);
x5=rand('binomial',(3/30), 30);
x6=rand('binomial',(1/30), 30);

d1=AMPECS_Services_DOLLARS;
d2=AMPECS_Communications_DOLLARS;
d3=AMPECS_Entertainment_DOLLARS;
d4=AMPECS_Merchandise_DOLLARS;
d5=AMPECS_Supplies_DOLLARS;
d6=AMPECS_Travel_DOLLARS;

output;
drop i;
end;
run;

data sq.SOWq_data;
set  simulate_trx;
array x(6)      x1-x6;
array d(6)      d1-d6;
array_01x(6)  _01x1-_01x6;
array_01xxd(6) _01xxd1-_01xxd6;
```

```
array xd(6)       xd1-xd6;
array xxd(6)   xxd1-xxd6;

do j=1 to 6;
_01x(j)=0;
if x(j) ne 0 then_01x(j)=1;
if d(j) le 0 then d(j)=uniform(12345)*100;
xd(j)=   x(j)*d(j);
_01xxd(j)= _01x(j)*xd(j);

sum_x =sum(of   x1-   x6);
sum_01x=sum(of_01x1-_01x6);

_01xxd(j)= _01x(j)*x(j)*d(j);
xxd(j) = x(j)*x(j)*d(j);

SUM_catgDOL=sum(of_01xxd1-_01xxd6);
SUM_trnxDOL=sum(of   xxd1-   xxd6);
drop j;
end;

SOW_q=SUM_catgDOL/SUM_trnxDOL;
label
sum_x='TOTAL_TRX'
sum_01x='TOTAL_CATGS'
x1='SERVICES_TRX(prob. expected)'
x2='COMMUNICATIONS_TRX(prob. expected)'
x3='ENTERTAINMENT_TRX(prob. expected)'
x4='MERCHANDISE_TRX(prob. expected)'
x5='SUPPLIES_TRX(prob. expected)'
x6='TRAVEL_TRX(prob. expected)'
xd1='SERVICES_DOL'
xd2='COMMUNICATIONS_DOL'
xd3='ENTERTAINMENT_DOL'
xd4='MERCHANDISE_DOL'
xd5='SUPPLIES_DOL'
xd6='TRAVEL_DOL'
xxd1='TOTAL SERVICES_DOL(prob. expected)'
xxd2='TOTAL COMMUNICATIONS_DOL(prob. expected)'
xxd3='TOTAL ENTERTAINMENT_DOL(prob. expected)'
xxd4='TOTAL MERCHANDISE_DOL (prob. expected)'
xxd5='TOTAL SUPPLIES_DOL(prob. expected)'
xxd6='TOTAL TRAVEL_DOL(prob. expected)'

SUM_catgDOL='SUM of catg-DOLLARS'
SUM_trnxDOL='SUM of DOLLAR WEIGHTS'
SOW_q='SOW_q';
run;
```

Appendix 11.B Seven Steps

```
libname sq 'c://0-SOW_q';
title2' BS=30000 BAL_TO_LIMIT PAY_AMOUNT_1 PAY_AMOUNT_2
PAY_AMOUNT_3';

data SSOWq_data;
set sq.SOWq_data (in = a) sq.SOWq_data (in = b);
if 0.00< SOW_q <0.05 then SOW_q=SOW_q*0.00;
if 0.05<= SOW_q <0.10 then SOW_q=SOW_q*0.05;
if 0.10<= SOW_q <0.20 then SOW_q=SOW_q*0.10;
if 0.20<= SOW_q <0.30 then SOW_q=SOW_q*0.20;
if 0.30<= SOW_q <0.40 then SOW_q=SOW_q*0.30;
if 0.40<= SOW_q <0.50 then SOW_q=SOW_q*0.40;
if 0.50<= SOW_q <0.60 then SOW_q=SOW_q*0.50;
if 0.60<= SOW_q <0.70 then SOW_q=SOW_q*0.60;
if 0.70<= SOW_q <0.80 then SOW_q=SOW_q*0.70;
if 0.80<= SOW_q <0.85 then SOW_q=SOW_q*0.80;
if 0.85<= SOW_q <0.90 then SOW_q=SOW_q*0.85;
if a then do;
Y = 1;
wt = SOW_q;
end;
if b then do;
Y = 0;
wt = 1 - SOW_q;
end;
run;

PROC LOGISTIC data = SSOWq_data nosimple des outest=coef;
model Y =
BAL_TO_LIMIT PAY_AMOUNT_1 PAY_AMOUNT_2 PAY_AMOUNT_3;
weight wt;
run;

PROC SCORE data=SSOWq_data predict type=parms score=coef out=score;
var BAL_TO_LIMIT PAY_AMOUNT_1 PAY_AMOUNT_2 PAY_AMOUNT_3;
run;

data score;
set score;
estimate=Y2;
label estimate='estimate';
wtt=1;
run;

data notdot;
```

```
set score;
if estimate ne .;

PROC MEANS data=notdot sum noprint; var wtt;
output out=samsize (keep=samsize) sum=samsize;
run;

data scoresam (drop=samsize);
set samsize score;
retain n;
if _n_=1 then n=samsize;
if _n_=1 then delete;
run;

PROC SORT data=scoresam; by descending estimate;
run;

data score;
set scoresam;
if estimate ne . then cum_n+wtt;
if estimate = . then dec=.;
else dec=floor(cum_n*10/(n+1));
prob_hat=exp(estimate)/(1 + exp(estimate));
run;

/* Bootstrapping score data */
data score (drop = i sample_size);
choice = int(ranuni(36830)*n) + 1;
set score point = choice nobs = n;
i+1;
sample_size=30000;
if i = sample_size + 1 then stop;
run;
/* End of bootstrapping */

PROC TABULATE data=score missing;
class dec;
var Y SOW_q;
table dec all, (Y*(mean*f=5.3 (n sum)*f=6.0) (SOW_q)*((mean min max)*f=5.3));
weight wt;
run;

PROC SUMMARY data=score missing;
class dec;
var SOW_q wtt;
output out=sum_dec sum=sum_can sum_wt;

data sum_dec;
set sum_dec;
```

```
avg_can=sum_can/sum_wt;
run;

data avg_rr;
set sum_dec;
if dec=.;
keep avg_can;
run;

data sum_dec1;
set sum_dec;
if dec=. or dec=10 then delete;
cum_n +sum_wt;
r =sum_can;
cum_r +sum_can;
cum_rr=(cum_r/cum_n)*100;
avg_cann=avg_can*100;
run;

data avg_rr;
set sum_dec1;
if dec=9;
keep avg_can;
avg_can=cum_rr/100;
run;

data scoresam;
set avg_rr sum_dec1;
retain n;
if _n_=1 then n=avg_can;
if _n_=1 then delete;
lift=(cum_rr/n);
if dec=0 then decc='top';
if dec=1 then decc='2';
if dec=2 then decc='3';
if dec=3 then decc='4';
if dec=4 then decc='5';
if dec=5 then decc='6';
if dec=6 then decc='7';
if dec=7 then decc='8';
if dec=8 then decc='9';
if dec=9 then decc='bottom';
if dec ne .;
run;

PROC PRINT data=scoresam d split='*' noobs;
var decc sum_wt r avg_cann cum_rr lift;
label decc='DECILE'
   sum_wt ='NUMBER OF*ACCOUNTS'
```

```
  r ='NUMBER OF*ACCOUNTS'
  cum_r ='CUM No. CUSTOMERS w/*SOW_q'
  avg_cann ='MEAN*SOW_q (%)'
  cum_rr ='C U M*SOW_q (%)'
  lift =' C U M*LIFT (%)';
sum sum_wt r;
format sum_wt r cum_n cum_r comma10.;
format avg_cann cum_rr 4.2;
format lift 3.0;
run;
```

References

1. Fowler, F.J., *Survey Research Methods,* 3rd ed., Sage, Thousand Oaks, CA, 2002.
2. Odom, J.G., *Validation of Techniques Utilized to Maximize Survey Response Rates,* Tech. ERIC document reproduction service no. ED169966, Education Resource Information Center, 1979.
3. Glady, N., and Croux, C., Predicting customer wallet without survey data, *Journal of Service Research,* 11(3), 219–231, 2009.
4. McCullagh, P., and Nelder. J.A., *Generalized Linear Models,* 2nd ed., Chapman & Hall/CRC, London, 1989.
5. Kieschnick, R. and McCullough, B., Regression analysis of variates observed on (0, 1): Percentages, proportions, and fractions, *Statistical Modeling,* 3, 193–213, 2003.
6. Papke, L., and Wooldridge, J.M., Econometric methods for fractional response variables with an application to 401(K) plan participation rates, *Journal of Applied Econometrics,* 11(6), 619–632, 1996.
7. Liu, W., and Xin, J., Modeling fractional outcomes with SAS, in *Proceedings of the SAS Global Forum 2014 Conference,* SAS Institute Inc., Cary, NC, 2014.

12

Ordinary Regression: The Workhorse of Profit Modeling

12.1 Introduction

Ordinary regression is the popular technique for predicting a quantitative outcome, such as profit and sales. It is considered the workhorse of *profit* modeling as its results are the gold standard. Moreover, the ordinary regression model serves as the benchmark for assessing the superiority of new and improved techniques. In a database marketing application, an individual's profit* to a prior solicitation is the quantitative dependent variable, and an ordinary regression model is built to predict the individual's profit to a future solicitation.

I provide a brief overview of ordinary regression and include the SAS© program for building and scoring an ordinary regression model. Then, I present a mini case study to illustrate that the data mining techniques presented in Chapter 10 carry over with minor modification to ordinary regression. Model builders, who are called on to provide statistical support to managers monitoring expected revenue from marketing campaigns, will find this chapter an excellent reference for profit modeling.

12.2 Ordinary Regression Model

Let Y be a quantitative dependent variable that assumes a continuum of values. The ordinary regression model, formally known as the ordinary least squares (OLS) regression model, predicts the Y value for an individual based on the values of the predictor (independent) variables $X_1, X_2, \ldots, X_n$ for that individual. The definition of the OLS model is in Equation 12.1:

$$Y = b_0 + b_1{}^*X_1 + b_2{}^*X_2 + \ldots + b_n{}^*X_n \quad (12.1)$$

Plugging in the values of the predictor variables for an individual in Equation 12.1 produces that individual's estimated (predicted) Y value. The b's are the OLS regression coefficients, determined by the calculus-based method of least squares estimation. The lead coefficient b_0 is the intercept, which has no corresponding X_0 as do the other coefficients.

* Profit is variously defined as any measure of an individual's valuable contribution to the bottom line of a business.

In practice, the quantitative dependent variable does not have to assume a progression of values that vary by minute degrees. It can assume just several dozens of discrete values and work quite well within the OLS methodology. When the dependent variable assumes only two values, the logistic regression model, not the ordinary regression model, is the appropriate technique. Even though logistic regression has been around for 60-plus years, there is some misunderstanding over the practical (and theoretical) weakness of using the OLS model for a binary response dependent variable. Briefly, an OLS model, with a binary dependent variable, produces some probabilities of response greater than 100% and less than 0% and often does not include important predictor variables.

12.2.1 Illustration

Consider dataset A, which consists of 10 individuals and 3 variables (Table 12.1). These are the quantitative variables PROFIT in dollars (Y), INCOME in thousands of dollars (X1), and AGE in years (X2). I regress PROFIT on INCOME and AGE using dataset A. The OLS output in Table 12.2 includes the ordinary regression coefficients and other columns of

TABLE 12.1

Dataset A

PROFIT ($)	INCOME ($000)	AGE (years)
78	96	22
74	86	33
66	64	55
65	60	47
64	98	48
62	27	27
61	62	23
53	54	48
52	38	24
51	26	42

TABLE 12.2

OLS Output: PROFIT with INCOME and AGE

Source	df	Sum of Squares	Mean Square	*F* Value	Pr > *F*
Model	2	460.3044	230.1522	6.01	0.0302
Error	7	268.0957	38.2994		
Corrected total	9	728.4000			
		Root MSE	6.18865	R-square	0.6319
		Dependent mean	62.60000	Adj R-Sq	0.5268
		Coeff var	9.88602		

Variable	df	Parameter Estimate	Standard Error	*t* Value	Pr > |*t*|
Intercept	1	52.2778	7.7812	7.78	0.0003
INCOME	1	0.2669	0.2669	0.08	0.0117
AGE	1	−0.1622	−0.1622	0.17	0.3610

information. The "Parameter Estimates" column contains the coefficients for INCOME and AGE variables and the Intercept. The coefficient b_0 for the Intercept variable serves as a "start" value given to all individuals, regardless of the specific values of the predictor variables in the model.

The estimated OLS PROFIT model is in Equation 12.2:

$$\text{PROFIT} = 52.2778 + 0.2667{*}\text{INCOME} - 0.1622{*}\text{AGE} \tag{12.2}$$

12.2.2 Scoring an OLS Profit Model

The SAS program (Figure 12.1) produces the OLS profit model built with dataset A and scores the external dataset B in Table 12.3. The SAS procedure REG produces the ordinary regression coefficients and puts them in the "ols_coeff" file, as indicated by the code "outest = ols_coeff." The ols_coeff file produced by SAS is in Table 12.4.

The SAS procedure SCORE scores the five individuals in dataset B using the OLS coefficients, as indicated by the code "score = ols_coeff." The procedure appends the predicted

```
/****** Building the OLS PROFIT Model on dataset A ************/
PROC REG data = A outest = ols_coeff;
pred_PROFIT: model PROFIT =
INCOME AGE;
run;

/****** Scoring the OLS PROFIT Model on dataset B ************/
PROC SCORE data = B predict type = parms score = ols_coeff
out = B_scored;
var INCOME AGE;
run;
```

FIGURE 12.1
SAS program for building and scoring OLS PROFIT model.

TABLE 12.3
Dataset B

INCOME ($000)	AGE (years)	Predicted PROFIT ($)
148	37	85.78
141	43	82.93
97	70	66.81
90	62	66.24
49	42	58.54

TABLE 12.4
OLS_Coeff File

OBS	_MODEL_	_TYPE_	_DEPVAR_	_RMSE_	INTERCEPT	INCOME	AGE	PROFIT
1	est_PROFIT	PARMS	PROFIT	6.18865	52.52778	0.26688	−0.16217	−1

PROFIT variable in Table 12.3 (called pred_PROFIT as indicated by "pred_PROFIT" in the second line of code in Figure 12.1) to the output file B_scored, as indicated by the code "out = B_scored."

12.3 Mini Case Study

I present a "big" discussion on ordinary regression modeling with the mini dataset A. I use this extremely small dataset not only to make the discussion of data mining techniques tractable but also to emphasize two aspects of data mining. First, data mining techniques of great service should work as well with small data as with big data, as explicitly stated in the definition of data mining in Chapter 1. Second, every fruitful effort of data mining on small data is evidence that big data are not always necessary to uncover structure in the data. This evidence is in keeping with the exploratory data analysis (EDA) philosophy that the data miner should work from simplicity until indicators emerge to go further. If predictions are not acceptable, then increase the data size.

The objective of the mini case study is as follows: to build an OLS profit model based on INCOME and AGE. The ordinary regression model (celebrating 200-plus years of popularity since the invention of the method of least squares on March 6, 1805) is the quintessential linear model, which implies the all-important assumption: The underlying relationship between a given predictor variable and the dependent variable is linear. Thus, I use the method of smoothed scatterplots, as described in Chapter 3, to determine whether the linear assumption holds for PROFIT with INCOME and with AGE. For the mini dataset, 10 slices each of size 1 define the smoothed scatterplot. Effectively, the smooth scatterplot is the simple scatterplot of 10 paired (PROFIT, Predictor Variable X_i) points. (Contrasting note: The logit plot as discussed with the logistic regression in Chapter 10 is neither possible nor relevant with OLS methodology. The quantitative dependent variable does not require a transformation, like converting logits into probabilities, as found in logistic regression.)

12.3.1 Straight Data for Mini Case Study

Before proceeding with the analysis of the mini case study, I clarify the use of the bulging rule when analysis involves OLS regression. The bulging rule states that the model builder should try reexpressing the predictor variables as well as the dependent variable. As discussed in Chapter 10, it is not possible to reexpress the dependent variable in a logistic regression analysis. However, in performing an ordinary regression analysis, reexpressing the dependent variable is possible, but the bulging rule needs modification. I discuss the modification in the illustration that follows.

Consider a hypothetical building of a profit model with the quantitative dependent variable Y and three predictor variables X_1, X_2, and X_3. Based on the bulging rule, the model builder determines that the powers of ½ and 2 for Y and X_1, respectively, produce an adequate straightening of the Y–X_1 relationship. Let us assume that the correlation between the square root Y (sqrt_Y) and the square of X_1 (sq_X_1) has a reliable r_{sqrt_Y, sq_X1} value of 0.85.

Continuing this scenario, the model builder determines that the powers of 0 and ½ and –½ and 1 for Y and X_2 and Y and X_3, respectively, also produce an adequate straightening of the Y–X_2 and Y–X_3 relationships, respectively. Let us assume the correlations between the log of Y (log_Y) and the square root of X_2 (sq_X_2) and between the negative square root of Y (negsqrt_Y) and X_3 have reliable r_{log_Y, sq_X1} and $r_{negsqrt\_Y, X3}$ values of 0.76 and 0.69, respectively. In sum, the model builder has the following results:

1. The best relationship between square root Y (p = ½) and square of X_1 has $r_{sqrt\_Y, sq\_X1} = 0.85$.
2. The best relationship between log of Y (p = 0) and square root of X_2 has $r_{log\_Y, sq\_X2} = 0.76$.
3. The best relationship between the negative square root of Y (p = –½) and X_3 has $r_{neg\_sqrt\_Y, X3} = 0.69$.

In pursuit of a good OLS profit model, the following guidelines have proven valuable when the bulging rule suggests several reexpressions of the quantitative dependent variable.

1. If there is a small range of *dependent variable powers* (powers used in reexpressing the dependent variable), then the best reexpressed dependent variable is the one with the noticeably largest correlation coefficient. In the illustration, the best reexpression of Y is the square root Y. Its correlation has the largest value: r_{sqrt_Y, sq_X1} equals 0.85. Thus, the data analyst builds the model with the square root Y and the square of X_1 and needs to reexpress X_2 and X_3 again on the square root of Y.
2. If there is a small range of dependent variable powers and the correlation coefficient values are comparable, then the best reexpressed dependent variable is defined by the *average power* among the dependent variable powers. In the illustration, if the data analyst were to consider the r values (0.85, 0.76, and 0.69) comparable, then the average power would be 0, which is one of the powers used. Thus, the data analyst builds the model with the log of Y and the square root of X_2 and needs to reexpress X_1 and X_3 again on the log of Y.

   If the average power were not one of the dependent variable powers used, then all predictor variables would need to be reexpressed again with the newly assigned reexpressed dependent variable, Y, raised to the average power.
3. When there is a large range of dependent variable powers, which is common when they are many predictor variables, the practical and productive approach to the bulging rule for building an OLS profit model consists of initially reexpressing only the predictor variables, leaving the dependent variable unaltered. Choose several handfuls of reexpressed predictor variables, which have the largest correlation coefficients with the unaltered dependent variable. Then, proceed as usual, invoking the bulging rule for exploring the best reexpressions of the dependent variable and the predictor variables. If reexpressing the dependent variable is appropriate, then apply Steps 1 or 2.

Meanwhile, there is an approach considered the most desirable for picking out the best reexpressed quantitative dependent variable. The approach is, however, neither practical nor easily assessable [3]. However, this approach is extremely tedious to perform by hand as is required because there is no commercially available software for its calculations. Its inaccessibility has no consequence to the model builders' quality of the model as the approach has not provided a noticeable improvement over the procedure in Step 3 for marketing applications where model implementation is at the decile level.

Now that I have examined all the issues surrounding the quantitative dependent variable, I return to a discussion of reexpressing the predictor variables of the mini case study, starting with INCOME and then AGE.

12.3.1.1 Reexpressing INCOME

I envision an underlying positively sloped straight line running through the 10 points in the PROFIT–INCOME smooth plot in Figure 12.2, even though the smooth trace reveals four severe kinks. Based on the general association test with the test statistic (TS) value of 6, which is *almost* equal to the cutoff score of 7, as presented in Chapter 3, I conclude there is an *almost noticeable* straight-line relationship between PROFIT and INCOME. The correlation coefficient for the relationship is a reliable $r_{PROFIT,\ INCOME}$ of 0.763. Notwithstanding these indicators of straightness, the relationship could use some straightening, but clearly, the bulging rule does not apply.

An alternative method for straightening data, especially characterized by nonlinearities, is the GenIQ Model, a machine-learning, genetic-based data mining method. As I extensively cover this model in Chapters 40 and 41, it suffices it to say that I use GenIQ to reexpress INCOME. The genetic structure, which represents the reexpressed INCOME variable, labeled gINCOME, is defined in Equation 12.3:

$$\text{gINCOME} = \sin(\sin(\sin(\sin(\text{INCOME})*\text{INCOME}))) + \log(\text{INCOME}) \qquad (12.3)$$

The structure uses the nonlinear reexpressions of the trigonometric sine function (four times) and the log (to base 10) function to loosen the "kinky" PROFIT–INCOME relationship. The relationship between PROFIT and INCOME (via gINCOME) is now smooth as the smooth trace reveals no serious kinks in Figure 12.3. Based on TS equal to 6, which

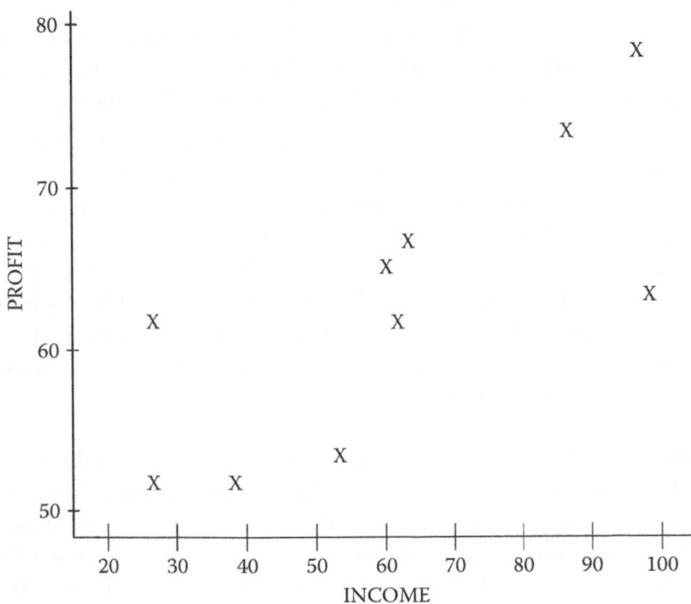

FIGURE 12.2
Plot of PROFIT and INCOME.

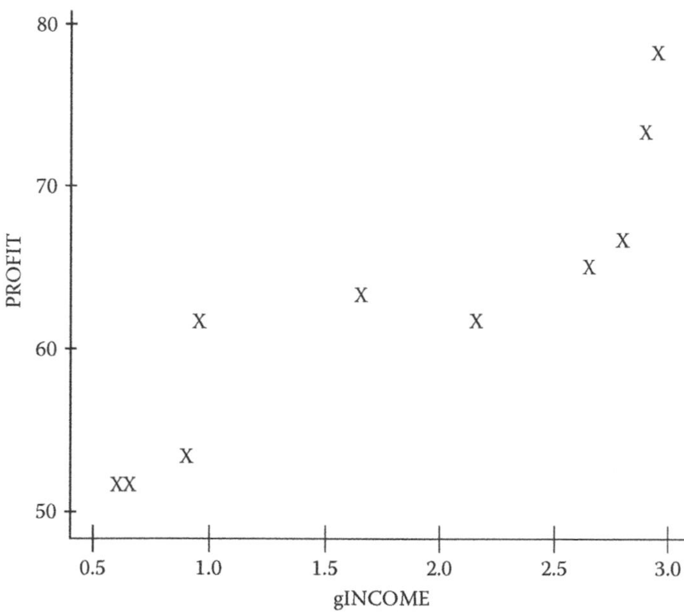

FIGURE 12.3
Plot of PROFIT and gINCOME.

again is almost equal to the cutoff score of 7, I conclude there is an almost noticeable straight-line PROFIT–gINCOME relationship, a nonrandom scatter about an underlying positively sloped straight line. The correlation coefficient for the reexpressed relationship is a reliable $r_{PROFIT,\ gINCOME} = 0.894$.

Visually, the effectiveness of the GenIQ procedure in straightening the data is obvious: the sharp peaks and valleys in the original PROFIT smooth plots versus the smooth wave of the reexpressed smooth plot. Quantitatively, the gINCOME-based relationship represents a noticeable improvement of 7.24% (= (0.894 – 0.763)/0.763) increase in correlation coefficient "points" over the INCOME-based relationship.

Two items are noteworthy: I previously invoked the statistical factoid that states a dollar-unit variable undergoes reexpression with the log function. Thus, it is not surprising that the genetically evolved structure gINCOME uses the log function. On logging the PROFIT variable, I concede that PROFIT could not benefit from a log reexpression due to the "mini" in the dataset (i.e., the small size of the data). So, I chose to work with PROFIT, not log of PROFIT, for the sake of simplicity (another EDA mandate, even for instructional purposes).

12.3.1.2 Reexpressing AGE

The stormy scatter of the 10-paired (PROFIT, AGE) points in the smooth plot in Figure 12.4 is an exemplary plot of no relationship between two variables. Not surprisingly, the TS value of 3 indicates there is no noticeable PROFIT–AGE relationship. Senselessly, I calculate the correlation coefficient for this nonexistent linear relationship: $r_{PROFIT,\ AGE} = -0.172$, which is clearly not meaningful. Obviously, the bulging rule does not apply.

I use GenIQ to reexpress AGE, labeled gAGE. The genetically based structure is in Equation 12.4:

$$gAGE = \sin(\tan(\tan(2*AGE) + \cos(\tan(2*AGE)))) \tag{12.4}$$

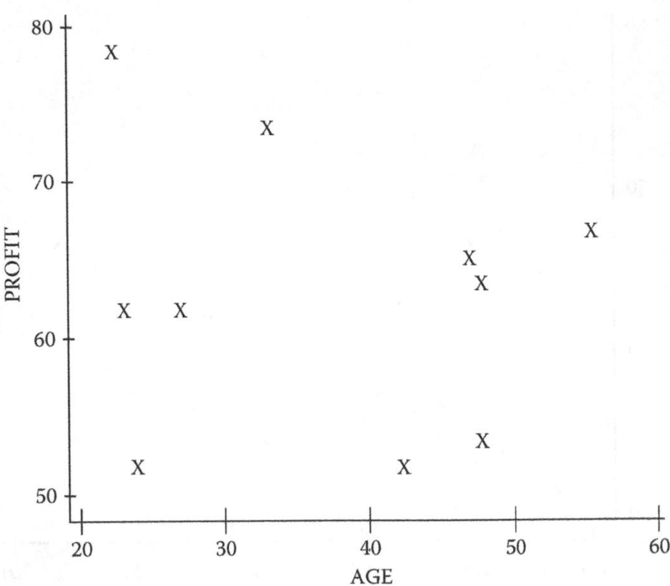

FIGURE 12.4
Plot of PROFIT and AGE.

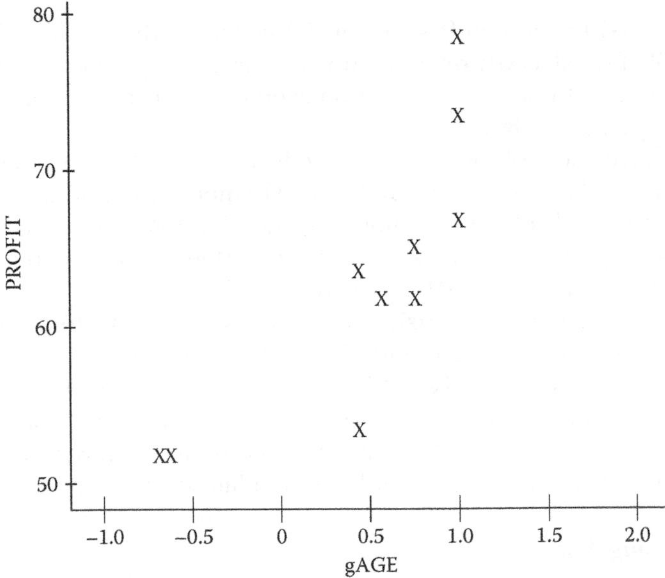

FIGURE 12.5
Plot of PROFIT and gAGE.

The structure uses the nonlinear reexpressions of the trigonometric sine, cosine, and tangent functions to calm the stormy-nonlinear relationship. The relationship between PROFIT and AGE (via gAGE) is indeed smooth, as the smooth trace reveals in Figure 12.5. There is an almost noticeable PROFIT–gAGE relationship with TS = 6, which favorably compares to the original TS of 3. The reexpressed relationship admittedly does not portray an exemplary straight line, but given its stormy origin, I see a beautiful positively sloped ray, not very straight, but trying to shine through. I consider the corresponding correlation coefficient $r_{PROFIT,\, gAGE} = 0.819$ as reliable and remarkable.

Visually, the effectiveness of the GenIQ procedure in straightening the data is obvious: the abrupt spikes in the original smooth plot of PROFIT and AGE versus the rising counterclockwise *wave* of the second smooth plot of PROFIT and gAGE. With enthusiasm and without quantitative restraint, the gAGE-based relationship represents a noticeable improvement—a whopping 376.2% (= (0.819 − 0.172)/0.172; disregarding the sign) improvement in correlation coefficient points over the AGE-based relationship. Because the original correlation coefficient is meaningless, the improvement percentage is also meaningless.

12.3.2 Plot of Smooth Predicted versus Actual

For a closer look at the detail of the strength (or weakness) of the gINCOME and gAGE structures, I construct the corresponding plots of PROFIT smooth predicted versus actual. The scatter about the 45° lines in the smooth plots for both gINCOME and gAGE in Figures 12.6 and 12.7, respectively, indicate a reasonable level of certainty in the reliability of the structures. In other words, both gINCOME and gAGE should be important variables for predicting PROFIT. The correlations between gINCOME-based predicted and actual smooth PROFIT values and between gAGE-based predicted and actual smooth PROFIT values have $r_{sm.PROFIT,\, sm.gINCOME}$ and $r_{sm.PROFIT,\, sm.gAGE}$ values equal to 0.894 and 0.819, respectively. (Why are these r values equal to $r_{PROFIT,\, INCOME}$ and $r_{PROFIT,\, AGE}$, respectively?)

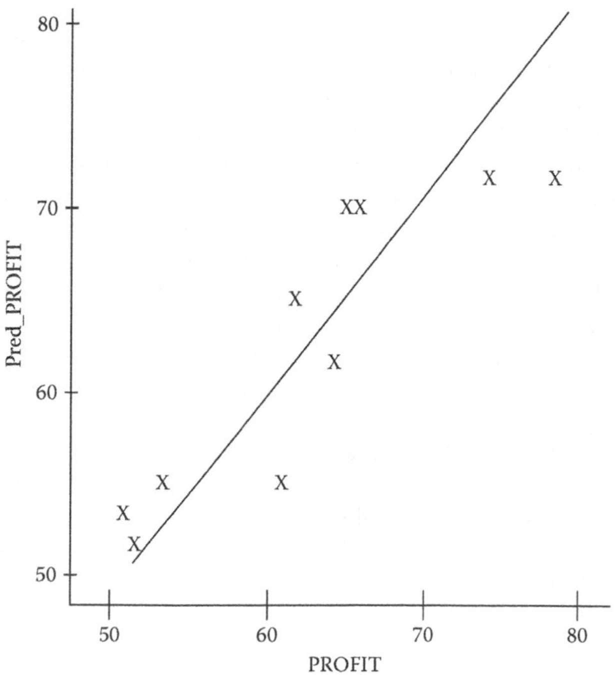

FIGURE 12.6
Smooth PROFIT predicted versus actual based on gINCOME.

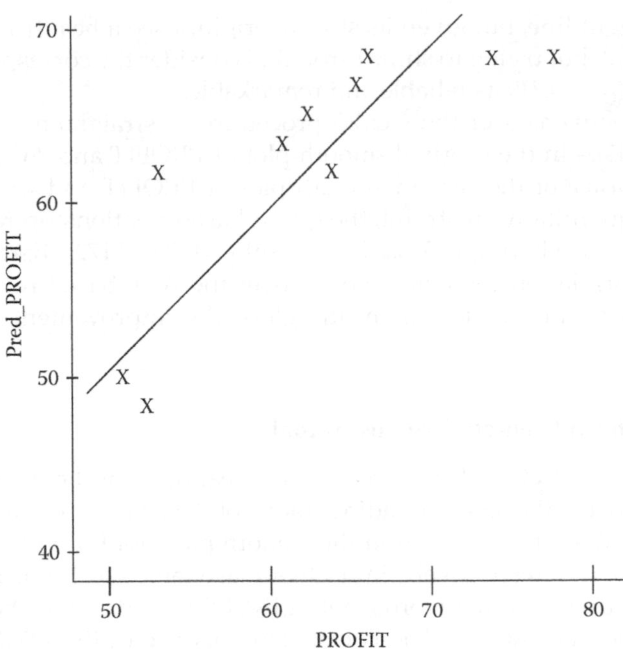

FIGURE 12.7
Smooth PROFIT predicted versus actual based on gAGE.

12.3.3 Assessing the Importance of Variables

As in the corresponding section of Chapter 10, the classical approach of assessing the statistical significance of a variable for model inclusion is the well-known null hypothesis significance testing procedure,* based on the reduction in prediction error (actual PROFIT minus predicted PROFIT) associated with the variable in question. The only difference between the discussions of the logistic regression in Chapter 10 is the apparatus used. The statistical apparatus of the formal testing procedure for ordinary regression consists of the sum of squares (total, due to regression, and due to error), the F statistic, degrees of freedom (df), and the p-value. The procedure uses the apparatus within a theoretical framework with weighty and untenable assumptions, which, from a purist's point of view, can cast doubt on findings of statistical significance. Even if findings of statistical significance are acceptably correct, this may not be of practical importance or have noticeable value to the study at hand. For the data miner with a pragmatist slant, the limitations and lack of scalability of the classical system of variable assessment cannot be overlooked, especially within big data settings. In contrast, the data mining approach uses the F statistic, R-squared, and degrees of freedom in an informal data-guided search for variables that suggest a noticeable reduction in prediction error. Note, the informality of the data mining approach calls for suitable change in terminology, from declaring a result as statistically significant to worthy of notice or noticeably important.

* "What If There Were No Significance Testing?" (on author's website, http://www.geniq.net/res/What-If-There-Were-No-Significance-Testing.html).

12.3.3.1 Defining the F Statistic and R-Squared

In data mining, the assessment of the importance of a subset of variables for predicting profit involves the notion of a noticeable reduction in prediction error due to the subset of variables. The assessment makes use of the F statistic, R-squared, and degrees of freedom always reported in the ordinary regression output. For the sake of reference, I provide their definitions and relationship with each other in Equations 12.5, 12.6, and 12.7.

$$\text{F} = \frac{\text{Sum of squares due to regression/df due to regression model}}{\text{Sum of squares due to error/df due to error in regression model}} \tag{12.5}$$

$$\text{R-squared} = \frac{\text{Sum of squares due to regression}}{\text{Total sum of squares}} \tag{12.6}$$

$$\text{F} = \frac{\text{R-squared/number of variables in model}}{(1 - \text{R-squared})/(\text{sample size} - \text{number of variables in model} - 1)} \tag{12.7}$$

For the sake of completion, I provide an additional statistic: the adjusted R-squared. R-squared is affected, among other things, by the ratio of the number of predictor variables in the model to the size of the sample. The larger the ratio, the greater the overestimation of R-squared is. Thus, the adjusted R-squared as defined in Equation 12.8 is not particularly useful in big data settings.

$$\text{Adjusted R-squared} = (1 - \text{R-squared}) = \frac{(\text{sample size} - 1)}{(\text{sample size} - \text{number of variables in model} - 1)} \tag{12.8}$$

In the following sections, I detail the decision rules for three scenarios for assessing the importance of variables (i.e., the likelihood the variables have some predictive power). In brief, the larger the F statistic, R-squared, and adjusted R-squared values, the more important the variables are in predicting profit.

12.3.3.2 Importance of a Single Variable

If X is the only variable considered for inclusion into the model, the decision rule for declaring X an important predictor variable is: If the F value due to X is greater than the *standard F value 4*, then X is an important predictor variable. Note, the decision rule only indicates that the variable has some importance, not how much importance. The decision rule implies: A variable X1 with an F value greater than variable X2's F value indicates that X1 has a greater *likelihood of some importance than does X1.* The decision rule does suggest X1 has greater importance than X2. If X is declared important, then testing X for entry into the model is conducted.

12.3.3.3 Importance of a Subset of Variables

When subset A consisting of k variables is the only subset considered for model inclusion, the decision rule for declaring subset A important is: If F/df* is greater than standard F value 4,

* df = number of predictor variables, k.

then subset A is an important subset of predictor variables and is worthy of inclusion in the model. As before, the decision rule only indicates that the subset has some importance, not how much importance.

12.3.3.4 Comparing the Importance of Different Subsets of Variables

Let subsets A and B consist of k and p variables, respectively. The number of variables in each subset does not have to be equal. If the number of variables is equal, then all but one variable can be the same in both subsets. Let F(k) and F(p) be the F values corresponding to the models with subsets A and B, respectively.

The decision rule for declaring which of the two subsets is more important (greater likelihood of some predictive power) in predicting profit is:

1. If F(k)/k is greater than F(p)/p, then subset A(k) is the more important predictor variable subset; otherwise, B(p) is the more important subset.
2. If F(k)/k and F(p)/p are equal or have comparable values, then both subsets are to be regarded tentatively as of comparable importance. The model builder should consider additional indicators to assist in the decision about which subset is better. It clearly follows from the decision rule that the model defined by the more important subset defines the better model. (Of course, the rule assumes that F/k and F/p are greater than the standard F value 4.)

Equivalently, the decision rule can use either R-squared or adjusted R-squared in place of F/df. The R-squared statistic is a friendly concept in that its values serve as indicators of the percentage of variation explained by the model.

12.4 Important Variables for Mini Case Study

I perform two ordinary regressions, regressing PROFIT on gINCOME and PROFIT on gAGE. The outputs are in Tables 12.5 and 12.6, respectively. The F values are 31.83 and 16.28, respectively, which are greater than the standard F value 4. Thus, both gINCOME and gAGE are declared important predictor variables of PROFIT.

TABLE 12.5

OLS Output: PROFIT with gINCOME

Source	df	Sum of Squares	Mean Square	*F* Value	Pr > *F*
Model	1	582.1000	582.1000	31.83	0.0005
Error	8	146.3000	18.2875		
Corrected total	9	728.4000			
		Root MSE	4.2764	*R*-square	0.7991
		Dependent mean	62.6000	Adj *R*-sq	0.7740
		Coeff var	6.8313		

Variable	df	Parameter Estimate	Standard Error	*t* Value	Pr > |*t*|
Intercept	1	47.6432	2.9760	16.01	<0.0001
gINCOME	1	8.1972	1.4529	5.64	0.0005

TABLE 12.6

OLS Output: PROFIT with gAGE

Source	df	Sum of Squares	Mean Square	*F* Value	Pr > *F*
Model	1	488.4073	488.4073	16.28	0.0038
Error	8	239.9927	29.9991		
Corrected total	9	728.4000			
		Root MSE	5.4771	*R*-square	0.6705
		Dependent mean	62.6000	Adj *R*-sq	0.6293
		Coeff var	8.7494		

Variable	df	Parameter Estimate	Standard Error	*t* Value	Pr > |*t*|
Intercept	1	57.2114	2.1871	26.16	<0.0001
gAGE	1	11.7116	2.9025	4.03	0.0038

12.4.1 Relative Importance of the Variables

Chapter 10 contains the same heading (Section 10.12), with only a minor variation on the statistic used. The *t statistic* as posted in ordinary regression output can serve as an indicator of the relative importance of a variable and for selecting the best subset and will be discussed next.

12.4.2 Selecting the Best Subset

The decision rules for finding the best subset of important variables are nearly the same as those discussed in Chapter 10; refer to Section 10.12.1. Point 1 remains the same for this discussion. However, the second and third points change as follows:

1. Select an initial subset of important variables.
2. For the variables in the initial subset, generate smooth plots and straighten the variables as required. The most noticeable handfuls of original and reexpressed variables make up the starter subset.
3. Perform the preliminary ordinary regression on the starter subset. Delete one or two variables with absolute t-statistic values less than the *t cutoff value* 2 from the model. Deletion of variables results in the first incipient subset of important variables. Note the changes to Points 4, 5, and 6 on the topic of this chapter.
4. Perform another ordinary regression on the incipient subset. Delete one or two variables with t values less than the t cutoff value 2 from the model. The data analyst can create an illusion of important variables appearing and disappearing with the deletion of different variables. The remainder of the discussion in Chapter 10 remains the same.
5. Repeat Step 4 until all retained predictor variables have comparable t values. This step often results in different subsets as the data analyst deletes judicially different pairings of variables.
6. Declare the best subset by comparing the relative importance of the different subsets using the decision rule in Section 12.3.3.4.

12.5 Best Subset of Variables for Case Study

I build a preliminary model by regressing PROFIT on gINCOME and gAGE. The output is in Table 12.7. The two-variable subset has an F/df value of 7.725 (= 15.45/2), which is greater than the standard F value 4. But, the t value for gAGE is 0.78 less than the t cutoff value (see bottom section of Table 12.7). If I follow Step 4, then I would have to delete gAGE, yielding a simple regression model with the lowly, albeit straight, predictor variable gINCOME. By the way, the adjusted R-squared is 0.7625 (after all, the entire mini sample is not big).

Before I dismiss the two-variable (gINCOME, gAGE) model, I construct the smooth residual plot in Figure 12.8 to determine the quality of the predictions of the model. The smooth residual plot is declared to be equivalent to the null plot based on the general association test (TS = 5). Thus, the overall quality of the predictions is considered good. That is, on average, the predicted PROFIT is equal to the actual PROFIT. Regarding the descriptive statistics for the smooth residual plot, for the smooth residual, the minimum

TABLE 12.7

OLS Output: PROFIT with gINCOME and gAGE

Source	df	Sum of Squares	Mean Square	*F* Value	Pr > *F*
Model	2	593.8646	296.9323	15.45	0.0027
Error	7	134.5354	19.2193		
Corrected total	9	728.4000			
		Root MSE	4.3840	*R*-squared	0.8153
		Dependent mean	62.6000	Adj *R*-sq	0.7625
		Coeff var	7.0032		

Variable	df	Parameter Estimate	Standard Error	*t* Value	Pr > |*t*|
Intercept	1	49.3807	3.7736	13.09	<0.0001
gINCOME	1	6.4037	2.7338	2.34	0.0517
gAGE	1	3.3361	4.2640	0.78	0.4596

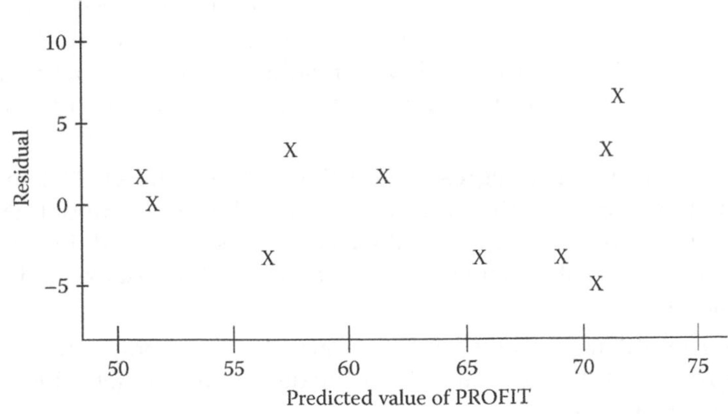

FIGURE 12.8
Smooth residual plot for (gINCOME, gAGE) model.

and maximum values and range are –4.567, 6.508, and 11.075, respectively; the standard deviation of the smooth residuals is 3.866.

12.5.1 PROFIT Model with gINCOME and AGE

With a vigilance in explaining the unexpected, I suspect the reason for the relative nonimportance of gAGE (i.e., gAGE is not important in the presence of gINCOME) is the strong correlation of gAGE with gINCOME: $r_{gINCOME,\, gAGE} = 0.839$. The strong correlation supports my contention but does not confirm it.

Given the preceding, I build another two-variable model regressing PROFIT on gINCOME and AGE. The output is in Table 12.8. The (gINCOME, AGE) subset has an F/df value of 12.08 (= 24.15/2), which is greater than the standard F value 4. Statistic happy, I see the t values for both variables are greater than the t cutoff value 2. I cannot overlook the fact that the raw variable AGE, which by itself is not important, now has relative importance in the presence of gINCOME. (More about this "phenomenon" is presented at the end of the chapter.) Thus, the evidence is that the subset of gINCOME and AGE is better than the original (gINCOME, gAGE) subset. By the way, the adjusted R-squared is 0.8373, representing a 9.81% (= (0.8373 – 0.7625)/0.7625) improvement in adjusted R-squared points over the original variable adjusted R-squared.

The smooth residual plot for the (gINCOME, AGE) model in Figure 12.9 is declared to be equivalent to the null plot based on the general association test with TS = 4. Thus, there is an indication that the overall quality of the predictions is good. Regarding the descriptive statistics for the two-variable smooth residual plot, for the smooth residual, the minimum and maximum values and range are –5.527, 4.915, and 10.442, respectively, and the standard deviation of the smooth residual is 3.200.

To obtain a further indication of the quality of the predictions of the model, I construct the plot of the smooth actual versus smooth predicted for the (gINCOME, AGE) model in Figure 12.10. The smooth plot is acceptable with minimal scatter of the 10 smooth points about the 45° line. The correlation between smooth actual versus predicted PROFIT based on gINCOME and AGE has $r_{sm.gINCOME,\, sm.AGE} = 0.93$. (This value is the square root of R-squared for the model. Why?)

For a point of comparison, I construct the plot of smooth actual versus smooth predicted for the (gINCOME, gAGE) model in Figure 12.11. The smooth plot is acceptable

TABLE 12.8

OLS Output: PROFIT with gINCOME and AGE

Source	df	Sum of Squares	Mean Square	*F* Value	Pr > *F*
Model	2	636.2031	318.1016	24.15	0.0007
Error	7	92.1969	13.1710		
Corrected Total	9	728.4000			
		Root MSE	3.6291	*R*-squared	0.8734
		Dependent mean	62.6000	Adj *R*-sq	0.8373
		Coeff var	5.79742		

Variable	df	Parameter Estimate	Standard Error	*t* Value	Pr > |*t*|
Intercept	1	54.4422	4.1991	12.97	< 0.0001
gINCOME	1	8.4756	1.2407	6.83	0.0002
AGE	1	–0.1980	0.0977	–2.03	0.0823

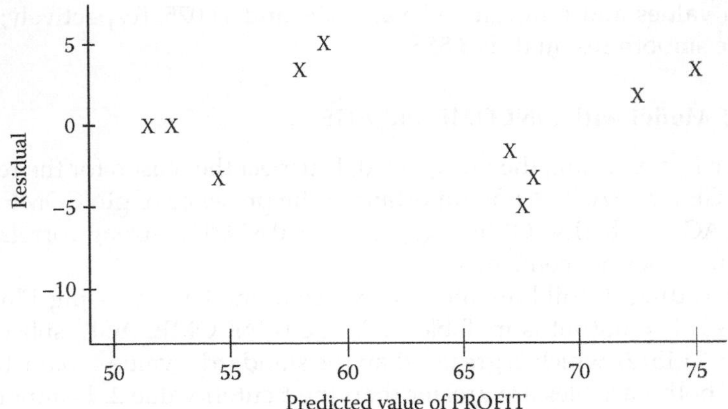

FIGURE 12.9
Smooth residual plot for (gINCOME, AGE) model.

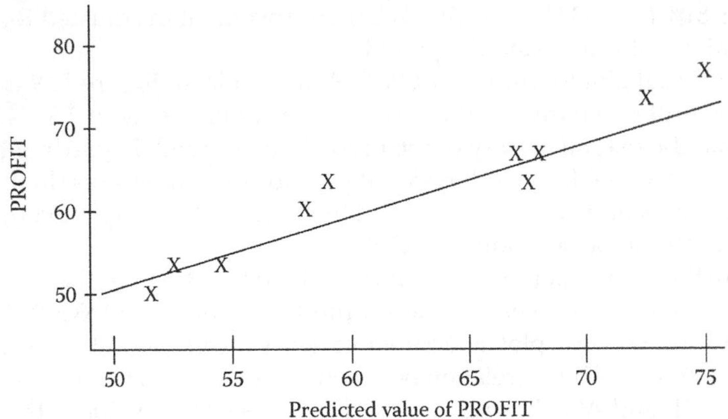

FIGURE 12.10
Smooth actual versus predicted plot for (gINCOME, AGE) model.

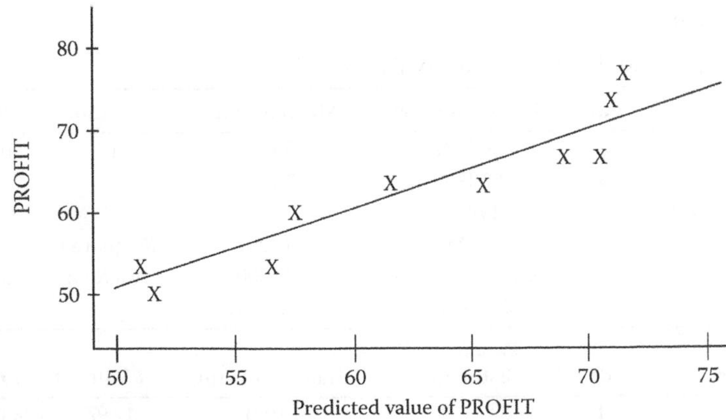

FIGURE 12.11
Smooth actual versus predicted plot for (gINCOME, gAGE) model.

TABLE 12.9

Comparison of Vital Statistics for Two PROFIT Models

				Smooth Residual		
Model	**Predictor Variables**	***F* Value**	***t* Value**	**Range**	**StdDev**	**Adjusted *R*-Square**
First	gINCOME, gAGE	Greater than cutoff value	Greater than cutoff value for only gINCOME	11.075	3.866	0.7625
Second	gINCOME, AGE	Greater than cutoff value	Greater than cutoff value for both variables	10.442	3.200	0.8373
Indication	Improvement of 2nd model over 1st model	NA	Because t value for gAGE is less than t cutoff value, gAGE contributes "noise" in model, as evidenced in range, StdDev, and adjusted R-square	–5.7%	–17.2%	9.8%

with minimal scatter of the 10 smooth points about the 45° line, with a noted exception of some wild scatters for PROFIT values greater than \$65. The correlation between smooth actual versus smooth predicted PROFIT based on gINCOME and gAGE has $r_{sm.gINCOME,\, sm.gAGE} = 0.90$. (This value is the square root of R-squared for the model. Why?)

12.5.2 Best PROFIT Model

To decide which of the two PROFIT models is better, I put the vital statistics of the preceding analyses in Table 12.9. Based on the consensus of a committee of one, I prefer the gINCOME–AGE model to the gINCOME–gAGE model because there are noticeable indications that the predictions of the gINCOME–AGE model are better than the gINCOME–gAGE model. The former model offers a 9.8% increase in the adjusted R-squared (less bias), a 17.2% decrease in the smooth residual standard deviation (more stable), and a 5.7% decrease in the smooth residual range (more stable).

12.6 Suppressor Variable AGE

A variable whose behavior is like that of AGE—poorly correlated with the dependent variable Y but becomes important by its inclusion in a model for predicting Y—is known as a *suppressor variable* [1,2]. The consequence of a suppressor variable is that it increases the R-squared of the model.

I explain the behavior of the suppressor variable within the context of the mini case study. The presence of AGE in the model removes or suppresses the information (variance) in gINCOME that is not related to the variance in PROFIT; that is, AGE suppresses the unreliable noise in gINCOME. The suppression due to AGE renders the AGE-adjusted variance in gINCOME more reliable or potent for predicting PROFIT.

I analyze the paired correlations among the three variables to exactly clarify what AGE is doing. Recall, squaring the correlation coefficient represents the shared variance between the two variables under consideration. The paired bits of information are in Table 12.10.

TABLE 12.10

Comparison of Pairwise Correlations among PROFIT, AGE, and gINCOME

Correlation Pair	Correlation Coefficient	Shared Variance (%)
PROFIT and AGE	–0.172	3
PROFIT and gINCOME	0.894	80
PROFIT and AGE in the presence of gINCOME	0.608	37
PROFIT and gINCOME in the presence of AGE	0.933	87

I know from the prior analysis that PROFIT and AGE have no noticeable relationship; their shared variance of 3% confirms this. I also know that PROFIT and gINCOME do have a noticeable relationship; their shared variance of 80% confirms this as well. In the presence of AGE, the relationship between PROFIT and gINCOME, specifically, the relationship between PROFIT and gINCOME adjusted for AGE has a shared variance of 87%. This represents an improvement of 8.75% (= (0.87 – 0.80/0.80) in shared variance. This "new" variance is now available for predicting PROFIT, increasing the R-squared (from 79.91% to 87.34%).

It is a pleasant surprise in several ways that AGE turns out to be a suppressor variable. First, suppressor variables occur most often in big data settings, not often with small data, and are truly unexpected with mini data. Second, the suppressor variable scenario serves as object lessons for the EDA paradigm: dig, dig, dig into the data, and you will find gold or some reward for your effort. Third, the suppressor variable scenario is a small reminder of a big issue. The model builder must not rely solely on predictor variables, highly correlated with the dependent variable, but also must consider the poorly correlated predictor variables as they are a great source of latent predictive importance.

12.7 Summary

The ordinary regression model serves as the workhorse of profit modeling as it has been in steady use for 200-plus years. As such, I illustrate in an orderly and detailed way the essentials of ordinary regression. Moreover, I show the enduring usefulness of this popular analysis and modeling technique as it works well within the EDA/data mining paradigm of today.

I first illustrate the rudiments of the ordinary regression model by discussing the SAS program for building and scoring an ordinary regression model. The program is a welcome addition to the toolkit of techniques used by model builders working on predicting a quantitative dependent variable.

Then, I discuss ordinary regression modeling with mini data. I use this extremely small dataset not only to make the discussion of the data mining techniques tractable but also to emphasize two aspects of data mining. First, data mining techniques of great service should work as well with big data as with small data. Second, every fruitful effort of data mining on small data is evidence that big data are not always necessary to uncover structure in the data.

I use the mini case study as the vehicle for introducing a host of data mining techniques, inspired by the EDA paradigm, tailored to ordinary regression modeling. The techniques involve the concept of importance, not significance, of individual predictor variables and subsets of predictor variables as well as the indispensable use of smooth plots. (The data mining techniques, discussed in the logistic regression framework of Chapter 10, carry over with minor modification to ordinary regression.)

Within my illustration of the case study is a pleasant surprise—the existence of a suppressor variable. A variable whose behavior is poorly correlated with the dependent variable but becomes important by its inclusion in a model for predicting the dependent variable is known as a suppressor variable. The consequence of a suppressor variable is that it increases the R-square of the model. A suppressor variable occurs most often in big data settings, not often with small data, and is truly unexpected with mini data. The suppressor variable scenario serves as an object lesson for the EDA paradigm: Dig deeply into the data and you will find a reward for your effort. And, the suppressor variable scenario is a small reminder of a bigger issue: The model builder must not rely solely on predictor variables, highly correlated with the dependent variable, but also should consider the poorly correlated predictor variables as they are a great source of latent predictive importance.

References

1. Horst, P., The role of predictor variables which are independent of the criterion, *Social Science Research Bulletin*, 48, 431–436, 1941.
2. Conger, A.J., A revised definition for suppressor variables: A guide to their identification and interpretation, *Educational and Psychological Measurement*, 34, 35–46, 1974.
3. Tukey and Mosteller., Data Analysis and Regression, *Graphical Fitting by Stages*, 271–279.

13

Variable Selection Methods in Regression: Ignorable Problem, Notable Solution

13.1 Introduction

Variable selection in regression—identifying the best subset of many variables to include in a model—is arguably the hardest part of the model-building process. Many variable selection methods exist. Many statisticians know them, but few know they produce poorly performing models. The variable selection methods are a miscarriage of statistics because they debase sound statistical theory into a misguided pseudo-theoretical foundation. The goal of this chapter is twofold: (1) resurface the scope of literature on the weaknesses of variable selection methods and (2) enliven a notable solution for defining a substantial performing regression model anew. To achieve my goal tactically, I divide the chapter into two objectives. First, I review the five frequently used variable selection methods. Second, I present Tukey's exploratory data analysis (EDA) relevant to the titled topic: the natural seven-step cycle of statistical modeling and analysis (previously discussed in Section 1.1). The seven-step cycle serves as a notable solution to variable selection in regression. I feel that newcomers to Tukey's EDA need the seven-step cycle introduced within the narrative of Tukey's analytic philosophy. Accordingly, I enfold the solution with front and back matter: the essence of EDA and the EDA school of thought, respectively. John W. Tukey (1915–2000) was a megacontributor to the field of statistics and was a humble, unpretentious man, as he always considered himself as a data analyst. Tukey's seminal book, *Exploratory Data Analysis* [1], is uniquely known by the book's initialed title—EDA.

13.2 Background

Classic statistics dictates that the statistician set about dealing with a given problem with a prespecified procedure designed for that problem. For example, solving the problem of predicting a continuous dependent variable (e.g., profit) uses the ordinary least squares (OLS) regression model *along with checking* the well-known underlying OLS assumptions [2]. At hand, there are *several* candidate predictor variables, allowing a workable task for the statistician to check assumptions (e.g., predictor variables are uncorrelated with errors). Likewise, the dataset has a *practicable* number of observations, making it also a workable task for the statistician to check assumptions (e.g., the errors are uncorrelated). As well, the statistician can perform the *well-regarded—yet often discarded*—EDA to examine and apply the appropriate remedies for individual records that contribute to sticky data characteristics (e.g., gaps, clumps, and outliers). It is important that EDA allows the statistician to assess whether a given variable, say, X needs a transformation/reexpression—for example, log(X), sin(X), or 1/X. The traditional variable selection methods cannot perform such transformations or a priori construction of new variables from the original variables.* The *inability* to construct new variables is a serious weakness of the variable selection methodology [1].

Today, building an OLS regression model or a logistic regression model (LRM; the dependent variable is binary) is problematic because of the size of the datasets used. Model builders work on *big data*—consisting of a teeming multitude of variables and an army of observations. The workable tasks are no longer feasible. Model builders cannot surefootedly use OLS regression and LRM on big data as the two statistical regression models were conceived, tested, and experimented within the small-data setting of more than 60 to 200-plus years ago for LRM and OLS regression, respectively. The theoretical regression foundation and the tool of significance testing† employed on big data are without statistical binding force. Thus, fitting big data to a prespecified small-framed model produces a skewed model with doubtful interpretability and questionable results.

In the late 1960s and early 1970s, according to folklore, the practice of variable selection methods began. During that period, data were small and slowly growing into the early size of big data today. With only a single bibliographic citation ascribing variable selection methods to unsupported notions, I believe a reasonable scenario of the genesis of the methods was as follows [3]: College statistics nerds (intelligent thinkers) and computer science geeks (intelligent doers) lay out the variable selection methodology using a *trinity of selection components:*

1. Statistical tests (e.g., F, chi-square, and t-test) and significance testing
2. Statistical criteria (e.g., R-squared [R-sq], adjusted R-sq, Mallows' C_p, and MSE [mean squared error]) [4]
3. Statistical stopping rules (e.g., p-value flags for variable entry/deletion/staying in a model)

The newly developed variable selection methods were on bearing soil of expertness and adroitness in computer-automated misguided statistics. The trinity distorts its components' original theoretical and inferential meanings when framed within the newborn methods. The statistician executing the computer-driven trinity of statistical apparatus in a seemingly intuitive and insightful way gave proof—*face validity*—that the problem of variable selection (also known as subset selection) was solved (at least to the uninitiated statistician).

The newbie subset selection methods initially enjoyed wide acceptance with extensive use and still do presently. Statisticians build *at-risk* accurate and stable models—either *unknowingly* using these unconfirmed methods or *knowingly* exercising these methods because *they know not what else to do*. It was not long before the weaknesses of these methods, some contradictory, generated many commentaries in the literature. I itemize nine ever-present weaknesses for two of the traditional variable selection methods,

* The variable selection methods do not include the new breed of methods that have data mining capability.

† "What If There Were No Significance Testing?" (on the author's website, http://www.geniq.net/res/What-If-There-Were-No-Significance-Testing.html).

all-subset and *stepwise* (SW). I concisely describe the five frequently used variable selection methods in the next section.

1. For all-subset selection with more than 40 variables [3]:
   a. The number of possible subsets can be huge.
   b. Often, there are several good models, although some are unstable.
   c. The best X variables may be no better than random variables if the sample size is relatively small compared to the number of all variables.
   d. The regression statistics and regression coefficients are biased.
2. All-subset selection regression can yield models that are too small [5].
3. The number of candidate variables and not the number in the final model is the number of degrees of freedom to consider [6].
4. The data analyst knows more than the computer—and failure to use that knowledge produces inadequate data analysis [7].
5. SW selection yields confidence limits that are far too narrow [8].
6. Regarding frequency of obtaining authentic and noise variables: The degree of correlation among the predictor variables affected the frequency with which authentic predictor variables found their way into the final model. The number of candidate predictor variables affected the number of noise variables that gained entry to the model [9].
7. SW selection will not necessarily produce the best model if there are redundant predictors (common problem) [10].
8. There are two distinct questions here: (a) When is SW selection appropriate? (b) Why is it so popular [11]?
9. Regarding Question 8b, there are two groups that are inclined to favor its usage. One group consists of individuals with little formal training in data analysis. This group confuses knowledge of data analysis with knowledge of the syntax of SAS, SPSS, and the like. This group believes that *if SW is in a program, it has to be good and better than actually thinking about what the data might look like.* The second group consists of right-thinking, well-trained data analysts. They believe in statistics to the extent a suitable computer program can objectively make substantive inferences without active consideration of the underlying hypotheses. *Stepwise selection* is the parent of this line of *blind data analysis*.*

Currently, there is burgeoning research that continues the original efforts of subset selection by shoring up its pseudo-theoretical foundation. It follows a line of examination that adds assumptions and makes modifications for eliminating the weaknesses. As the traditional methods undergo constant revision, there are innovative approaches with starting points far afield from their traditional counterparts. Under development are freshly minted methods such as the *enhanced variable selection method* of the GenIQ Model [12–15].

* Comment without an attributed citation: Frank Harrell, professor of biostatistics and department chair, Department of Biostatistics, Vanderbilt University School of Medicine, 2009.

13.3 Frequently Used Variable Selection Methods

Variable selection in regression—identifying the best subset of many variables to include in a model—is arguably the hardest part of the model-building process. Many variable selection methods exist because they provide a solution to one of the most important problems in statistics [16].* Many statisticians know them, but few know they produce poorly performing models. The deficient variable selection methods are a miscarriage of statistics, debasing sound statistical theory into a misguided pseudo-theoretical foundation. Execution of the methods depends on computer-intensive search heuristics guided by rules of thumb. Each method uses a unique trio of elements, one from each component of the trinity of selection components.† Different sets of elements typically produce different subsets. The number of variables in common with the different subsets is small, and the sizes of the subsets can vary considerably.

An alternative view of the problem of variable selection is to examine certain subsets and select the best subset that either maximizes or minimizes an appropriate criterion. Two subsets are obvious: the best single variable and the complete set of variables. The problem lies in selecting an intermediate subset that is better than both of these extremes. Therefore, the issue is how to find the *necessary variables* among the complete set of variables by deleting both *irrelevant variables* (variables not affecting the dependent variable) and *redundant variables* (variables not adding anything to the prediction of the dependent variable) [17].

I review five frequently used variable selection methods found in major statistical software packages.‡ The test statistic (TS) for the first three methods uses either the F statistic for a continuous dependent variable or the G statistic for a binary dependent variable. The TS for the fourth method is either R-sq for a continuous dependent variable or the score statistic for a binary dependent variable. The last method uses one of the following criteria: R-sq, adjusted R-sq, or Mallows' C_p.

1. *Forward selection (FS)*: This method adds variables to the model until no remaining variable (outside the model) can add anything significant to the dependent variable. FS begins with no variable in the model. For each variable, the TS, a measure of the contribution of the variable to the model, is calculated. The variable with the largest TS value that is greater than a preset value, C, goes into the model. Then, TS for the variables remaining is calculated, and the evaluation process repeats. Thus, variables are added to the model one by one until no remaining variable produces a TS value that is greater than C. Once a variable is in the model, it remains there.
2. *Backward elimination (BE)*: This method deletes variables one by one from the model until all remaining variables contribute something significant to the dependent variable. BE begins with a model that includes all variables. Variables are then deleted from the model one by one until all the variables remaining in the model have TS values greater than C. At each step, the variable showing the smallest

* Comment without an attributed citation: In 1996, Tim C. Hesterberg, research scientist at Insightful Corporation, asked Brad Efron for the most important problems in statistics, fully expecting the answer to involve the bootstrap given Efron's status as inventor. Instead, Efron named a single problem, variable selection in regression. This entails selecting variables from among a set of candidate variables, estimating parameters for those variables, and inference—hypotheses tests, standard errors, and confidence intervals.

† Other criteria are based on information theory and Bayesian rules.

‡ SAS/STAT Manual. See PROC REG and PROC LOGISTIC, support.sas.com, 2011.

contribution to the model (i.e., with the smallest TS value that is less than C) is subject to deletion.

3. *Stepwise (SW)*: This method is a modification of the FS approach and differs in that variables already in the model do not necessarily stay. As in FS, SW adds variables to the model one at a time. Variables that have a TS value greater than C go into the model. After a variable enters a model, however, SW looks at all the variables already included to delete any variable that does not have a TS value greater than C.
4. *R-squared (R-sq)*: This method finds several subsets of different sizes that best predict the dependent variable. R-sq finds subsets of variables that best predict the dependent variable based on the appropriate TS. The best subset of size k has the largest TS value. For a continuous dependent variable, TS is the popular measure R-sq, the coefficient of multiple determination, which measures the proportion of the *explained* variance in the dependent variable by the multiple regression. For a binary dependent variable, TS is the theoretically correct but less-known Score statistic.* R-sq finds the best one-variable model, the best two-variable model, and so forth. However, it is unlikely that one subset will stand out as clearly the best because TS values often bunch together. For example, they are equal in value when rounding at the, say, third place after the decimal point.† R-sq generates some subsets of each size, which allows the user to select a subset, possibly using nonstatistical criteria.
5. *All-possible subsets*: This method builds all one-variable models, all two-variable models, and so on until the last creation of the all-variable model. The method requires intensive computer power as it produces many models and the selection of any one of the criteria: R-sq, adjusted R-sq, or Mallows' C_p.

13.4 Weakness in the Stepwise

An ideal variable selection method for regression models would find one or more subsets of variables that produce an *optimal* model‡. The objective of the ideal method states that the resultant models include the following elements: accuracy, stability, parsimony, interpretability, and lack of bias in drawing inferences. Needless to say, the methods enumerated do not satisfy most of these elements. Each method has at least one drawback specific to its selection criterion. In addition to the nine weaknesses mentioned, I itemize a compiled list of weaknesses of the *most popular SW* method.§

1. It yields R-sq values that are badly biased toward high.
2. The F and chi-squared statistics quoted next to each variable on the printout do not have the claimed distribution.

* R-squared theoretically is not the appropriate measure for a binary dependent variable. However, many analysts use it with varying degrees of success.

† For example, consider two TS values: 1.934056 and 1.934069. These values are equal when rounding occurs at the third place after the decimal point: 1.934.

‡ Even if there were a perfect variable selection method, it is unrealistic to believe there is a unique best subset of variables.

§ Comment without an attributed citation: Frank Harrell, professor of biostatistics and department chair, Department of Biostatistics, Vanderbilt University School of Medicine, 2010.

3. The method yields confidence intervals for effects and predicted values that are falsely narrow.
4. It yields p-values that do not have the proper meaning, and the proper correction for them is a very difficult problem.
5. It gives biased regression coefficients that need shrinkage (the coefficients for remaining variables are too large).
6. It has severe problems in the presence of collinearity.
7. SW uses methods (e.g., F statistic) intended to be used to test prespecified hypotheses.
8. Increasing the sample size does not help very much.
9. It allows us not to think about the problem.
10. It uses a lot of paper. (This item is no longer relevant as the output can go to the cloud.)
11. The number of candidate predictor variables affect the number of noise variables that gain entry to the model.

I add to the tally of weaknesses by stating common weaknesses in regression models as well as those specifically related to the OLS regression model and LRM:

> The everyday variable selection methods in the regression model result typically in models having too many variables, an indicator of overfitting. The prediction errors, which are inflated by outliers, are not stable. Thus, model implementation results in unsatisfactory performance. For OLS regression, it is well-known that in the absence of normality or absence of linearity assumption or outlier presence in the data, variable selection methods perform poorly. For logistic regression, the reproducibility of the computer-automated variable selection models is poor. The variables selected as predictor variables in the models are sensitive to unaccounted sample variation in the data.

Given the litany of weaknesses cited, the lingering question is: Why do statisticians use variable selection methods to build regression models? To paraphrase Mark Twain: "Get your [data] first, and then you can distort them as you please" [18]. My answer is: "Modeler builders use variable selection methods every day because they can." As a counterpoint to the absurdity of "because they can," I enliven anew Tukey's solution of the natural seven-step cycle to define a substantial performing regression model. I feel that newcomers to Tukey's EDA need the seven-step cycle introduced within the narrative of Tukey's analytic philosophy. Accordingly, I enfold the solution with front and back matter—the essence of EDA and the EDA school of thought, respectively. I delve into the trinity of Tukey's masterwork. But, first I discuss an enhanced variable selection method for which I might be the only exponent for appending this method to the current baseless arsenal of variable selection.

13.5 Enhanced Variable Selection Method

In lay terms, the statement of the variable selection problem in regression is:

> Find the best combination of the original variables to include in a model. The variable selection method neither states nor implies that it has an attribute to concoct new variables stirred up by mixtures of the original variables.

The attribute—data mining—is either overlooked, perhaps because it is reflective of the simple-mindedness of the problem solution at the onset, or currently sidestepped because the problem is too difficult to solve. A variable selection method without a data mining attribute obviously hits a wall, beyond which it would otherwise increase the predictiveness of the technique. In today's terms, the variable selection methods are *without* data mining capability. They cannot dig the data for the mining of potentially important new variables. (This attribute, which has never surfaced in my literature search, is a partial mystery to me.) Accordingly, I put forth a definition of an enhanced variable selection method:

> An enhanced variable selection method is one that identifies a subset that consists of the original variables *and* data-mined variables, whereby *the latter are a result of the data mining attribute of the method itself.*

The following five discussion points clarify the attribute weakness and illustrate the concept of an enhanced variable selection method:

1. Consider the complete set of variables X_1, X_2, …, X_{10}. Any of the current variable selection methods finds the best combination of the original variables (say X_1, X_3, X_7, X_{10}), but it can never automatically transform a variable (say transform X_1 to log X_1) if it were needed to increase the *information content* (*predictive power*) of that variable. Furthermore, none of the methods can generate a reexpression of the original variables (perhaps X_3/X_7) if the constructed variable (*structure*) were to offer more predictive power than the original component variables combined. In other words, current variable selection methods cannot find an *enhanced subset*, which needs, say, to include transformed and newly constructed variables (possibly X_1, X_3, X_7, X_{10}, log X_1, X_3/X_7). A subset of variables without the potential of new structure offering more predictive power limits the model builder in building the best model.
2. Specifically, the current variable selection methods fail to identify a data structure of the type discussed here: *transformed variables* with a *preferred* shape. A variable selection procedure should have the ability to transform an individual variable, if necessary, to induce a symmetric distribution. Symmetry is the preferred shape of an individual variable. For example, the workhorse of statistical measures—the mean and variance—is based on a symmetric distribution. A skewed distribution produces inaccurate estimates for means, variances, and related statistics, such as the correlation coefficient. Symmetry facilitates the interpretation of the effect of the variable in an analysis. A skewed distribution is difficult to examine because most of the observations are bunched together at one end of the distribution. Modeling and analyses based on skewed distributions typically provide a model with doubtful interpretability and questionable results.
3. The current variable selection method also should have the ability to *straighten* nonlinear relationships. A linear or straight-line relationship is the *preferred* shape when considering two variables. A straight-line relationship between independent and dependent variables is an assumption of the popular statistical linear regression models (e.g., OLS regression and LRM). (Recall, a linear model

is defined as a sum of weighted variables, such as $Y = b_0 + b_1{*}X_1 + b_2{*}X_2 + b_3{*}X_3$.*) Moreover, straight-line relationships *among all* the independent variables constitute a *desirable property* [19]. In brief, straight-line relationships are easy to interpret: A unit of increase in one variable produces an expected constant increase or decrease in a second variable.

4. *Constructed variables* are mathematical mixtures of original variables and simple arithmetic functions. A variable selection method should have the ability to construct simple reexpressions of the original variables. Sum, difference, ratio, or product variables potentially offer more information than the original variables themselves. For example, when analyzing the efficiency of an automobile engine, two important variables are miles traveled and fuel used (gallons). However, it is well-known that the ratio variable of miles per gallon is the best variable for assessing the performance of the engine.
5. *Constructed variables* are mathematical mixtures of original variables using a set of functions (e.g., arithmetic, trigonometric, or Boolean functions). A variable selection method should have the ability to construct complex reexpressions with mathematical functions to capture the complex relationships in the data and offer potentially more information than the original variables themselves. In an era of data warehouses and the Internet, big data consisting of hundreds of thousands to millions of individual records and hundreds to thousands of variables are commonplace. Relationships among many variables produced by so many individuals are sure to be complex and beyond the simple straight-line pattern. Discovering the mathematical expressions of these relationships, although difficult with theoretical guidance, should be the hallmark of a high-performance variable selection method. For example, consider the well-known relationship among three variables: the lengths of the three sides of a right triangle. A powerful variable selection procedure would identify the relationship among the sides, even in the presence of measurement error: The longer side (diagonal) is the square root of the sum of squares of the two shorter sides.

In sum, the attribute weakness implies that a variable selection method should have the ability to generate an enhanced subset of candidate predictor variables.

13.6 Exploratory Data Analysis

I present the trinity of Tukey's EDA that is relevant to the titled topic: (1) the essence of EDA, (2) the natural seven-step cycle of statistical modeling and analysis, serving as a notable solution to variable selection in regression, and (3) the EDA school of thought.

1. The essence of EDA is best described in Tukey's own words: "Exploratory data analysis is detective work—numerical detective work—or counting detective work—or graphical detective work. … [It is] about looking at data to see what

* The weights or coefficients (b_0, b_1, b_2, and b_3) are derived to satisfy some criterion, such as minimize the mean squared error used in ordinary least squares regression or minimize the joint probability function used in logistic regression.

it seems to say. "[1, pp.1] It concentrates on simple arithmetic and easy-to-draw pictures. It regards whatever appearances we have recognized as partial descriptions and tries to look beneath them for new insights." EDA includes the following characteristics:

a. *Flexibility*—Techniques with greater flexibility to delve into the data
b. *Practicality*—Advice for procedures of analyzing data
c. *Innovation*—Techniques for interpreting results
d. *Universality*—Use all statistics that apply to analyzing data
e. *Simplicity*—Above all, the belief that simplicity is the golden rule

The computational strength of the personal computer (PC) has empowered the statistician. Without the PC, the statistician would not be able to perform the natural seven-step cycle of statistical modeling and analysis. The PC and the analytical cycle comprise the perfect pairing as long as the steps follow the prescribed order. The information obtained from one step is used in the next succeeding step. Unfortunately, statisticians are human and succumb to taking shortcuts through the seven-step cycle. They ignore the cycle and focus solely on the sixth step. However, diligent statistical endeavor requires steady execution of the sequential tasks, as described in the originally outlined seven-step cycle.*

2. The natural seven-step cycle of statistical modeling and analysis consists of the following analytical techniques and criteria:

   a. *Definition of the problem*: Determining the best way to tackle the problem is not always obvious. Management objectives are often expressed qualitatively, in which case the selection of the outcome or target (dependent) variable is subjectively biased. When the objectives are clearly stated, the appropriate dependent variable is often not available, in which case a surrogate must be used.
   b. *Determining technique*: The technique first selected is often the one with which the data analyst is most comfortable; it is not necessarily the best technique for solving the problem.
   c. *Use of competing techniques*: Applying alternative techniques increases the odds that a thorough analysis is conducted.
   d. *Rough comparisons of efficacy*: Comparing variability of results across techniques can suggest additional techniques or the deletion of alternative techniques.
   e. *Comparison in terms of a precise (and thereby inadequate) criterion*: An explicit criterion is difficult to define. Therefore, precise surrogates are often used.
   f. *Optimization in terms of a precise and inadequate criterion*: An explicit criterion is difficult to define. Therefore, precise surrogates are often used.
   g. *Comparison in terms of several optimization criteria*: This constitutes the final step in determining the best solution.

3. The EDA school of thought

Tukey's book is more than a collection of new and creative rules and operations. It defines EDA as a discipline that holds that data analysts fail if only they fail to try many things. It further espouses the belief that data analysts

* The seven steps are Tukey's. The annotations are mine.

are especially successful if their detective work forces them to notice the unexpected. In other words, the philosophy of EDA is a trinity of *attitude* and *flexibility* to do whatever it takes to refine the analysis and *sharp-sightedness* to observe the unexpected when it does appear. EDA is thus a self-propagating theory; each data analyst adds his or her contribution, thereby contributing to the discipline.

The sharp-sightedness of EDA warrants more attention as it is a very important feature of the EDA approach. The data analyst should be a keen observer of those indicators that are capable of being dealt with successfully and use them to paint an analytical picture of the data. In addition to the ever-ready visual graphical displays as indicators of what the data reveal, there are numerical indicators, such as counts, percentages, averages, and the other classical descriptive statistics (e.g., standard deviation, minimum, maximum, and missing values). The data analyst's personal judgment and interpretation of indicators are not considered a bad thing because the goal is to draw informal inferences rather than those statistically significant inferences that are the hallmark of statistical formality.

In addition to visual and numerical indicators, there are the *indirect messages* in the data that force the data analyst to take notice, prompting responses such as "The data look like …," or, "It appears to be …" Indirect messages may be vague, but their importance is to help the data analyst draw informal inferences. Thus, indicators do not include any of the hard statistical apparatus, such as confidence limits, significance tests, or standard errors.

With EDA, a new trend in statistics was born. Tukey and Mosteller quickly followed up in 1977 with the second EDA book, *Data Analysis and Regression* (EDA II), which recasts the basics of classical inferential procedures of data analysis and regression. EDA II takes a fresh view of the classics as an assumption-free, nonparametric approach guided by "(a) a sequence of philosophical attitudes … for effective data analysis, and (b) a flow of useful and adaptable techniques that make it possible to put these attitudes to work" [20].

Hoaglin, Mosteller, and Tukey, in 1983, succeeded in advancing EDA with, *Understanding Robust and Exploratory Data Analysis,* which provides an understanding of how badly the classical methods behave when their restrictive assumptions do not hold and offers alternative robust and exploratory methods to broaden the effectiveness of statistical analysis [21]. It includes a collection of methods to cope with data in an informal way, guiding the identification of data structures relatively quickly and easily and trading off optimization of objectives for the stability of results.

Hoaglin, Mosteller, and Tukey, in 1991, continued their fruitful EDA efforts with, *Fundamentals of Exploratory Analysis of Variance* [22]. They recast the basics of the analysis of variance with the classical statistical apparatus (e.g., degrees of freedom, F ratios, and p-values) in a host of numerical and graphical displays. These displays often give insight into the structure of the data, such as size effects, patterns and interaction, and behavior of residuals.

EDA set off a burst of activity in the visual portrayal of data. *Graphical Methods for Data Analysis* (1983) presented new and old methods—some of which require a computer, while others only paper and a pencil—but all are powerful data

```
Problem  ==>  Model  ===>    Data     ===>  Analysis  ===>  Results/interpretation (Classical)
Problem  <==>  Data  <===>  Analysis  <===>   Model   ===>  Results/interpretation (EDA)

              Attitude, flexibility, and sharp-sightedness (EDA trinity)
```

FIGURE 13.1
EDA paradigm.

analysis tools to learn more about data structure [23]. In 1986, du Toit, Steyn, and Stumpf came out with *Graphical Exploratory Data Analysis*, providing a comprehensive, yet simple presentation of the topic [24]. Jacoby, with *Statistical Graphics for Visualizing Univariate and Bivariate Data* (1997) and *Statistical Graphics for Visualizing Multivariate Data* (1998), carries out his objective to obtain pictorial representations of quantitative information by elucidating histograms, one-dimensional and enhanced scatterplots, and nonparametric smoothing [25,26]. Also, he successfully transfers graphical displays of multivariate data on a single sheet of paper, a two-dimensional space.

EDA presents a major paradigm shift, depicted in Figure 13.1, in the starting point of the model-building process. With the mantra "Let your data be your guide," EDA offers a view that is a complete reversal of the classical principles that govern the usual steps of the model building. EDA declares the model must always follow the data, not the other way around, as in the classical approach.

In the classical approach, the problem is stated and formulated regarding an outcome variable Y. The working assumption is the *true* model explaining all the variation in Y is known. Specifically, all the structures (predictor variables, X_i's) affecting Y and their forms are known and present in the model. For example, if Age affects Y, but the log of Age reflects the true relationship with Y, then the log of Age must be present in the model. Once the model is specified, the data undergo model-specific analysis, which provides the results regarding numerical values associated with the structures or estimates of the true predictor variables' coefficients. Then, the data analyst interprets the specified model by declaring the importance of X_i, assessing how X_i affects the prediction of Y, and ranking X_i in order of predictive importance.

Of course, the data analyst never knows the true model. So, familiarity with the content domain of the problem is used to put forth explicitly the true *surrogate* model, which yield good predictions of Y. According to Box, "All models are wrong, but some are useful" [27]. In this case, the model selected provides serviceable predictions of Y. Regardless of the model used, the assumption of knowing the truth about Y sets the statistical logic in motion to cause likely bias in the analysis, results, and interpretation.

In the EDA approach, the only assumption is having some prior experience with the content domain of the problem. The right attitude, flexibility, and sharp-sightedness are the forces behind the data analyst, who assesses the problem and lets the data guide the analysis, which then suggests the structures and their forms of the model. If the model passes the validity check, then it is considered final and ready for final results and interpretation. If not, with the force still behind the data analyst, the analysis or data are revisited until new structures produce a sound and validated model. The model's final results and interpretation are available for review. Take a second look at Figure 13.1. Without exposure to assumption violations, the EDA paradigm offers a degree of confidence that its prescribed exploratory efforts are not biased, at least in the manner of the classical approach. Of course, no analysis is bias free as all analysts admit their bias into the equation.

With all its strengths and determination, EDA as originally developed had two minor weaknesses that could have hindered its wide acceptance and great success. One is of a subjective or psychological nature, and the other is a misconceived notion. Data analysts know that failure to look into a multitude of possibilities can result in a flawed analysis; thus, they find themselves in a competitive struggle against the data itself. So, EDA can foster data analysts with insecurity that their work is never complete. The PC can assist data analysts in being thorough with their analytical due diligence but bears no responsibility for the arrogance EDA engenders.

The belief that EDA, originally developed for the small-data setting, does not work as well with large samples is a misconception. Indeed, some of the graphical methods, such as the stem-and-leaf plots, and some of the numerical and counting methods, such as folding and binning, do break down with large samples. However, the majority of the EDA methodology is unaffected by data size. The manner by which the methods are carried out and the reliability of the results remain in place. In fact, some of the most powerful EDA techniques scale up quite nicely but do require the PC to do the serious number crunching of the big data [28]. For example, techniques such as ladder of powers, reexpressing, and smoothing are valuable tools for large sample or big data applications.

13.7 Summary

Finding the best possible subset of variables to put in a model has been a frustrating exercise. Many variable selection methods exist. Many statisticians know them, but few know they produce poorly performing models. The deficient variable selection methods are a miscarriage of statistics, debasing sound statistical theory into a misguided pseudo-theoretical foundation. I review the five widely used variable selection methods, itemize some of their weaknesses, and answer why they are still in use. Then, I present the notable solution to variable selection in regression: the natural seven-step cycle of statistical modeling and analysis. I feel that newcomers to Tukey's EDA need the seven-step cycle introduced within the narrative of Tukey's analytic philosophy. Accordingly, I enfolded the solution with front and back matter—the essence of EDA and the EDA school of thought, respectively.

References

1. Tukey, J.W., *The Exploratory Data Analysis*, Addison-Wesley, Reading, MA, 1977.
2. Classical underlying assumptions, 2009. http://en.wikipedia.org/wiki/Regression_analysis.
3. Miller, A.J., *Subset Selection in Regression*, Chapman and Hall, New York, 1990, pp. iii–x.

4. Statistica-Criteria-Supported-by-SAS.pdf, 2010. http://www.geniq.net/res/Statistical-Criteria-Supported-by-SAS.pdf.
5. Roecker, E.B., Prediction error and its estimation for subset-selected models, *Technometrics*, 33, 459–468, 1991.
6. Copas, J.B., Regression, prediction and shrinkage (with discussion), *Journal of the Royal Statistical Society B*, 45, 311–354, 1983.
7. Henderson, H.V., and Velleman, P.F., Building multiple regression models interactively, *Biometrics*, 37, 391–411, 1981.
8. Altman, D.G., and Andersen, P.K., Bootstrap investigation of the stability of a Cox regression model, *Statistics in Medicine*, 8, 771–783, 1989.
9. Derksen, S., and Keselman, H.J., Backward, forward and stepwise automated subset selection algorithms, *British Journal of Mathematical and Statistical Psychology*, 45, 265–282, 1992.
10. Judd, C.M., and McClelland, G.H., *Data Analysis: A Model Comparison Approach*, Harcourt Brace Jovanovich, New York, 1989.
11. Bernstein, I.H., *Applied Multivariate Analysis*, Springer-Verlag, New York, 1988.
12. Kashid, D.N., and Kulkarni, S.R., A more general criterion for subset selection in multiple linear regression, *Communication in Statistics–Theory & Method*, 31(5), 795–811, 2002.
13. Tibshirani, R., Regression shrinkage and selection via the Lasso, *Journal of the Royal Statistical Society B*, 58(1), 267–288, 1996.
14. Ratner, B., *Statistical Modeling and Analysis for Database Marketing: Effective Techniques for Mining Big Data*, CRC Press, Boca Raton, FL, 2003, Chapter 15, which presents the GenIQ Model. http://www.GenIQModel.com.
15. Chen, S.-M., and Shie, J.-D., *A New Method for Feature Subset Selection for Handling Classification Problems, Journal Expert Systems with Applications: An International Journal*, Volume 37 Issue 4, Pergamon Press, Inc. Tarrytown, NY, April, 2010.
16. SAS Proc Reg Variable Selection Methods.pdf, support.sas.com, SAS/STAT(R) 9.3 User's Guide, 2011.
17. Dash, M., and Liu, H., Feature selection for classification, *Intelligent Data Analysis*, 1, 131–156, 1997.
18. Twain, M., Get your facts first, then you can distort them as you please, 2011. http://thinkexist.com/quotes/mark_twain/.
19. Fox, J., *Applied Regression Analysis, Linear Models, and Related Methods*, Sage, Thousand Oaks, CA, 1997.
20. Mosteller, F., and Tukey, J.W., *Data Analysis and Regression*, Addison-Wesley, Reading, MA, 1977.
21. Hoaglin, D.C., Mosteller, F., and Tukey, J.W., *Understanding Robust and Exploratory Data Analysis*, Wiley, New York, 1983.
22. Hoaglin, D.C., Mosteller, F., and Tukey, J.W., *Fundamentals of Exploratory Analysis of Variance*, Wiley, New York, 1991.
23. Chambers, M.J., Cleveland, W.S., Kleiner, B., and Tukey, P.A., *Graphical Methods for Data Analysis*, Wadsworth & Brooks/Cole, Pacific Grove, CA, 1983.
24. du Toit, S.H.C., Steyn, A.G.W., and Stumpf, R.H., *Graphical Exploratory Data Analysis*, Springer-Verlag, New York, 1986.
25. Jacoby, W.G., *Statistical Graphics for Visualizing Univariate and Bivariate Data*, Sage, Thousand Oaks, CA, 1997.
26. Jacoby, W.G., *Statistical Graphics for Visualizing Multivariate Data*, Sage, Thousand Oaks, CA, 1998.
27. Box, G.E.P., Science and statistics, *Journal of the American Statistical Association*, 71, 791–799, 1976.
28. Weiss, S.M., and Indurkhya, N., *Predictive Data Mining*, Morgan Kaufman, San Francisco, CA, 1998.

14

*CHAID for Interpreting a Logistic Regression Model**

14.1 Introduction

The logistic regression model is the standard technique for building a response model. Its theory is well established, and its estimation algorithm is available in all major statistical software packages. The literature on the theoretical aspects of logistic regression is large and rapidly growing. However, the literature is seemingly empty on the interpretation of the logistic regression model. The purpose of this chapter is to present a *data mining* method based on chi-squared automatic interaction detection (CHAID) for interpreting a logistic regression model, specifically to provide a complete assessment of the effects of the predictor variables, defining a logistic regression model, on a binary response variable.

14.2 Logistic Regression Model

I state the definition of the logistic regression model briefly. Let Y be a binary response (dependent) variable, which takes on yes/no values (typically coded 1/0, respectively), and X_1, X_2, ..., X_n be the predictor (independent) variables. The logistic regression model estimates the *logit of Y*—the log of the odds of an individual responding yes—defined in Equation 14.1. The logit yields an individual's probability of responding yes from the transformation formula in Equation 14.2:

$$\text{Logit Y} = b_0 + b_1{*}X_1 + b_2{*}X_2 + \ldots + b_n{*}X_n \tag{14.1}$$

$$\text{Prob}(Y = 1) = \frac{\exp(\text{Logit Y})}{1 + \exp(\text{Logit Y})} \tag{14.2}$$

Calculating an individual's predicted probability of responding yes is performed by plugging in the values of the predictor variables for that individual in Equations 14.1 and 14.2. The b's are the logistic regression coefficients. The coefficient b_0 is referred to as the intercept and has no corresponding X_0 like the other b's.

* This chapter is based on an article with the same title in *Journal of Targeting, Measurement and Analysis for Marketing*, 6, 2, 1997. Used with permission.

The *odds ratio* is the traditional measure of assessing the effect of a predictor variable on the response variable, actually on the odds of response = 1, given that the other predictor variables are "held constant." The phrase "given that …" implies that the odds ratio is the average effect of a predictor variable on response when the effects of the other predictor variables are "partialled out" of the relationship between response and the predictor variable. Thus, the odds ratio does not explicitly reflect the variation of the other predictor variables. Exponentiating the coefficient of the predictor variable defines the odds ratio for a predictor variable. That is, the odds ratio for X_i equals $\exp(b_i)$, where exp is the exponential function and b_i is the coefficient of X_i.

14.3 Database Marketing Response Model Case Study

A woodworker's tool supplier, who wants to increase response to her catalog in an upcoming campaign, needs a model for generating a list of her most responsive customers. The response model, built on a sample drawn from a recent catalog mailing with a 2.35% response rate, is defined by the following variables:

1. The response variable is RESPONSE, which indicates whether a customer made a purchase (yes = 1, no = 0) from the recent catalog mailing.
2. There are three predictor variables consisting of
   a. CUST_AGE, the customer age in years
   b. LOG_LIFE, the log of total purchases in dollars since the customer's first purchase; that is, the log of lifetime dollars
   c. PRIOR_BY, the dummy variable indicating a purchase in the 3 months before the recent catalog mailing (yes = 1, no = 0).

The logistic regression analysis on RESPONSE with the three predictor variables produces the output in Table 14.1. The "Parameter Estimate" column contains the logistic regression coefficients, which defines the RESPONSE model in Equation 14.3.

$$\begin{aligned}\text{Logit RESPONSE} &= -8.43 + 0.02{*}\text{CUST\_AGE} \\ &\quad + 0.74{*}\text{LOG\_LIFE} + 0.82{*}\text{PRIOR\_BY}\end{aligned} \tag{14.3}$$

TABLE 14.1

Logistic Regression Output

Variable		Parameter Estimate	Standard Error	Wald Chi-square	Pr > Chi-square	Odds Ratio
Intercept	1	−8.4349	0.0854	9760.7175	1.E + 00	
CUST_AGE	1	0.0223	0.0004	2967.8450	1.E + 00	1.023
LOG_LIFE	1	0.7431	0.0191	1512.4483	1.E + 00	2.102
PRIOR_BY	1	0.8237	0.0186	1962.4750	1.E + 00	2.279

14.3.1 Odds Ratio

The following discussion illustrates two weaknesses of the odds ratio. First, the odds ratio is regarding *odds of responding yes*: It is an unintuitive awkward measure due to the necessary exponentiating of the coefficient of the predictor variable. Even the mathematically adept feel slightly queasy when interpreting it. Second, the odds ratio provides a static assessment of the effect of a predictor variable as its value is constant regardless of the relationship among the other predictor variables. The odds ratio is part of the standard output of the logistic regression analysis. For the case study, the odds ratio is in the rightmost column in Table 14.1.

1. For PRIOR_BY with a coefficient value of 0.8237, the odds ratio is 2.279 (= exp(0.8237)). The odds ratio indicates that the odds of an individual who has made a purchase within the prior 3 months (PRIOR_BY = 1) is 2.279 times the odds of an individual who has *not* made a purchase within the prior 3 months (PRIOR_BY = 0)—given CUST_AGE and LOG_LIFE are held constant.*
2. CUST_AGE has an odds ratio of 1.023. The odds ratio indicates that for every one-year increase in a customer's age, the odds increase by 2.3%—given PRIOR_BY and LOG_LIFE are held constant.
3. LOG_LIFE has an odds ratio of 2.102, which indicates that for every one log-lifetime-dollar-unit increase, the odds increase by 110.2%—given PRIOR_BY and CUST_AGE are held constant.

The proposed CHAID-based data mining method supplements the odds ratio for interpreting the effects of a predictor on RESPONSE. It provides treelike displays regarding non-threatening probability units, everyday values ranging from 0% to 100%. Moreover, the CHAID-based graphics provide a complete assessment of the effect of a predictor variable on RESPONSE. They bring forth the simple, unconditional relationship between a given predictor variable and RESPONSE as well as the conditional relationship between a given predictor variable and RESPONSE shaped by the relationships between the other Xs and a given predictor variable and the other Xs and RESPONSE.

14.4 CHAID

Briefly stated, CHAID is a technique that recursively partitions a population into separate and distinct subpopulations or segments such that the variation of the dependent variable is minimized within the segments and maximized among the segments. A CHAID analysis results in a tree-like diagram, commonly called a CHAID tree. In 1980, CHAID was the first technique originally developed for finding "combination" or interaction variables. In database marketing today, CHAID primarily serves as a market segmentation technique. The use of CHAID in the proposed application for interpreting a

* For given values of CUST_AGE and LOG_LIFE, say, a and b, respectively, the odds ratio for PRIOR_BY is defined as:

$$= \frac{\text{odds (PRIOR\_BY = 1 given CUST\_AGE = a and LOG\_LIFE = b)}}{\text{odds (PRIOR\_BY = 0 given CUST\_AGE = a and LOG\_LIFE = b)}}$$

logistic regression model is possible because of the salient data mining features of CHAID. CHAID is eminently good in uncovering structure, in this application, within the conditional and unconditional relationships among RESPONSE and predictor variables. Moreover, CHAID is excellent in graphically displaying multivariable relationships. The CHAID output is a tree like figure that branches from a single root. The CHAID tree is easy to read and interpret.

It is worth emphasizing that the proposed application of CHAID does not offer an alternative method for analyzing or modeling the data at hand. CHAID serves as a visual aid for depicting the statistical mechanics of an already built logistic regression model: addressing how the variables of the model work together in contributing to the predicted probability of response.

It is important to note the versatility of the proposed method as it can be applied to any built model—not necessarily logistic regression. As such, the proposed method provides a complete assessment of the effects of the predictor variables, defining any model—statistical or machine-learning, on the dependent variable.

14.4.1 Proposed CHAID-Based Method

When performing an ordinary CHAID analysis, the model builder selects both the binary response variable and a set of predictor variables. For the proposed CHAID-based data mining method, the CHAID response variable is the already-built logistic response model's *estimated probability of response*. The CHAID set of predictor variables consists of the predictor variables that defined the logistic response model in their original units not in reexpressed units (if reexpressing was necessary). Reexpressed variables are invariably in units that hinder the interpretation of the CHAID-based analysis as well as the logistic regression model. Moreover, to facilitate the analysis, the continuous predictor variables are categorized into meaningful intervals based on the content domain of the problem under study.

For the current study, the CHAID response variable is the estimated probability of RESPONSE, called Prob_est, which is obtained from Equations 14.2 and 14.3 and defined in Equation 14.4:

$$\text{Prob\_est} = \frac{\exp(-8.43 + 0.02 * \text{CUST\_AGE} + 0.74 * \text{LOG\_LIFE} + 0.82 * \text{PRIOR\_BY})}{1 + \exp(-8.43 + 0.02 * \text{CUST\_AGE} + 0.74 * \text{LOG\_LIFE} + 0.82 * \text{PRIOR\_BY})} \tag{14.4}$$

The CHAID set of predictor variables is CUST_AGE categorized into two classes, PRIOR_BY and the original variable LIFETIME DOLLARS (log-lifetime-dollar units are hard to understand), categorized into three classes. The woodworker's tool supplier views her customers regarding the following classes:

1. The two CUST_AGE classes are "less than 35 years" and "35 years and up." CHAID uses the bracket and parenthesis symbols in its display of intervals, denoting two customer age intervals: [18, 35) and [35, 93], respectively. CHAID defines intervals as a closed interval and a left-closed right-open interval. The former is denoted by [a, b], indicating all values between and including a and b. The latter is denoted by [a, b), indicating all values greater than/equal to a and less than b. Minimum and maximum ages in the sample are 18 and 93, respectively.

2. The three LIFETIME DOLLARS classes are: less than \$15,000, \$15,001 to \$29,999, and equal to or greater than \$30,000. CHAID denotes the three lifetime dollar intervals as [12, 1500), [1500, 30000), and [30000, 675014], respectively. Minimum and maximum lifetime dollars in the sample are \$12 and \$675,014, respectively.

The CHAID trees in Figures 14.1 to 14.3 are based on the Prob_est variable with three predictor variables and are read as follows:

1. All CHAID trees have a top box (root node), which represents the sample under study: sample size and response rate. For the proposed CHAID application, the top box reports the sample size and the average estimated probability (AEP)

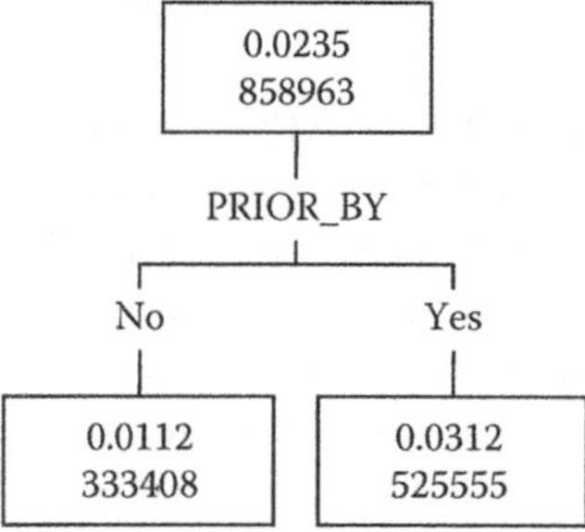

FIGURE 14.1
CHAID tree for PRIOR_BY.

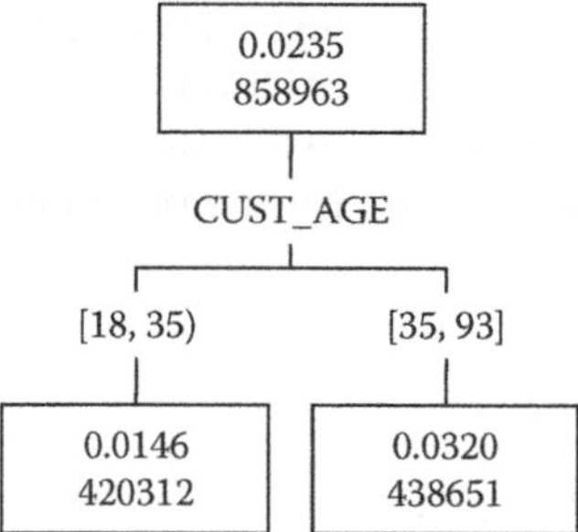

FIGURE 14.2
CHAID tree for CUST_AGE.

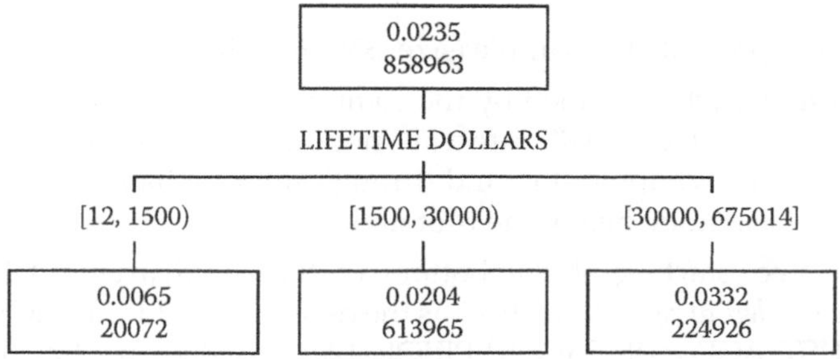

FIGURE 14.3
CHAID tree for LIFETIME DOLLARS.

of response. For the case study, the sample size is 858,963 and the AEP of response is 0.0235.*

2. The CHAID tree for PRIOR_BY is shown in Figure 14.1. The left leaf node represents a segment (size: 333,408) defined by PRIOR_BY = no. These customers have *not* made a purchase in the prior 3 months; their AEP of response is 0.0112. The right leaf node represents a segment (size: 525,555) defined by PRIOR_BY = yes. These customers have made a purchase in the prior 3 months; their AEP of response is 0.0312.
3. The CHAID tree for CUST_AGE is shown in Figure 14.2. The left leaf node represents a segment (size: 420,312) defined by customers whose ages are in the interval [18, 35); their AEP of response is 0.0146. The right leaf node represents a segment (size: 438,651) defined by customers whose ages are in the interval [35, 93]; their AEP of response is 0.0320.
4. The CHAID tree for LIFETIME DOLLARS is in Figure 14.3. The left leaf node represents a segment (size: 20,072) defined by customers whose lifetime dollars are in the interval [12, 1500); their AEP of response is 0.0065. The middle leaf node represents a segment (size: 613,965) defined by customers whose lifetime dollars are in the interval [1500, 30000); their AEP of response is 0.0204. The left leaf node represents a segment (size: 224,926) defined by customers whose lifetime dollars are in the interval [30000, 675014]; their AEP of response is 0.0332.

At this point, the single-predictor-variable CHAID tree shows the effect of the predictor variable on RESPONSE. The values of the leaf nodes from left to the right clearly reveal the pattern of increasing RESPONSE and predictor variable values. Although the single-predictor-variable CHAID tree is easy to interpret with probability units, it does not reveal the effects of a predictor variable on the variation of the other predictor variables in the model. A *multivariable CHAID tree* serves as a complete visual display of the effect of a predictor variable on response accounting for the presence of other predictor variables.

14.5 Multivariable CHAID Trees

The multivariable CHAID tree in Figure 14.4 shows the effects of LIFETIME DOLLARS on RESPONSE on the variation of CUST_AGE and PRIOR_BY = *no*. The LIFETIME DOLLARS-PRIOR_BY = no CHAID tree is read and interpreted as follows:

1. The root node represents the sample (size: 858,963); the AEP of response is 0.0235.
2. The tree has six *branches*, defined by the combination or interaction of the CUST_AGE and LIFETIME DOLLARS intervals/nodes. Branches are read from an end leaf node (bottom box) upward to and through intermediate leaf nodes, stopping at the first-level leaf node below the root node.
3. Reading the tree, starting at the bottom of the multivariable CHAID tree in Figure 14.4, from *left to right*: Branch 1 has the description of LIFETIME DOLLARS = [12, 1500), CUST_AGE = [18, 35), and PRIOR_BY = no. Branch 2 has the description

* The average estimated probability or response rate is always equal to the true response rate.

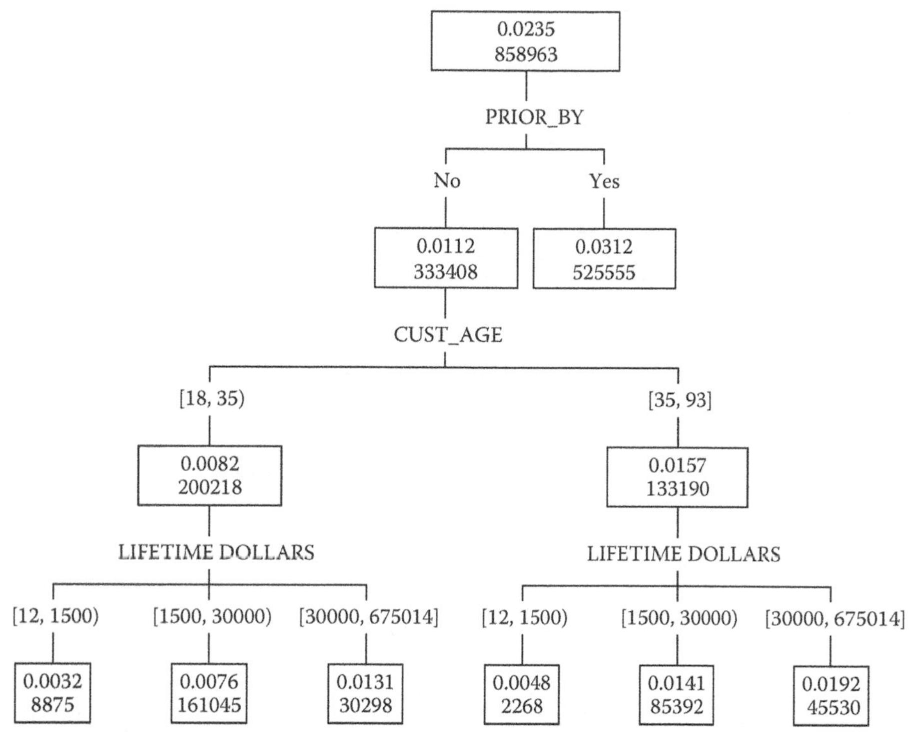

FIGURE 14.4
Multivariable CHAID tree for effects of LIFETIME DOLLARS, accounting for CUST_AGE and PRIOR_BY = no.

of LIFETIME DOLLARS = [1500, 30000), CUST_AGE = [18, 35), and PRIOR_BY = no. Branches 3 to 5 are similarly described. The rightmost branch, Branch 6, has the description of LIFETIME DOLLARS = [30000, 675014], CUST_AGE = [35, 93], and PRIOR_BY = no.

4. Branches 1 to 3 show the effects on RESPONSE as LIFETIME DOLLARS increase from the multivariable CHAID tree. The AEP of RESPONSE goes from 0.0032 to 0.0076 to 0.0131 for customers whose ages are in the interval [18, 35) *and* have not purchased in the prior 3 months.
5. Branches 4 to 6 show the effects on RESPONSE as LIFETIME DOLLARS increase from the multivariable CHAID tree. The AEP of RESPONSE ranges from 0.0048 to 0.0141 to 0.0192 for customers whose ages are in the interval [35, 93] *and* have not purchased in the prior 3 months.

The multivariable CHAID tree in Figure 14.5 shows the effect of LIFETIME DOLLARS on RESPONSE on the variation of CUST_AGE and PRIOR_BY = yes. Briefly, the LIFETIME DOLLARS-PRIOR_BY = yes CHAID tree is interpreted as follows:

1. Branches 1 to 3 show the effects on RESPONSE as LIFETIME DOLLARS increase from the multivariable CHAID tree. The AEP of response goes from 0.0077 to 0.0186 to 0.0297 for customers whose ages are in the interval [18, 35) *and* have purchased in the prior 3 months.

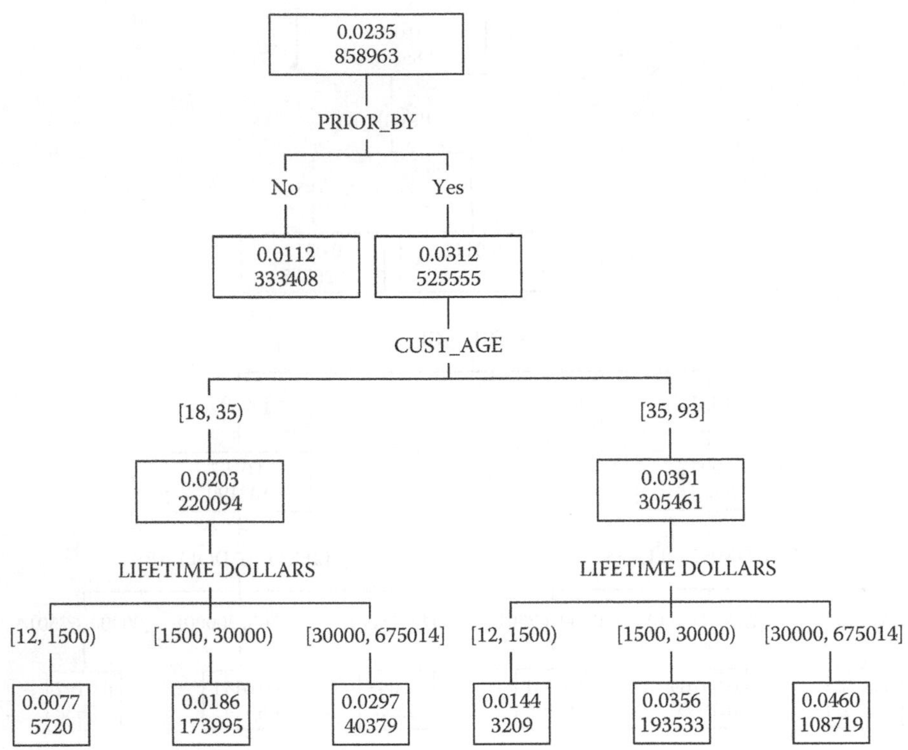

FIGURE 14.5
Multivariable CHAID tree for effects of LIFETIME DOLLARS, accounting for CUST_AGE and PRIOR_BY = yes.

2. Branches 4 to 6 show the effects on RESPONSE as LIFETIME DOLLARS increase from the multivariable CHAID tree. The AEP of RESPONSE goes from 0.0144 to 0.0356 to 0.0460 for customers whose ages are in the interval [35, 93] *and* have purchased in the prior 3 months.

The multivariable CHAID trees for the effects of PRIOR_BY on RESPONSE on the variation of CUST_AGE = [18, 35) and LIFETIME DOLLARS and on the variation of CUST_AGE = [35, 93] and LIFETIME DOLLARS are in Figures 14.6 and 14.7, respectively. They are similarly read and interpreted as the LIFETIME DOLLARS multivariable CHAID trees in Figures 14.4 and 14.5.

The multivariable CHAID trees for the effects of CUST_AGE on RESPONSE on the variation of PRIOR_BY = no and LIFETIME DOLLARS and on the variation of PRIOR_BY = yes and LIFETIME DOLLARS are in Figures 14.8 and 14.9, respectively. They are similarly read and interpreted as the LIFETIME DOLLARS multivariable CHAID trees.

14.6 CHAID Market Segmentation

I take this opportunity to use the analysis (so far) to illustrate CHAID as a market segmentation technique. A closer look at the full CHAID trees in Figures 14.4 and 14.5 identifies three market segment pairs, which show three levels of response performance,

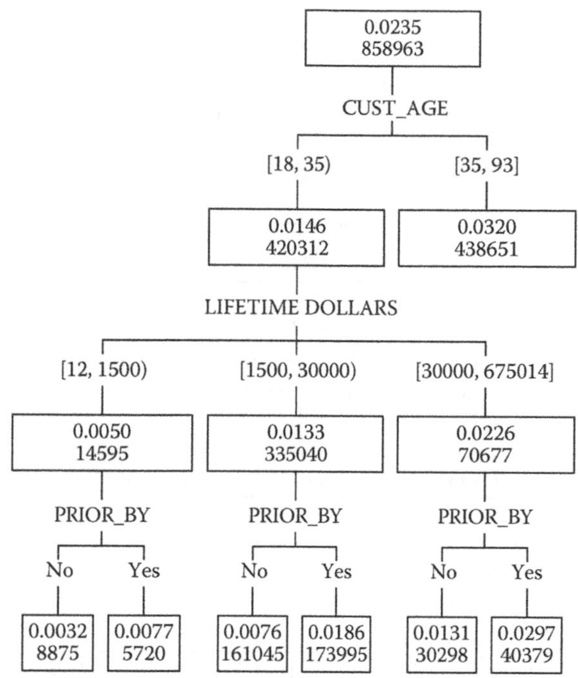

FIGURE 14.6
Multivariable CHAID tree for effect of PRIOR_BY, accounting for LIFETIME DOLLARS and CUST_AGE = [18, 35].

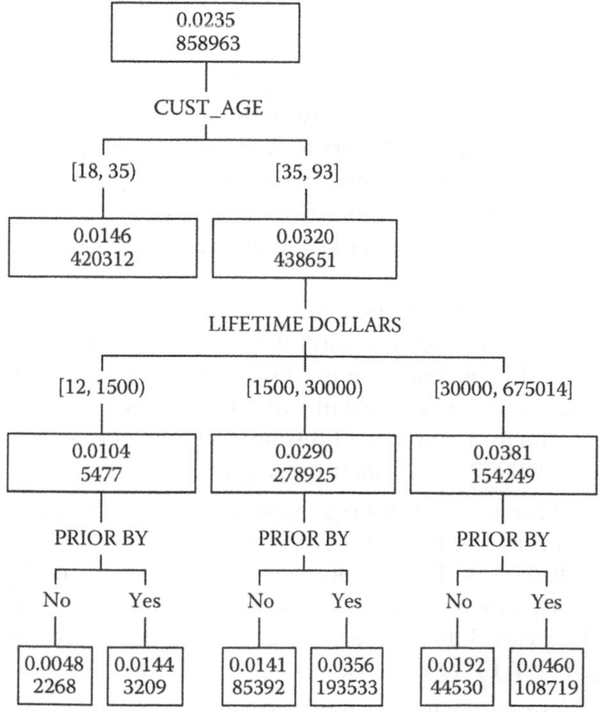

FIGURE 14.7
Multivariable CHAID tree for effect of PRIOR_BY, accounting for LIFETIME DOLLARS and CUST_AGE = [35, 93].

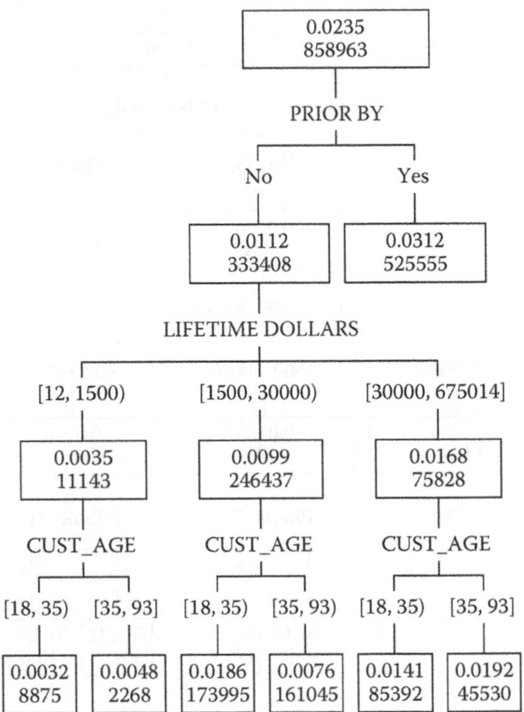

FIGURE 14.8
Multivariable CHAID tree for effect of CUST_AGE, accounting for LIFETIME DOLLARS and PRIOR_BY = no.

0.76%/0.77%, 0.48%/4.60%, and 1.41%/1.44%. CHAID provides the catalogue with marketing intelligence for high-, medium-, and low-performing segments. Marketing strategy can be developed to stimulate the high performers with techniques such as cross-selling, or to pique interest in the medium performers with new products, as well as to prod the low performers with incentives and discounts.

The descriptive profiles of the three market segments are as follows:

Market Segment 1: These are customers whose ages are in the interval [18, 35), who have *not* purchased in the prior 3 months and who have lifetime dollars in the interval [1500, 30000). The AEP of response is 0.0076. See Branch 2 in Figure 14.4. Also, there are customers whose ages are in the interval [18, 35), who have purchased in the prior 3 months and who have lifetime dollars in the interval [12, 1500). The AEP of response is 0.0077. See Branch 1 in Figure 14.5.

Market Segment 2: These are customers whose ages are in the interval [35, 93], who have *not* purchased in the prior 3 months and who have lifetime dollars in the interval [12, 1500). The AEP of response is 0.0048. See Branch 4 in Figure 14.4. Also, there are customers whose ages are in the interval [35, 93], who have purchased in the prior 3 months and who have lifetime dollars in the interval [30000, 675014]. The AEP of response is 0.0460. See Branch 6 in Figure 14.5.

Market Segment 3: These are customers whose ages are in the interval [35, 93], who have *not* purchased in the prior 3 months and who have lifetime dollars in the interval [1500, 30000). The AEP of response is 0.0141. See Branch 5 in Figure 14.4. Also, there

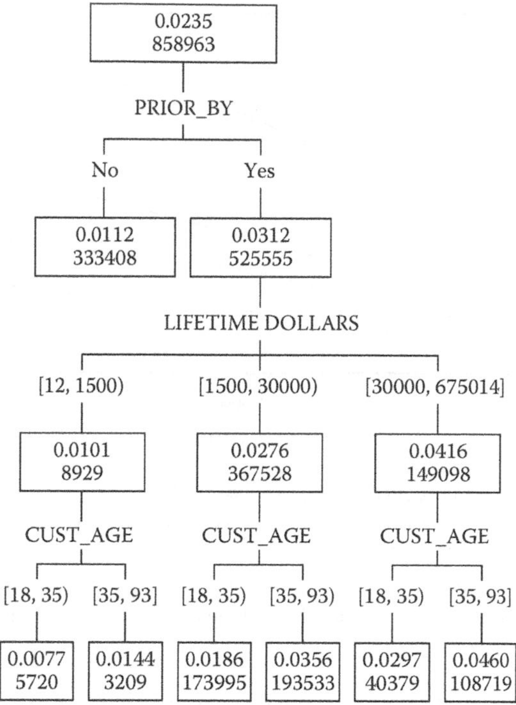

FIGURE 14.9
Multivariable CHAID tree for effect of CUST_AGE, accounting for LIFETIME DOLLARS and PRIOR_BY = yes.

are customers whose ages are in the interval [35, 93], who have purchased in the prior 3 months and who have lifetime dollars in the interval [12, 1500). The AEP of response is 0.0144. See Branch 4 in Figure 14.5.

14.7 CHAID Tree Graphs

Displaying the multivariable CHAID trees in a *single* graph provides the desired displays of a *complete* assessment of the effects of the predictor variables on RESPONSE. Construction and interpretation of the *CHAID tree graph* for a given predictor are as follows:

1. Collect the set of multivariable CHAID trees for a given predictor variable. For example, for PRIOR_BY there are two trees, which correspond to the two values of PRIOR_BY, yes and no.
2. For each branch, plot the AEP of response values (Y-axis) and the minimum values of the end leaf nodes of the given predictor variable (X-axis).*

* The minimum value is one of several values that can be used; alternatives are the mean or median of each predefined interval.

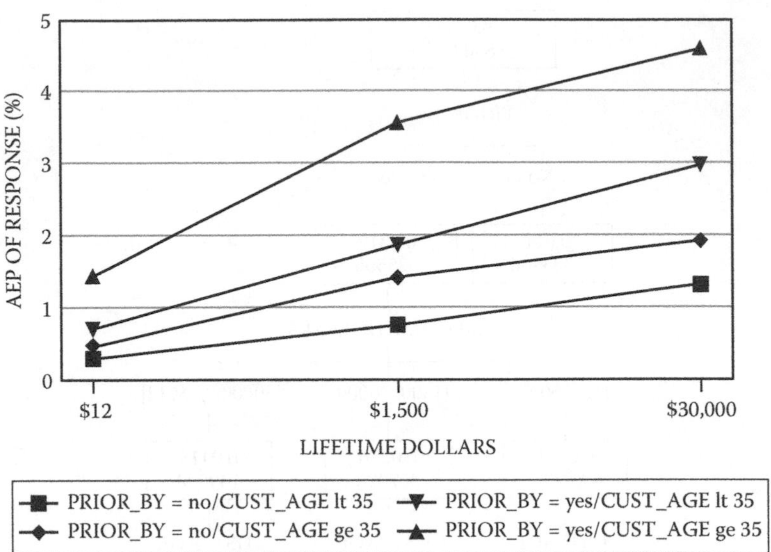

FIGURE 14.10
CHAID tree graph for effects of LIFETIME on RESPONSE by CUST_AGE and PRIOR_BY.

3. For each branch, connect the *nominal points* (AEP response value, minimum value). The resultant *trace line* segment represents a market or customer segment defined by the branch's intermediate leaf intervals/nodes.
4. The shape of the trace line indicates the effect of the predictor variable on RESPONSE for that segment. A comparison of the trace lines provides a total view of how the predictor variable affects response, accounting for the presence of the other predictor variables.

The LIFETIME DOLLARS CHAID tree graph in Figure 14.10 is based on the multivariable LIFETIME DOLLARS CHAID trees in Figures 14.4 and 14.5. The top trace line with a noticeable bend corresponds to older customers (age 35 years and older) who have made purchases in the prior 3 months. The implication is Lifetime Dollars has a nonlinear effect on RESPONSE for this customer segment. As LIFETIME DOLLARS goes from a nominal $12 to a nominal $1,500 to a nominal $30,000, RESPONSE increases at a nonconstant rate as depicted in the tree graph.

The other trace lines are straight lines* with various slopes. The implication is LIFETIME DOLLARS has various constant effects on response across the corresponding customer segments. As LIFETIME DOLLARS goes from a nominal $12 to a nominal $1,500 to a nominal $30,000, RESPONSE increases at various constant rates as depicted in the tree graph.

The PRIOR_BY CHAID tree graph in Figure 14.11 has its foundation from the multivariable PRIOR_BY CHAID trees in Figures 14.6 and 14.7. I focus on the slopes of the trace lines.† The evaluation rule is as follows: The steeper the slope is, the greater the constant effect on RESPONSE. Among the six trace lines, the top trace line for the "top" customer seg-

* The segment that is defined by PRIOR_BY = no and CUST_AGE greater than or equal to 35 years appears to have a very slight bend. However, I treat its trace line as straight because the bend is very slight.

† The trace lines are necessarily straight because two points (PRIOR_BY points no and yes) always determine a straight line.

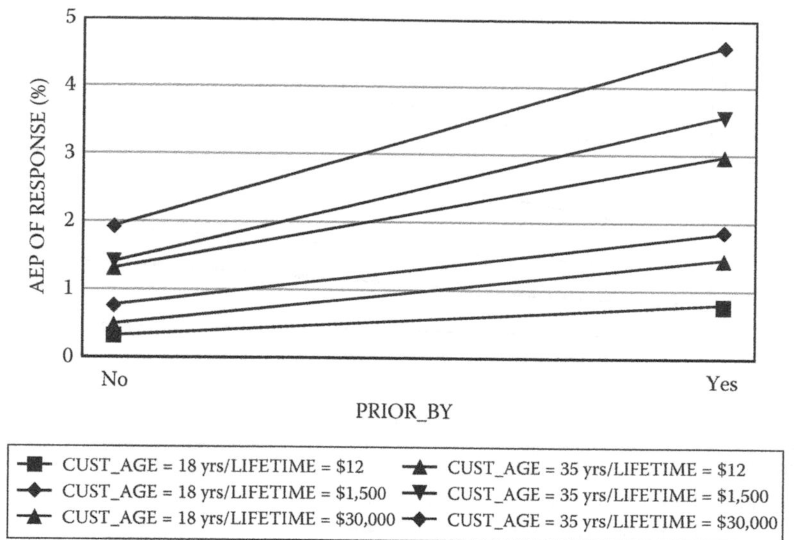

FIGURE 14.11
CHAID tree graph for effects of PRIOR_BY on RESPONSE by CUST_AGE and LIFETIME DOLLARS.

ment of older customers with LIFETIME DOLLARS equal to or greater than $30,000 has the steepest slope. The implications are as follows: (1) PRIOR_BY has a rather noticeable constant effect on RESPONSE for the top customer segment and (2) the size of the PRIOR_BY effect for the top customer segment is greater than the PRIOR_BY effect for the remaining five customer segments. As PRIOR_BY goes from no to yes, RESPONSE increases at a rather noticeable constant rate for the top customer segment as depicted in the tree graph.

The remaining five trace lines have slopes of varying steepness. The implication is PRIOR_BY has various constant effects on RESPONSE across the corresponding five customer segments. As PRIOR_BY goes from no to yes, RESPONSE increases at various constant rates as depicted in the tree graph.

The CUST_AGE CHAID tree graph in Figure 14.12 has its foundation from the multivariable CUST_AGE CHAID trees in Figures 14.8 and 14.9. There are two sets of parallel trace lines with different slopes. The first set of the top two parallel trace lines corresponds to two customer segments defined by

1. PRIOR_BY = no and LIFETIME DOLLARS in the interval [1500, 30000)
2. PRIOR_BY = yes and LIFETIME DOLLARS in the interval [30000, 675014]

The implication is CUST_AGE has the same constant effect on RESPONSE for the two customer segments. As CUST_AGE goes from a nominal age of 18 years to a nominal age of 35 years, RESPONSE increases at a constant rate as depicted in the tree graph.

The second set of the next three parallel trace lines (two trace lines virtually overlap each other) corresponds to three customer segments defined by

1. PRIOR_BY = no and LIFETIME DOLLARS in the interval [30000, 675014]
2. PRIOR_BY = yes and LIFETIME DOLLARS in the interval [1500, 30000)
3. PRIOR_BY = yes and LIFETIME DOLLARS in the interval [12, 1500]

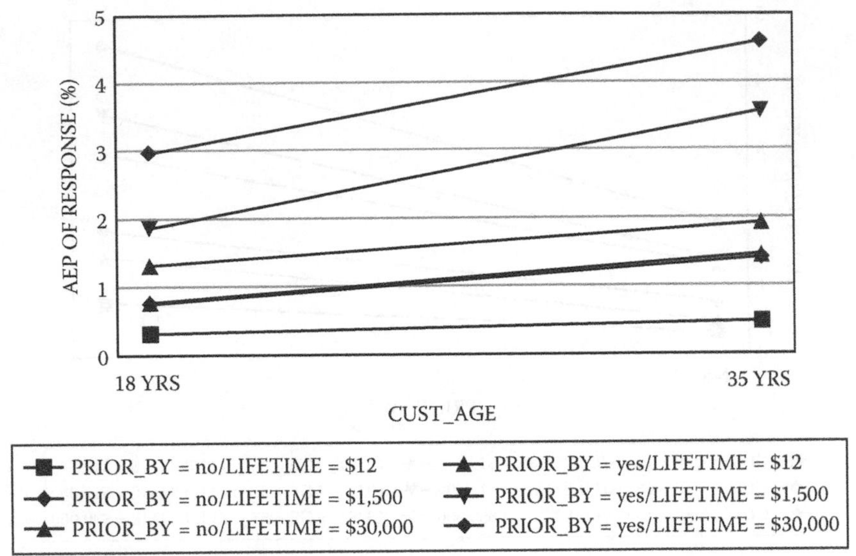

FIGURE 14.12
CHAID tree graph for effects of CUST_AGE on RESPONSE by PRIOR_BY and LIFE_DOLLARS.

The implication is CUST_AGE has the same constant effect on RESPONSE for the three customer segments. As CUST_AGE goes from a nominal age of 18 years to a nominal age of 35 years, RESPONSE increases at a constant rate as depicted in the tree graph. Note that the constant CUST_AGE effect for the three customer segments is less than the CUST_AGE effect for the former two customer segments as the slope of the former segments is less steep than that of the latter segments.

Last, the bottom trace line, which corresponds to the customer segment defined by PRIOR_BY = no and LIFETIME DOLLARS in the interval [12, 1500), has virtually no slope because it is nearly horizontal. The implication is CUST_AGE has no effect on RESPONSE for the corresponding customer segment.

14.8 Summary

After a brief introduction of the logistic regression model as the standard technique for building a binary response model, I focus on its interpretation, an area in the literature that has not received noticeable attention. I discuss the traditional approach to interpreting a logistic regression model using the odds ratio statistic, which measures the effect of a predictor variable on the odds of response.

I propose a CHAID-based data mining method to supplement the odds ratio, which has two discussed weaknesses. The CHAID-based method adds a visual touch to the original concept of the odds ratio. I illustrate the new method, which exports the information of the odds ratio into CHAID trees, visual displays regarding simple probability values. More important, the CHAID-based method makes possible the desired complete assessment of the effect of a predictor variable on response explicitly reflecting the relationship of the other predictor variables. I illustrate the new method, which combines individual predictor variable CHAID trees into a multivariable CHAID tree graph as a complete visual assessment of a predictor variable.

Furthermore, I note the versatility of the proposed method because it can be applied to any built model not necessarily logistic regression. As such, the proposed method provides a complete assessment of the effects of the predictor variables, defining any model—statistical or machine-learning—on the dependent variable.

15

The Importance of the Regression Coefficient

15.1 Introduction

Interpretation of the ordinary regression model—the most popular technique for making predictions of a single continuous variable Y—focuses on the model's coefficients with the aid of three concepts: the statistical p-value, *variables held constant*, and the standardized regression coefficient. The purpose of this chapter is to take a closer look at these widely used, yet often misinterpreted, concepts. I demonstrate that the statistical p-value as the sole measure for declaring predictor variable X important is sometimes problematic. The concept of variables held constant is critical for reliable assessment of how predictor variable X affects the prediction of Y. And, the standardized regression coefficient provides the correct ranking of variables in order of predictive importance—only under special circumstances.

15.2 The Ordinary Regression Model

The ordinary regression model, formally known as the ordinary least squares multiple linear regression model, is the most popular technique for making predictions of a single continuous variable. Its theory is well-established, and the estimation algorithm is available in all statistical computer software packages. The model is relatively easy to build and usually produces usable results.

Let Y be a continuous dependent variable (e.g., sales) and the set of X_1, X_2, ..., X_n comprises the predictor variables. The regression model (prediction equation) is in Equation 15.1:

$$Y = b_0 + b_1 * X_1 + b_2 * X_2 + \ldots + b_n * X_n \tag{15.1}$$

The b's are the regression coefficients,* estimated by the method of ordinary least squares. Once the coefficients are estimated, the calculation of an individual's predicted Y value (estimated sales) involves plugging in the values of the predictor variables for that individual into Equation 15.1.

* b_0 is called the intercept, which serves as a mathematical necessity for the regression model. However, b_0 can be considered the coefficient of $X_0 = 1$.

15.3 Four Questions

Interpretation of the model focuses on the regression coefficient with the aid of three concepts: the statistical p-value, the average change in Y associated with a unit change in X_i when the other Xs* are held constant, and the standardized regression coefficient. The following four questions apply universally to any discussion of the regression coefficient. They are discussed in detail to provide a better understanding of why the regression coefficient is important.

1. Is X_i important for making good predictions? The usual answer is that X_i is an important predictor variable if its associated p-value is less than 5%. This answer is correct for experimental studies but may not be correct for big data† applications.
2. How does X_i affect the prediction of Y? The usual answer is that Y experiences an average change of b_i with a unit increase in X_i when the other Xs are held constant. This answer is an honest one, which often is not accompanied by a caveat.
3. Which variables in the model have the *greatest effect* on the prediction of Y, in ranked order? The usual answer is that the variable with the largest regression coefficient has the greatest effect; the variable with the next largest regression coefficient has the next greatest effect, and so on. This answer is usually incorrect.
4. Which variables in a model are the *most important* predictors, in ranked order? The usual answer is that the variable with the largest standardized regression coefficient is the most important variable; the variable with the next largest standardized regression coefficient is the next important variable, and so on. This answer is usually incorrect.

15.4 Important Predictor Variables

X_i is declared an important predictor variable if it significantly reduces the prediction error of (actual Y – predicted Y) the regression model. The size of reduction in prediction error due to X_i can be tested for significance with the null hypothesis (NH) significance testing procedure. Briefly, I outline the procedure‡ as follows:

1. The NH and alternative hypothesis (AH) are defined as follows:

   NH: The change in mean squared prediction error due to X_i (cMSE_X_i) is equal to zero

   AH: cMSE_X_i is not equal to zero.
2. Significance testing for cMSE_X_i is equivalent to significance testing for the regression coefficient for X_i, b_i. Therefore, the hypotheses NH and AH are alternatively stated:

   NH: b_i is equal to zero

   AH: b_i is not equal to zero.

* The other Xs consist of n – 1 variables without X_i, namely, $X_1, X_2, \ldots, X_{i-1}, X_{i+1}, \ldots, X_n$.

† Big data are defined in Section 1.6.

‡ See any good textbook on the null hypothesis-significance testing procedure, such as Chow, S.L., *Statistical Significance*, Sage, Thousand Oaks, CA, 1996.

3. The working assumptions* of the testing procedure are that the sample size is correct, and the sample accurately reflects the relevant population. (Well-established procedures for determining the correct sample size for experimental studies are readily available in intermediate-level statistics textbooks.)
4. The decision to reject or fail to reject NH relies on the statistical p-value.† The statistical p-value is the probability of observing a value of the sample statistic (cMSE or b_i) as extreme or more extreme than the observed value (sample evidence) given NH is true.‡
5. The decision rule:
   a. If the p-value is *not* very small, typically greater than 5%, then the sample evidence supports the decision to fail to reject NH.§ The conclusion is that b_i is zero, and X_i does not significantly contribute to the reduction in prediction error. Thus, X_i is not an important predictor variable.
   b. If the p-value is very small, typically less than 5%, then the sample evidence supports the decision to reject NH for accepting AH. The conclusion is that b_i (or cMSE_X_i) has some nonzero value, and X_i contributes a significant reduction in prediction error. Thus, X_i is an important predictor variable.

The decision rule makes it clear that the p-value is an indicator of the *likelihood* that the variable has *some* predictive importance—*not* an indicator of how *much* importance (AH does not specify a value of b_i). Thus, *a smaller p-value implies a greater likelihood of some predictive importance not a greater predictive importance.* The implication is contrary to the common misinterpretation of the p-value: The smaller the p-value is, the greater the predictive importance of the associated variable.

15.5 p-Values and Big Data

Relying solely on the p-value for declaring important predictor variables is problematic in big data applications, which are characterized by large-to-big samples drawn from populations with the unknown spread of the Xs. The p-value is affected by the sample size (as sample size increases, the p-value decreases given NH is true) and affected by the spread of X_i (as the X_i spread increases, the p-value decreases) [1].¶ Accordingly, a small p-value may be due to a large sample or a large spread of the Xs. Thus, *in big data applications, a small p-value is only an indicator of a potentially important predictor variable.*

The issue of how the p-value is affected by moderating sample size is currently unresolved. Big data are nonexperimental data for which there are no procedures for determining the correct large sample size. A large sample produces many small p-values—a spurious result. The associated variables are often declared important when in fact they

* A suite of classical assumptions is required for the proper testing of the least squares estimate of b_i. See any good mathematical statistics textbook, such as Ryan, T.P., *Modern Regression Methods*, Wiley, New York, 1997.

† Failure to reject NH is not equivalent to accepting NH.

‡ The p-value is a conditional probability.

§ The choice of "very small" is arbitrary, but convention sets it at 5% or less.

¶ The size of b_i also affects the p-value: As b_i increases, the p-value decreases. This factor cannot be controlled by the analyst.

are not important; this reduces the stability of a model [2].[*,†] A procedure that adjusts the p-values when working with big data is needed.[‡] Until the development of such procedures, the recommended ad hoc approach is as follows: *In big data applications, variables with small p-values must undergo a final assessment of importance based on their actual reduction in prediction error. Variables associated with the greatest reduction in prediction error can be declared important predictors.* Elimination of problematic cases helps in moderating the effects of the spread of the Xs. For example, if the relevant population consists of 35- to 65-year olds and the big data include 18- to 65-year olds, then simply excluding the 18- to 34-year olds eliminates the spurious effects of spread.

15.6 Returning to Question 1

Is X_i important for making good predictions? The usual answer is that X_i is an important predictor variable if it has an associated p-value less than 5%.

This answer is correct for experimental studies, which have pre-determined sample sizes and samples have presumably known spread. For big data applications, in which large samples with unknown spread adversely affect the p-value, a small p-value is only an indicator of a potentially important predictor variable. The associated variables must go through an ad hoc evaluation of their actual reduction in the prediction errors before ultimately being declared important predictors.

15.7 Effect of Predictor Variable on Prediction

An assessment of the effect of predictor variable X_i on the prediction of Y focuses on the regression coefficient b_i. The common interpretation is that b_i is the average change in the predicted Y value associated with a unit change in X_i when the other Xs are held constant. A detailed discussion of what b_i measures and how to calculate b_i shows this is an honest interpretation that must be accompanied by a caveat, discussed in Section 15.8.

The regression coefficient b_i, also known as a partial regression coefficient, is a measure of the linear relationship between Y and X_i when the influences of the other Xs are partialled out or held constant. The expression *held constant* implies that the calculation of b_i involves the removal of the effects of the other Xs. Although the details of the calculation are beyond the scope of this chapter, it suffices to outline the steps involved, as delineated next.[§]

The calculation of b_i uses a method of statistical control in a three-step process:

1. The removal of the linear effects of the other Xs *from Y* produces a new variable Y-adj (= Y adjusted linearly for the other Xs).

[*] Falsely reject NH.

[†] Falsely fail to reject NH. The effect of a "small" sample is that a variable can be declared unimportant when, in fact, it is important.

[‡] This is a procedure similar to the Bonferroni method, which adjusts p-values downward because repeated testing increases the chance of incorrectly declaring a relationship or coefficient significant.

[§] The procedure for statistical control can be found in most basic statistics textbooks.

2. The removal of the linear effects of the other Xs *from* X_i produces a new variable X_i-adj (= X_i adjusted linearly for the other Xs).
3. The regression of Y-adj on X_i-adj produces the desired partial regression coefficient b_i.

The partial regression coefficient b_i is an honest estimate of the relationship between Y and X_i (with the effects of the other Xs partialled out) because its foundation is the concept of statistical control not experimental control. The statistical control method estimates b_i without data for the relationship between Y and X_i when the other Xs are held constant. The rigor of this method ensures the estimate is an honest one.

In contrast, the experimental control method involves collecting data for the observed relationship between X_i and Y when the other Xs are held constant. The resultant partial regression coefficient is directly measured and therefore yields a true estimate of b_i. Regrettably, experimental control data are difficult and expensive to collect.

15.8 The Caveat

There is one caveat to ensure the proper interpretation of the partial regression coefficient. *It is not enough to know the variable being multiplied by the regression coefficient; the other Xs must also be known* [3]. Recognition of the other Xs regarding their values in the sample ensures that interpretation is valid. Specifically, the average change b_i in Y is valid for each and every unit change in X_i, within the range of X_i values in the sample when the other Xs are held constant within the ranges or region of the values of the other Xs in the sample.* The clarification of this point is in the following illustration.

I regress SALES (in dollar units) on EDUCATION (abbreviated EDUC, in year units), AGE (in year units), GENDER (in female gender units; 1 = female and 0 = male), and INCOME (in thousand-dollar units). The regression equation is in Equation 15.2:

$$\text{SALES} = 68.5 + 0.75{*}\text{AGE} + 1.03{*}\text{EDUC} + 0.25{*}\text{INCOME} + 6.49{*}\text{GENDER} \qquad (15.2)$$

Interpretation of the regression coefficients is as follows:

1. The individual variable ranges in Table 15.1 suffice to mark the boundary of the region of the values of the other Xs.
2. For AGE, the average change in SALES is 0.75 dollar for each one-year increase in AGE within the range of AGE values in the sample when EDUC, INCOME, and GENDER (E–I–G) are held constant within the E–I–G region in the sample.
3. For EDUC, the average change in SALES is 1.03 dollars for each one-year increase in EDUC within the range of EDUC values in the sample when AGE, INCOME, and GENDER (A–I–G) are held constant within the A–I–G region in the sample.
4. For INCOME, the average change in SALES is 0.25 dollar for each $1,000 increase in INCOME within the range of income values in the sample when AGE, EDUC, and GENDER (A–E–G) are held constant within the A–E–G region in the sample.

* The region of the values of the other Xs is defined as the values in the sample common to the entire individual variable ranges of the other Xs.

TABLE 15.1
Descriptive Statistics of Sample

Variable	Mean	Ranges (min, max)	StdDev	H-spread
SALES	30.1	(8, 110)	23.5	22
AGE	55.8	(44, 76)	7.2	8
EDUC	11.3	(7, 15)	1.9	2
INCOME	46.3	(35.5, 334.4)	56.3	28
GENDER	0.58	(0, 1)	0.5	1

5. For GENDER, the average change in SALES is 6.49 dollars for a female-gender unit increase (a change from male to female) when AGE, INCOME, and EDUC (A–I–E) are held constant within the A–I–E region in the sample.

To further the discussion on the proper interpretation of the regression coefficient, I consider a composite variable (e.g., $X_1 + X_2$, X_1*X_2, or X_1/X_2), which often gets into regression models. I show that the regression coefficients of the composite variable and the variables defining the composite variable are not interpretable.

I add a composite variable (defined by multiplication) to the original regression model: EDUC_INC (= EDUC*INCOME, in years*thousand-dollar units). The resultant regression model is in Equation 15.3:

$$\begin{aligned} \text{SALES} = 72.3 + 0.77\text{*AGE} + 1.25\text{*EDUC} + 0.17\text{*INCOME} \\ + 6.24\text{*GENDER} + 0.006\text{*EDUC\_INC} \end{aligned} \tag{15.3}$$

Interpretation of the new regression model and its coefficients is as follows:

1. The coefficients of the original variables have changed. This expected change is because the value of the regression coefficient for X_i depends not only on the relationship between Y and X_i but also on the relationships between the other Xs and X_i and the other Xs and Y.
2. Coefficients for AGE and GENDER changed from 0.75 to 0.77 and from 6.49 to 6.24, respectively.
3. For AGE, the average change in SALES is 0.77 dollar for each one-year increase in AGE within the range of AGE values in the sample when EDUC, INCOME, GENDER, and EDUC_INC (E–I–G–E_I) are held constant within the E–I–G–E_I region in the sample.
4. For GENDER, the average change in SALES is 6.24 dollars for a female-gender unit increase (a change from male to female) when AGE, EDUC, INCOME, and EDUC_INC (A–E–I–E_I) are held constant within the A–E–I–E_I region in the sample.
5. Unfortunately, the inclusion of EDUC_INC in the model compromises the interpretation of the regression coefficients for EDUC and INCOME—two variables whose interpretation is not possible. Consider the following:
   a. For EDUC, the usual interpretation is that the average change in SALES is 1.25 dollars for each one-year increase in EDUC within the range of EDUC values in the sample when AGE, INCOME, GENDER, and EDUC_INC (A–I–G–E_I) are held constant within the A–I–G–E_I region in the sample. *This statement is meaningless.* It is not possible to hold constant EDUC_INC for any one-year increase in EDUC. As EDUC would vary, so EDUC_INC would vary. Thus, no meaningful interpretation is possible for the regression coefficient for EDUC.

b. Similarly, no meaningful interpretations are possible for the regression coefficients for INCOME and EDUC_INC. It is not possible to hold constant EDUC_INC for INCOME. And, for EDUC_INC, it is not possible to hold constant EDUC and INCOME.

15.9 Returning to Question 2

How does the X_i affect the prediction of Y? The usual answer is that Y experiences an average change of b_i with a unit increase in X_i when the other Xs are held constant.

This answer is an honest one (because of the statistical control method that estimates b_i) *only with mention* of the values of the X_i range and the region of the other Xs. Unfortunately, the effects of a composite variable and the variables defining the composite variable are undeterminable because their regression coefficients are not interpretable.

15.10 Ranking Predictor Variables by Effect on Prediction

I return to the first regression model in Equation 15.2.

$$\text{SALES} = 68.5 + 0.75{*}\text{AGE} + 1.03{*}\text{EDUC} + 0.25{*}\text{INCOME} + 6.49{*}\text{GENDER} \tag{15.2}$$

A common misinterpretation of the regression coefficient is that GENDER has the greatest effect on SALES, followed by EDUC, then AGE, and lastly INCOME because the coefficient ranking is in that order. The problem with this interpretation is in the following discussions. I also present the correct rule for ranking predictor variables regarding their effects on the prediction of dependent variable Y.

This regression model illustrates the difficulty in relying on the regression coefficient for ranking predictor variables. The regression coefficients are incomparable because different units are involved. No meaningful comparison can be made between AGE and INCOME because the variables have different units (years and thousand-dollar units, respectively). Comparing GENDER and EDUC is another mixed-unit comparison between female gender and years, respectively. Even a comparison between AGE and EDUC, whose units are the same (i.e., years), is problematic because the variables have unequal spreads (e.g., standard deviation, denoted StdDev) in Table 15.1.

The correct ranking of predictor variables regarding their effects on the prediction of Y is the ranking of the variables by the magnitude of the standardized regression coefficient. The sign of the standardized coefficient is not relevant as it only indicates direction. (Exception to this rule is discussed further in the chapter.) The standardized regression coefficient (also known as the beta regression coefficient) is the product of the original regression coefficient (also called the raw regression coefficient) and a conversion factor (CF). The standardized regression coefficient is unitless, just a plain number allowing meaningful comparisons among the variables. The transformation equation that converts a unit-specific raw regression coefficient into a unitless standardized regression coefficient is in Equation 15.4:

$$\begin{aligned}&\text{Standardized regression coefficient for } X_i\\ &\quad= \text{CF}{*}\text{Raw regression coefficient for } X_i\end{aligned} \tag{15.4}$$

TABLE 15.2
Raw and Standardized Regression Coefficients

		Standardized Coefficient (unitless) Based on	
Variable (unit)	**Raw Coefficients (unit)**	**StdDev**	**H-Spread**
AGE (years)	0.75 (dollars/years)	0.23	0.26
EDUC (years)	1.03 (dollars/years)	0.09	0.09
INCOME (000 dollars)	0.25 (dollars/000 dollars)	0.59	0.30
GENDER (female gender)	6.49 (dollars/ female-gender)	0.14	0.27

The CF is the ratio of a unit measure of Y variation to a unit measure of X_i variation. The StdDev is the usual measure used. However, if the distribution of a variable is not bell-shaped, then the StdDev is not reliable, and the resultant standardized regression coefficient is questionable. An alternative measure, one that is not affected by the shape of the variable, is the H-spread. The H-spread is the difference between the 75th percentile and 25th percentile of the variable distribution. Thus, there are two popular CFs and two corresponding transformations in Equations 15.5 and 15.6.

$$\text{Standardized regression coefficient for } X_i \\ = [\text{StdDev of } X_i / \text{StdDev of Y}]^* \text{Raw regression coefficient for } X_i \quad (15.5)$$

$$\text{Standardized regression coefficient for } X_i \\ = [\text{H-spread of } X_i / \text{H-spread of Y}]^* \text{Raw regression coefficient for } X_i \quad (15.6)$$

Returning to the illustration of the first regression model, I note the descriptive statistics in Table 15.1. AGE has equivalent values for both StdDev and H-spread: 23.5 and 22, respectively. EDUC has equivalent values for both StdDev and H-spread: 7.2 and 8, respectively. INCOME, which is typically not bell-shaped, has quite different values for an unreliable StdDev and a reliable H-spread: 56.3 and 28, respectively.

Dummy variables have no meaningful measure of variation. StdDev and H-spread, often reported for dummy variables as a matter of course, have no value at all.

The correct ranking of the predictor variables regarding their effects on SALES—based on the magnitude of the standardized regression coefficients in Table 15.2—puts INCOME first, with the greatest effect, followed by AGE and EDUC. This ordering depends on either the StdDev or the H-spread. However, INCOME's standardized coefficient should be based on the H-spread as INCOME typically has a sizeable skewness. Because GENDER is a dummy variable with no meaningful CF, its effect on the prediction of Y cannot be ranked.

15.11 Returning to Question 3

Which variables in the model have the greatest effect on the prediction of Y, in ranked order? The usual answer is that the variable with the largest regression coefficient has the greatest effect; the variable with the next largest regression coefficient has the next greatest effect, and so on. This answer is usually incorrect.

The correct ranking of predictor variables regarding their effects on the prediction of Y is the ranking of the variables by the magnitude of the standardized regression coefficient. Predictor variables that cannot be ranked are (1) dummy variables, which have no meaningful measure of variation and (2) composite variables and the variables defining them, which have both uninterpretable raw regression coefficients and standardized regression coefficients.

15.12 Returning to Question 4

Which variables in the model are the most important predictors, in ranked order? The usual answer is that the variable with the largest standardized regression coefficient is the most important variable; the variable with the next largest standardized regression coefficient is the next important variable, and so on.

This answer is correct only when the predictor variables are uncorrelated, a rare occurrence in most models. With uncorrelated predictor variables in a regression model, there is a rank order correspondence between the magnitude of the standardized coefficient and reduction in prediction error. Thus, the magnitude of the standardized coefficient can rank the predictor variables in order of most to least important. Unfortunately, there is no rank order correspondence with correlated predictor variables. Thus, the magnitude of the standardized coefficient of correlated predictor variables cannot rank the variables regarding predictive importance. The proof of these facts is beyond the scope of this chapter [4].

15.13 Summary

It should now be clear that the common misconceptions of the regression coefficient lead to an incorrect interpretation of the ordinary regression model. This exposition puts to rest these misconceptions, promoting an appropriate and useful presentation of the regression model.

Common misinterpretations of the regression coefficient are problematic. The use of the statistical p-value as a sole measure for declaring predictor variable X important is sometimes problematic because of its sensitivity to sample size and spread of the X values. In experimental studies, the model builder must ensure that the study design takes into account these sensitivities to allow drawing valid inferences. In big data applications, variables with small p-values must undergo a final assessment of importance based on their actual reduction in prediction error. Variables associated with the greatest reduction in prediction error can be declared important predictors.

When assessing how X_i affects the prediction of Y, the model builder must report the other Xs regarding their values. Moreover, the analyst must not attempt to assess the effects of a composite variable and the variables defining the composite variable because their regression coefficients are not interpretable.

By identifying the variables in the model that have the greatest effect on the prediction of Y—in rank order—the superiority of the standardized regression coefficient for providing a correct ranking of predictor variables, rather than the raw regression coefficient, should be apparent. Furthermore, it is important to recognize that the standardized regression coefficient ranks variables in order of predictive importance (most to least) for only uncorrelated predictor variables. This ranking is not true for correlated predictor variables. It is imperative that the model builder does not use the coefficient to rank correlated predictor variables despite the allure of its popular misuse.

References

1. Kraemer, H.C., and Thiemann, S., *How Many Subjects?* Sage, Thousand Oaks, CA, 1987.
2. Dash, M., and Liu, H., *Feature Selection for Classification*, Intelligent Data Analysis, Elsevier Science, New York, 1997.
3. Mosteller, F., and Tukey, J., *Data Analysis and Regression*, Addison-Wesley, Reading, MA, 1977.
4. Hayes, W.L., *Statistics for the Social Sciences*, Holt, Rinehart and Winston, Austin, TX, 1972.

16

The Average Correlation: A Statistical Data Mining Measure for Assessment of Competing Predictive Models and the Importance of the Predictor Variables

16.1 Introduction

The purpose of this chapter is to introduce the number one statistic, the mean, and the runner-up, the correlation coefficient, which when used together as discussed here yield the *average correlation*, providing a fruitful statistical data mining measure. The average correlation, along with the correlation coefficient, provides a quantitative criterion for assessing (1) competing predictive models and (2) the importance of the predictor variables. I provide an SAS© subroutine for calculating the average correlation in Chapter 44.

16.2 Background

Two essential characteristics of a predictive model are its *reliability* and *validity*, terms that are often misunderstood and misused. *Reliability* refers to the model* yielding consistent results.† For a predictive model, the proverbial question is how dependable (reliable) the model is for predictive purposes. A predictive model is reliable to the extent the model is producing repeatable predictions. It is normal for an individual to vary in performance because chance influences are always in operation, but performance is expected to fall within narrow limits. Thus, for a predictive model, predictions for the same individual, obtained from repeated implementations of the (reliable) model, are expected to fluctuate closely.

Model *validity* refers to the extent to which the model measures what it is intended to measure on a given criterion (e.g., a predictive model criterion is small prediction errors). One indispensable element of a valid model is it has high reliability. If the reliability of the

* A "model" can be predictive (statistical regression), explanatory (principal components analysis, PCA), or predictive and explanatory (structural equation).

† A model should be monitored for consistent results after each implementation. If the results of the model show signs of degradation, then the model should either be recalibrated (update the regression coefficients: keep the same variables in the model but use fresh data) or retrained (update the model: add new variables to the original variables of the model and use fresh data).

model is low, the validity of the model is low. Reliability is a necessary, but not sufficient, condition for a valid model. Thus, a predictive model is valid to the extent the model is *efficient* (precise and accurate) in predicting at a given time. I give a clarifying explanation of efficiency in the next section.

The common question—Is the model valid?—is not directly answerable. A model does not possess a standard level of validity because it may be highly valid at one point in time but not at another because the environment about the initial modeling process is likely to change. The inference is that models have to be kept in a condition of good efficiency.

There are two other aspects of model validity: face validity and content validity. *Face validity* is a term used to characterize the notion of a model "looking like it is going to work." It is a subjective criterion valuable to model users. Face validity gives users, especially those who may not have the specialized background to build a model but do have practical knowledge of how models work, what they want of a model. Thus, if the model does not look right for the required objective, the confidence level of the model utility drops, as does the acceptance and implementation of the model.

Content validity refers to the variables in a model in that the content of the individual variables and the ensemble of the variables are relevant to the purpose of the model. As a corollary, the model should not have irrelevant, unnecessary variables. Such variables are subject to elimination by assessing the correlation between the dependent variable and each variable preliminarily. The correlation coefficient for this elimination process is not foolproof because at times it allows variables with nonrelevant content to creep spuriously into the model. More important, taking out the indicated creepy variables is best achieved by determining the *quality* of the variable selection used for defining the best subset of the predictor variables. In other words, the model builder must assess *the content of the variables subjectively*. An *objective* discussion of content validity is beyond the scope of this chapter.*

The literature† on validity does not address the left side of the model equation: the dependent variable. The dependent variable often reflects an expression in management terms (i.e., nonoperational terms). The model builder needs the explicit backstory about the management objective to create a valid definition of the dependent variable. Consider the management objective to build a lifetime value (LTV) model to estimate the expected LTV over the next 5 years (LTV5). Defining the dependent variable LTV5 is not always straight away. For most modeling projects, there are not enough historical data to build an LTV5 model with a literal definition of dependent variable LTV5 using, say, the current 5 years of data. That is, if using a full 5-year window of data for the definition of LTV5, there are not enough remaining years of data for a sufficient pool of candidate predictor variables. The model builder has to shorten the LTV5 window, say, using LTV2, defined by a window of 2 years, and then define LTV5 = LTV2*2.5. Such a legitimate arithmetic adjustment yields a sizable amount of data for a promising set of candidate predictor variables along with an acceptable content-valid dependent variable. As a result, the building of an LTV5 model is in a favorable situation for excellent valid and reliable performance.

* An excellent reference is Carmines, E.G., and Zeller, R.A., *Reliability and Viability Assessment*, Sage, Thousand Oaks, CA, 1991.

† The literature—to my knowledge—does not address the left side of the model.

16.3 Illustration of the *Difference* between Reliability and Validity

Reliability does not imply validity. For example, a reliable regression model is predicting consistently (precisely) but may not be predicting what it is intended to predict. Regarding accuracy and precision, reliability is analogous to precision, while validity is analogous to accuracy.

An example often used to illustrate the difference between reliability and validity involves the bathroom scale. If someone weighing 118 pounds steps on the same scale say, five consecutive times, yielding varied readings of, say, 115, 125, 195, 140, and 136, then the scale is not reliable or precise. If the scale yields consistent readings of, say, 130, then the scale is reliable but not valid or accurate. If the scale readings are 118 all five times, then the scale is both reliable and valid.

16.4 Illustration of the *Relationship* between Reliability and Validity*

I may have given the impression that reliability and validity are separate concepts. In fact, they complement each other. I discuss a common example of the relationship between reliability and validity. Think of the center of the bull's-eye target as the concept that you seek to measure. You take a shot at the target. If you measure the concept perfectly, you hit the center of the target. If you do not, you miss the center. The more you are off target, the further you are from the center.

There are four situations involving reliability and validity, which are depicted in Figure 16.1:

1. In the first bull's-eye target, you hit the target consistently, but you miss the center of the target. You are producing consistently poor performances among all hits. Your target practice is reliable but not valid.
2. The second display shows that your hits are spread randomly about the target. You seldom hit the center of the target, but you get good performance among all hits. Your target practice is not reliable but valid. Here, you can clearly see that reliability is directly related to the *variability* of your target practice performance.

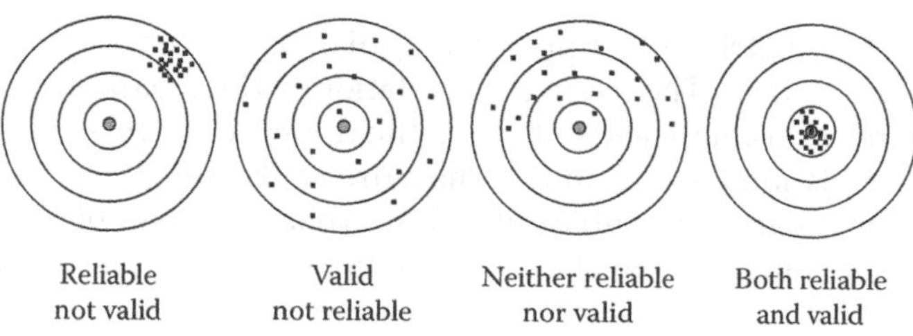

FIGURE 16.1
Relationship between reliability and validity.

* Consult http://www.socialresearchmethods.net/kb/relandval.php, 2010.

3. The third display shows that your hits spread across the target, and you are consistently missing the center. Your target practice, in this case, is neither reliable nor valid.
4. Finally, there is the "Wild Bill Hickok" fanfare*: You hit the center of the target consistently. You (and Wild Bill) are both reliable and valid as sharpshooters.

Note: Reliability and validity concepts themselves are independent of each other. But in the context of a predictive model, reliability is necessary but not sufficient for validity.

16.5 The Average Correlation

The average correlation, along with the correlation coefficient, provides a quantitative criterion for assessing (1) competing predictive models† and (2) the importance of the predictor variables defining a model. About Item 2, numerous statistics textbooks are amiss in explaining which variables in the model have the greatest effect on the prediction of Y. The usual answer is that the variable with the largest regression coefficient has the greatest effect; the variable with the next largest regression coefficient has the next greatest effect, and so on. This answer is usually incorrect. It is correct only when the predictor variables are uncorrelated, quite a rare occurrence in practice.‡ An illustration is the best route to make the average correlation understandable and, hopefully, part of the model builders' toolkit for addressing Items 1 and 2.

16.5.1 Illustration of the Average Correlation with an LTV5 Model

Let us consider the following objective: to build an LTV5 model. The previous discussion of defining the LTV5-dependent variable serves as background to this illustration. The LTV5 model consists of 11 predictor variables (VAR1–VAR11), whose content domain is the nature of sales performance, sales incentives, sales program enrollment, number of transactions, and the like. The correlation matrix of VAR1–VAR11 is in Table 16.1.

For ease of presentation, I replace the "backslash" diagonal of correlation coefficients of Value 1 with "dots" (see Table 16.2). (The correlation coefficient of a variable with itself is obviously equal to 1.) The purpose of replacing the 1s is to show where within the matrix the calculations are performed. The correlation coefficients are all positive—*not always* found in practice.

The dots divide the matrix into upper triangular and lower triangular correlation matrices, which are identical. The average correlation is the mean of the absolute values of the pairwise correlation coefficients of either the upper- or lower-triangular matrix but not both. As the correlation coefficient is a measure of the reliability or closeness of the relationship between a pair of variables, it follows that the mean of all pairs is pleasing to the mind as an *honest* measure of closeness among the predictor variables in a model.

* James Butler Hickok (May 27, 1837–August 2, 1876), better known as Wild Bill Hickok, was a figure in the American Old West. He fought in the Union Army during the American Civil War and gained publicity after the war as a sharpshooter.

† See "A Dozen Statisticians, a Dozen Outcomes" on my website: http://www.geniq.net/res/A-Dozen-Statisticians-with-a-Dozen-Outcomes.html.

‡ I am mindful that if the model builder uses principal component (PC) variables, the PCs are uncorrelated.

TABLE 16.1

LTV Model: Pearson Correlation Coefficients/Prob > |r| under HO: Rho = 0

	VAR1	VAR2	VAR3	VAR4	VAR5	VAR6	VAR7	VAR8	VAR9	VAR10	VAR11
VAR1	1.00000	−0.69609	0.26903	0.30443	0.35499	0.35166	0.37812	0.31297	0.35020	0.29410	0.25545
		<0.0001	<0.0001	<0.0001	<0.0001	<0.0001	<0.0001	<0.0001	<0.0001	<0.0001	<0.0001
VAR2	−0.69689	1.00000	−0.18834	0.19207	−0.22438	−0.21200	−0.23549	−0.20007	−0.20309	−0.17179	−0.14809
	<0.0001		<0.0001	<0.0001	<0.0001	<0.0001	<0.0001	<0.0001	<0.0001	<0.0001	<0.0001
VAR3	0.26903	−0.10034	1.00000	0.22467	0.37111	0.31629	0.39288	0.38376	0.42024	0.29891	0.26452
	<0.0001	<0.0001		<0.0001	<0.0001	<0.0001	<0.0001	<0.0001	<0.0001	<0.0001	<0.0001
VAR4	0.30443	−0.19207	0.22467	1.00000	0.34542	0.29531	0.33502	0.26600	0.24341	0.23202	0.26305
	<0.0001	<0.0001	<0.0001		<0.0001	<0.0001	<0.0001	<0.0001	<0.0001	<0.0001	<0.0001
VAR5	0.35499	−0.22438	0.37111	0.34542	1.00000	0.50653	0.46107	0.39413	0.41061	0.44856	0.33444
	<0.0001	<0.0001	<0.0001	<0.0001		<0.0001	<0.0001	<0.0001	<0.0001	<0.0001	<0.0001
VAR6	0.35166	−0.21200	0.31629	0.29531	0.50653	1.00000	0.46999	0.35111	0.44730	0.48984	0.34348
	<0.0001	<0.0001	<0.0001	<0.0001	<0.0001		<0.0001	<0.0001	<0.0001	<0.0001	<0.0001
VAR7	0.37812	−0.23549	0.39288	0.33582	0.46107	0.46999	1.00000	0.40503	0.44634	0.41615	0.33916
	<0.0001	<0.0001	<0.0001	<0.0001	<0.0001	<0.0001		<0.0001	<0.0001	<0.0001	<0.0001
VAR8	0.31297	−0.20007	0.38376	0.26688	0.39413	0.35111	0.40503	1.00000	0.39891	0.29346	0.31058
	<0.0001	<0.0001	<0.0001	<0.0001	<0.0001	<0.0001	<0.0001		<0.0001	<0.0001	<0.0001
VAR9	0.35020	−0.20389	0.42024	0.24341	0.41061	0.44730	0.44634	0.39891	1.00000	0.44746	0.35423
	<0.0001	<0.0001	<0.0001	<0.0001	<0.0001	<0.0001	<0.0001	<0.0001		<0.0001	<0.0001
VAR10	0.29410	−0.17179	0.29891	0.23282	0.44856	0.48984	0.41615	0.29346	0.44746	1.00000	0.35845
	<0.0001	<0.0001	<0.0001	<0.0001	<0.0001	<0.0001	<0.0001	<0.0001	<0.0001		<0.0001
VAR11	0.25545	−0.14809	0.26452	0.26305	0.33444	0.34348	0.33916	0.31058	0.35423	0.35845	1.00000
	<0.0001	<0.0001	<0.0001	<0.0001	<0.0001	<0.0001	<0.0001	<0.0001	<0.0001	<0.0001	

TABLE 16.2

The Correlation Matrix of VAR1–VAR11 Divided by Replacing the 1's with Dots (.).

	VAR1	VAR2	VAR3	VAR4	VAR5	VAR6	VAR7	VAR8	VAR9	VAR10	VAR11
VAR1	.	0.69689	0.26903	0.30443	0.35499	0.35166	0.37812	0.31297	0.35020	0.29410	0.25545
VAR2	0.69689	.	0.18834	0.19207	0.22438	0.21200	0.23549	0.20007	0.20389	0.17179	0.14809
VAR3	0.26903	0.18834	.	0.22467	0.37111	0.31629	0.39288	0.38376	0.42024	0.29891	0.26452
VAR4	0.30443	0.19207	0.22467	.	0.34542	0.29531	0.33582	0.26688	0.24341	0.23282	0.26305
VAR5	0.35499	0.22438	0.37111	0.34542	.	0.50653	0.46107	0.39413	0.41061	0.44856	0.33444
VAR6	0.35166	0.21200	0.31629	0.29531	0.50653	.	0.46999	0.35111	0.44730	0.48984	0.34348
VAR7	0.37812	0.23549	0.39288	0.33582	0.46107	0.46999	.	0.40503	0.44634	0.41615	0.33916
VAR8	0.31297	0.20007	0.38376	0.26688	0.39413	0.35111	0.40503	.	0.39891	0.29346	0.31058
VAR9	0.35020	0.20389	0.42024	0.24341	0.41061	0.44730	0.44634	0.39891	.	0.44746	0.35423
VAR10	0.29410	0.17179	0.29891	0.23282	0.44856	0.48984	0.41615	0.29346	0.44746	.	0.35845
VAR11	0.25545	0.14809	0.26452	0.26305	0.33444	0.34348	0.33916	0.31058	0.35423	0.35845	.

I digress to justify my freshly minted measure, the average correlation. One familiar with Cronbach's alpha may think the average correlation is a variant of alpha; it is. I did not realize the relationship between the two measures until the writing of this chapter. My background, in part, which includes the field of psychometrics, was in action without my knowing it. Cronbach developed alpha for a measure of reliability within the field of psychometrics. Cronbach started with a variable X, actually a test item score defined as X = t + e, where X is an observed score, t is the true score, and e is the random error [1]. Alpha does not take absolute values of the correlation coefficients; the correlations of alpha are always positive. I developed the average correlation for a measure of closeness among predictor variables *in a model*. "In a model" implies that the average correlation is a conditional measure. That is, the average correlation is a *relative* measure for the reason that it anchors on a dependent variable. Alpha measures reliability among variables belonging to a test; there is no "anchor" variable. In practice, the psychometrician seeks large alpha values, whereas the model builder seeks small average correlation values.

A rule of thumb for a desirable value of the average correlation: small positive values.

1. A small value indicates the predictor variables are not highly correlated.* That is, they do not suffer from the condition of multicollinearity. Multicollinearity makes it virtually impossible to assess the true contribution or importance the predictor variables have on the dependent variable.
   a. Statistics textbooks refer to multicollinearity as a "data problem, not a weakness in the model." Multicollinearity is a data problem because its only affects a clearly assigned contribution of each predictor variable to the dependent. The assigned contribution of each predictor variable is muddy. The muddied contribution is not a weakness in the model and not consequential of model performance, only model interpretation. The textbooks do not address the practical implication of multicollinearity on model performance.

      In practice the impact of multicollinearity on model performance is consequential. Model performance is *not affected* by the condition of multicollinearity *as long as* the condition of multicollinearity found while building the initial model remains the same in the immediate future. If the condition is the same, then implementation of the model yields good performance. However, for every reimplementation of the model after the first, in practice, the condition of multicollinearity *has* shown *not* to remain the same, I uphold and posit that multicollinearity is a data problem, and multicollinearity *does affect* model performance.
2. Average correlation values in the range of 0.35 or less are desirable. In this situation, the assessment of the contributions of the predictor variables to the performance of the model is *soundly* honest.
3. Average correlation values that are greater than 0.35 and less than 0.55 are moderately desirable. In this situation, the assessment of the contributions of the predictor variables to the performance of the model is *fairly* honest.
4. Average correlation values that are greater than 0.55 are not desirable because they indicate the predictor variables are excessively redundant. In this situation, the assessment of the contributions of the predictor variables to the performance of the model is *dubiously* honest.

* Actually highly linearly correlated.

As long as the average correlation value is acceptable (less than 0.40), the second proposed item of assessing competing models (every modeler builds several models and must choose the best one) is in play. If a project session brings forth models within the acceptable range of average correlation values, the model builder uses both the average correlation value *and the set* of the individual correlations of predictor variable with the dependent variable. The individual correlations indicate the content *validity* of the model. Rules of thumb for the values of the individual correlation coefficients are as follows:

1. Values between 0.0 and 0.3 (0.0 and –0.3) indicate poor validity.
2. Values between 0.3 and 0.7 (–0.3 and –0.7) indicate moderate validity.
3. Values between 0.7 and 1.0 (–0.7 and –1.0) indicate a strong validity.

In sum, the model builder uses the average correlation and the individual correlations to assess competing predictive models and the importance of the predictor variables. I continue with the illustration of the LTV5 model to make sense of these discussions and rules of thumb in the next section.

16.5.2 Continuing with the Illustration of the Average Correlation with an LTV5 Model

The average correlation of the LTV5 model is 0.33502. The individual correlations of the predictor variables with LTV5 (Table 16.3) indicate the variables have moderate to strong validity, except for VAR2. The combination of 0.33502 and values of Table 16.3 is compelling for any modeler to be pleased with the reliability and validity of the LTV5 model.

16.5.3 Continuing with the Illustration with a Competing LTV5 Model

Assessing competing models is a task of the model builder who can compare two sets of measures (the average correlation and the set of individual correlations). The model builder must have a keen skill to balance the metrics. The continuing illustration makes the balancing of the metrics clear.

TABLE 16.3

Correlations of Predictors (VAR) with LTV5

Predictors	CORR_COEFF_with_LTV5
VAR8	0.71472
VAR7	0.70277
VAR6	0.68443
VAR5	0.67982
VAR9	0.65602
VAR3	0.61076
VAR10	0.59583
VAR11	0.59087
VAR4	0.52794
VAR1	0.49827
VAR2	–0.27945

TABLE 16.4
Correlations of Predictors (VAR_) with LTV5

Predictors	CORR_COEFF_with_LTV5
VAR_2	0.80377
VAR_3	0.70945
VAR_1	0.62148

I build a second model, the competing model, whose average correlation equals 0.37087. The individual correlations of the predictor variables with LTV5 (Table 16.4) indicate the predictor variables have strong validity, except with VAR_1, which has a moderate degree. Clearly, the difference between the average correlations of 0.33502 and 0.37087, for the first and second models, respectively, offers a quick thought for the modeler to contemplate as the difference is minimal. From there, I examine the individual correlations in Tables 16.3 and 16.4. The competing model has large-valued correlation coefficients and has few (three) variables, an indicator of a reliable model. The first model has individual correlations with mostly strong validity, except for VAR2, but there are too many predictor variables (for my taste). My determination of the better model is the competing model because not only are the metrics pointing to that decision, but also I always prefer a model with no more than a handful of predictor variables.

16.5.3.1 The Importance of the Predictor Variables

As for the importance of the predictor variables, if the average correlation value is acceptable, the importance of the predictor variables is readily apparent from the individual correlations table. Thus, for the better model discussed, the most important predictor variable is VAR_2, followed by VAR_3 and VAR_1, successively.

16.6 Summary

I introduce the fruitful statistical data mining measure, the average correlation, which provides, along with the correlation coefficient, a quantitative criterion for assessing both competing predictive models and the importance of the predictor variables. After providing necessary background, which includes model reliability and validity, I chose an illustration with an LTV5 model as the vehicle to make the average correlation understandable and, hopefully, part of the model builders' toolkit for assessing both competing predictive models and the importance of the predictor variables. I provide an SAS subroutine for calculating the average correlation in Chapter 44.

Reference

1. Cronbach, L.J., Coefficient alpha and the internal structure of tests, *Psychometrika*, 16(3), 297–334, 1951.

17

CHAID for Specifying a Model with Interaction Variables

17.1 Introduction

To increase the predictive power of a model beyond that provided by its components, data analysts create an interaction variable, which is the product of two or more component variables. The purpose of this chapter is to propose chi-squared automatic interaction detection (CHAID) as an alternative data mining method for specifying a model, thereby justifying the omission of the component variables under certain circumstances. Database marketing provides an excellent example of the proposed data mining method. I illustrate the alternative method with a response model case study.

17.2 Interaction Variables

Consider variables X_1 and X_2. The product of these variables, denoted by X_1X_2, is called a *two-way* or *first-order interaction variable*. An obvious property of this interaction variable is that X_1 and X_2 share their information or variance. In other words, X_1X_2 has an inherent high correlation with both X_1 and X_2.

If there is a third variable (X_3), then the product of the three variables ($X_1X_2X_3$) is called a three-way or second-order interaction variable. It is also highly correlated with each of its component variables. Simply multiplying the component variables can create higher-order variables. However, interaction variables of an order greater than 3 are rarely justified by theory or empirical evidence.

When data have highly correlated variables, they have a condition known as multicollinearity. When high correlation exists because of the specific relationship between variables, the multicollinearity assumes *essential ill-conditioning*. An example of this is the correlation between gender and income in the current workforce. The fact that males earn more than their female counterparts creates essential ill-conditioning. However, when the high correlation is due to an interaction variable, the multicollinearity assumes *nonessential ill-conditioning* [1].

When multicollinearity exists, it is difficult to assess reliably the statistical significance, as well as the noticeable informal importance, of the highly correlated variables. Accordingly, multicollinearity makes it difficult to define a strategy for excluding variables.

A sizable literature has developed for essential ill-conditioning, which has produced several approaches for specifying models [2]. In contrast, there is a modest collection of articles for nonessential ill-conditioning [3,4].

17.3 Strategy for Modeling with Interaction Variables

The popular strategy for modeling with interaction variables is the *principle of marginality*, which states that a model including an interaction variable should also include the component variables that define the interaction [5,6]. A cautionary note accompanies this principle: *Neither testing the statistical significance (or noticeable importance) nor interpreting the coefficients of the component variables should be performed* [7]. A significance test, which requires a unique partitioning of the dependent variable variance regarding the interaction variable and its components, is not possible due to the multicollinearity.

An unfortunate by-product of this principle is that models with *unnecessary* component variables go undetected. Such models are prone to overfit, which results in either unreliable predictions or deterioration of performance, as compared to a well-fit model with the necessary component variables.

Nelder's *notion of a special point* offers an alternative strategy [8]. Nelder's strategy relies on understanding the functional relationship between the component variables and the dependent variable. When a theory or prior knowledge about the relationship is limited or unavailable, Nelder suggests using exploration data analysis to determine the relationship. However, Nelder provides no general procedure or guidelines for uncovering the relationship between the variables.

I propose using CHAID as the data mining method for uncovering the functional relationship among the variables. Higher-order interactions are seldom found to be significant, at least in database marketing. Therefore, I limit this discussion only to first-order interactions. If higher-order interactions are required, an extension of the proposed method is straightforward.

17.4 Strategy Based on the Notion of a Special Point

For simplicity, consider the full model in Equation 17.1:

$$Y = b_0 + b_1X_1 + b_2X_2 + b_3X_1X_2 \tag{17.1}$$

$X_1 = 0$ is a *special point* on the scale if when $X_1 = 0$, there is no relationship between Y and X_2. If $X_1 = 0$ is a special point, then omit X_2 from the full model; otherwise, X_2 should not be omitted.

Similarly for X_2, $X_2 = 0$ is a special point on the scale if when $X_2 = 0$, there is no relationship between Y and X_1. If $X_2 = 0$ is a special point, then omit X_1; otherwise, X_1 should not be omitted.

If both X_1 and X_2 have special points, then X_1 and X_2 are omitted from the full model, and the model reduces to $Y = b_0 + b_3X_1X_2$.

If the component variable does not assume a zero value, then no special point exists, and the procedure is not applicable.

17.5 Example of a Response Model with an Interaction Variable

Database marketing provides an excellent example of the proposed data mining method. I illustrate the method with a response model case study, but it applies as well to a profit model. A music continuity club requires a model to increase response to its solicitations. Based on a random sample (size 299,214) of a recent solicitation with a 1% response, I conduct a logistic regression analysis of RESPONSE on two available predictor variables, X_1 and X_2. The variables are defined as:

1. RESPONSE is the indicator of a response to the solicitation: 0 indicates nonresponse, 1 indicates a response.
2. X_1 is the number of months since the last inquiry. A zero month value indicates an inquiry within the month of the solicitation.
3. X_2 is a measure of an individual's affinity for the club based on the number of previous purchases and the listening interest categories of the purchases.

The output of the logistic analysis is in Table 17.1. X_1 and X_2 have Wald chi-square values of 10.5556 and 2.9985, respectively. Using a Wald cutoff value of 4, as outlined in Section 10.12.1, there is an indication that X_1 is an important predictor variable, whereas X_2 is not quite important. The classification accuracy of the base response model is displayed in Table 17.2. The "Total" column represents the actual number of nonresponders and responders: there are 296,120 nonresponders and 3,094 responders. The "Total" row represents the predicted or the classified number of nonresponders and responders: There are 144,402 individuals classified as nonresponders and 154,812 classified as responders. The diagonal cells indicate the correct classifications of the model. The upper-left cell (actual = 0 and classified = 0) indicates the model correctly classified 143,012 nonresponders. The lower-right cell (actual = 1

TABLE 17.1

Logistic Regression of Response on X_1 and X_2

Variable	df	Parameter Estimate	Standard Error	Wald Chi-Square	Pr > Chi-Square
Intercept	1	−4.5414	0.0389	13613.7923	0.E + 00
X_1	1	−0.0338	0.0104	10.5556	0.0012
X_2	1	0.0145	0.0084	2.9985	0.0833

TABLE 17.2

Classification Table of Model with X_1 and X_2

		Classified		
		0	1	Total
Actual	0	143,012	153,108	296,120
	1	1,390	1,704	3,094
Total		144,402	154,812	299,214
			TCCR	48.37%

TABLE 17.3

Logistic Regression of Response on X_1, X_2, and X_1X_2

Variable	df	Parameter Estimate	Standard Error	Wald Chi-Square	Pr > Chi-Square
Intercept	1	−4.5900	0.0502	8374.8095	0.E + 00
X_1	1	−0.0092	0.0186	0.2468	0.6193
X_2	1	0.0292	0.0126	5.3715	0.0205
X_1X_2	1	−0.0074	0.0047	2.4945	0.1142

TABLE 17.4

Classification Table of Model with X_1, X_2, and X_1X_2

		Classified		
		0	1	Total
Actual	0	164,997	131,123	296,120
	1	1,616	1,478	3,094
Total		166,613	132,601	299,214
			TCCR	55.64%

and classified = 1) indicates the model correctly classified 1,704 responders. The total correct classification rate (TCCR) is equal to 48.37% (= (143,012 + 1,704)/299,214).

After creating the interaction variable X_1X_2 (= X_1*X_2), I conduct another logistic regression analysis of RESPONSE with X_1, X_2, and X_1X_2. The output is in Table 17.3. I observe the Wald chi-square values for X_1 and X_2 are reverse in magnitude. The Wald chi-square value for X_1X_2 is close to the value of X_2 in the base model. These observations are moot in that no direct statistical assessment is possible as per the cautionary note.

The classification accuracy of this full response model is displayed in classification Table 17.4. TCCR(X_1, X_2, X_1X_2) equals 55.64%, which represents a 15.0% improvement over the TCCR(X_1, X_2) of 48.37% for the model without the interaction variable. These two TCCR values provide a benchmark for assessing the effects of omitting—if possible under the notion of a special point—X_1 or X_2.

Can component variable X_1 or X_2 be omitted from the full RESPONSE model in Table 17.3? To omit a component variable, say X_2, no relationship between RESPONSE and X_2—when X_1 is a special point—can exist. CHAID can be used to determine whether the relationship exists. Following a brief review of the CHAID technique, I illustrate how to use CHAID for this new approach.

17.6 CHAID for Uncovering Relationships

CHAID is a technique that recursively partitions (or splits) a population into separate and distinct segments. These segments, called *nodes*, are split in such a way that the variation of the dependent variable (categorical or continuous) is minimized within the segments and maximized among the segments. After the initial splitting of the population into two or more nodes (defined by values of an independent or predictor variable), the splitting process repeats on each of the nodes. Each node serves as a new subpopulation. The node then splits into two or more nodes (defined by the values of another predictor variable) such that the variation of the dependent variable is minimized within the nodes and maximized among the nodes. The splitting process repeats until stopping rules are satisfied. The output of CHAID is a *tree* display, where the root is the population and the branches are the connecting segments such that the variation of the dependent variable is minimized within all the segments and maximized among all the segments.

In 1980, CHAID was the first technique originally developed for finding "combination" or interaction variables. In database marketing, CHAID is primarily used today as a market segmentation technique. Here, I utilize CHAID as a data mining method for uncovering the relationship among component variables and the dependent variable to provide the information needed to test for a special point.

For this application of CHAID, the RESPONSE variable is the dependent variable, and the component variables are predictor variables X_1 and X_2. The CHAID analysis is forced to produce a tree, in which the initial splitting of the population is on the component variable to serve as a special point. A zero value must define one of the nodes. Then, the "zero" node is split by the other component variable, producing response rates for testing for a special point.

17.7 Illustration of CHAID for Specifying a Model

Can X_1 be omitted from the full RESPONSE model? If when $X_2 = 0$, there is no relationship between RESPONSE and X_1, then $X_2 = 0$ is a special point, and X_1 can be omitted from the model. The relationship between RESPONSE and X_1 can be assessed easily by the RESPONSE CHAID tree (Figure 17.1), which is read and interpreted as follows:

1. The top box (root node) indicates that for the sample of 299,214 individuals, there are 3,094 responders and 296,120 nonresponders. The response rate is 1%, and nonresponse rate is 99%.
2. The left leaf node (of the first level) of the tree represents 208,277 individuals whose X_2 values are not equal to zero. The response rate among these individuals is 1.1%.
3. The right leaf node represents 90,937 individuals whose X_2 values are equal to zero. The response rate among these individuals is 1.0%.
4. Tree notation: Trees for a continuous predictor variable denote the continuous values in intervals—a closed interval or a left-closed right-open interval. The former is denoted by [a, b], indicating all values between and including a and b. The latter is denoted by [a, b), indicating all values greater than or equal to a and less than b.
5. I reference the bottom row of five branches (defined by the intersection of X_2 and X_1 intervals/nodes), from left to right: 1 through 5.
6. Branch 1 represents 40,995 individuals whose X_2 values are equal to zero *and* X_1 values lie in [0, 2). The response rate among these individuals is 1.0%.
7. Branch 2 represents 17,069 individuals whose X_2 values are equal to zero *and* X_1 values lie in [2, 3). The response rate among these individuals is 0.9%.

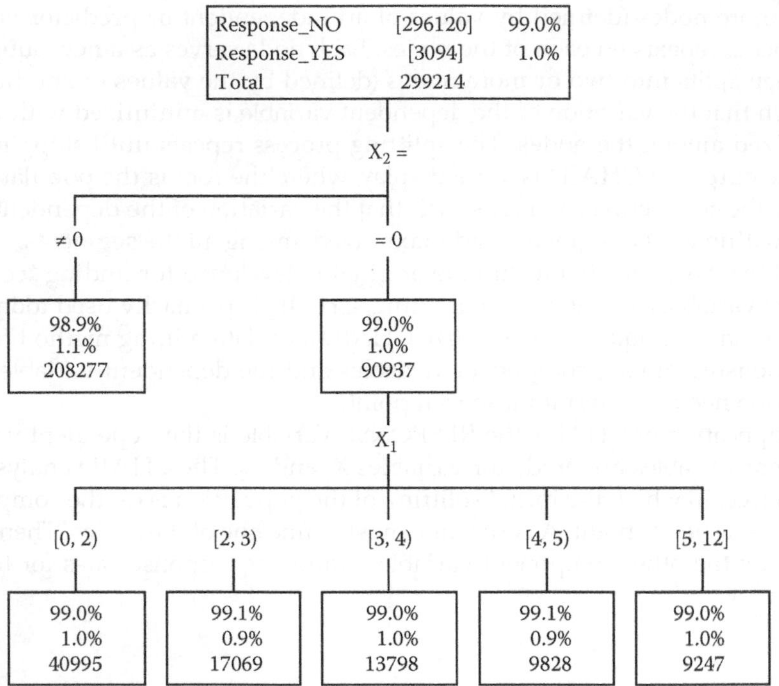

FIGURE 17.1
CHAID tree for testing X_2 for a special point.

8. Branch 3 represents 13,798 individuals whose X_2 values are equal to zero *and* X_1 values lie in [3, 4). The response rate among these individuals is 1.0%.
9. Branch 4 represents 9,828 individuals whose X_2 values are equal to zero *and* X_1 values lie in [4, 5). The response rate for these individuals is 0.9%.
10. Node 5 represents 9,247 individuals whose X_2 values are equal to zero *and* X_1 values lie in [5, 12]. The response rate among these individuals is 1.0%.
11. The pattern of response rates (1.0%, 0.9%, 1.0%, 0.9%, 1.0%) across the five branches reveals there is no relationship between response and X_1 when $X_2 = 0$. Thus, $X_2 = 0$ is a special point.
12. The implication is the model builder can omit X_1 from the response model.

Can X_2 be omitted from the full RESPONSE model? If when $X_1 = 0$ there is no relationship between RESPONSE and X_2, then $X_1 = 0$ is a special point and X_2 can be omitted from the model. The relationship between RESPONSE and X_2 is in the CHAID tree in Figure 17.2, which is read and interpreted as follows.

1. The top box indicates that the sample of 299,214 individuals consists of 3,094 responders and 296,120 nonresponders. The response rate is 1%, and nonresponse rate is 99%.
2. The left leaf node represents 229,645 individuals whose X_1 values are not equal to zero. The response rate among these individuals is 1.0%.
3. The right leaf node represents 69,569 individuals whose X_1 values are equal to zero. The response rate among these individuals is 1.1%.

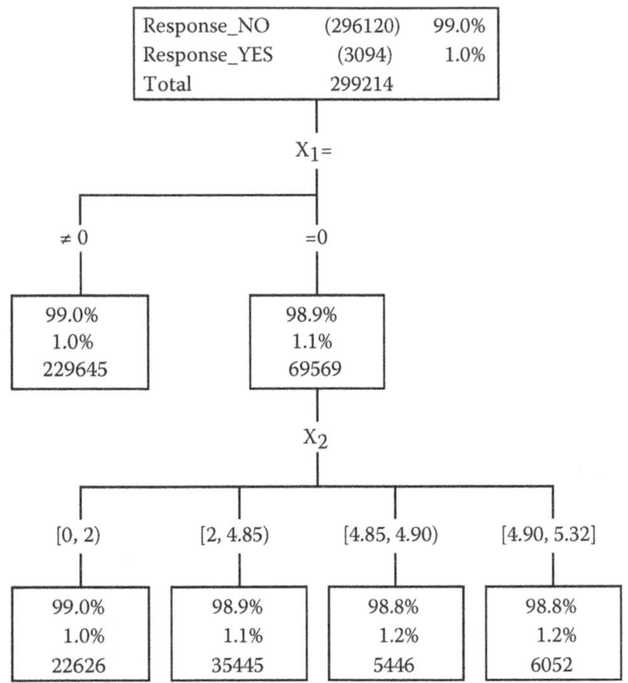

FIGURE 17.2
CHAID tree for testing X_1 for a special point.

4. I reference the bottom row of four branches (defined by the intersection of X_1 and X_2 intervals/nodes), from left to right: 1 through 4.
5. Branch 1 represents 22,626 individuals whose X_1 values are equal to zero *and* X_2 values lie in [0, 2). The response rate among these individuals is 1.0%.
6. Branch 2 represents 35,445 individuals whose X_1 values are equal to zero *and* X_2 values lie in [2, 4.85). The response rate among these individuals is 1.1%.
7. Branch 3 represents 5,446 individuals whose X_1 values are equal to zero *and* X_2 values lie in [4.85, 4.90). The response rate among these individuals is 1.2%.
8. Branch 4 represents 6,052 individuals whose X_1 values are equal to zero *and* X_2 values lie in [4.90, 5.32]. The response rate among these individuals is 1.2%.
9. The pattern of response rates across the four nodes, best observed in the smooth plot of response rates by the minimum values* of the intervals for the X_2 branches, is in Figure 17.3. There appears to be a positive straight-line relationship between RESPONSE and X_2 when $X_1 = 0$. Thus, $X_1 = 0$ is *not* a special point.
10. The implication is the model builder cannot omit X_2 from the response model.

Because I choose to omit X_1, I perform a logistic regression analysis on RESPONSE with X_2 and X_1X_2. The output is in Table 17.5. The classification accuracy of this RESPONSE model is in Table 17.6: TCCR(X_2, X_1X_2) equals 55.64%. TCCR for the full model,

* The minimum value is one of the several values that can be used; alternatives are the means or medians of the predefined ranges.

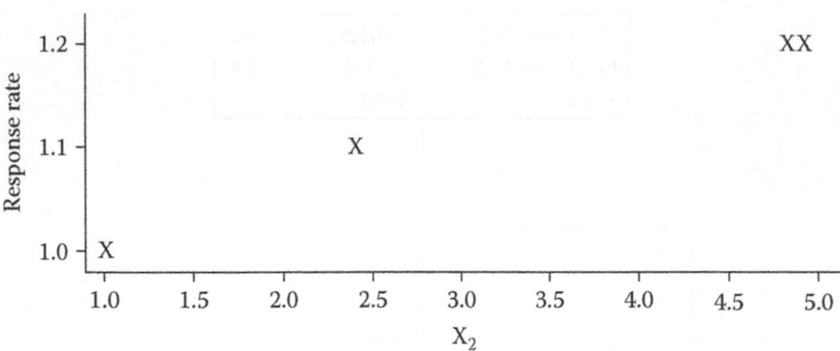

FIGURE 17.3
Smooth plot of response and X_2.

TABLE 17.5
Logistic Regression of Response on X_2 and X_1X_2

Variable	df	Parameter Estimate	Standard Error	Wald Chi-Square	Pr > Chi-Square
Intercept	1	–4.6087	0.0335	18978.8487	0.E + 00
X_1X_2	1	–0.0094	0.0026	12.6840	0.0004
X_2	1	0.0332	0.0099	11.3378	0.0008

TABLE 17.6
Classification Table of Model with X_2 and X_1X_2

		Classified		
		0	1	Total
Actual	0	165,007	131,113	296,120
	1	1,617	1,477	3,094
Total		166,624	132,590	299,214
			TCCR	55.64%

TCCR(X_1, X_2, X_1X_2), is also equal to 55.64%. The implication is that X_1 is an unnecessary variable in the model.

In sum, the parsimonious (best-so-far) RESPONSE model is defined by X_2 and X_1X_2. The CHAID method for specifying a model with interaction variables justifies the omission of the component variables in situations like the one illustrated.

17.8 An Exploratory Look

A closer look at the plot in Figure 17.3 seems to indicate that the relationship between RESPONSE and X_2 bends slightly in the upper-right portion, which would imply that there is a quadratic component in the relationship. Because trees are always exploratory in

TABLE 17.7

Logistic Regression of Response on X_2, X_1X_2, and X_2_SQ

Variable	df	Parameter Estimate	Standard Error	Wald Chi-Square	Pr > Chi-Square
Intercept	1	−4.6247	0.0338	18734.1156	0.E + 00
X_1X_2	1	−0.0075	0.0027	8.1118	0.0040
X_2	1	0.7550	0.1151	43.0144	5.E−11
X_2_SQ	1	−0.1516	0.0241	39.5087	3.E−10

TABLE 17.8

Classification Table of Model with X_2, X_1X_2, and X_2_SQ

		Classified		
		0	1	Total
Actual	0	191,998	104,122	296,120
	1	1,838	1,256	3,094
Total		193,836	105,378	299,214
			TCCR	64.59%

nature, I choose to test the X_2 squared term (X_2*X_2), denoted by X_2_SQ, in the model. (Note: the bulging rule discussed in Chapter 10, which seeks to straighten unconditional data, does not apply here because the relationship between RESPONSE and X_2 is a conditional one as it is based on individuals with the "condition $X_1 = 0$.")

The logistic regression analysis on RESPONSE with X_2, X_1X_2, and X_2_SQ is in Table 17.7. The classification accuracy of this model is shown in Table 17.8. TCCR(X_2, X_1X_2, X_2_SQ) equals 64.59%, which represents a 16.1% improvement over the best-so-far model with TCCR(X_2, X_1X_2) equals 55.64%.

I conclude that the relationship is quadratic, and the corresponding model is a good fit of the data. Thus, X_2, X_1X_2, and X_2_SQ define the best RESPONSE model.

17.9 Database Implication

Database marketers are among those who use response models to identify individuals most likely to respond to their solicitations and thus place more value on the information in the Actual = 1, and Classified = 1 entry—the number of responders correctly classified—than in the TCCR. Table 17.9 indicates the number of responders correctly classified for the models tested. The model that *appears* to be the best is not the best for a database marketer because it identifies the least number of responders (1,256).

I summarize the modeling process as follows: The base RESPONSE model with the two original variables, X_1 and X_2, produces TCCR(X_1, X_2) = 48.37%. The interaction variable X_1X_2, added to the base model, produces the full model with TCCR(X_1, X_2, X_1X_2) = 55.64%, for a 15.03% [=(55.64% − 48.37%)/48.37%] classification improvement over the base model.

TABLE 17.9

Summary of Model Performance

Model		TCCR	Number of Responder	RCCR
Type	Defined by	(%)	Correct Classification	(%)
Base	X_1, X_2	48.37	1,704	1.10
Full	X_1, X_2, X_1X_2	55.64	1,478	1.11
Best so Far	X_2, X_1X_2	55.64	1,477	1.11
Best	X_2, X_1X_2, X_2_SQ	64.59	1,256	1.19

Using the new CHAID-based data mining approach to determine whether a component variable is subject to omission, I observe that X_2 (but not X_1) is subject to the omission from the full model. Thus, the best-so-far model—with X_2 and X_1X_2—has no loss of performance over the full model: TCCR(X_2, X_1X_2) = TCCR(X_1, X_2, X_1X_2) = 55.64%.

A closer look at the smooth plot in RESPONSE and X_2 suggests that X_2_SQ should enter to the best-so-far model. The inclusion of X_2_SQ produces the best model with a TCCR(X_2, X_1X_2, X_2_SQ) of 64.59%, which indicates a 16.01% (=(64.59% – 55.64%)/55.64%) classification improvement over the best-so-far model.

Database marketers assess the performance of a response model by how well the model correctly classifies responders among the total number of individuals classified as responders. That is, the percentage of responders correctly classified, or the responder correct classification rate (RCCR), is the pertinent measure. For the base model, the "Total" row in Table 17.2 indicates the model classifies 154,812 individuals as responders, among whom there are 1,704 correctly classified: RCCR is 1.10% in Table 17.9. RCCR values for the best, best-so-far, and full models are 1.19%, 1.11%, and 1.11%, respectively, in Table 17.9. Accordingly, the best RCCR-based model is still the best model, originally based on TCCR.

It is interesting to note that the performance improvement based on RCCR is not as large as the improvement based on TCCR. The best model compared to the best-so-far model has a 7.2% (= 1.19%/1.11%) RCCR improvement versus 16.1% (= 64.59%/55.64%) TCCR improvement.

17.10 Summary

After briefly reviewing the concepts of interaction variables and multicollinearity and the relationship between the two, I restate the popular strategy for modeling with interaction variables. The principle of marginality states that a model including an interaction variable should also include the component variables that define the interaction. I reinforce the cautionary note that accompanies this principle: The data analyst should neither test the statistical significance (or noticeable importance) nor interpret the coefficients of the component variables. Moreover, I point out that an unfortunate by-product of this principle is that models with unnecessary component variables go undetected, resulting in either unreliable predictions or deterioration of performance.

Then, I present an alternative strategy, based on Nelder's notion of a special point. I define the strategy for first-order interaction, X_1*X_2, as higher-order interactions are rare in database marketing applications. Predictor variable $X_1 = 0$ is a special point on the scale if when $X_1 = 0$, there is no relationship between the dependent variable and a second predictor variable X_2. If $X_1 = 0$ is a special point, then omit X_2 from the model; otherwise, X_2 should not be omitted. I propose using CHAID as a data mining method for determining whether there is a relationship between the dependent variable and X_2.

I present a case study involving the building of a database marketing response model to illustrate the special point CHAID data mining method. The resultant response model, which omitted one component variable, clearly demonstrated the utility of the new method. Then, I took advantage of the full-bodied case study, as well as the mantra of data mining (never stop digging into the data), to improve the model. I determine that the additional term, the square of the included component variable, adds 16.2% improvement over the original response model regarding the traditional measure of model performance, the TCCR.

Digging a little deeper, I emphasize the difference between the traditional and database measures of model performance. Database marketers are more concerned about models with a larger RCCR than a larger TCCR. As such, the improved response model, which initially appeared not to have the largest RCCR, was the best model regarding RCCR as well as TCCR.

References

1. Marquardt, D.W., You should standardize the predictor variables in your regression model, *Journal of the American Statistical Association*, 75, 87–91, 1980.
2. Aiken, L.S., and West, S.G., *Multiple Regression: Testing and Interpreting Interactions*, Sage, Thousand Oaks, CA, 1991.
3. Chipman, H., Bayesian variable selection with related predictors, *Canadian Journal of Statistics*, 24, 17–36, 1996.
4. Peixoto, J.L., Hierarchical variable selection in polynomial regression models, *The American Statistician*, 41, 311–313, 1987.
5. Nelder, J.A., Functional marginality is important (letter to editor), *Applied Statistics*, 46, 281–282, 1997.
6. McCullagh, P.M., and Nelder, J.A., *Generalized Linear Models*, Chapman & Hall, London, 1989.
7. Fox, J., *Applied Regression Analysis, Linear Models, and Related Methods*, Sage, Thousand Oaks, CA, 1997.
8. Nelder, J.A., The selection of terms in response-surface models—How strong is the weak-heredity principle? *The American Statistician*, 52, 315–318, 1998.

18

*Market Segmentation Classification Modeling with Logistic Regression**

18.1 Introduction

Logistic regression analysis is a recognized technique for classifying individuals into two groups. Perhaps less known but equally important, polychotomous logistic regression (PLR) analysis is another method for performing classification. The purpose of this chapter is to present PLR analysis as a multigroup classification technique. I illustrate the technique using a cellular phone market segmentation study to build a market segmentation classification model as part of a customer relationship management (better known as CRM) strategy.

I start the discussion by defining the typical two-group (binary) logistic regression model. After introducing necessary notation for expanding the binary logistic regression (BLR) model, I define the PLR model. For readers uncomfortable with such notation, the PLR model provides several equations for classifying individuals into one of many groups. The number of equations is one less than the number of groups. Each equation looks like the BLR model.

After a brief review of the estimation and modeling processes used in PLR, I illustrate PLR analysis as a multigroup classification technique with a case study based on a survey of cellular phone users. The survey data were used initially to segment the cellular phone market into four groups. I use PLR analysis to build a model for classifying cellular users into one of the four groups.

18.2 Binary Logistic Regression

Let Y be a binary dependent variable that assumes two outcomes or classes, typically labeled 0 and 1. The BLR model classifies an individual into one of the classes based on the values of the predictor (independent) variables $X_1, X_2, \ldots, X_n$ for that individual. BLR estimates the *logit of Y*—the log of the odds of an individual belonging to Class 1; the logit is in Equation 18.1. The logit easily converts into the probability of an individual belonging to Class 1, Prob(Y = 1), defined in Equation 18.2.

* This chapter is based on an article with the same title in *Journal of Targeting Measurement and Analysis for Marketing*, 8, 1, 1999. Used with permission.

$$\text{logit Y} = b_0 + b_1{*}X_1 + b_2{*}X_2 + \ldots + b_n{*}X_n \tag{18.1}$$

$$\text{Prob}(Y = 1) = \frac{\exp(\text{logit Y})}{1 + \exp(\text{logit Y})} \tag{18.2}$$

An individual's predicted probability of belonging to Class 1 is calculated by plugging in the values of the predictor variables for that individual in Equations 18.1 and 18.2. The b's are the logistic regression coefficients, determined by the calculus-based method of maximum likelihood. Note, unlike the other coefficients, b_0 (referred to as the intercept) has no corresponding predictor variable. Needless to say, the probability of an individual belonging to Class 0 is 1 – Prob(Y = 1).*

18.2.1 Necessary Notation

I introduce notation for the discussion in the next section. There are several explicit restatements in Equations 18.3, 18.4, 18.5, and 18.6 of the logit of Y of Equation 18.1. They are superfluous when Y takes on only two values, 0 and 1:

$$\text{logit Y} = b_0 + b_1{*}X_1 + b_2{*}X_2 + \ldots + b_n{*}X_n \tag{18.3}$$

$$\text{logit}(Y = 1) = b_0 + b_1{*}X_1 + b_2{*}X_2 + \ldots + b_n{*}X_n \tag{18.4}$$

$$\text{logit}(Y = 1 \text{ vs. } Y = 0) = b_0 + b_1{*}X_1 + b_2{*}X_2 + \ldots + b_n{*}X_n \tag{18.5}$$

$$\text{logit}(Y = 0 \text{ vs. } Y = 1) = -\,[b_0 + b_1{*}X_1 + b_2{*}X_2 + \ldots + b_n{*}X_n] \tag{18.6}$$

Equation 18.3 is the standard notation for the BLR model. Y implies the class modeled is 1. Equation 18.4 explicitly states the class modeled is 1. Equation 18.5 formally indicates that the class modeled is 1 versus 0. Equation 18.6 is the reverse of Equation 18.5, in that class modeled is 0 versus 1. Specifically, Equation 18.6 is the negative of Equation 18.5, as the negative sign on the right-hand side of the equation indicates.

18.3 Polychotomous Logistic Regression Model

When the categorical dependent variable takes on more than two outcomes or classes, the PLR model, an extension of the BLR model, can be used to predict the class membership. For ease of presentation, I discuss Y with three categories, coded 0, 1, and 2.

Three binary logits are in Equations 18.7, 18.8, and 18.9.†

$$\text{logit\_10} = \text{logit}(Y = 1 \text{ vs. } Y = 0) \tag{18.7}$$

$$\text{logit\_20} = \text{logit}(Y = 2 \text{ vs. } Y = 0) \tag{18.8}$$

$$\text{logit\_21} = \text{logit}(Y = 2 \text{ vs. } Y = 1) \tag{18.9}$$

* Because Prob(Y = 0) + Prob(Y = 1) = 1.

† It can be shown that from any pair of logits, the remaining logit can be obtained.

I use the first two logits (because of the similarity with the standard expression of the BLR) to define the PLR model in Equations 18.10, 18.11, and 18.12:

$$\text{Prob}(Y=0) = \frac{1}{1+\exp(\text{logit\_10})+\exp(\text{logit\_20})} \tag{18.10}$$

$$\text{Prob}(Y=1) = \frac{\exp(\text{logit\_10})}{1+\exp(\text{logit\_10})+\exp(\text{logit\_20})} \tag{18.11}$$

$$\text{Prob}(Y=2) = \frac{\exp(\text{logit\_20})}{1+\exp(\text{logit\_10})+\exp(\text{logit\_20})} \tag{18.12}$$

The PLR model is easily extended when there are more than three classes. When Y = 0, 1, 2, ..., k (i.e., k + 1 outcomes), the model is defined in Equations 18.13, 18.14, and 18.15:

$$\text{Prob}(Y=0) = \frac{1}{1+\exp(\text{logit\_10})+\exp(\text{logit\_20})+\ldots+\exp(\text{logit\_k0})} \tag{18.13}$$

$$\text{Prob}(Y=1) = \frac{\exp(\text{logit\_10})}{1+\exp(\text{logit\_10})+\exp(\text{logit\_20})+\ldots+\exp(\text{logit\_k0})} \tag{18.14}$$

$$\ldots,$$

$$\text{Prob}(Y=k) = \frac{\exp(\text{logit\_k0})}{1+\exp(\text{logit\_10})+\exp(\text{logit\_20})+\ldots+\exp(\text{logit\_k0})} \tag{18.15}$$

where

$$\begin{aligned} \text{logit\_10} &= \text{logit}(Y=1 \text{ vs. } Y=0) \\ \text{logit\_20} &= \text{logit}(Y=2 \text{ vs. } Y=0) \\ \text{logit\_30} &= \text{logit}(Y=3 \text{ vs. } Y=0) \\ &\ldots \\ \text{logit\_k0} &= \text{logit}(Y=k \text{ vs. } Y=0) \end{aligned}$$

Note, there are k logits for a PLR with k + 1 classes.

18.4 Model Building with PLR

The estimation method for PLR is maximum likelihood estimation, the same method for BLR estimation. The theory of stepwise variable selection, model assessment, and validation has been worked out for PLR. Some theoretical problems remain. For example, a variable can be declared significant for all but, say, one logit. Because there is no theory for estimating a PLR model with the constraint of setting a coefficient equal to zero for a given logit, the PLR model may produce unreliable classifications.

Choosing the best set of predictor variables is the toughest part of modeling and is perhaps more difficult with PLR because there are k logit equations to consider. The traditional stepwise procedure is the popular variable selection process for PLR. (See Chapter 41 for a discussion of the traditional stepwise procedure and recall Chapter 13.) Without arguing the pros and cons of the stepwise procedure, its use as the determinant of the final model is questionable.* The stepwise approach is best as a rough-cut method for boiling down many variables—about 50 or more—to a manageable set of about 10. I prefer a methodology based on chi-squared automatic interaction detection (CHAID) as the variable selection process for the PLR because it fits well into the data mining paradigm of digging into the data to find unexpected structure. In the next section, I illustrate CHAID as the variable selection procedure for building a market segmentation classification model for a cellular phone.

18.5 Market Segmentation Classification Model

In this section, I describe the cellular phone user study. Using the four user groups derived from a cluster analysis (not shown) of the survey data, I build a four-group classification model with PLR. I use CHAID to identify the final set of candidate predictor variables, interaction terms, and variable structures (i.e., reexpressions of original variables defined with functions such as a log and square root) for inclusion in the model. After a detailed discussion of the CHAID analysis, I define the market segmentation classification model and assess the total classification accuracy of the resultant model.

18.5.1 Survey of Cellular Phone Users

A survey of 2,005 past and current users of cellular phones from a wireless carrier was conducted to gain an understanding of customer needs as well as the variables that affect churn (cancellation of cellular service) and long-term value. The survey data were used to segment this market of consumers into homogeneous groups so that group-specific marketing programs, namely, CRM strategies, can then be developed to maximize the individual customer relationship.

The cluster analysis produces four segments. Segment names and sizes are in Table 18.1. The hassle-free segment is concerned with the ability of the customer to design the contract and rate plan. The service segment is focused on quality of the call, such as no dropped calls and call clarity. The price segment values discounts, such as offering 10% off the

TABLE 18.1

Cluster Analysis Results

Name	Size
Hassle-free	13.2% (265)
Service	24.7% (495)
Price	38.3% (768)
Features	23.8% (477)
Total	100% (2,005)

* Briefly, a stepwise approach is misleading because all possible subsets are not considered; the final selection is also data dependent and sensitive to influential observations. Also, it does not automatically check for model assumptions and does not automatically test for interaction terms. Moreover, the stepwise approach does not guarantee finding the globally best subset of the variables.

monthly base charges and 30 free minutes of use. Lastly, the features segment represents the latest technology, such as long-lasting batteries and free phone upgrades. A model is needed to divide the entire database of the wireless carrier into these four actionable segments. Cellular carriers who sponsored the study develop marketing programs tailored to the specific needs of these predefined groups and then implement the programs.

The cellular carrier provides its billing record information for appending to the survey. For all respondents, now classified into one of four segments, there are 10 usage variables, such as the number of mobile phones, minutes of use, peak and off-peak calls, airtime revenue, base charges, roaming charges, and free minutes of use (yes/no).

18.5.2 CHAID Analysis

Briefly, CHAID is a technique that recursively partitions a population into separate and distinct subpopulations or nodes such that the variation of the dependent variable is minimized within the nodes and maximized among the nodes. The dependent variable can be binary (dichotomous), polychotomous, or continuous. The nodes, defined by independent variables, pass through an algorithm for partitioning. The independent variables can be categorical or continuous.

To perform a CHAID analysis, I define the dependent variable and the set of independent variables. For this application of CHAID, the set of independent variables is the set of usage variables appended to the survey data. The dependent variable is the class variable Y identifying the four segments from the cluster analysis. Specifically, I define the dependent variable as follows:

$$
\begin{aligned}
Y &= 0 \text{ if segment is Hassle-free} \\
&= 1 \text{ if segment is Service} \\
&= 2 \text{ if segment is Price} \\
&= 3 \text{ if segment is Features}
\end{aligned}
$$

The CHAID analysis identifies* four important predictor variables:†

1. NUMBER OF MOBILE PHONES: the number of mobile phones a customer has
2. MONTHLY OFF-PEAK CALLS: the number of off-peak calls averaged over a 3-month period
3. FREE MINUTES, yes/no: free first 30 minutes of use per month
4. MONTHLY AIRTIME REVENUE: total revenue excluding monthly charges averaged over a 3-month period

The CHAID tree for NUMBER OF MOBILE PHONES, shown in Figure 18.1, is read as follows:

1. The top box indicates that for the sample of 2,005 customers, the sizes (and incidences) of the segments are 264 (13.2%), 495 (24.7%), 767 (38.3%), and 479 (23.9%) for Hassle-free, Service, Price, and Features segments, respectively.

* Based on the average chi-square value per degrees of freedom (number of nodes).

† I do not consider interaction variables identified by CHAID because the sample is too small.

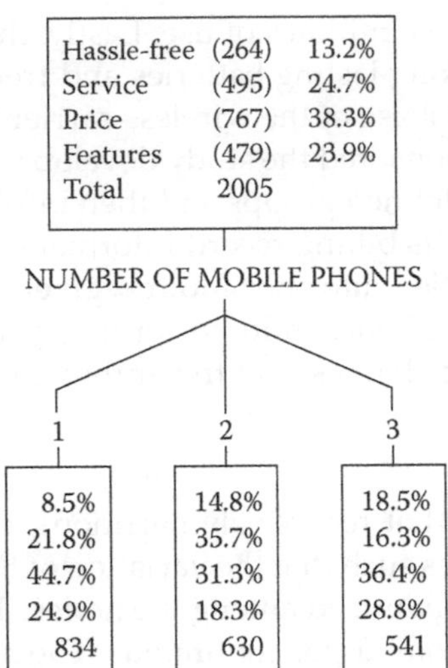

FIGURE 18.1
CHAID tree for NUMBER OF MOBILE PHONES.

2. The left node represents 834 customers who have one mobile phone. Within this subsegment, the incidence rates of the four segments are 8.5%, 21.8%, 44.7%, and 24.9% for Hassle-free, Service, Price, and Features segments, respectively.
3. The middle node represents 630 customers who have two mobile phones, with incidence rates of 14.8%, 35.7%, 31.3%, and 18.3% for Hassle-free, Service, Price, and Features segments, respectively.
4. The right node represents 541 customers who have three mobile phones, with incidence rates of 18.5%, 16.3%, 36.4%, and 28.8% for Hassle-free, Service, Price, and Features segments, respectively.

The CHAID tree for MONTHLY OFF-PEAK CALLS, shown in Figure 18.2, is read as follows:

1. The top box is the sample breakdown of the four segments; it is identical to the top box for NUMBER OF MOBILE PHONES.
2. There are three nodes: (a) The left node is the number of calls in the left-closed right-open interval [0,1); this means zero calls. (b) The middle node is the number of calls in the left-closed right-open interval [1,2); this means one call. (c) The right node is the number of calls in the closed interval [2, 270]; this means calls between greater than or equal to 2 and less than or equal to 270.
3. The left node represents 841 customers who have zero off-peak calls. Within this subsegment, the incidence rates of the four segments are 18.1%, 26.4%, 32.6%, and 22.9% for Hassle-free, Service, Price, and Features, respectively.
4. The middle node represents 380 customers who have one off-peak call, with incidence rates of 15.3%, 27.1%, 43.9%, and 13.7% for hassle-free, service, price, and features, respectively.
5. The right node represents 784 customers who have off-peak calls inclusively between 2 and 270, with incidence rates of 6.9%, 21.7%, 41.6%, and 29.8% for Hassle-free, Service, Price, and Features, respectively.

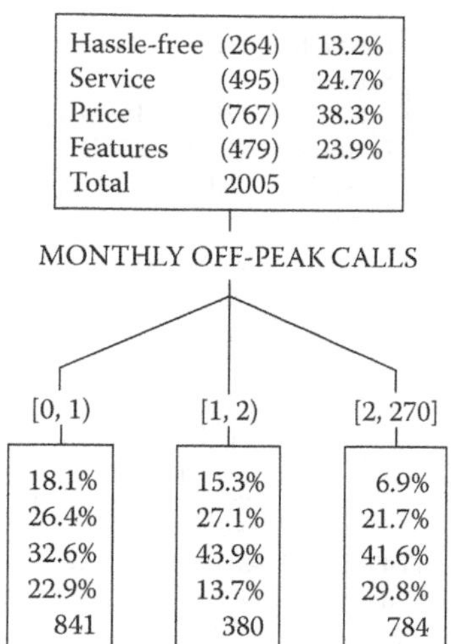

FIGURE 18.2
CHAID tree for MONTHLY OFF-PEAK CALLS.

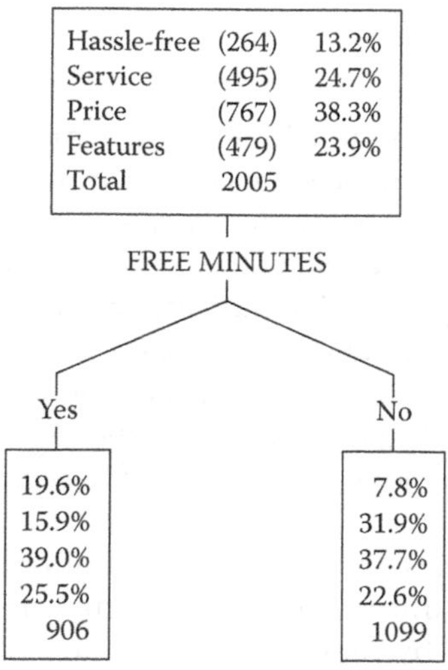

FIGURE 18.3
CHAID tree for FREE MINUTES.

Similar readings apply to the other predictor variables, FREE MINUTES and MONTHLY AIRTIME REVENUE, identified by CHAID. Their CHAID trees are in Figures 18.3 and 18.4, respectively.

Analytically, CHAID declaring a significant variable means the segment incidence rates (as a column array of rates) differ significantly across the nodes. For example, MONTHLY OFF-PEAK CALLS has three column arrays of segment incidence rates, corresponding to the three nodes: {18.1%, 26.4%, 32.6%, 22.9%}, {15.3%, 27.1%, 43.9%, 13.7%}, and {6.9%, 21.7%, 41.6%, 29.8%}. These column arrays are significantly different from each other. The column arrays represent a complex concept that has no interpretive value, at least in the context of identifying variables with classification power. However, the CHAID tree can help evaluate the potential predictive power of a variable when the variable lies in a tree graph.

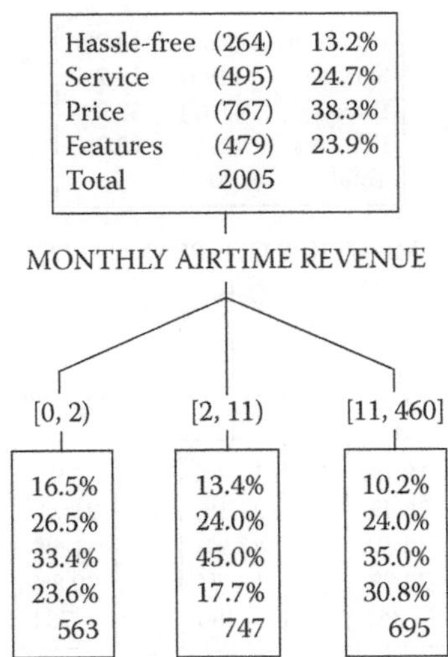

FIGURE 18.4
CHAID tree for MONTHLY AIRTIME REVENUE.

18.5.3 CHAID Tree Graphs

Displaying a CHAID tree in a graph facilitates the evaluation of the potential predictive power of a variable. I plot the incidence rates by the minimum values* of the intervals for the nodes and connect the *smooth* points to form a *trace line*, one for each segment. The shape of the trace line indicates the effect of the predictor variable on identifying individuals in a segment. The baseline plot, which indicates a predictor variable with no classification power, consists of all the segment trace lines being horizontal or flat. The extent to which the segment trace lines are not flat indicates the potential predictive power of the variable for identifying an individual belonging to the segments. A comparison of all trace lines (one for each segment) provides a total view of how the variable affects classification *across* the segments.

The following discussion relies on an understanding of the basics of reexpressing data as discussed in Chapter 10. It suffices it to say that sometimes the predictive power offered by a variable can be increased by reexpressing or transforming the original form of the variable. The final forms of the four predictor variables identified by CHAID variables are in the next section.

The PLR is a linear model,† which requires a linear or straight-line relationship between the predictor variable and each implicit binary segment dependent variable.‡ The tree graph suggests the appropriate reexpression when the empirical relationships between predictor and binary segment variables are not linear.§

* The minimum value is one of several values that can be used; alternatives are the mean or median of each predefined interval.
† That is, each individual logit is a sum of weighted predictor variables.
‡ For example, for segment hassle-free (Y = 0), binary hassle-free segment variable = 1 if Y = 0; otherwise, hassle-free segment variable = 0.
§ The suggestions are determined from the ladder of powers and the bulging rule discussed in Chapter 10.

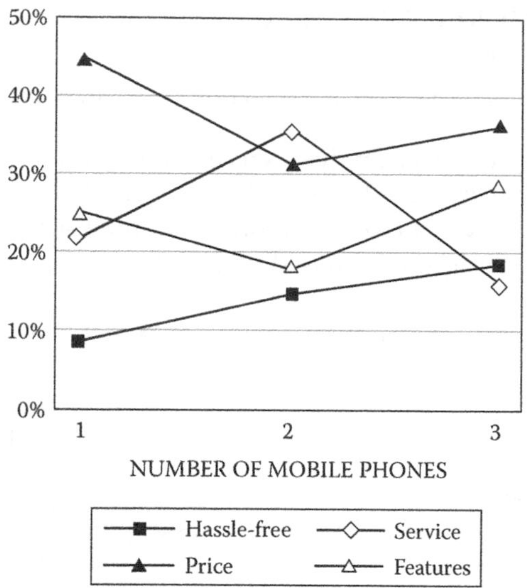

FIGURE 18.5
CHAID tree graph for NUMBER OF MOBILE PHONES.

The CHAID tree graph for NUMBER OF MOBILE PHONES in Figure 18.5 indicates the following:

1. There is a positive and nearly linear relationship* between NUMBER OF MOBILE PHONES and the identification of customers in the hassle-free segment. This relationship implies that only the variable NUMBER OF MOBILE PHONES in its raw form, no reexpression, may be needed.
2. The relationship for the features segment also has a positive relationship but with a bend from below.† This relationship implies that the variable NUMBER OF MOBILE PHONES in its raw form and its square may be required.
3. Price segment has a negative effect with a bend from below. This relationship implies that the variable NUMBER OF MOBILE PHONES itself in raw form and its square root may be required.
4. The service segment has a negative relationship but with a bend from above. This relationship implies that the variable NUMBER OF MOBILE PHONES itself in raw form and its square may be required.

The CHAID tree graphs for the other predictor variables, MONTHLY OFF-PEAK CALLs, MONTHLY AIRTIME REVENUE, and FREE MINUTES are in Figures 18.6, 18.7, and 18.8, respectively. Interpreting the graphs, I diagnostically determine the following:

1. MONTHLY OFF-PEAK CALLS: This variable in its raw, square, and square root forms may be required.

* A relationship is assessed by determining the slope between the left and right node smooth points.
† The position of a bend is determined by the middle-node smooth point.

2. MONTHLY AIRTIME REVENUE: This variable in its raw, square, and square root forms may be required.
3. FREE MINUTES: This variable as is, in its raw form, may be needed.

The preceding CHAID tree graph analysis serves as the initial variable selection process for PLR. The final selection criterion for including a variable in the PLR model is that the variable must be significantly/noticeably important on no less than three of the four logit equations based on the techniques discussed in Chapter 10.

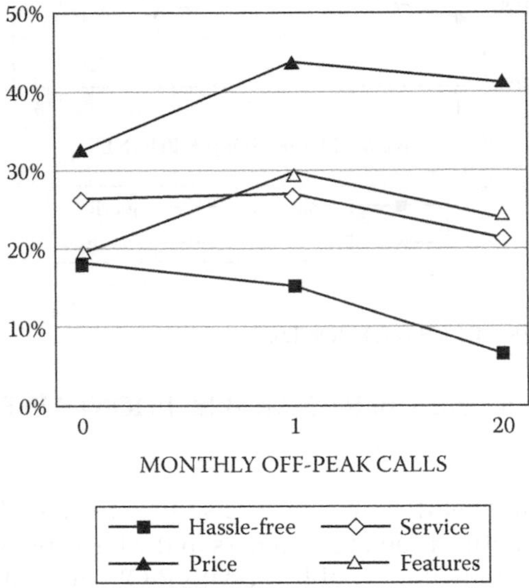

FIGURE 18.6
CHAID tree graph for MONTHLY OFF-PEAK CALLS.

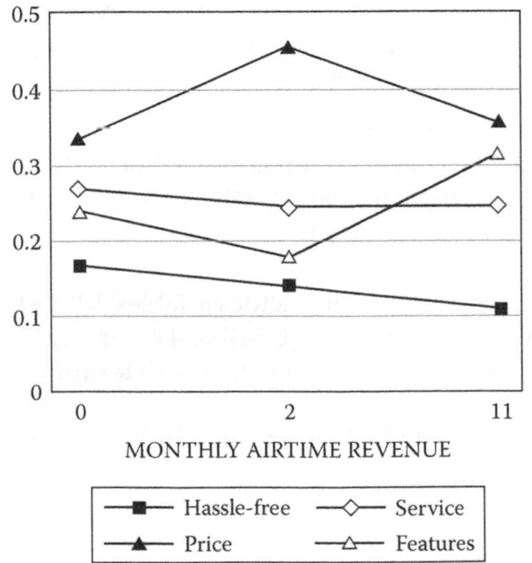

FIGURE 18.7
CHAID tree graph for MONTHLY AIRTIME REVENUE.

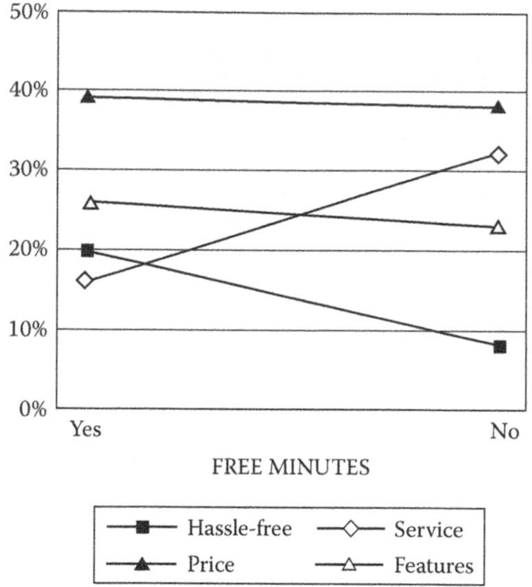

FIGURE 18.8
CHAID tree graph for FREE MINUTES.

18.5.4 Market Segmentation Classification Model

The final PLR model for classifying customers into one of the four cellular phone behavioral market segments has the following variables:

1. Number of Mobile Phones (NMP)
2. Square of NMP
3. Square root of NMP
4. Monthly Off-Peak Calls
5. Monthly Airtime Revenue (MAR)
6. Square of MAR
7. Square root of MAR
8. Free Minutes

Without arguing the pros and cons of validation procedures, I draw a fresh holdout sample of size 5,000 to assess the total classification accuracy of the model.* The classification results of the model in Table 18.2 read as follows:

1. The row totals are the actual counts in the sample. The sample consists of 650 Hassle-free customers, 1,224 Service customers, 1,916 Price customers, and 1,210 Features customers. The percentage figures are the percentage compositions of the segments on the total sample. For example, 13.0% of the sample consists of actual Hassle-free customers.

* Here is a great opportunity to debate and test various ways of calibrating and validating a "difficult" model under the best of conditions.

TABLE 18.2

Market Segment Model: Classification Table of Cell Counts

		PREDICTED				
		Hassle-free	Service	Price	Features	Total
ACTUAL	Hassle-free	326 (50.0%)	68	158	98	650 (13.0%)
	Service	79	460 (36.3%)	410	275	1,224 (24.5%)
	Price	147	431	922 (49.0%)	416	1,916 (38.3%)
	Features	103	309	380	418 (34.6%)	1,210 (24.2%)
	Total	655 (13.1%)	1,268 (25.4%)	1,870 (37.4%)	1,207 (24.1%)	5,000 (100%)

2. The column totals are *predicted* counts. The model predicts 655 Hassle-free customers, 1,268 Service customers, 1,870 Price customers, and 1,207 Features customers. The percentage figures are the percentage compositions on the predicted counts. For example, the model predicts 13.1% of the sample as Hassle-free customers.
3. Given that the sample consists of 13.0% Hassle-free customers, and the model predicts 13.1% Hassle-free customers, the model has no *bias* on classifying Hassle-free customers. Similarly, the model shows no bias in classifying the other groups: for Service, the actual incidence is 24.5% versus the predicted 25.4%; for Price, the actual incidence is 38.3% versus the predicted 37.4%; and for Features, the actual incidence is 24.2% versus the predicted 24.1%.
4. Although the model shows no bias, the big question is how accurate are the predictions. Among those customers predicted to be in the Hassle-free segment, how many are? Among those customers predicted to be in the Service segment, how many are Service in the segment? Similarly, for Price and Features segments, among those customers predicted to be in the segments, how many are? The percentages of the table cells provide the answer. For Hassle-free customers, the model correctly classifies 50.0% (= 326/655) of the time. Without a model, I would expect 13.0% correct classifications of Hassle-free customers. Thus, the model has a lift of 385 (50.0%/13.0%). This lift means: The model provides 3.85 times the number of correct classifications of Hassle-free customers obtained by chance.
5. For Service, the model has a lift of 148 (36.3%/24.5%); for Price, the model has a lift of 128 (49.0%/38.3%); and, for Features, the model has a lift of 143 (34.6%/24.2%).
6. As a summary measure of how well the model makes correct classifications, I look at the total correct classification rate (TCCR). Simply put, the TCCR is the total number of correct classifications across all groups divided by total sample size. Accordingly, I have 326 + 460 + 922 + 418 = 2,126 divided by 5,000, which yields TCCR = 42.52%.
7. To assess the improvement of total correct classification provided by the model, I must compare it to the total correct classification provided by the model chance.

TCCR(chance model) is the sum of the squared actual group incidence. For the data at hand, TCCR(chance model) is 28.22% (= (13.0%*13.0%) + (24.5%*24.5%) + (38.3%*38.3%) + (24.2%*24.2%)).

8. Thus, the model lift is 151 (= 42.52%/28.22%). That is, the model provides 51% more total correct classifications across all groups than obtained by chance.

18.6 Summary

I cast the multigroup classification technique of PLR as an extension of the most familiar two-group (binary) logistic regression model. I derive the PLR model for a dependent variable assuming k + 1 groups by "piecing together" k individual binary logistic models. I propose a CHAID-based data mining methodology as the variable selection process for the PLR—as it fits well into the data mining paradigm—to dig into the data to find important predictor variables and unexpected structure.

To bring into practice the PLR, I illustrate the building of a classification model based on a four-group market segmentation of cellular phone users. For the variable selection procedure, I demonstrate how CHAID is a valued technique. For this application of CHAID, the dependent variable is the variable identifying the four market segments. CHAID trees were used to identify the starter subset of predictor variables. Then, I generate CHAID tree graphs from the CHAID trees. The CHAID tree graphs offer potential reexpressions—or identification of structures—of the starter variables. The final market segmentation classification model has a few reexpressed variables (involving square roots and squares).

Last, I assess the performance of the final four-group/market segment classification model regarding TCCR. TCCR for the final model is 42.52%, which represents a 51% improvement over the TCCR (28.22%) of the chance model.

19

Market Segmentation Based on Time-Series Data Using Latent Class Analysis

19.1 Introduction

Market segmentation is an often-used marketing model for efficient allocation of a company's resources. Market segmentation divides a population of customers into subpopulations—segments of customers. Customers within a segment are similar in their products and services, and customers across segments are dissimilar in their products and services. Market segmentation model implementation allows for effectively applying resources by targeting customers within their assigned segments. There are many statistical methods for market segmentation. A traditional and popular method is k-means clustering. A not-as-well-known method is latent class analysis (LCA). The purpose of this chapter is to present a novel approach to building a market segmentation model based on time-series data using LCA. I provide SAS© subroutines so data miners can perform segmentations similar to those presented, and I offer a unique way of incorporating time-series data in an otherwise cross-sectional dataset. The subroutines are also available for downloading from my website: http://www.geniq.net/articles.html#section9.

19.2 Background

I pedagogically outline this chapter. First, I concisely describe k-means clustering. Second, I cursorily review principal component analysis (PCA). I need to revisit PCA (from Chapter 7) because PCA and factor analysis (FA) are often confused, and FA is quite helpful in explaining LCA. Third, I present the LCA. Fourth, I compare LCA and k-means clustering. Lastly, I illustrate building a market segmentation model based on times-series data using LCA.

19.2.1 K-Means Clustering*

K-means clustering generates k mutually exclusive groups by distances computed from one or more quantitative variables (Xs) [1]. Every observation belongs to one and only

* This section draws on https://support.sas.com/rnd/app/stat/procedures/fastclus.html.

one group of the k clusters. Because the number of clusters k is unknown, the model builder performs as many cluster solutions as desired. The typical k-means procedure uses Euclidean distance (i.e., least-squares calculations) that yields cluster means (of the observations in a cluster for the Xs). If groups exist, then all distances among observations in a group are less than all distances among observations in a different group.

The k-means algorithm is heuristic, which involves the following steps:

1. Diversely select k initial seeds (i.e., random points in the X-space of the observations).
2. These points represent initial cluster means.
3. Assign each observation to the cluster whose mean is the closest to the observation.
4. When all observations are assigned, recalculate the k means.
5. Repeat Steps #2 and #3, until the k means are stable.

19.2.2 PCA

PCA transforms a set of p variables* X1, X2, …, Xj, ..., Xp into p linear combination variables PC1, PC2, ..., PCj, ..., PCp (PCj denotes the j-th principal component). The essential objective of PCA is to establish a smaller set of the new PCj variables that represents most of the information (variation) in the original set of variables. An attractive analytic feature of the PCs is that they are uncorrelated with each other. The PCs are defined as:

PC1 = a11*X1 + a12*X2 +… + a1j*Xj + …+ a1p*Xp

PC2 = a21*X1 + a22*X2 +…+ a2j*Xj + …+ a2p*Xp

⋮

PCi = ai1*X1 + ai2*X2 +…+ aij*Xj + …+ aip*Xp

⋮

PCp = ap1*X1 + ap2*X2 +…+ apj*Xj + …+ app*Xp

where the aij's are the PC coefficients.

19.2.3 FA

There is confusion among too many statistics practitioners who do not understand the difference between PCA and FA. The reasons for the confusion is perhaps because:

1. PCA is often mentioned in textbooks as a particular case of FA.
2. Statistical computer packages treat PCA as an option in FA modules.

* For ease of presentation, the Xs are standardized.

3. PCA and FA both aim to reduce the dimensionality of the given dataset: PCA and FA are data reduction techniques. For example, a dataset with, say, 1,000 variables can be reduced to a statistically equivalent dataset with, say, only 150 variables.
4. PCA has been used extensively as a part of the FA solution.

19.2.3.1 FA Model

The FA model is defined as: p variables X1, X2, …, Xj, ..., Xp can be expressed (up to an error term) as a linear combination of m latent (unobservable) *continuous* variables or factor F1, F2, …, Fj, ..., Fm. Fs are defined as:

$$X1 = c11*F1 + c12*F2 + ... + c1j*Fj + ... + c1m*Fm$$

$$X2 = c21*F1 + c22*F2 + ... + c2j*Fj + ... + c2m*Fm$$

$$\vdots$$

$$Xi = ci1*F1 + ci2*F2 + ... + cij*Fj + ... + cim*Fm$$

$$\vdots$$

$$Xp = cp1*F1 + cp2*F2 + ... + cpj*Fj + ... + cpm*Fm$$

where cij's are like regression coefficients, called factor loadings.

The difference between PCA and FA is immediately apparent. Focus on the i-th PC and Xi:

$$PCi = ai1*X1 + ai2*X2 + \ldots + aij2*Xij + \ldots + aip*Xp$$

$$Xi = ci1*F1 + ci2*F2 + ... + cij*Fj + \ldots + cim*Fm$$

The PCs are a linear combination of the original *observed* Xs. Each Xi is a linear combination of *unobserved* latent factors F1, F2, …, Fj, ..., Fm.

FA attempts to achieve a reduction from p to m dimensions by postulating a model relating p Xs to m latent factors. In contrast, PCA directly transforms the Xs into PCs, which possess the desirable properties cited in Chapter 7.

19.2.3.2 FA Model Estimation

At first sight, the FA model (with simplified subscripts) looks like a standard ordinary least squares (OLS) regression model.

$$\text{FA: } X = c1*F1 + c2*F2 + ... + cj*Fj + ... + cm*Fm \tag{19.1}$$

$$\text{OLS: } Y = b1*X1 + b2*X2 + \ldots bj*Xj + ... + bm*Xm \tag{19.2}$$

However, a closer inspection reveals a substantial difference. With FA, c's and Fs are both *unknown*. With OLS, b's are *unknown*, and the Xs are known.

The PCA solution to FA or the so-called principal FA involves the following steps:

1. Obtain initial estimates of the c's by taking the first m PCs from a PCA.
2. Obtain initial estimates of the Fs. X and c's are known. Thus, F can be initially determined.
3. Initial estimates are then fine-tuned, looping within Step #2, until optimal.

19.2.3.3 FA versus OLS Graphical Depiction

The typical graphical depictions of the FA and OLS models clearly indicate the theoretical framework of these two models in Figures 19.1 and 19.2.

The two models illustrated with three Xs are now visibly different in conceptual ways. The FA model has line arrows from the factor F to the Xs. The direction of the arrows indicates that the factor, the unobserved latent variable, affects the independent variables X1, X2, and X3. The lower case e's indicate the unknown errors associated with the measurement of the observed corresponding Xs. The OLS regression model has line arrows from the independent variables X1, X2, and X3 to the dependent variable Y. The direction of the arrows indicates the independent variables X1, X2, and X3 affect the dependent variable Y. There are no measurement errors assumed in OLS that influence Xs.

19.2.4 LCA versus FA Graphical Depiction

As a proper introduction to LCA is in the next section, I bring out the traditional setting of LCA, which is often considered similar to FA—with one difference. When the LCA model (Figure 19.3) is compared to the FA model (Figure 19.1), the similarity is visible. The significant and useful difference between the methods is the factor F of FA is a *continuous* latent variable, while the latent variable LC of LCA is *categorical.*

The usual graphic display for LCA is Figure 19.3. However, this display does not clearly portray the categorical nature of LC. I offer a refocused visual for an LCA 2-Class model in Figure 19.4, which clearly indicates LC is categorical. LC(1) and LC(2) are the two categories of LC. The crisscross-like arrows indicate each LC affects the independent variables X1, X2, and X3. Important to note, LCA nomenclature for an independent variable is *indicator.*

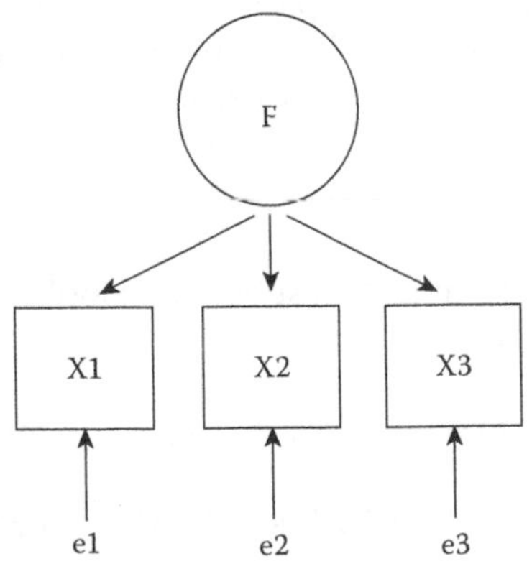

FIGURE 19.1
FA model graphic.

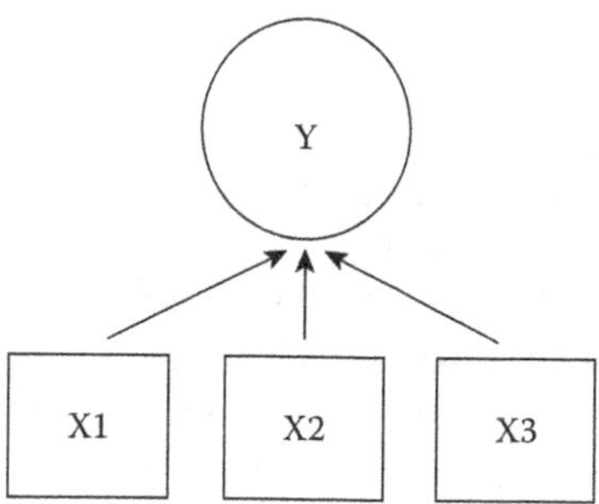

FIGURE 19.2
OLS model graphic.

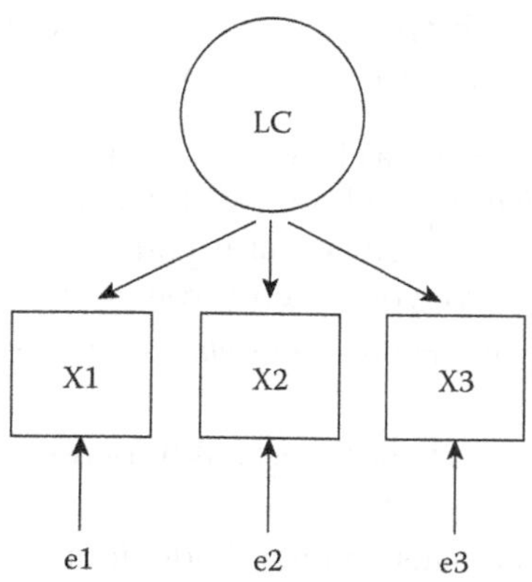

FIGURE 19.3
LCA model graphic.

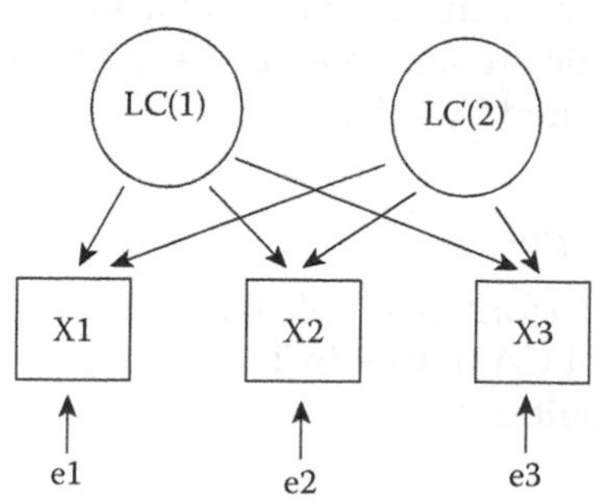

FIGURE 19.4
LCA 2-Class model graphic.

19.3 LCA

The concept of LCA, a statistical technique for identifying unobservable subpopulations within a population, was conceived in 1950 by Lazarsfeld, the father of this concept [2]. In 1968, Lazarsfeld (by then the *grandfather* of LCA) published the first comprehensive treatment of LCA with only categorical indicators, but he did not provide a reliable method for parameter estimation [3]. In 1974, Goodman solved the problem of obtaining maximum likelihood estimates* of LCA model parameters [4]. The traditional LCA was generalized by Vermunt in 1998 to include all scales of data—categorical, continuous, count, and ordinal [5].

19.3.1 LCA of Universal and Particular Study

As an illustration of how LCA works along with its output, I present a well-known and often revisited "Role Conflict and Personality" article, often referred to as the Universal and Particular Study, conducted in 1950 [6], which Goodman examined in 1974 [7]. In the 1950 study, 216 Harvard/Radcliffe undergraduates were asked how they would respond in four role-conflict situations. The premise of the role conflict study is "What right has your friend to expect you to protect him?" The four conflict scenarios† are:

1. You are riding in a car driven by a close friend, and he hits a pedestrian. You know that he is going at least 35 miles an hour (MPH) in a 20 MPH speed zone. There are no other witnesses. His lawyer says that if you testify under oath that the speed was only 20 MPH, it may save your friend from serious consequences.
   a. Universalistic response: "He has *no right* as a friend to expect me to testify to the lower figure."
   b. Particularistic response: "He has *a right* as a friend to expect me to give evidence to the lower figure."
2. As a physician, your friend asks you to "shade doubts" about a physical examination for an insurance policy.
3. As a drama critic, your friend asks you to "go easy on a review" of a bad play in which all of his savings are invested.
4. As a member of the board of directors, your friend intimates that you will "tip him off" about financially ruinous, though secret, company information.

The response patterns of the 216 respondents generate the 16 (= 2 × 2 × 2 × 2) row patterns in Table 19.1.‡ For each of the conflict scenarios (A, B, C, D), responses tending toward one that is universalistic are indicated by "+ ", and responses tending toward one that is particularistic are indicated by "–" [7].

19.3.1.1 Discussion of LCA Output

Hagenaars and McCutcheon perform a simple two-class LCA on the Universal-Particular study [8]. The vital statistics of LCA output in Table 19.2 consist of (1) latent class probabilities and (2) conditional probabilities.

* Maximum likelihood estimation, specifically, the expectation–maximization (EM) process.
† The descriptions of the scenarios are taken from the original study.
‡ Table is taken from Goodman 1974, p. 216.

TABLE 19.1

Response Patterns of the Four Conflict Scenarios

A	B	C	D	Observed Frequency	A	B	C	D	Observed Frequency
+	+	+	+	42	–	+	+	+	1
+	+	+	–	23	–	+	+	–	4
+	+	–	+	6	–	+	–	+	1
+	+	–	–	25	–	+	–	–	6
+	–	+	+	6	–	–	+	+	2
+	–	+	–	24	–	–	+	–	9
+	–	–	+	7	–	–	–	+	2
+	–	–	–	38	–	–	–	–	20

TABLE 19.2

LCA Output of Study

Observed Indicators	Universal	Particular
Auto passenger friend	0.993	0.714
Insurance doctor friend	0.939	0.329
Drama critic friend	0.929	0.354
Board of directors friend	0.769	0.132
Latent Class Size	0.280	0.720

The latent class probabilities are the sizes of the two latent classes, which are 0.280 and 0.720 for Universal and Particular, respectively. These probabilities show 28% and 72% of the study population are in the Universal and Particular classes, respectively.

Within the similarity of LCA and FA, a conditional probability is comparable to the factor loading in FA. Large conditional probabilities for a latent class imply the corresponding indicators are highly associated with the latent class and therefore define the latent class.

The large probabilities of the first class column, ranging from 0.993 to 0.769, confirm the correct labeling of the column as Universal. For the second class column, the large (0.714), moderate (0.329 and 0.354), and low (0.132) conditional probabilities reasonably define the Particular class. In other words, latent class Universal is reliably identified respondents, whereas Particular is moderately determined.

Within LCA, the conditional probability is the probability of an individual in a given class responding at a given level of the indicator. The conditional probability is 99.3% for a respondent in class Universal responding that there is no right to lie for the driver friend. The conditional probability is 71.4% for a respondent in Particular class responding that there is a right to lie for the driver friend.

Similarly, for a respondent who is in the Universal class and answers that there is no reason for doctor friend to lie, the conditional probability is 93.9%. And, the conditional probability is 32.9% for a respondent in Particular class responding that there is a reason for doctor friend to lie.

19.3.1.2 Discussion of Posterior Probability

The relevant statistic noticeably missing in Table 19.2 is the posterior probability—the probability of class membership *given* a respondent's answers to a given scenario. This statistic allows for probabilistically classifying a respondent into one of the latent classes. The posterior probability is notationally represented as Prob (LC = c | A = i, B = j, C = k, D = l), where c = 1, 2; i = yes, no; j = yes, no; k = yes, no; and l = yes, no. The symbol | represents the word *given*.

The calculation of posterior probability needs the LCA joint probability, Prob (A = i, B = j, C = k, D = l, CL = c), defined as the product of:

- Prob (individual at CL = c)
- Prob (individual at A = i | CL = c)
- Prob (individual at B = j | CL = c)
- Prob (individual at C = k | CL = c)
- Prob (individual at D = l | CL = c)

Thus, the posterior probability of an individual belonging to latent class c (CL = c) given A = i, B = j, C = k, D = l is in Equation 19.3:

$$\text{Prob}\left(\text{LC}=\text{c}|\text{A}=\text{i},\text{B}=\text{j},\text{C}=\text{k},\text{D}=\text{l}\right)$$
$$=\text{Prob}\left(\text{A}=\text{i},\text{B}=\text{j},\text{C}=\text{k},\text{D}=\text{l},\text{CL}=\text{c}\right)/\,\text{sum}\left[\text{Prob}\left(\text{A}=\text{i},\text{B}=\text{j},\text{C}=\text{k},\text{D}=\text{l},\text{CL}=\text{c}\right)\right],\text{across c} \tag{19.3}$$

Thus far, the discussion of LCA regards only categorical indicators. LCA with continuous and categorical indicators generates output virtually identical to that of LCA with only categorical indicators. Expectantly, the statistical computations of LCA with continuous variables add a level of detail and knowledge of the mathematical statistics underpinnings of LCA beyond the scope of this chapter. Accordingly, there is no discussion of the extended LCA, for which Vermunt (1998) is the definitive source.

19.4 LCA versus k-Means Clustering

LCA and k-means clustering have the same objective of dividing a population of individuals into k disjoint and exhaustive subpopulations (clusters) such that individuals within a group are as similar as possible, and the individuals among the groups are as dissimilar as possible. In statistics-speak, cluster construction seeks to maximize the between-cluster differences and minimize the within-cluster differences.

The fundamental underlying difference between LCA and k-means methodologies is that LCA is model-based, whereas k-means is a heuristic (technique). A heuristic is any approach to solving a problem that uses intuition based on the problem domain. Heuristics do not provide optimal solutions but rather good solutions. Markedly, LCA as a model-based statistical technique means that it posits a theoretical equation that represents the data-generating process of the population and drawn sample data. Moreover, the probability distribution of the statistical model is the distinguishing feature between model-based versus data-based heuristics.

K-means clustering is a popular, practical, and useful tool, especially for market segmentation. Similar to all techniques, k-means has strengths and weaknesses. The strengths (Items 1–4) and weaknesses (Items 5–12) are listed as follows:

1. K-means is easy to use, understand, and implement.
2. K-means performs best with many variables, which it can accommodate because it is computationally efficient.
3. K-means always produces a cluster solution.
4. K-means tends to produce tighter clusters than alternative techniques.
5. K-means can create a cluster with one or a handful of individuals.
6. K-means cannot suggest the optimal number of clusters.
7. K-means is sensitive to outliers. (When few individuals define a cluster, they are outliers.)
8. K-means is not a robust method in that pseudorandom seeds yield different solutions. Or, a holdout dataset can produce a different cluster solution.
9. The k-means algorithm does not reveal which variables are relevant.
10. The k-means algorithm is affected by variables with large variances. Accordingly, k-means often use standardized data.
11. K-means always produces a cluster solution. Always-a-solution gives a false sense of statistical security, in that k-means finds a good solution.
12. An objective criterion by which to assess the quality of the cluster solution does not exist.

LCA as a statistical model has the apparatus that indicates its strengths (first four items in the list that follows). Unfortunately, the apparatus does not eliminate weaknesses (last seven items in the list that follows).

1. There are objective criteria by which to assess the goodness-of-fit of the cluster solution.
2. Statistical testing exists for comparison between two or more candidate cluster solutions.
3. LCA solutions are not affected by the different scales and unequal or large variances of the indicators.
4. LCA, like all statistical models, produces residuals. Residual analysis is crucial in assessing remedies of lack-of-fit.
5. LCA cannot suggest the optimal number of clusters.
6. A serious problem with LCA is the assumption of conditional dependence, also known as local independence. Local independence means the indicators within a cluster class are independent of each other. The local independence is sometimes not a tenable assumption. The standard LCA model must be modified to account for this (http://john-uebersax.com/stat/faq.htm).
7. Methods for relaxing the conditional independence assumption are in-progress in recent years.
8. LCA uses maximum likelihood estimation like many statistical models. Thus, LCA solutions are subject to local maxima not necessarily the global maxima.

9. There are many goodness-of-fitness criteria, such as the basics: likelihood (L), log-likelihood (LL), and L-square (L^2).
10. Some criteria weight the fit and the parsimony of a model based on sample size and degrees of freedom. They are:
    a. Akaike Information Criteria (AIC):* AIC, AIC(L^2), AIC3(L^2), AIC(LL), and AIC3(LL).
    b. Bayesian Information Criteria (BIC):† BIC, BIC2(L^2), and BIC(LL).
    c. None of these criteria are universally deemed superior to another.
    d. The behavior of these statistics creates confusion and uncertainty as to which is the best cluster solution. For example, when comparing two models, it is not known how much difference in BIC is significant in order to choose one model over another [9].
11. Restrictions when assessing goodness-of-fit complicate building the model. Hypotheses are tested by imposing restrictions and determining how these limitations affect the fit of the model to the data. Two such restrictions include (1) equality constraint (e.g., parallel indicators or equal error rate) and (2) deterministic (e.g., setting conditional probability to a particular value—usually 1 or 0).
12. Sparseness, due to many indicators with many response options, leads to difficulties in model evaluation (e.g., determining the degrees of freedom).

19.5 LCA Market Segmentation Model Based on Time-Series Data

I present the building of a market segmentation model based on times-series data using LCA.

The proposed model is not to be confused with partitioning of a times series, producing a sequence of discrete segments to reveal the underlying structure of input time-series data. A typical method of a times-series segmentation is a piecewise linear regression, which forecasts, say, stock market trading, inputting time series into k straight lines, each of an equal length. The proposed LCA market segmentation model is a novel and efficient segmentation technique, based on time-series data that are used to create indicators, as required by LCA.

I outline the intended LCA time-series market segmentation model pedagogically in a step-by-step manner to show (1) the components of the time-series data preparation with the SAS subroutines used and (2) building the LCA segmentation. The LCA program used (not provided here) is among several commercial software packages available.

19.5.1 Objective

Hi-tech company PirSQ wants to build a market segmentation as an efficient way to target their best customers. The segmentation, providing segments or clusters, allows PirSQ to develop marketing strategies to optimize future unit orders.

* AIC= 2*Npar – 2*ln(L). AIC(L^2) = L^2 – 2*df. AIC3(L^2) = L^2 – 3*df. AIC(LL) = 2 log L + 2*Npar. AIC3(LL) = 2 log L + 3 Npar. Npar = number of parameters.

† BIC= -2*ln(L) + Npar*ln(N). BIC(L^2) =L^2 – log(N)*df. BIC(LL) = 2*log L + log(N)*Npar. Npar = number of parameters, and N = sample size.

The data available are only a times series, consisting of 14 years of quarterly *units* from 2001 to 2014. There are 2,403 customers. The data array, ranging from 2001Q1 to 2001Q4, 2002Q1 to 2002Q4, …, 2014Q1 to 2014Q4, is presented for the first five customers (Comp_ID #1–#5) in Table 19.3.

Steps for constructing the indicators required for a unit LCA market segmentation are given here. The subroutine in Appendix 19.A constructs the indicators.

1. Correlation coefficients are the measures used to capture the trends and serve as the basis for the construction of the indicators of unit orders (UNITS).
2. TREND_COEFF_2001 is the correlation coefficient between the year 2001 quarterly UNITS and Time.
3. TREND_COEFF_2002 is the correlation coefficient between *two* years of 2001–2002 quarterly UNITS and Time.
4. TREND_COEFF_2003 is the correlation coefficient between *three* years of 2001–2003 quarterly UNITS and Time.
5. And so on, for TREND_COEFF_2004, …, TREND_COEFF_2013.
6. Lastly, TREND_COEFF_2014 is the correlation coefficient between *14* years of 2001–2014 quarterly UNITS and Time.

The essence of the 14 TREND_COEFF_ variables is that they represent a unique metric of the time series UNITS by constructing *running* trends by years, 2001, 2001–2002, 2001–2003, …, 2001–2014. The results of the construction of the 14 indicators are in Table 19.4.

TABLE 19.3

Fourteen Years of Quarterly UNITS from 2001 to 2014

Comp_ID	UNITS_2001Q1	UNITS_2001Q2	UNITS_2001Q3	UNITS_2001Q4
1	0	0	0	0
2	346,472	520,161	428,341	142,186
3	0	0	0	0
4	0	6,960	3,186	0
5	0	0	0	0

Comp_ID	UNITS_2002Q1	UNITS_2002Q2	UNITS_2002Q3	UNITS_2002Q4
1	0	0	0	0
2	444,892	204,730	347,460	537,252
3	0	0	0	0
4	15,918	31,399	−8,029	141,595
5	0	0	0	0
	…	…	…	…

Comp_ID	UNITS_2014Q1	UNITS_2014Q2	UNITS_2014Q3	UNITS_2014Q4
1	177	2,510	418	0
2	632,076	1,204,691	1,657,926	2,035,833
3	0	0	0	13,832
4	175,989	108,697	679,535	22,248
5	27,283	1,535	7,047	10,918

TABLE 19.4
TREND_COEFF_2001 to TREND_COEFF_2014

Comp_ID	TREND_ COEFF_ 2001	TREND_ COEFF_ 2002	TREND_ COEFF_ 2003	TREND_ COEFF_ 2004	TREND_ COEFF_ 2005	TREND_ COEFF_ 2006	TREND_ COEFF_ 2007
1	0.00000	0.00000	0.00000	0.52854	0.34932	0.34618	0.31359
2	−0.56439	0.02142	0.17152	0.16059	0.01171	0.37555	0.64201
3	0.00000	0.00000	0.39304	0.14003	0.01989	−0.04518	−0.08339
4	−0.14722	0.60432	0.56916	0.37449	0.23192	0.09085	0.02582
5	0.00000	0.00000	0.61358	0.28335	0.10846	0.31078	0.43425

Comp_ID	TREND_ COEFF_ 2008	TREND_ COEFF_ 2009	TREND_ COEFF_ 2010	TREND_ COEFF_ 2011	TREND_ COEFF_ 2012	TREND_ COEFF_ 2013	TREND_ COEFF_ 2014
1	0.17177	0.10581	0.15980	0.22325	0.31579	0.23446	0.18805
2	0.68702	0.72033	0.59574	0.61859	0.61541	0.63827	0.66565
3	0.15319	0.08146	0.02981	0.23478	0.33829	0.36037	0.37021
4	0.01991	−0.03729	−0.09386	−0.14927	−0.18942	−0.21025	−0.03164
5	0.46905	0.52976	0.53785	0.52653	0.42322	0.31415	0.26732

Steps for the building of the sought-after *five* indicators for the UNITS-LCA are given here. The subroutine in Appendix 19.B builds the five indicators. Note, there is no AVG_ZER_TREND indicator corresponding to N_ZER_TRENDS. By definition, AVG_ZER_TREND values are always zero, have no variance, cannot add any information to any model, and obviously are not in the model. However, excluding AVG_ZER_TREND does not preclude using N_ZER_COEFF_2014 defined as follows.

1. N_POS_TRENDS – number of positive-valued TREND_COEFF_2001, ..., TREND_COEFF_2014
2. AVG_POS_TREND – average of positive-valued TREND_COEFF_2001, ..., TREND_COEFF_2014
3. N_NEG_TRENDS – number of negative-valued TREND_COEFF_2001, ..., TREND_COEFF_2014
4. AVG_NEG_TREND – average of negative-valued TREND_COEFF_2001, ..., TREND_COEFF_2014
5. N_ZER_TRENDS – number of zero-valued TREND_COEFF_2001, ..., TREND_COEFF_2014

The results of constructing the five indicators are in Table 19.5.

19.5.2 Best LCA Models

I build four LCA models, one-cluster, two-cluster, three-cluster, and four-cluster. As the number of clusters is not known, I start from the simplest to increasingly complex models. The goodness-of-fit measures are LL, AIC(LL), AIC3(LL), and BIC(LL),* along with the important statistic number of parameters in the model (Npar), which plays a prominent

* AIC(LL), AIC(LL)3, and BIC(LL) are information criteria that weigh the fit and the parsimony of a model by sample size and degrees of freedom.

TABLE 19.5

Five Indicators of the UNITS-LCA

Comp_ID	N_POS_ TRENDS	AVG_POS_ TREND	N_NEG_ TRENDS	AVG_NEG_ TREND	N_ZER_ TRENDS
1	11	0.26696	0	0.00000	3
2	13	0.45568	1	−0.56439	0
3	10	0.21211	2	−0.06429	2
4	7	0.27378	7	−0.12271	0
5	12	0.40153	0	0.00000	2

TABLE 19.6

1–4 Cluster Models with Measures of Fit

		LL	BIC(LL)	AIC(LL)	AIC3(LL)	Npar	Class. Err.
Model 1	1-cluster	−39705.7405	79468.4058	79425.4809	79432.4809	7	0.0000
Model 2	2-cluster	−29232.8119	58587.6056	58495.6239	58510.6239	15	0.0154
Model 3	3-cluster	−26026.9807	52241.0002	52099.9614	52122.9614	23	0.0420
Model 4	4-cluster	−24911.3300	50074.7558	49884.6601	49915.6601	31	0.0610

TABLE 19.7

Classification of Latent_Clus_ by Posterior Probability

LATENT_Clus_	POSTERIOR_PROB_Clus_			
	1	2	3	Total (%)
1	1,348	54	44	1,446 42.52
2	54	939	0	993 29.20
3	44	0	918	962 28.29
Total (%)	1,446 42.52	993 29.20	962 28.29	3,401 100.00

_3_CLUSTER_ TCCR	CHANCE_ TCCR	IMPROV_ OVER_ CHANCE
94.24%	34.60%	172.3%

role in determining the best model. For LL, the larger its value, the better the model. In contrast, the other three goodness-of-fit measures work in reverse order: the smaller the value, the better the model.

The couple of statistical fitness measure, say, LL and Npar is the typical approach for selecting the best model. There must be a trade-off between LL and Npar. Based on the information in Table 19.6, I declare Model 3 from the three-cluster solution the best model, which is further supported by a small 0.0420 classification error (Class. Err.).

Another measure reaffirming the selected LCA model is the classification of the three-cluster latent variable (LANTENT_Clus_) by posterior probability (POSTERIOR_PROB_Clus_) in Table 19.7. The three-cluster LCA model has a 94.24% total correct classification rate (TCCR), which represents a 172.3% (=(94.24% – 34.60%)/34.60%) improvement over the corresponding chance model TCCR of 34.60%). In a consequential manner, the model fit statistics (small values of LL and p-value, along with a reasonable Npar) and the remarkable welcome improvement (172.43%) over the chance classification confirm that local independence is satisfied.

19.5.2.1 Cluster Sizes and Conditional Probabilities/Means

Cluster sizes and conditional probabilities/means* of the three-cluster LCA are in Table 19.8. Cluster 1 is the largest cluster, which accounts for 42.52% of the population. Clusters 2 and 3 are roughly equivalent to 29.20% and 28.28% of the population, respectively. Cluster analysis with a disproportionally large cluster size is not uncommon, but sometimes the large cluster presents a problem in achieving an acceptable TCCR.

I originally conducted LCA with continuous variables with AVG_POS_TREND and AVG_NEG_TREND as well as the other count variables, N_POS_TRENDS, N_NEG_TRENDS, and N_ZER_TRENDS. The resultant table was virtually indecipherable because there were 10 rows for both AVG_POS_TRENDS and AVG_NEG_TRENDS. Of the 20 rows for average trends, many are filled with too many small or conspicuous in importance conditional probabilities. To render an easily read and understood table of conditional probabilities, I create variables AVG_POS_TRENDP and AVG_NEG_TRENDN, collapsed versions of AVG_POS_TREND and AVG_NEG_TREND with an appended P and N as the last letter, respectively, into the typical three categories of correlation values as defined in the if–then rules that follow.

1. If 0 ≤ AVG_POS_TREND < 0.333 then AVG_POS_TRENDP='+low'.
2. If 0.333 ≤ AVG_POS_TREND < 0.667 then AVG_POS_TRENDP='+med'.

TABLE 19.8

Cluster Sizes and Conditional Probabilities/Means

		Cluster 1	Cluster 2	Cluster 3
Cluster Size		0.4252	0.2920	0.2828
Indicators				
AVG_POS_TRENDP				
	+hig	0.0014	0.0513	0.0000
	+low	0.7676	0.8096	0.9473
	+med	0.2310	0.1390	0.0527
AVG_NEG_TRENDN				
	–hig	0.0179	0.0031	0.0000
	–low	0.8500	0.7061	0.9999
	–med	0.1321	0.2908	0.0000
N_POS_TRENDS				
	Mean	9.7360	2.7002	5.5642
N_NEG_TRENDS				
	Mean	3.2056	11.1382	0.1102
N_ZER_TRENDS				
	Mean	1.0584	0.1617	8.3253

* Conditional probabilities and conditional means for categorical and continuous/count variables, respectively.

3. If 0.667 ≤ AVG_POS_TREND ≤ 1 then AVG_POS_TRENDP='+high'.
4. If 0 ≥ AVG_NEG_TREND > −0.333 then AVG_NEG_TRENDN='-low'.
5. If −0.333 ≥ AVG_NEG_TREND > −0.667 then AVG_NEG_TRENDN='-med'.
6. If −0.667 ≥ AVG_NEG_TREND ≥ −1 then AVG_NEG_TRENDN='-high'.

I present the collapsed Table 19.8, which allows for the easy naming of the latent classes. Recall, an FA loading is the correlation coefficient of an item and a factor. Loadings with high values are given greater importance when labeling the factor; loadings with low values do not add any substantive understanding of the meaning of the factor. Similarly, LCA has conditional probabilities/means, which are treated like loadings, in that their values carry greater or lesser importance for naming the latent classes.

Clearly, Table 19.8 is readable for labeling the clusters. However, I take simplicity one step further by only including substantial conditional probabilities/means in Table 19.9.

Note, if conditional probabilities are equal to 0 or 1, then there is no error; if all p's are equal to 0.05, then there is noise.

For Cluster 1, significant indicator levels, in order of extent, are:

1. AVG_NEG_TRENDN–low has a conditional probability of 85.00%. Cluster 1 individuals in this row have low AVG_NEG_TREND values on the left-closed right-open interval [0, −0.333).
2. AVG_POS_TRENDP+low has a conditional probability of 76.76%. Cluster 1 individuals in this row have low AVG_POS_TREND values on the left-closed right-open interval [0, 0.333).
3. N_POS_TRENDS has an average of 9.7360.
4. N_NEG_TRENDS has an average of 3.2056.
5. N_ZER_TRENDS has an average of 1.0584.

TABLE 19.9

Important Conditional Probabilities/Means

		Cluster 1	Cluster 2	Cluster 3
Cluster Size		0.4252	0.2920	0.2828
Indicators				
AVG_POS_TRENDP				
	+low	0.7676	0.8096	0.9473
AVG_NEG_TRENDN				
	–low	0.8500	0.7061	0.9999
N_POS_TRENDS				
	Mean	9.7360	2.7002	5.5642
N_NEG_TRENDS				
	Mean	3.2056	11.1382	0.1102
N_ZER_TRENDS				
	Mean	1.0584	0.1617	8.3253

Cluster 1 consists of individuals who experience about 10 low-positive trends in UNITS, much less low-negative trends in UNITS (cf. 9.360 versus 3.2056), and one zero trend in UNITS.

For Cluster 2, significant indicator levels, in order of extent, are:

1. AVG_POS_TRENDP+low has a conditional probability of 80.96%. Cluster 2 individuals in this row have low AVG_POS_TREND values on the left-closed right-open interval [0, 0.333).
2. AVG_NEG_TRENDN–low has a conditional probability of 70.61%. Cluster 2 individuals in this row have low AVG_NEG_TREND values on the left-closed right-open interval [0, –0.333).
3. N_NEG_TRENDS has an average of 11.1382.
4. N_POS_TRENDS has an average of 2.7002.
5. N_ZER_TRENDS has an average of 0.1617.

Cluster 2 consists of individuals who experience about 11 low-negative trends in UNITS, much less low-positive trends in UNITS, and no zero trends in UNITS.

For Cluster 3, important indicator levels, in order of extent, are:

1. AVG_NEG_TRENDN–low has a conditional probability of 99.99%. Cluster 3 individuals in this row have low AVG_NEG_TREND values on the left-closed right-open interval [0, –0.333).
2. AVG_POS_TRENDP+low has a conditional probability of 94.73%. Cluster 3 individuals in this row have low AVG_POS_TREND values on the left-closed right-open interval [0, 0.333).
3. N_ZER_TRENDS has an average of 8.3253.
4. N_POS_TRENDS has an average of 5.5642.
5. N_NEG_TRENDS has an average of 0.1102.

Cluster 3 consists of individuals who experience about eight zero trends in UNITS, slightly less (5.5) low-positive trends in UNITS, and no zero low-negative trends in UNITS.

It is interesting to note that the intensity levels (conditional probabilities) of naming Clusters 1 and 2 are equivalent, although the reverse for negative and positive trends. A closer look at the intensity levels of the clusters reveals the reliability of naming Cluster 3 is greater because its levels are virtually optimal, although the levels of Clusters 1 and 2 are completely firm in importance.

Taking simplicity one additional step further, I create a summary of the conditional probabilities/means in Table 19.10. The table facilitates the naming of the clusters.

TABLE 19.10

Summary of Conditional Probabilities/Means of Three-Cluster LCA

Cluster	AVG_NEG_TRENDN [0, –0.333] N_NEG_TRENDS	AVG_POS_TRENDP [0, +0.333] N_POS_TRENDS	N_ZER_TRENDS
1	3	10	1
2	11	3	0
3	8	6	0

- Cluster 1 is labeled *Prominent Low-Positive/Moderate Low-Negative UNITS Trends.* The implication is that Cluster 1 needs marketing strategies to stimulate the positive sales.
- Cluster 2 is labeled *Prominent Low-Negative /Moderate Low-Positive UNITS Trends.* The implication is that Cluster 2 needs marketing strategies to reverse the negative sales.
- Cluster 3 is labeled *Prominent Zero/Moderate Low-Positive UNITS Trends.* The implication is that Cluster 3 needs marketing strategies to increase sales.

19.5.2.2 Indicator-Level Posterior Probabilities

Cluster sizes are prior probabilities, indicating the likelihood of individuals belonging to each of the clusters *before* observing the data. In contrast, posterior probabilities indicate the likelihood of individuals belonging to each of the clusters *after* observing the data.

If a comparison of posterior and prior probabilities yields similarly reasonable values, then the model is called weakly identifiable (poorly estimated). The prior probabilities (top row) and the posterior probabilities of the three-cluster model (all subsequent rows) are in Table 19.11. The prior and posterior probabilities are very different. Thus, the model

TABLE 19.11

Indicator-Level Posterior Probabilities

		Cluster 1	Cluster 2	Cluster 3
Overall		0.4252	0.2920	0.2828
Indicators				
AVG_POS_TRENDP				
	+hig	0.0377	0.9623	0.0000
	+low	0.3929	0.2846	0.3225
	+med	0.6390	0.2640	0.0969
AVG_NEG_TRENDN				
	−hig	0.8935	0.1065	0.0000
	−low	0.4250	0.2424	0.3326
	−med	0.3981	0.6019	0.0000
N_POS_TRENDS				
	0–2	0.0078	0.7011	0.2911
	3–5	0.1131	0.5627	0.3242
	6–7	0.3540	0.2204	0.4257
	8–10	0.6710	0.0065	0.3225
	11–14	0.9394	0.0000	0.0606
N_NEG_TRENDS				
	0–0	0.2155	0.0000	0.7845
	1–1	0.7761	0.0000	0.2239
	2–5	0.9652	0.0007	0.0340
	6–9	0.5127	0.4873	0.0000
	10–14	0.0003	0.9997	0.0000
N_ZER_TRENDS				
	0–0	0.4778	0.5222	0.0000
	1–1	0.7179	0.2813	0.0008
	2–6	0.5429	0.0509	0.4062
	7–14	0.0072	0.0000	0.9928

is well estimated. Note, the posterior probabilities, like the prior probabilities, summed across the clusters are equal to 100%.

The utility of indicator-level posterior probabilities is that the marketer can know the likelihood of cluster membership for an individual at a given level of an indicator. The table reads as follows. For example, for a given individual in, say, Cluster 1, the probability of experiencing:

1. AVG_POS_TRENDP+med = 63.90%
2. AVG_NEG_TRENDN–hig = 89.35%
3. N_POS_TRENDS (11–14) = 93.94%
4. N_NEG_TRENDS (2–5) = 96.52%
5. N_ZER_TRENDS (1–1) = 71.79%

The profile based on the five items here is not intended to represent a complete identification for the individual. I picked the largest prior probabilities within each indicator level.

19.6 Summary

I present a novel approach, imbued with high practical value, for building a market segmentation model. The model-based LCA, instead of the popular heuristic k-means, is the preferred technique. I provide background on LCA and k-means listing advantages and disadvantages of each. Also, I provide a concise treatment of PCA and FA because LCA is often considered a categorical FA. I illustrate all discussion points with a case study for a hi-tech company who wants an efficient way to target their best customers, allowing it to develop marketing strategies to optimize future unit orders. The only data available are 14 years of unit sales. Thus, the proposed technique offers times-series market segmentation modeling. The proposed method is unique and not found in the literature. I provide SAS subroutines so data miners can perform similar segmentations as presented, along with a unique way of incorporating time-series data in an otherwise cross-sectional dataset.

Appendix 19.A Creating Trend3 for UNITS

```
libname lcat 'c:\0-LCA-t';

data UNITS_vars;
set lcat.UNITS_2001_14;
keep Comp_ID
UNITS_2001Q1
UNITS_2001Q2   UNITS_2001Q3   UNITS_2001Q4   UNITS_2002Q1   UNITS_2002Q2
UNITS_2002Q3   UNITS_2002Q4
```

```
UNITS_2003Q1 UNITS_2003Q2 UNITS_2003Q3 UNITS_2003Q4 UNITS_2004Q1
UNITS_2004Q2 UNITS_2004Q3
UNITS_2004Q4 UNITS_2005Q1 UNITS_2005Q2 UNITS_2005Q3 UNITS_2005Q4
UNITS_2006Q1 UNITS_2006Q2
UNITS_2006Q3 UNITS_2006Q4 UNITS_2007Q1 UNITS_2007Q2 UNITS_2007Q3
UNITS_2007Q4 UNITS_2008Q1
UNITS_2008Q2 UNITS_2008Q3 UNITS_2008Q4 UNITS_2009Q1 UNITS_2009Q2
UNITS_2009Q3 UNITS_2009Q4
UNITS_2010Q1 UNITS_2010Q2 UNITS_2010Q3 UNITS_2010Q4 UNITS_2011Q1
UNITS_2011Q2 UNITS_2011Q3
UNITS_2011Q4 UNITS_2012Q1 UNITS_2012Q2 UNITS_2012Q3 UNITS_2012Q4
UNITS_2013Q1 UNITS_2013Q2
UNITS_2013Q3 UNITS_2013Q4 UNITS_2014Q1 UNITS_2014Q2 UNITS_2014Q3
UNITS_2014Q4;
run;

PROC TRANSPOSE data=UNITS_vars out =outtrans;
id Comp_ID;
run;

data trend1;
set outtrans;
TIME+1;
TREND_COEFF_2001=TIME;
TREND_COEFF_2002=TIME;
TREND_COEFF_2003=TIME;
TREND_COEFF_2004=TIME;
TREND_COEFF_2005=TIME;
TREND_COEFF_2006=TIME;
TREND_COEFF_2007=TIME;
TREND_COEFF_2008=TIME;
TREND_COEFF_2009=TIME;
TREND_COEFF_2010=TIME;
TREND_COEFF_2011=TIME;
TREND_COEFF_2012=TIME;
TREND_COEFF_2013=TIME;
TREND_COEFF_2014=TIME;

if TREND_COEFF_2001 gt 4 then TREND_COEFF_2001=.;
if TREND_COEFF_2002 gt 8 then TREND_COEFF_2002=.;
if TREND_COEFF_2003 gt 12 then TREND_COEFF_2003=.;
if TREND_COEFF_2004 gt 16 then TREND_COEFF_2004=.;
if TREND_COEFF_2005 gt 20 then TREND_COEFF_2005=.;
if TREND_COEFF_2006 gt 24 then TREND_COEFF_2006=.;
if TREND_COEFF_2007 gt 28 then TREND_COEFF_2007=.;
if TREND_COEFF_2008 gt 32 then TREND_COEFF_2008=.;
if TREND_COEFF_2009 gt 36 then TREND_COEFF_2009=.;
if TREND_COEFF_2010 gt 40 then TREND_COEFF_2010=.;
```

```
if TREND_COEFF_2011 gt 44 then TREND_COEFF_2011=.;
if TREND_COEFF_2012 gt 48 then TREND_COEFF_2012=.;
if TREND_COEFF_2013 gt 52 then TREND_COEFF_2013=.;
if TREND_COEFF_2014 gt 56 then TREND_COEFF_2014=.;
drop TIME;
run;

title1 'TREND_COEFFs trend1';
PROC CORR data=trend1 outp=trend2; with _1 - _3402;
var TREND_COEFF_2001 - TREND_COEFF_2014;
run;

data trend3;
set trend2;
if _TYPE_='MEAN' then delete;
if _TYPE_='STD' then delete;
if _TYPE_='N' then delete;
drop _TYPE_;
rename _NAME_=Comp_ID;

data lcat.trend3;
set trend3;
Comp_ID=substr(Comp_ID,2);
array num(*) _numeric_;
do j = 1 to dim(num);
if missing(num(j)) then num(j)=0;
end;
drop j;
run;

PROC CONTENTS;
run;

PROC PRINT data=lcat.trend3 (obs=5);
var Comp_ID TREND_COEFF_2001 - TREND_COEFF_2007;
run;

PROC PRINT data=lcat.trend3 (obs=5);
var Comp_ID TREND_COEFF_2008 - TREND_COEFF_2014;
title3 ' trends for UNITS ';
run;
```

Appendix 19.B POS-ZER-NEG Creating Trend4

```
libname lcat 'c:\0-LCA-t';

data lcat.trend4_data;
set lcat.trend3;
```

```
array tr(14) trend_coeff_2001 - trend_coeff_2014;
array pos(14) ptrend_coeff_2001 - ptrend_coeff_2014;
array neg(14) ntrend_coeff_2001 - ntrend_coeff_2014;
array zer(14) ztrend_coeff_2001 - ztrend_coeff_2014;

array ppos(14) pptrend_coeff_2001 - pptrend_coeff_2014;
array nneg(14) nntrend_coeff_2001 - nntrend_coeff_2014;
array nzer(14) nztrend_coeff_2001 - nztrend_coeff_2014;

do i=1 to 14;
if tr(i) gt 0 then pos(i)=tr(i); else pos(i)=.;
AVG_POS_TREND=mean(of ptrend_coeff_2001 - ptrend_coeff_2014);
if AVG_POS_TREND=. then AVG_POS_TREND=0;

if tr(i) lt 0 then neg(i)=tr(i); else neg(i)=.;
AVG_NEG_TREND=mean(of ntrend_coeff_2001 - ntrend_coeff_2014);
if AVG_NEG_TREND=. then AVG_NEG_TREND=0;

if tr(i) eq 0 then zer(i)=tr(i); else zer(i)=.;
AVG_ZER_TREND=mean(of ztrend_coeff_2001 - ztrend_coeff_2014);
if AVG_ZER_TREND=. then AVG_ZER_TREND=0;

if tr(i) gt 0 then ppos(i)=1; else ppos(i)=0;
N_POS_TRENDS=sum(of pptrend_coeff_2001 - pptrend_coeff_2014);
if tr(i) lt 0 then nneg(i)=1; else nneg(i)=0;
N_NEG_TRENDS=sum(of nntrend_coeff_2001 - nntrend_coeff_2014);
if tr(i) eq 0 then nzer(i)=1; else nzer(i)=0;
N_ZER_TRENDS=sum(of nztrend_coeff_2001 - nztrend_coeff_2014);
end;
drop i;
run;

proc print data= lcat.trend4_data (obs=5);
var Comp_ID
N_POS_TRENDS AVG_POS_TREND
N_NEG_TRENDS AVG_NEG_TREND
N_ZER_TRENDS;
run;
```

References

1. MacQueen, J., Some methods for classification and analysis of multivariate observations, in *Proceedings of the Fifth Berkeley Symposium on Mathematical Statistics and Probability*, Vol. 1, pp. 281–297, University of California Press, 1967.
2. Lazarsfeld, P.F., The logical and mathematical foundation of latent structure analysis & the interpretation and mathematical foundation of latent structure analysis, in Stouffer, S.A., et al., Eds., *Measurement and Prediction*, pp. 362–472, Princeton University Press, Princeton, NJ, 1950.

3. Lazarsfeld, P.F., and Henry, N.W., *Latent Structure Analysis*, Houghton Mifflin, Boston, MA, 1968.
4. Goodman, L.A., The analysis of systems of qualitative variables when some of the variables are unobservable. Part I: A modified latent structure approach, *American Journal of Sociology*, 79, 1179–1259, 1974.
5. Vermunt, J.D., The regulation of constructive learning processes, *British Journal of Educational Psychology*, 68, 149–171, 1998.
6. Stouffer, S.A., and Toby, J., Role conflict and personality, *American Journal of Sociology*, 56, 395–406, 1951.
7. Goodman, L.A., Exploratory latent structure analysis using both identifiable and unidentifiable models, *Biometrika*, 61, 215–231, 1974.
8. Hagenaars, J.A., and McCutcheon, A.L., *Applied Latent Class Analysis*, Cambridge University Press, MA.
9. Neath, A.A., and Cavanaugh, J.E., The Bayesian information criterion: Background, derivation, and applications, *WIREs Computational Statistics*, 4, 199–203, 2012.

20

Market Segmentation: An Easy Way to Understand the Segments

20.1 Introduction

Market segmentation modeling is performed to gain insight into a company's customers' needs and wants to effectively market to their customers. A business database consists of heterogeneous segments. A cluster analysis of a business database identifies the segments. After a fruitful cluster analysis uncovers the segments, the next step is to understand the nature of the segments. The purpose of this compact chapter is to present an easy way to understand the segments to develop effective marketing strategies for targeting customers within segments and across segments. I illustrate the proposed technique and provide the corresponding SAS© subroutines so data miners can add this worthy statistical technique to their toolkits. The subroutines are also available for downloading from my website: http://www.geniq.net/articles.html#section9.

20.2 Background

Market segmentation modeling—technically, cluster analysis—is the commonly used statistical method to gain insight into a company's customers' needs and wants to effectively market-n-sell customers. A customer database assuredly consists of heterogeneous segments. The simplest segmentation structure is the two-group solution: the best and the not-so-best segments of customers. Initially, the statistician conducts cluster analyses for two- through four-cluster solutions. Based on the primary statistics, goodness-of-fit measures, and number of parameters, the statistician determines the best cluster solution. If an acceptable cluster solution does not unfold, then the statistician increases the number of clusters until the primary statistics indicate the best solution [1]. After a fruitful cluster analysis uncovers the segments, the next step is to understand the nature of the segments. (Segments and clusters are interchangeable terms.)

The literature is replete with clustering methodologies. The notion of a *cluster* cannot be precisely defined, which is one of the reasons why there are so many clustering algorithms (https://en.wikipedia.org/wiki/Clustering_algorithm) [2]. In contrast, an area of segmentation modeling that has not received much attention is how to interpret the uncovered segments. The purpose of this compact chapter is to present an easy way to understand the segments to develop effective marketing strategies for targeting customers within segments and across segments.

20.3 Illustration

Consider the input dataset SAMPLE in Table 20.1, which consists of five customers and their values on three variables, the segmentors of the segmentation, AGE (years), INCOME ($000), and EDUC (years). The subroutine to create SAMPLE is in Appendix 20.A.

Next, I will discuss the steps of the proposed method for understanding the revealed segments of a market segmentation:

1. I perform a cluster analysis (not shown) on SAMPLE. The output of the segmentation model is the dataset SAMPLE_CLUSTERED in Table 20.2, which has the appended SEGMENT variable.
2. I run subroutine in Appendix 20.B, Segmentor-Means, to obtain the Segmentor-Mean by Cluster in Table 20.3.
3. Next, I use the proposed easy method of *indexing* the means of the segmentors over the "All" *base means* of the corresponding segmentors.
   a. For AGE and Clus1, base mean AGE_mean is 50.6 and Clus1 mean is 45.0. AGE indexed for Clus1 is Clus1 mean minus AGE_mean divided by AGE_mean, producing an AGE indexed (for Clus1) value of –11.1% (=(45.0 – 50.6)/50.6).
      i. The calculations for AGE indexed for Clus2 and Clus3 are similar to that of AGE indexed for clus1.
   b. For the remaining segmentors, INCOME and EDUC, the indexing calculations are similarly performed.
   c. I run the subroutine in Appendix 20.C to obtain desired indexed profiles of segments Clus1, Clus2, and Clus3 for AGE, EDUC, and INCOME, respectively, in Table 20.4.

TABLE 20.1
SAMPLE

obs	AGE	INCOME	EDUC
1	43	130	10
2	47	140	12
3	52	250	14
4	44	230	14
5	67	390	19

TABLE 20.2
SAMPLE_CLUSTERED

AGE	INCOME	EDUC	SEGMENT
43	130	10	Clus1
47	140	12	Clus1
52	250	14	Clus2
44	230	14	Clus2
67	390	19	Clus3

TABLE 20.3

Segmentor-Mean by Cluster

	AGE	INCOME	EDUC
SEGMENT	**Mean**	**Mean**	**Mean**
Clus1	45.0	135.0	11.0
Clus2	48.0	240.0	14.0
Clus3	67.0	390.0	19.0
All	50.6	228.0	13.8

TABLE 20.4

Indexed Profiles of Clusters

SEGMENT	AGE Indexed over AGE_mean	INCOME Indexed over INCOME_mean	EDUC Indexed over EDUC_mean
Clus1	(11.1%)	(40.8%)	(20.3%)
Clus2	(5.1%)	5.3%	1.4%
Clus3	32.4%	71.1%	37.7%
	AGE_mean	**INCOME_mean**	**EDUC_mean**
	50.6	228.0	13.8

20.4 Understanding the Segments

The following narrative yields a clear understanding of the segments.

1. Clus3 has an indexed profile of all positive values. The implications are:
   a. Clus3 customers are older and more experienced than the average customer in the database by 32.4%.
   b. Clus3 customers draw a significantly greater income, 71.1% of the average customer.
   c. Clus3 customers have more education, 37.7% more years over the average customer.
2. Clus1 has an indexed profile of all negatives values. The implications are:
   a. Clus1 customers are younger than the average customer, 11.1% younger.
   b. Clus1 customers bring home much less income, 40.8% less income than the average customer.
   c. Clus1 customers have less education, 20.3% fewer years of education.
3. Yielding a *flat* note, Clus2 customers have a mixture of positive and negative indexed values of modest magnitude. The implications are:
   a. Clus2 customers are slightly younger; their average age is 5.1% less than the average customer.

b. Clus2 customers bring home somewhat more income, 5.3% over the average customer.
c. Clus2 customers have a tad more education, 1.4% years of education over the average customer.

The segment names are based on the indexed profiles. I leave it to the reader to create mnemonics for the three-cluster solution. Worthy of note, the easiness of the illustration should not belie the power of the new method.

20.5 Summary

The literature is replete with clustering methodologies, of which any one can serve for conducting a market segmentation. In contrast, the literature is virtually sparse in the area of how to interpret the segmentation results. I propose an easy way to understand the discovered customer segments. I illustrate the new method with an admittedly simple example that should not belie the power of the approach with real studies. I provide SAS subroutines for conducting the proposed technique so data miners can add this worthy statistical technique to their toolkits.

Appendix 20.A Dataset SAMPLE

```
data SAMPLE;
input AGE INCOME EDUC;
cards;
43 130 10
47 140 12
52 250 14
44 230 14
67 390 19
;
run;

PROC PRINT data=SAMPLE;
run;

data SAMPLE_CLUSTERED;
input AGE INCOME EDUC SEGMENT $5.;
cards;
43 130 10 clus1
47 140 12 clus1
52 250 14 clus2
44 230 14 clus2
67 390 19 clus3
```

```
;
run;

PROC PRINT;
run;
```

Appendix 20.B Segmentor-Means

```
PROC TABULATE data=SAMPLE_CLUSTERED;
class SEGMENT;
var AGE INCOME EDUC;
table segment all, ((AGE INCOME EDUC)*((mean)*f=7.1));
run;
```

Appendix 20.C Indexed Profiles

```
PROC SUMMARY data=SAMPLE_CLUSTERED;
class SEGMENT;
var AGE INCOME EDUC;
output out=VAR_means mean=;
run;

data BASE_MEANS;
set VAR_MEANS;
if _type_=0;
drop _freq_ _type_;
k=1;
rename
AGE = AGE_mean
INCOME = INCOME_mean
EDUC = EDUC_mean;
format AGE_mean INCOME_mean EDUC_mean 5.1;
run;

data VAR_means;
set VAR_means;
k=1;
run;

PROC SORT data=BASE_MEANS; by k;
PROC SORT data=VAR_means; by k;
run;
```

```
PROC PRINT data=BASE_MEANS;
format AGE_mean INCOME_mean EDUC_mean 5.1;
run;

data VAR_means;
set VAR_means;
k=1;
run;

PROC SORT data=BASE_MEANS; by k;
PROC SORT data=VAR_means; by k;
run;

data INDEX;
merge
BASE_MEANS VAR_means; by k;
array CLUS_MEANS AGE INCOME EDUC;
array BASE_MEANS AGE_mean INCOME_mean EDUC_mean;
array INDEX AGEx INCOMEx EDUCx;
do over INDEX;
index=(CLUS_MEANS-BASE_MEANS)/BASE_MEANS;
end;
label
AGEx= 'AGE Indexed over AGE_mean'
INCOMEx='INCOME Indexed over INCOME_mean'
EDUCx= 'EDUC Indexed over EDUC_mean';
if segment=' ' then delete;
run;

PROC PRINT data=INDEX label;
var segment AGEx INCOMEx EDUCx;
format AGEx INCOMEx EDUCx PERCENT8.1;
run;
```

References

1. Ratner, B., *Statistical and Machine-Learning Data Mining: Techniques for Better Predictive Modeling, Analysis of Big Data*, 2nd edition, pp. 79–84, 2012.
2. Estivill-Castro, V., Why so many clustering algorithms—A position paper, *ACM SIGKDD Explorations Newsletter* 4(1), 65–75, 2002.

21

The Statistical Regression Model: An Easy Way to Understand the Model

21.1 Introduction

The purpose of this short and insightful chapter is to present an easy way to understand the statistical regression model—that is, ordinary least squares (OLS) and logistic regression (LR) models—as an extension of the method of understanding a market segmentation (MS) presented in Chapter 20. I illustrate the proposed method with an LR model. The illustration brings out the power of the method in that it imparts supplementary information, making up for a deficiency in the ever-relied-upon regression coefficient for understanding a statistical regression model. I provide the SAS© subroutines, which serve as a valued addition to any bag of statistical methods. The subroutines are also available for downloading from my website: http://www.geniq.net/articles.html#section9.

21.2 Background

From Chapter 23, I use the LR model: the RESPONSE model based on predictor variables X11, X12, X13, X19, and X21, which are renumbered X10, X11, X12, X13, and X14, respectively, for ease of presentation. I repost the corresponding decile analysis in Table 21.1. Due to the nondescriptive nature of the predictor variables, an easily flowing narrative similar to the MS of Chapter 20 is not impossible. However, the nondescriptive predictor variables provide a narrative of a different sort, which yields another valuable feature of the proposed *EZ-method* for the statistical regression model.

The displayed data structures of the market segmentation (MS) EZ-method are Segmentor-Mean by Cluster (Table 20.3) and Indexed Profiles of Clusters (Table 20.4). The extension of the MS EZ-method subroutines is all but identical for the LR EZ-method. Carrying the MS EZ-method to the LR EZ-method, there are two required modifications. The first is obvious: predictor variables and deciles replace segmentors and clusters, respectively. The second is not obvious, and unfortunately, the discussion of its justification is in Chapter 42. Briefly, to affect the indexed profiles of deciles regarding the concept of *holding constant all but one predictor variable,* say, X1 holding constant X2, X3, and X4, the construction of an *M-spread common region* is required. For example, the M50-spread common region consists of the observations {X1, X2, X3, X4}

TABLE 21.1

Decile Analysis of RESPONSE Based on X10, X11, X12, X13, and X14

Decile	Number of Individuals	Number of Responses	Response Rate (%)	Cum Response Rate (%)	Cum Lift (%)
top	1,600	1,118	69.9	69.9	314
2	1,600	637	39.8	54.8	247
3	1,601	332	20.7	43.5	195
4	1,600	318	19.9	37.6	169
5	1,600	165	10.3	32.1	144
6	1,601	165	10.3	28.5	128
7	1,600	158	9.88	25.8	116
8	1,601	256	16.0	24.6	111
9	1,600	211	13.2	23.3	105
bottom	1,600	199	12.4	22.2	100
	16,003	3,559			

whose values are common to the individual M50-spreads (the middle 50% of the values*) for each of the four variables X1, X2, X3, and X4. Similarly, the M65-spread common region consists of the observations {X1, X2, X3, X4} whose values are common to the individual M65-spreads (the middle 65% of the values) for each of the four variables X1, X2, X3, and X4.

Note, the Indexed Profiles of Clusters do not require an M-spread common region construction. The effects of one segmentor are not contingent upon holding constant the remaining segmentors because the segmentors are not tied together by a regression model.

21.3 EZ-Method Applied to the LR Model

The four-step process of the EZ-method applied to the LR model consists of the following:

1. Establish an M-spread common region of the dataset used for the decile analysis, denoted Mx-spread, where x = size of spread. Typically, M50-spread is the starting point. If the resultant means, hereafter referred to as *base means*, of the predictor variables on the *reliable section* of the original dataset are problematic (e.g., zeros), then incrementally increase or decrease the spread, say, by 5%. The effects of the size of the spread is discussed in Chapter 42.
2. Create a Mx-spread common region to calculate the base means of the predictor variables defining the LR model.
   a. For the LR illustration, I use M65-spread because M50-spread generates base means of zero for X12–X14. Zero base means are problematic because the

* M50 is the more familiar term for H-spread.

required calculations involve division by the base means: Resultant quotients are not defined. The base means for X10, X11, X12, X13, and X14, labeled *X10_mean, X11_mean, …, X14_mean,* are in Table 21.2.

   b. The subroutine for this step is in Appendix 21.A.

3. Create 10 datasets, one for each decile based on the M65-spread dataset.
   a. For the LR illustration, the decile datasets are dec0, decl, dec2, dec3, dec4, dec5, dec6, dec7, dec8, and dec9.
   b. The subroutine for this step is in Appendix 21.A.
4. Calculate the indexed profiles of deciles based on the M65-spread decile datasets.
   a. Recall from MS EZ-method
      i. $\text{Xi indexed for Decile j} = \dfrac{\text{Xi\_mean(Decile j)} - \text{Xi\_mean}}{\text{Xi\_mean}}$

      where i = 10 to 14, j = 0 to 9, and Xi_mean(Decile j) denotes the mean of predictor variable Xi at Decile j.
   b. The subroutines for calculating predictors' means by decile and indexed profiles of deciles are in Appendix 21.C.
   c. For the LR illustration, the predictors' means by decile are in Table 21.3.
   d. For the LR illustration, the indexed profiles of deciles are in Table 21.4.

TABLE 21.2

Base Means for X10, X11, X12, X13, and X14 Based on M65-spread

X10_mean	X11_mean	X12_mean	X13_mean	X14_mean
138525.4	1.6	−0.2	−0.2	−0.2

TABLE 21.3

Predictors' Means by Decile Based on M65-spread

Decile	X10 Mean	X11 Mean	X12 Mean	X13 Mean	X14 Mean
top	71196.8	1.5	2.0	2.0	1.6
2	90918.1	1.5	1.0	1.4	0.7
3	57815.5	1.0	0.0	0.0	0.0
4	91136.1	1.4	0.0	0.0	0.0
5	77287.1	2.0	0.0	0.0	0.0
6	146690.6	2.0	0.0	0.0	0.0
7	254069.1	2.0	0.0	0.0	0.0
8	148158.3	1.2	−1.0	−0.8	−0.8
9	261946.5	1.6	−1.0	−1.0	−1.0
bottom	257318.5	1.5	−2.0	−2.0	−2.0

TABLE 21.4
Indexed Profiles of Deciles Based on M65-spread

Decile	X10 Indexed over X10_mean	X11 Indexed over X11_mean	X12 Indexed over X12_mean	X13 Indexed over X13_mean	X14 Indexed over X14_mean
top	(48.6%)	(5.2%)	(937.2%)	(953.0%)	(783.9%)
2	(34.4%)	(0.8%)	(518.6%)	(691.6%)	(380.1%)
3	(58.3%)	(35.8%)	(100.0%)	(100.0%)	(100.0%)
4	(34.2%)	(11.5%)	(100.0%)	(100.0%)	(100.0%)
5	(44.2%)	28.4%	(100.0%)	(100.0%)	(100.0%)
6	5.9%	28.4%	(100.0%)	(100.0%)	(100.0%)
7	83.4%	28.4%	(100.0%)	(100.0%)	(100.0%)
8	7.0%	(23.7%)	318.6%	254.4%	217.7%
9	89.1%	2.7%	318.6%	326.5%	318.3%
bottom	85.8%	(2.3%)	737.2%	753.0%	736.7%
	X10_mean	**X11_mean**	**X12_mean**	**X13_mean**	**X14_mean**
	138525.4	1.6	−0.2	−0.2	−0.2

21.4 Discussion of the LR EZ-Method Illustration

The nondescriptive predictor variables of the LR model provide a narrative of a different sort from that of the MS EZ-method illustration, which has well-defined demographic variables. The undefined predictor variables yield another valuable feature of the proposed EZ-method: indexed profiles of deciles with symbols, which are discussed later in this section. At first blush, Table 21.4 itself represents a disquietude because there are many large absolute indexed values due to the small base mean value for X12_mean, X13_mean, and X14_mean (i.e., −0.2). These indexed values are not problematic: They are just the facts of the profiles.

1. There are six large negative indexed values, ranging from −953.0% to −380.1%.
2. There are 15 indexed values equal to −100%.
3. There are nine large positive indexed values, ranging from 217.7% to 753.0%.
4. The remaining indexed values corresponding to X10_mean and X11_mean, whose values are 138525.4 and 1.6, respectively, are in a typical range, from −0.8% to 89.1%.

The interpretation of a positive (negative) indexed value: the percent Xi_mean(Decile j) is greater (less) than the corresponding Xi_mean.

1. For the top decile and X11, the X11 indexed value −5.2% (=(1.5–1.6)/1.6)* indicates the typical individual in the top decile has an X11 value 5.2% less than X11_mean.

* This expression suffers from rounding error. An accurate expression is obtained with values reported to four decimal places: X11 index is equal to −5.21% (=(1.4767–1.5579)/1.5579).

2. For the fifth decile and X11, the X11 indexed value 28.4% (=(2.0–1.6)/1.6)* indicates the typical individual in the fifth decile has an X11 value 28.4% greater than X11_mean.
3. An indexed value equal to –100% (or 100%) is due to the corresponding Xi_mean(Decile j) equal to 0.0. The interpretation of an indexed value –100% is as follows:
   a. For the third decile and X12, the X12 indexed value –100.0% (=(0.0 + 0.2)/–0.2) indicates the typical individual in the third decile has an X12 value –100% less than X12_mean. In other words, the typical value of 0.0 is 0.2 (= –100%*–0.2) greater than X12_mean value –0.2.

Instead of mechanically reading the indexed profiles for each decile using the values of undefined predictor variables, I generate a visual version of Table 21.4, in which symbols replace the indexed values. The indexed profiles of deciles with symbols in Table 21.5 reveals patterns of the magnitudes of the indexed profile values. The profile patterns allow for an easy read of the understanding of the LR model. The symbols are:

1. Triplet negative sign (– – –) for the six large negative indexed values, ranging from –953.0% to –380.1%.
2. Single zero sign (0) for the 15 nominally indexed values equal to –100%.
3. Triplet positive sign (+++) for the nine large positive indexed values, ranging from 217.7% to 753.0%.
4. Couplet negative sign (– –) for the eight moderate (somewhat) negative indexed values, ranging from 11.5% to 58.3%.
5. Couplet positive sign (++) for the six moderate positive (somewhat) indexed values, ranging from 28.4% to 89.1%.

TABLE 21.5

Indexed Profiles of Deciles with Symbols Based on M65-spread

Decile	X10 Indexed over X10_mean	X11 Indexed over X11_mean	X12 Indexed over X12_mean	X13 Indexed over X13_mean	X14 Indexed over X14_mean	Decile Response Rate (%)	Cum Lift (%)
top	(– –)	(–)	(– – –)	(– – –)	(– – –)	69.9	314
2	(– –)	(–)	(– – –)	(– – –)	(– – –)	39.8	247
3	(– –)	(– –)	(0)	(0)	(0)	20.7	195
4	(– –)	(– –)	(0)	(0)	(0)	19.9	169
5	(– –)	+ +	(0)	(0)	(0)	10.3	144
6	+	+ +	(0)	(0)	(0)	10.3	128
7	+ +	+ +	(0)	(0)	(0)	9.88	116
8	+	(– –)	+ + +	+ + +	+ + +	16.0	111
9	+ +	+	+ + +	+ + +	+ + +	13.2	105
bottom	+ +	(–)	+ + +	+ + +	+ + +	12.4	100
	X10_mean	**X11_mean**	**X12_mean**	**X13_mean**	**X14_mean**		
	138525.4	1.6	–0.2	–0.2	–0.2		

* This expression also suffer from rounding error. An accurate expression is obtained with values reported to four decimal places: X11 index is equal to 28.4% (=(2.0000–1.5579)/1.5579).

6. Single negative sign (–) for the one small (slightly) negative indexed value, –2.3%.
7. Single positive sign (+) for the two small (slightly) positive indexed values, 2.7% and 7.0%.
8. Additionally, appended to Table 21.5 are the "Decile Response Rate" and "Cum Lift" columns from the LR model's decile analysis in Table 21.1.

The decile-based profile patterns allow for a well-situated narrative of the understanding of the LR model.

1. The top and second deciles consist of individuals for which all predictor variables' values are less than the base means. Specifically, X12, X13, and X14 are largely less than the corresponding base means; X10 is somewhat less than the corresponding base mean; and, X11 is slightly less than the corresponding base mean. The responsiveness of the top 20% of the file, response rates 69.9% and 39.8% for top and second deciles, respectively, is significant with a Cum Lift of 247 at a 20% depth-of-file.
2. The third and fourth deciles consist of individuals for which all predictor variables' values are less than the corresponding base means but with a different pattern than that of the top two deciles. Specifically, X12, X13, and X14 are nominally less than the corresponding base means (as the means of X12, X13, and X14 are zero); and, X10 and X11 are somewhat less than the corresponding base mean. The lower responsiveness of the third and fourth deciles, response rates 20.7% and 19.9%, respectively, compared to the top two deciles, lessens the Cum Lift to 169 for a 40% depth-of-file.
3. The fifth decile consists of individuals for which three predictor variables' values are nominally less than the corresponding base means, one predictor variable's values are less than the corresponding base means, and one predictor variable's values are greater than the corresponding base mean. Specifically, X12, X13, and X14 are nominally less than the corresponding base means; X10 is somewhat less than the corresponding base mean; and X11 is somewhat greater than the corresponding base mean. The fifth decile, with a response rate of 10.3%, is on the cusp of actionable/nonactionable depth-of-file for model implementation. The Cum Lift is 144 at a 50% depth-of-file.
4. The lower five deciles (sixth through the bottom) are typically nonactionable deciles because of their minimal impact on the Cum Lift: The average Cum Lift for the lower five deciles is 112. However, these deciles are worthy of discussion as they exhibit a concerning pattern.
   a. For the sixth and seventh deciles, there is a block of the indexed symbol (0) for X12, X13, and X14.
   b. For the eighth through the bottom deciles, there is a block of the indexed symbol (+++) for X12, X13.
   c. For the lower five deciles, for X10 and X12, there is no apparent pattern of symbols.
   d. The lower five decile pattern is without value due to its position. However, it would not be unreasonable to see such a pattern hold importance in upper deciles.

The implication of this discussion is that the EZ-method is a powerful analytical tool that imparts supplementary information, making up for a deficiency in the ever-relied-upon regression coefficient for understanding a statistical regression model. The traditional use of the regression coefficient is conveying the singular effects of the corresponding predictor variable, holding constant the other variables, on the dependent variable. In contrast, the EZ-method imparts the combined effects of the full array of predictor variables, accounting for the concept of holding constant the other variables, on the dependent variable.

21.5 Summary

I present an EZ-method to understand the statistical regression model (i.e., OLS and LR models). I illustrate the EZ-method with an LR model, which brings out the power of the EZ-method: It imparts supplementary information, making up for a deficiency in the ever-relied-upon regression coefficient for understanding a statistical regression model. The traditional use of the regression coefficient is conveying the singular effects of the corresponding predictor variable, holding constant the other variables, on the dependent variable. In contrast, the EZ-method imparts the combined effects of the full array of predictor variables, accounting for the concept of holding constant the other variables, on the dependent variable. I provide the SAS subroutines, which serve as a valued addition to any bag of statistical methods.

Appendix 21.A M65-Spread Base Means X10–X14

```
libname c15c 'c:\0-chap15c';

%let spread=65;
title "Base means with M-spread&spread";

PROC RANK data=c15c.ezway_LRM groups=100 out=OUT;
var X10-X14;
ranks X10r X11r X12r X13r X14r;
run;

data spread&spread._X10;
set out;
rhp=(100-&spread)/2;
if X10r=> (rhp-1) and X10r<=(99-rhp);
keep ID X10 X10r;
run;

data spread&spread._X11;
set out;
rhp=(100-&spread)/2;
```

```
if X11r=> (rhp-1) and X11r<=(99-rhp);
keep ID X11 X11r;
run;

data spread&spread._X12;
set out;
rhp=(100-&spread)/2;
if X12r=> (rhp-1) and X12r<=(99-rhp);
keep ID X12 X12r;
run;

data spread&spread._X13;
set out;
rhp=(100-&spread)/2;
if X13r=> (rhp-1) and X13r<=(99-rhp);
keep ID X13 X13r;
run;

data spread&spread._X14;
set out;
rhp=(100-&spread)/2;
if X14r=> (rhp-1) and X14r<=(99-rhp);
keep ID X14 X14r;
run;

PROC SORT DATA=spread&spread._X10; by ID;
PROC SORT DATA=spread&spread._X11; by ID;
PROC SORT DATA=spread&spread._X12; by ID;
PROC SORT DATA=spread&spread._X13; by ID;
PROC SORT DATA=spread&spread._X14; by ID;
run;

data spread&spread._X10X11X12X13X14;
merge
spread&spread._X10 (in=var_X10)
spread&spread._X11 (in=var_X11)
spread&spread._X12 (in=var_X12)
spread&spread._X13 (in=var_X13)
spread&spread._X14 (in=var_X14);
by ID;
if var_X10=1 and var_X11=1 and var_X12=1 and var_X13=1 and var_X14=1;
run;

PROC MEANS data=spread&spread._X10X11X12X13X14 mean n maxdec=4;
var X10-X14;
run;
```

Appendix 21.B Create Ten Datasets for Each Decile

```
libname c15c 'c:\0-chap15c';
title' X10-X14 ';

PROC LOGISTIC data=c15c.ezway_LRM nosimple des outest=coef;
model RESPONSE = X10-X14;
run;

PROC SCORE data=c15c.ezway_LRM predict type=parms score=coef out=score;
var X10-X14;
run;

data score;
set score;
estimate=response2;
run;

data notdot;
set score ;
if estimate ne .;
PROC MEANS data=notdot noprint sum; var wt;
output out=samsize (keep=samsize) sum=samsize;
run;

data scoresam (drop=samsize);
set samsize score;
retain n;
if _n_=1 then n=samsize;
if _n_=1 then delete;
run;

PROC SORT data=scoresam; by descending estimate;
run;

data score;
set scoresam;
if estimate ne . then cum_n+wt;
if estimate = . then dec=.;
else dec=floor(cum_n*10/(n+1));
prob_hat=exp(estimate)/(1+ exp(estimate));
logit=estimate;
run;

data c15c.ezway_probs;
set scoresam;
if estimate ne . then cum_n+wt;
if estimate = . then dec=.;
else dec=floor(cum_n*10/(n+1));
```

```
prob_complete=exp(estimate)/(1+ exp(estimate));
keep ID response estimate dec X10-X14 wt;
run;

data c15c.dec0 c15c.dec1 c15c.dec2 c15c.dec3 c15c.dec4
c15c.dec5 c15c.dec6 c15c.dec7 c15c.dec8 c15c.dec9;
set c15c.ezway_probs;
if dec=0 then output c15c.dec0;
if dec=1 then output c15c.dec1;
if dec=2 then output c15c.dec2;
if dec=3 then output c15c.dec3;
if dec=4 then output c15c.dec4;
if dec=5 then output c15c.dec5;
if dec=6 then output c15c.dec6;
if dec=7 then output c15c.dec7;
if dec=8 then output c15c.dec8;
if dec=9 then output c15c.dec9;
run;
```

Appendix 21.C Indexed Profiles of Deciles

```
libname c15c 'c:\0-chap15c';
options pageno=1;

%macro doMIDSPREAD;
%do dec= 0 %to 9;
%let spread=65;

PROC RANK data=c15c.dec&dec. groups=100 out=OUT;
var X10-X14;
ranks X10r X11r X12r X13r X14r;
run;

title1 "dec=&dec ";
title2 "midspread=&spread";
run;

data midspread&spread._X10;
set out;
dec=&dec;
rhp=(100-&spread)/2;
if X10r=> (rhp-1) and X10r<=(99-rhp);
keep ID dec X10 X10r;
run;

data midspread&spread._X11;
set out;
dec=&dec;
```

```
rhp=(100-&spread)/2;
if X11r=> (rhp-1) and X11r<=(99-rhp);
keep ID dec X11 X11r;
run;

data midspread&spread._X12;
set out;
dec=&dec;
rhp=(100-&spread)/2;
if X12r=> (rhp-1) and X12r<=(99-rhp);
keep ID dec X12 X12r;
run;

data midspread&spread._X13;
set out;
dec=&dec;
rhp=(100-&spread)/2;
if X13r=> (rhp-1) and X13r<=(99-rhp);
keep ID dec X13 X13r;
run;

data midspread&spread._X14;
set out;
dec=&dec;
rhp=(100-&spread)/2;
if X14r=> (rhp-1) and X14r<=(99-rhp);
keep ID X14 X14r;
run;

PROC SORT data=midspread&spread._X10; by ID;
PROC SORT data=midspread&spread._X11; by ID;
PROC SORT data=midspread&spread._X12; by ID;
PROC SORT data=midspread&spread._X13; by ID;
PROC SORT data=midspread&spread._X14; by ID;
run;

data midspread&spread._dec&dec._X10X11X12X13X14;
merge
midspread&spread._X10 (in=var_X10)
midspread&spread._X11 (in=var_X11)
midspread&spread._X12 (in=var_X12)
midspread&spread._X13 (in=var_X13)
midspread&spread._X14 (in=var_X14);
by ID;
if var_X10=1 and var_X11=1 and var_X12=1 and var_X13=1 and var_X14=1;
run;

PROC MEANS data=midspread&spread._dec&dec._X10X11X12X13X14
mean n MAXDEC=4;
var dec X10-X14;
%end;
```

```
%mend;
%doMIDSPREAD
quit;

%let spread=65;
data midspread&spread._X10X11X12X13X14;
set midspread&spread:;
Decile=dec;
keep ID Decile X10-X14;
run;

PROC FORMAT;
value Decile
  0 = 'top'
  1 = ' 2 '
  2 = ' 3 '
  3 = ' 4 '
  4 = ' 5 '
  5 = ' 6 '
  6 = ' 7 '
  7 = ' 8 '
  8 = ' 9 '
  9 = 'bot';
run;

title ' ';
*21.3 Predictor Means by Deciles;
PROC TABULATE data=midspread&spread._X10X11X12X13X14;
class Decile;
var X10-X14;
table Decile, ((X10-X14) *((mean)*f=12.1));
format Decile Decile.;
run;

*21.4 Indexed Profiles of Deciles;
PROC SUMMARY data=midspread&spread._X10X11X12X13X14;
class DECILE;
var X10-X14;
output out=DECILE_means mean=;
run;

data DECILE_means;
set DECILE_means;
k=1;
run;

PROC SUMMARY data=spread&spread._X10X11X12X13X14;
var X10-X14;
output out=BASE_means mean=;
run;
```

```
data BASE_means;
set BASE_means;
drop _TYPE_ _FREQ_;
k=1;
rename
X10 = X10_mean
X11 = X11_mean
X12 = X12_mean
X13 = X13_mean
X14 = X14_mean;

PROC PRINT;
title' BASE_MEANS';
format X10_mean X11_mean X12_mean X13_mean X14_mean 12.1;
run;

PROC SORT data=BASE_MEANS; by k;
PROC SORT data=DECILE_MEANS; by k;
run;

data INDEX;
merge
BASE_MEANS DECILE_means; by k;
array DECILE_MEANS X10-X14;
array BASE_MEANS X10_mean X11_mean X12_mean X13_mean X14_mean;
array INDEX X10X X11X X12X X13X X14X;
do over INDEX;
INDEX=(DECILE_MEANS-BASE_MEANS)/BASE_MEANS;
end;

label
X10X= 'X10 indexed over X10_mean'
X11X= 'X11 indexed over X11_mean'
X12X= 'X12 indexed over X12_mean'
X13X= 'X13 indexed over X13_mean'
X14X= 'X14 indexed over X14_mean';
if DECILE=' ' then delete;
run;

PROC PRINT data=INDEX label;
var DECILE X10X X11X X12X X13X X14X;
format X10X X11X X12X X13X X14X PERCENT8.1;
format DECILE DECILE.;
title' Indexed Profiles of Deciles ';
run;
```

22

CHAID as a Method for Filling in Missing Values

22.1 Introduction

The problem of analyzing data with missing values is well known to data analysts. Data analysts know that almost all standard statistical analyses require complete data for reliable results. These analyses performed with incomplete data assuredly produce biased results. Thus, data analysts make every effort to fill in the missing data values in their datasets. The popular solutions to the problem of handling missing data belong to the collection of imputation or fill-in techniques. This chapter presents chi-squared automatic interaction detection (CHAID) as a data mining method for filling in missing data.

22.2 Introduction to the Problem of Missing Data

Missing data are a pervasive problem in data analysis. It is the rare exception when the data at hand have no missing data values. The objective of filling in missing data is to recover or minimize the loss of information due to the incomplete data. I introduce the problem of handling missing data briefly.

Consider a random sample of 10 individuals in Table 22.1. The individuals in the sample have three demographic characteristics, the variables AGE, GENDER, and INCOME. There are missing values, which are denoted by a dot (.). Eight of 10 individuals provide their age; 7 of 10 individuals provide their gender and income.

Two common solutions for handling missing data are *available case analysis* and *complete case analysis*.* The available case analysis uses only the cases for which the variable of interest is available. Consider the calculation of the mean AGE: The available sample size (number of nonmissing values) is 8, not the original sample size of 10. The calculation for the means of INCOME and GENDER† uses two different available samples of size 7. The calculation on different samples points to a weakness in available case analysis. Unequal sample sizes create practical problems. Comparative analysis between variables is difficult because the subsamples of the original sample are different. Also, estimates of multivariable statistics are prone to illogical values.‡

* Available case analysis is also known as pairwise deletion. Complete case analysis is also known as listwise deletion or casewise deletion.

† The mean of GENDER is the incidence of females in the sample.

‡ Consider the correlation coefficient of X1 and X2. If the available sample sizes for X1 and X2 are unequal, it is possible to obtain a correlation coefficient value that lies outside the theoretical [−1, 1] range.

TABLE 22.1

Random Sample of 10 Individuals

Individual	AGE (years)	GENDER (0 = male, 1 = female)	INCOME
1	35	0	$50,000
2	.	.	$55,000
3	32	0	$75,000
4	25	1	$100,000
5	41	.	.
6	37	1	$135,000
7	45	.	.
8	.	1	$125,000
9	50	1	.
10	52	0	$65,000
Total	317	4	$605,000
Number of nonmissing values	8	7	7
Mean	39.6	57%	$86,429

Dot (.) denotes missing value.

TABLE 22.2

Complete-Case Version

Individual	AGE (years)	GENDER (0 = male, 1 = female)	INCOME
1	35	0	$50,000
3	32	0	$75,000
4	25	1	$100,000
6	37	1	$135,000
10	52	0	$65,000
Total	181	2	$425,000
Number of nonmissing values	5	5	5
Mean	36.2	40%	$85,000

The popular complete case analysis uses examples for which all variables are present. A complete case analysis of the original sample in Table 22.1 includes only five cases, as reported in Table 22.2. The advantage of this type of analysis is simplicity because conducting standard statistical analysis is performed without modification to incomplete data. Comparative analysis between variables is not complicated because of the use of one common subsample of the original sample. The disadvantage of discarding incomplete cases is the resulting loss of information.

Another solution is dummy variable adjustment [1]. For a variable X with missing data, two new variables are used in its place. X_filled and X_dum are defined as follows:

1. X_filled = X if X is not missing; X_filled = 0 if X is missing.
2. X_dum = 0 if X is not missing; X_dum = 1 if X is missing.

The advantage of this solution is its simplicity of use without having to discard cases. The disadvantage is that the analysis can become unwieldy when there are many variables with missing data. Also, filling in the missing value with a zero is arbitrary, which is unsettling for some data analysts.

The definition of an imputation method is any process that fills in missing data to produce a complete dataset. The simplest and most popular imputation method is *mean value imputation*. The mean of the nonmissing values of the variable of interest is used to fill in the missing data. Consider Individuals 2 and 8 in Table 22.1. The mean AGE of the file, namely, 40 years (rounded from 39.6) replaces the missing ages with Individuals 2 and 8. The advantage of this method is undoubtedly its ease of use. The calculation of means, as required, is performed within classes, predefined by other variables related to the study at hand.

Another popular method is *regression-based imputation*. The predicted values from a regression analysis replace missing values. The dependent variable Y is the variable whose missing values need to be imputed. The predictor variables, the Xs, are the *matching variables*. Regression of Y on the Xs uses a complete case analysis dataset. If Y is continuous, then ordinary least squares (OLS) regression is appropriate. If Y is categorical, then the logistic regression model (LRM) is used. For example, I wish to impute AGE for Individual 8 in Table 22.1. I regress AGE on GENDER and INCOME (the matching variables) based on the complete-case dataset consisting of five individuals (IDs 1, 3, 4, 6, and 10). The OLS regression imputation model is in Equation 22.1:

$$\text{AGE\_imputed} = 25.8 - 20.5{*}\text{GENDER} + 0.0002{*}\text{INCOME} \quad (22.1)$$

Plugging in the values of GENDER (= 1) and INCOME (= $125,000) for Individual 8, the imputed AGE is 53 years.

22.3 Missing Data Assumption

Missing data methods presuppose that the missing data are *missing at random* (MAR). Rubin formalized this condition into two separate assumptions [2]:

1. Missing at random (MAR) means that what is missing does not depend on the missing values but may depend on the observed values.
2. Missing completely at random (MCAR) means that what is missing does not depend on either the observed values or the missing values. When this assumption is satisfied for all variables, the reduced sample of individuals with only complete data can be regarded as a simple random subsample from the original data. Note that the second assumption of MCAR represents a stronger condition than the MAR assumption.

The missing data assumptions are problematic. To some extent, by comparing the information from complete cases to the information from incomplete cases, the testing of the MCAR assumption is possible. A procedure often used is to compare the distribution of the variable of interest, say Y, based on nonmissing data with the distribution of Y based on missing data. If there are significant differences, then the assumption is considered not met. If no significant differences exist, then the test offers no direct evidence of assumption violation. In this case, the assumption is considered cautiously to be satisfied.* The MAR assumption is impossible to test for validity. (Why?)

It is accepted wisdom that missing data solutions at best perform satisfactorily, even when the amount of missing data is moderate, and the missing data assumptions are tenable. The potential of the two new imputation methods, maximum likelihood and multiple imputations, which offer substantial improvement over the complete case analysis, is questionable as their assumptions are usually untenable. Moreover, there has been no evaluation of the utility of the new methods in big data applications.

Nothing can take the place of the missing data. Allison noted, "The best solution to the missing data problem is not to have any missing data" [3]. Dempster and Rubin warned that "Imputation is both seductive and dangerous," seductive because it gives a false sense of confidence that the data are complete and dangerous because it can produce misguided analyses and untrue models [4].

The aforementioned admonitions are without reference to the impact of big data on filling in missing data. In big data applications, the problem of missing data is severe because it is common for at least one variable to have 30%–90% of its values missing. Thus, I strongly argue to exercise restraint for imputation of big data applications and judicious evaluation of the findings.

In the spirit of the exploratory data analysis (EDA) tenet—that failure is when one fails to try—I advance the proposed data mining/EDA CHAID imputation method as a hybrid mean-value/regression-based imputation method that explicitly accommodates missing data without imposing additional assumptions. The salient features of the new method are the EDA characteristics:

1. *Flexibility*: Assumption-free CHAID work especially well with big data containing large amounts of missing data.
2. *Practicality*: A descriptive CHAID tree provides analysis of data.
3. *Innovation*: The CHAID algorithm defines imputation classes.
4. *Universality*: Blending of two diverse traditions occurs—traditional imputation methods and a machine-learning algorithm for data structure identification.
5. *Simplicity*: CHAID tree imputation estimates are easy to use.

22.4 CHAID Imputation

I introduce a couple of terms required for the discussion of CHAID imputation. Imputation methods require the sample to be divided into groups or classes, called *imputation classes*, which are defined by variables called *matching variables*. The formation of imputation classes is an important step to ensure the reliability of the imputation estimates: As the homogeneity of the classes increases, the accuracy and stability of the

* This test proves the necessary condition for MCAR. It remains to be shown that there is no relationship between missingness on a given variable and the values of that variable.

estimates increase. The latter relationship is true given that the variance (on the variable whose missing values are to be imputed) within each class is small.

CHAID is a technique that recursively partitions a population into separate and distinct groups, defined by the predictor variables, such that the variance of the dependent variable is minimized within the groups and maximized across the groups. In 1980, CHAID was the first technique originally developed for finding "combination" or interaction variables. In database marketing today, CHAID primarily serves as a market segmentation technique. Here, I propose CHAID as an alternative data mining method for mean value/regression-based imputation.

The justification for CHAID as a method of imputation is as follows: By definition, CHAID creates optimal homogeneous groups, which can be used effectively as trusty imputation classes.* Accordingly, CHAID provides a reliable method of mean value/regression-based imputation.

The CHAID methodology provides the following:

- CHAID is a tree-structured, assumption-free modeling alternative to OLS regression. It provides reliable estimates without the assumption of specifying the true structural form of the model (i.e., knowing the correct independent variables and their correct reexpressed forms) and without regard to the weighty classical assumptions of the underlying OLS model. Thus, CHAID, with its trusty imputation classes, provides reliable *regression tree* imputation for a continuous variable.
- CHAID serves as a nonparametric tree-based alternative to the binary and polychotomous LRM without the assumption of specifying the true structural form of the model. Thus, CHAID, with its trusty imputation classes, provides reliable *classification tree* imputation for a categorical variable.
- CHAID potentially offers more reliable imputation estimates due to its ability to use most of the analysis sample. The analysis sample for CHAID is not as severely reduced by the pattern of missing values in the matching variables, as is the case for regression-based models, because CHAID can accommodate missing values for the matching variables in its analysis.† The regression-based imputation methods cannot make such an accommodation.

22.5 Illustration

Consider a sample of 29,132 customers from a cataloguer's database. The following is known about the customers: their ages (AGE_CUST), GENDER, total lifetime dollars (LIFE_DOL), and whether a purchase was made within the past 3 months (PRIOR_3). Missing values for each variable are denoted by "???."

The counts and percentages of missing and nonmissing values for the variables are in Table 22.3. For example, there are 691 missing values for LIFE_DOL, resulting in a 2.4% missing rate. It is interesting to note that the complete case sample size for analysis with

* Some analysts may argue about the optimality of the homogeneous groups but not the trustworthiness of the imputation classes.

† Missing values are allowed to "float" within the range of the matching variable and rest at the position that optimizes the homogeneity of the groups.

TABLE 22.3
Counts and Percentages of Missing and Nonmissing Values

	Missing		Nonmissing	
Variable	**Count**	**%**	**Count**	**%**
AGE_CUST	1,142	3.9	27,990	96.1
GENDER	2,744	9.4	26,388	90.6
LIFE_DOL	691	2.4	28,441	97.6
PRIOR_3	965	3.3	28,167	96.7
All Variables	2,096	7.2	27,025	92.8

all four variables is 27,025. The complete case sample size represents a 7.2% (= 2,096/29,132) loss of information from discarding incomplete cases from the original sample.

22.5.1 CHAID Mean-Value Imputation for a Continuous Variable

I wish to impute the missing values for LIFE_DOL. I perform a mean value imputation with CHAID using LIFE_DOL as the dependent variable and AGE_CUST as the predictor (matching) variable. The AGE_CUST CHAID tree is in Figure 22.1.

I set some conventions to simplify the discussions of the CHAID analyses:

1. The left-closed right-open interval [x, y) indicates values between x and y, including x and excluding y.
2. The closed interval [x, y] indicates values between x and y, including both x and y.
3. There is a distinction between nodes and imputation classes. Nodes are the visual displays of the CHAID groups. Imputation classes are the nodes.
4. Nodes are referenced by numbers (1, 2, …) from left to right as they appear in the CHAID tree.

The AGE_CUST CHAID tree, in Figure 22.1, is read as follows:

1. The top box indicates a mean LIFE_DOL of $27,288.47 for the available sample of 28,441 (nonmissing) observations for LIFE_DOL.
2. The CHAID creates four nodes on AGE_CUST. Node 1 consists of 6,499 individuals whose ages are in the interval [18, 30) with a mean LIFE_DOL of $14,876.75. Node 2 consists of 7,160 individuals whose ages are in the interval [30, 40) with a mean LIFE_DOL of $29,396.02. Node 3 consists of 7,253 individuals whose ages are in the interval [40, 55) with a mean LIFE_DOL of $36,593.81. Node 4 consists of 7,529 individuals whose ages are either in the interval [55, 93] or missing. The mean LIFE_DOL is $27,033.73.
3. Note: CHAID positions the missing values of AGE_CUST in the oldest age missing node.

The set of AGE_CUST CHAID mean value imputation estimates for LIFE_DOL is the mean values of nodes 1 to 4: $14,876.75, $29,396.02, $36,593.81, and $27,033.73, respectively. The AGE_CUST distribution of the 691 individuals with missing LIFE_DOL is in Table 22.4.

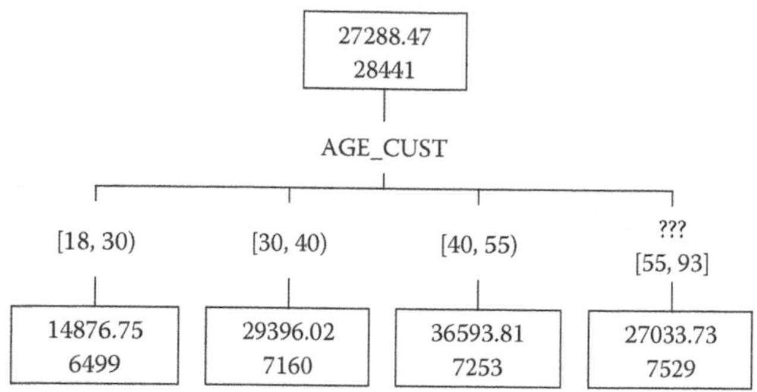

FIGURE 22.1
AGE_CUST CHAID tree for LIFE_DOL.

TABLE 22.4
CHAID Imputation Estimates for Missing Values of LIFE_DOL

AGE_CUST Class	Class Size	Imputation Estimate
[18, 30)	0	$14876.75
[30, 40)	0	$29396.02
[40, 55)	0	$36593.81
[55, 93] or ???	691	$27033.73

All the 691 individuals belong to the last class, and their imputed LIFE_DOL value is $27,033.73. Of course, if any of the 691 individuals were in the other AGE_CUST classes, the corresponding mean values would be used.

22.5.2 Many Mean-Value CHAID Imputations for a Continuous Variable

CHAID provides many mean-value imputations—as many as there are matching variables—as well as a measure to determine which imputation estimates to use. The assessment of the goodness or quality of a CHAID tree is by the measure percentage variance explained (PVE). The imputation estimates based on the matching variable with the largest PVE value are often selected as the preferred estimates. Note, however, a large PVE value does not necessarily guarantee reliable imputation estimates, and the largest PVE value does not necessarily guarantee the best imputation estimates. The data analyst may have to perform the analysis at hand with imputations based on several of the large PVE value-matching variables.

Continuing with imputation for LIFE_DOL, I perform two additional CHAID mean-value imputations. The first CHAID imputation, in Figure 22.2, uses LIFE_DOL as the dependent variable and GENDER as the matching variable. The second imputation, in Figure 22.3, uses LIFE_DOL as the dependent variable and PRIOR_3 as the matching variable. The PVE values for the matching variables AGE_CUST, GENDER, and PRIOR_3 are 10.20%, 1.45%, and 1.52%, respectively. Thus, the preferred imputation estimates for LIFE_DOL are based on AGE_CUST because the AGE_CUST-PVE value is noticeably largest.

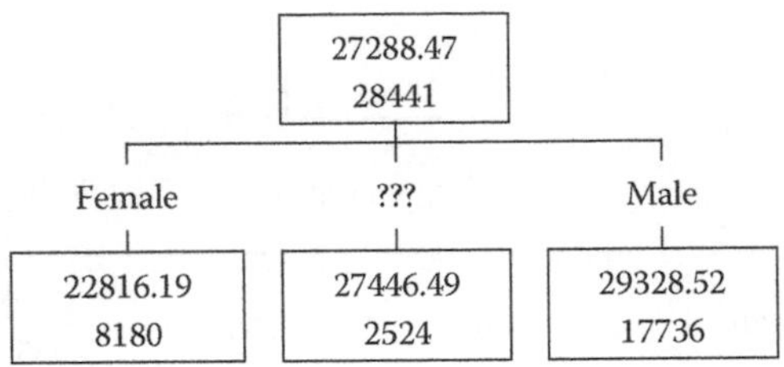

FIGURE 22.2
GENDER CHAID tree for LIFE_DOL.

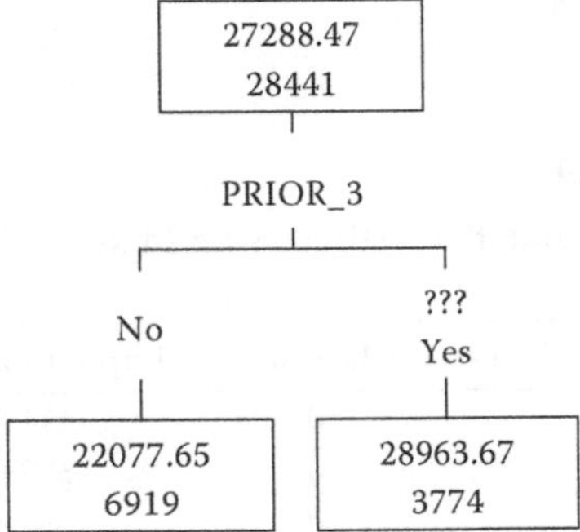

FIGURE 22.3
PRIOR_3 CHAID tree for LIFE_DOL.

Comparative note: Unlike CHAID mean value imputation, traditional mean value imputation provides no guideline for selecting an all case continuous matching variable (e.g., AGE_CUST), whose imputation estimates are preferred.

22.5.3 Regression Tree Imputation for LIFE_DOL

I can selectively add matching variables* to the preferred single-variable CHAID tree—generating a regression tree—to increase the reliability of the imputation estimates (i.e., to increase the PVE value). Adding GENDER and PRIOR_3 to the AGE_CUST tree, I obtain a PVE value of 12.32%, which represents an increase of 20.8% (= 2.12%/10.20%) over the AGE_CUST-PVE value. The AGE_CUST–GENDER–PRIOR_3 regression tree is in Figure 22.4.

The AGE_CUST–GENDER–PRIOR_3 regression tree reads as follows:

1. Extending the AGE_CUST CHAID tree, I obtain a regression tree with 13 end nodes.
2. Node 1 (two levels deep) consists of 2,725 individuals whose ages are in the interval [18, 30) *and* have not made a purchase in the past 3 months (PRIOR_3 = no). The mean LIFE_DOL is $13,353.11.

* Here is a great place to discuss the relationship between increasing the number of variables in a (tree) model and its effects on bias and stability of the estimates provided by the model.

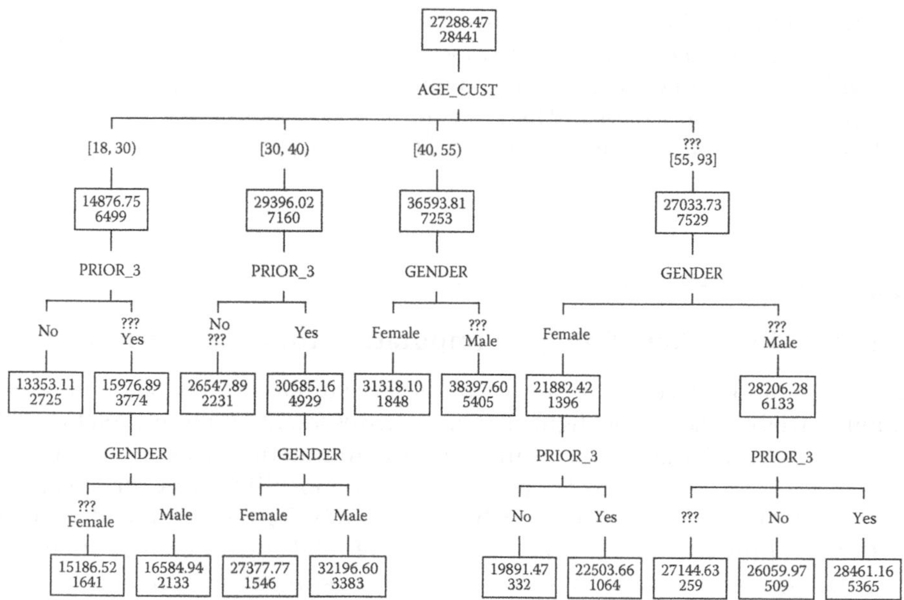

FIGURE 22.4
AGE_CUST, GENDER, and PRIOR_3 regression tree for LIFE_DOL.

TABLE 22.5

Regression Tree Imputation Estimates for Missing Values of LIFE_DOL

AGE_CUST Class	GENDER	PRIOR_3	Class Size	Imputation Estimates
??? or [55, 93]	Female	No	55	$19891.47
??? or [55, 93]	Female	Yes	105	$22503.66
??? or [55, 93]	Male	No	57	$26059.97
??? or [55, 93]	Male	Yes	254	$28461.16
??? or [55, 93]	???	No	58	$26059.97
??? or [55, 93]	???	Yes	162	$28461.16
Total			691	

3. Node 2 (three levels deep) consists of 1,641 individuals whose ages are in the interval [18, 30) *and* PRIOR_3 = ??? or yes *and* whose GENDER = ??? or female. The mean LIFE_DOL is $15,186.52.
4. The remaining nodes have a similar interpretation.

The AGE_CUST–GENDER–PRIOR_3 regression tree imputation estimates for LIFE_DOL are the mean values of 13 end nodes. An individual's missing LIFE_DOL value is replaced with the mean value of the imputation class to which the individual matches. The distribution of the 691 individuals with missing LIFE_DOL values regarding the three matching variables is in Table 22.5. All the missing LIFE_DOL values come from the five rightmost end nodes (Nodes 9 to 13). As before, if any of the 691 individuals were in the other eight nodes, the corresponding mean values would be used.

Comparative note: Traditional OLS regression–based imputation for LIFE_DOL based on four matching variables—AGE_CUST, two dummy variables for GENDER ("missing" GENDER is considered a category), and one dummy variable for PRIOR_3—results in a complete case sample size of 27,245. The complete case sample size represents a 4.2% (= 1,196/28,441) loss of information from the CHAID analysis sample.

22.6 CHAID Most Likely Category Imputation for a Categorical Variable

CHAID for imputation of a categorical variable is very similar to CHAID for imputation of a continuous variable, except for slight changes in assessment and interpretation. CHAID with a continuous variable assigns a mean value to an imputation class. In contrast, CHAID with a categorical variable assigns the predominant or most likely category to an imputation class. CHAID with a continuous variable provides PVE. In contrast, with a categorical variable, CHAID provides the measure proportion of total correct classifications (PTCC)* to identify the matching variable(s) whose imputation estimates are preferred.

As noted for CHAID with a continuous variable, there is a similar note that a large PTCC value does not necessarily guarantee reliable imputation estimates, and the largest PTCC value does not necessarily guarantee the best imputation estimates. The data analyst may have to perform the analysis with imputations based on several of the large PTCC value-matching variables.

22.6.1 CHAID Most Likely Category Imputation for GENDER

I wish to impute the missing values for GENDER. I perform a CHAID most likely category imputation using GENDER as the dependent variable and AGE_CUST as the matching variable. The AGE_CUST CHAID tree is in Figure 22.5. The PTCC value is 68.7%.

The AGE_CUST CHAID tree reads as follows:

1. The top box indicates the incidences of females and males are 31.6% and 68.4%, respectively, based on the available sample of 26,388 (nonmissing) observations for GENDER.
2. The CHAID creates five nodes on AGE_CUST. Node 1 consists of 1,942 individuals whose ages are in the interval [18, 24). The incidences of females and males are 52.3% and 47.7%, respectively. Node 2 consists of 7,203 individuals whose ages are in the interval [24, 35). The incidences of female and male are 41.4% and 58.6%, respectively.
3. The remaining nodes have a similar interpretation.
4. Interesting note: CHAID places the missing values in the middle-age node. Compare the LIFE_DOL CHAID: The missing ages are in the oldest age/missing node.

* PTCC is calculated with the percentage of observations in each of the end nodes of the tree that fall in the modal category. The weighted sum of these percentages over all end nodes of the tree is PTCC. A given node is weighted by the number of observations in the node relative to the total size of the tree.

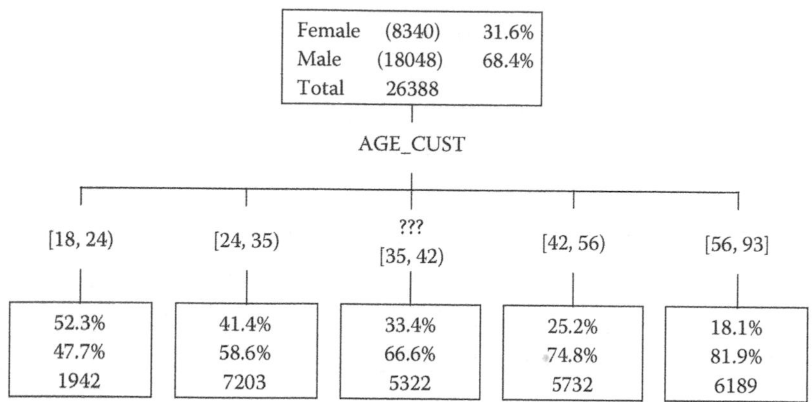

FIGURE 22.5
AGE_CUST CHAID tree for GENDER.

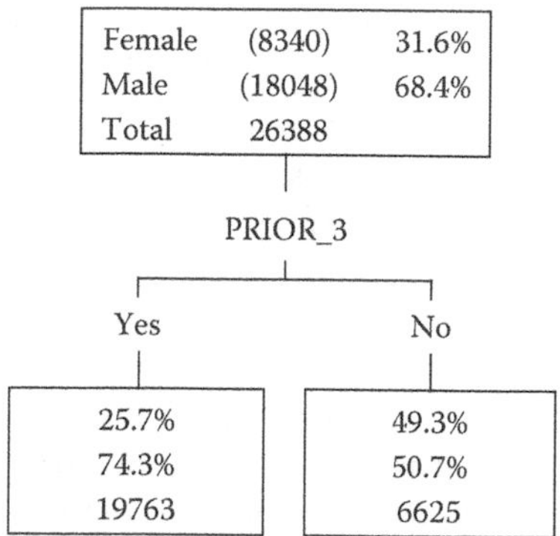

FIGURE 22.6
PRIOR_3 CHAID tree for GENDER.

I perform two additional CHAID most likely category imputations (Figures 22.6 and 22.7) for GENDER using the individual matching variables PRIOR_3 and LIFE_DOL, respectively. The PTCC values are identical, 68.4%. Thus, I select the imputation estimates for GENDER based on AGE_CUST because its PTCC value is the largest (68.7%), albeit not noticeably different from the other PTCC values.

The AGE_CUST CHAID most likely category imputation estimates for GENDER are the most likely categories of the nodes, i.e., categories with the largest percentage. The largest percentage categories of the nodes in Figure 22.5: female (52.3%), male (58.6%), male (66.6%), male (74.8%), and male (81.9%). I replace an individual's missing GENDER value with the predominant category of the imputation class to which the individual matches. There are 2,744 individuals with GENDER missing. Their AGE_CUST distribution is in Table 22.6. The individuals whose ages are in the interval [18, 24) are classified as female because the females have the largest percentage. The classification of all other individuals is male. It is not surprising that most of the classifications are males, given the large (68.4%) incidence of males in the sample.

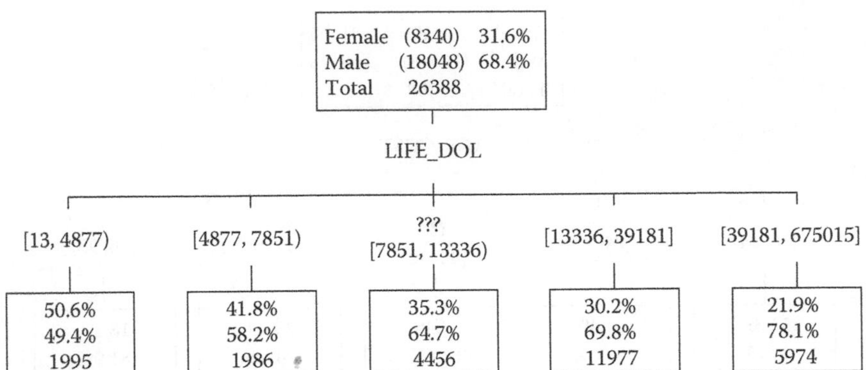

FIGURE 22.7
LIFE_DOL CHAID tree for GENDER.

TABLE 22.6
CHAID Imputation Estimates for Missing Values of GENDER

AGE_CUST Class	Class Size	Imputation Estimate
[18, 24)	182	Female
[24, 35)	709	Male
??? or [35, 42)	628	Male
[42, 56)	627	Male
[56, 93]	598	Male
Total	2,744	

Comparative note: Traditional mean imputation performs with only matching variable PRIOR_3. (Why?) CHAID most likely category imputation conveniently offers three choices of imputation estimates and a guideline to select the best.

22.6.2 Classification Tree Imputation for GENDER

I can selectively add matching variables* to the preferred single-variable CHAID tree—generating a classification tree—to increase the reliability of the imputation estimates (i.e., increase the PTCC value). Extending the AGE_CUST tree, I obtain the classification tree (Figure 22.8) for GENDER based on AGE_CUST, PRIOR_3, and LIFE_DOL. The PTCC value for this tree is 79.3%, which represents an increase of 15.4% (= 10.6%/68.7%) over the AGE_CUST-PTCC value.

The AGE_CUST-PRIOR_3-LIFE_DOL classification tree reads as follows:

1. Extending the GENDER tree with the addition of PRIOR_3 and LIFE_DOL, I obtain a classification tree with 12 end nodes.
2. Node 1 consists of 1,013 individuals whose ages are in the interval [18, 24) *and* who have *not* made a prior purchase in the past 3 months (PRIOR_3 = no). The female and male incidences are 59.0% and 41.0%, respectively.

* Here again is a great place to discuss the relationship between increasing the number of variables in a (tree) model and its effects on bias and stability of the estimates determined by the model.

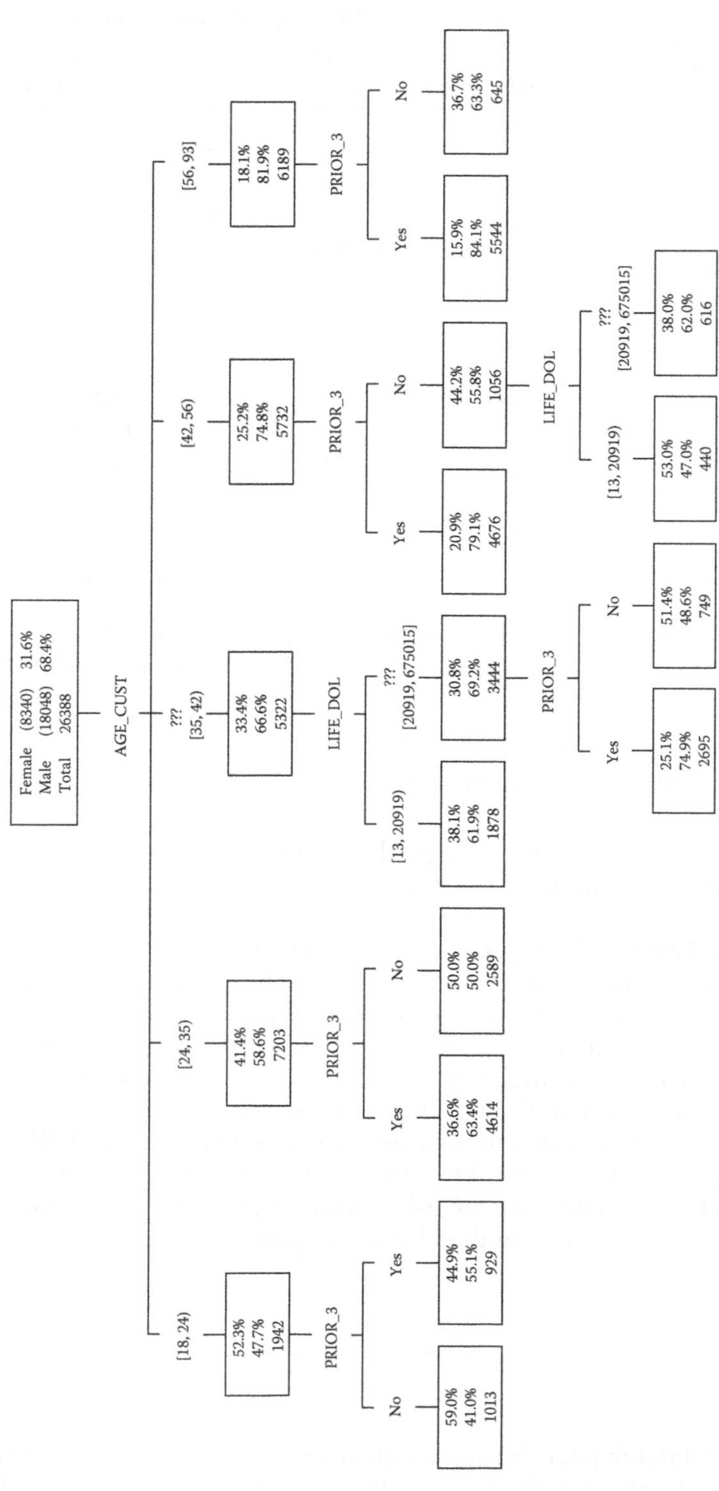

FIGURE 22.8
AGE_CUST-PRIOR_3-LIFE_DOL classification tree for GENDER.

TABLE 22.7

Classification Tree Imputation Estimates for Missing Values of GENDER

Node	AGE_CUST Class	PRIOR_3	LIFE_DOL Class	Class Size	Imputation Estimates
1	[18, 24)	No	–	103	Females
2	[18, 24)	Yes	–	79	Males
3	[24, 35)	Yes	–	403	Males
4	[24, 35)	No	–	306	Females/males
5	??? or [35, 42)	–	[13, 20919)	169	Males
6	??? or [35, 42)	Yes	??? or [20919, 675015]	163	Males
7	??? or [35, 42)	No	??? or [20919, 675015]	296	Females
8	[42, 56)	Yes	–	415	Males
9	[42, 56)	No	[13, 20919)	70	Females
10	[42, 56)	No	[20919, 675015]	142	Males
11	[56, 93]	Yes	–	449	Males
12	[56, 93]	No	–	149	Males
Total				2,744	

3. Node 2 consists of 929 individuals whose ages are in the interval [18, 24) *and* who have made a prior purchase in the past 3 months (PRIOR_3 = yes). The female and male incidences are 44.9% and 55.1%, respectively.
4. The remaining nodes have a similar interpretation.
5. Node 4 has no predominant category because the GENDER incidences are equal to 50%.
6. Nodes 1, 7, and 9 have female as the predominant category. All remaining nodes have male as the predominant category.

The AGE_CUST-PRIOR_3-LIFE_DOL classification tree imputation estimates for GENDER are the most likely categories of the nodes. I replace an individual's missing GENDER value with the predominant category of the imputation class to which the individual belongs. The distribution of missing GENDER values (Table 22.7) falls within all 12 nodes. The classification of individuals in Nodes 1, 7, and 9 is female. For individuals in Node 4, I flip a coin. The classification of all other individuals is male.

Comparative note: Traditional logistic regression-based imputation for GENDER based on three matching variables (AGE_CUST, LIFE_DOL, and one dummy variable for PRIOR_3) results in a complete case sample size of 26,219, which represents a barely noticeable 0.6% (= 169/26,388) loss of information from the CHAID analysis sample.

22.7 Summary

It is rare to find a dataset that has no missing data values. A given is that the data analyst first tries to recover or minimize the loss of information from the incomplete data. I illustrate briefly the popular missing data methods, which include complete case and available case analyses and mean value and regression-based imputation methods. All these methods have at least one version of the missing data assumptions: MAR and MCAR, which are difficult and impossible to test for validity, respectively.

I remark that the conventional wisdom of missing data solutions is that their performance is, at best, satisfactory, especially for big data applications. Experts in missing data admonish us to remember that imputation is seductive and dangerous. Therefore, the best solution to the missing data problem is not to have missing data. Thus, I strongly argue to exercise restraint for imputation of big data applications and judicious evaluation of the findings.

Then, I recover from the naysayer to advance the proposed CHAID imputation method. I present CHAID as an alternative method for mean value/regression-based imputation. The justification of CHAID for missing data is that CHAID creates optimal homogeneous groups that serve as trusty imputation classes, which ensure the reliability of the imputation estimates. The trusty imputation classes render CHAID as a reliable method of mean value/regression-based imputation. Moreover, the CHAID imputation method has salient features commensurate with the best of what EDA offers.

I illustrate the CHAID imputation method with a database catalogue case study. I show how CHAID—for both a continuous and a categorical variable to be imputed—offers imputations based on several individual matching variables and rules for selecting the preferred CHAID mean value imputation estimates and the preferred CHAID regression tree imputation estimates.

References

1. Cohen, J., and Cohen, P., *Applied Multiple Regression and Correlation Analysis for the Behavioral Sciences*, Erlbaum, Hillsdale, NJ, 1987.
2. Rubin, D.B., Inference and missing data, *Biometrika*, 63, 581–592, 1976.
3. Allison, P.D., *Missing Data*, Sage, Thousand Oaks, CA, 2002, p. 2.
4. Dempster, A.P., and Rubin, D.B., Overview, in Madow, W.G., Okin, I., and Rubin, D.B., Eds., *Incomplete Data in Sample Surveys, Vol. 2: Theory and Annotated Bibliography*, Academic Press, New York, 1983, p. 3.

23

Model Building with Big Complete and Incomplete Data

23.1 Introduction

"You can't win unless you learn how to accept me," said missing data to the statistician.

Statisticians cannot avoid missing data. All but every dataset has some missing data, causing concern about how to accept them. All big datasets have lots of missing data, causing greater concern. Traditional data-based methods (complete case analysis), predating big data, are known to be problematic with virtually all datasets. These methods now open a greater concern as to their unknown ineffectiveness on big data. The purpose of this chapter is to present a new data-based approach, in the face of known effeteness of data-based methods for model building with big complete and incomplete data. I illustrate the approach with a small dataset study for ease of presentation, which gives evidence the proposed procedure is viable for all sizes of datasets. I provide SAS© subroutines for the proposed method, which should become a utile technique for the statistical model builder. The subroutines are also available for downloading from my website: http://www.geniq.net/articles.html#section9.

23.2 Background

This chapter can be considered part two of Chapter 22. I refer the reader to review, as required, Section 22.2, where I put forth CHAID as an imputation method despite the accepted wisdom that missing data solutions at best perform satisfactorily, even when the amount of missing data is moderate, and the missing data assumption is not satisfied.

In this chapter, I put forth a data-based method for big complete and incomplete data with the use of principal component analysis (PCA) against compelling recognition of the faulty complete case analysis (CCA) when the MAR and MCAR assumptions are not satisfied. I am mindful of the rich body of good statistical practices in earshot of missing data's admonition (in next paragraph). I strongly argue exercised restraint on imputation application of big data, and urge to judiciously accept their findings. To that end, after extensive testing, the proposed method, referred to as CCA-PCA, is a well-tested heuristic that has great value for the missing problem with big data.

As in Chapter 22, I invoke the spirit of exploratory data analysis (EDA) by heeding the adage of missing data to the statistician, "You can't win unless you learn how to accept me." I advance CCA-PCA as a hybrid of *a justified CCA** and PCA's data mining mightiness that explicitly accommodates missing data without imposing additional assumptions.

The salient features of the new method are the EDA characteristics:

1. *Flexibility*: Data mining PCA is assumption-free and works especially well with big data containing large amounts of missing data.
2. *Practicality*: Model performance is established in a tabular display of the decile analysis.
3. *Innovation*: Treatment of complete and incomplete data separately uniquely accounts for all the information within the dataset.
4. *Universality*: Blending of two diverse traditions—upgrading the maligned traditional CCA and the force of data mining PCA.
5. *Simplicity*: CCA-PCA models are self-explanatory.

23.3 The CCA-PCA Method: Illustration Details

Per an engagement of a database marketer, who sought a model to identify most likely responders to a general merchandise solicitation, I proceed with obtaining a recent solicitation file from which I draw a random sample. The sample dataset GENMERCH consists of 30,000 individuals. The dependent variable is RESPONSE (yes = 1, no = 0) and candidate predictor variables are X2–X24. There are 6,636 responders, yielding a 22.1% RESPONSE rate. I generically rename the original variables' names only to prevent the usual series of comments of keen statisticians, who love to detail output. I prefer to focus the reader's attention on the new CCA-PCA method over sidebar commentary.

The first 10 records of GENMERCH are in Table 23.1. The first missing value is at ID #10 for X10. The next missing values are: ID #3 for X14; ID #6 for X15; and IDs #1, #7, #9, and #10 for X16.

23.3.1 Determining the Complete and Incomplete Datasets

I run the simple subroutine in Appendix 23.A to determine the number of missing observations for X2–X24. I run the subroutine in Appendix 23.B, Testing CCA Samsizes, to calculate the missing percentage per variable. The percentage per variable provides the CCA sample sizes for various sets of Xs. The CCA size is the *frequency* corresponding to a CCA_SAMSIZE = 0 as obtained from a frequency procedure.

I determine the best balance of variable subset size and CCA size. The selected set of variables consists of X11, X12, X13, X19, and X21, which produce the largest sample size. CCA (complete) sample size is 16,003, representing 53.34% of the sample. The *incomplete (ICA) dataset* size is 13,997, representing the 46.66% of the otherwise discarded data. See Table 23.2.

* As the MCAR and MAR assumptions are difficult-to-impossible to test, the rule of thumb for examining the assumptions is the test of reasonableness (i.e., much leeway is given these assumptions in practice). Hence, the concern of using CCA for building a logistic model should not be as problematic as the literature warns (http://art-artificial-evolution.dei.uc.pt/preface.htm).

TABLE 23.1

Dataset GENMERCH with Its Variables

ID	Response	X2	X3	X4	X5	X6	X7	X 8	X9	X10
1	1	24	3,913	3,102	689	0	0	0	2	20,000
2	1	26	2,682	1,725	2,682	3,272	3,455	3,261	2	1,20,000
3	1	30	65,802	67,369	65,701	66,782	36,137	36,894	2	70,000
4	1	24	15,376	18,010	17,428	18,338	17,905	19,104	1	20,000
5	1	39	316	316	316	0	632	316	2	1,20,000
6	1	26	41,087	42,445	45,020	44,006	46,905	46,012	2	70,000
7	1	40	5,512	19,420	1,473	560	0	0	1	4,50,000
8	1	27	−109	−425	259	−57	127	−189	1	60,000
9	1	33	30,518	29,618	22,102	22,734	23,217	23,680	2	50,000
10	1	25	0	780	0	0	0	0	1	.

ID	X11	X12	X13	X14	X15	X16	X17	X18	X19	X20	X21	X22	X23	X24
1	2	2	1	−1	−1	.	−2	0	689	0	0	0	0	2
2	−1	2	2	0	0	0	2	0	1,000	1,000	1,000	0	2,000	2
3	1	2	2	.	0	0	2	3,200	0	3,000	3,000	1,500	0	1
4	0	0	2	2	2	2	2	3,200	0	1,500	0	1,650	0	1
5	−1	1	1	−1	−1	−1	−1	316	316	0	632	316	0	2
6	2	0	2	0	.	2	2	2,007	3,582	0	3,601	0	1,820	2
7	−2	2	1	−2	−2	.	−2	19,428	1,473	560	0	0	1,128	2
8	1	2	2	−1	−1	−1	−1	.	1,000	0	500	0	1,000	1
9	2	0	2	0	0	.	0	1,718	1,500	1,000	1,000	1,000	716	1
10	1	1	2	−1	−2	.	−2	780	0	0	0	0	0	1

TABLE 23.2

CCA Sample Size (CCA = 1) based on X11–X13, X19, and X21. Incomplete Data Sample Size (CCA = 0)

CCA_SAMSIZE_ X11_X13X19X21	Frequency	Percent	Cumulative Frequency	Cumulative Percent
0	16,003	53.34	16,003	53.34
1	10,541	35.14	26,544	88.48
2	2,991	9.97	29,535	98.45
3	429	1.43	29,964	99.88
4	36	0.12	30,000	100.00

CCA	Frequency	Percent	Cumulative Frequency	Cumulative Percent
0	13,997	46.66	13,997	46.66
1	16,003	53.34	30,000	100.00

The subroutine for the construction of the CCA and ICA datasets is in Appendix 23.C.

23.4 Building the RESPONSE Model with Complete (CCA) Dataset

Based on the CCA dataset, I build a logistic regression of RESPONSE based on X11, X12, X13, X19, and X21. The essential output of the logistic regression on CCA is the maximum likelihood estimates in Table 23.3.

The CCA RESPONSE model is defined directly by the logit in Equation 23.1 and indirectly by transforming logits into probabilities in Equation 23.2:

$$\begin{aligned}\text{Logit(RESPONSE} = 1 \mid \text{CCA)} = & -0.8387 \\ & - 1.595\text{E}{-6}{*}\text{X11} \\ & - 0.1920{*}\text{X12} \\ & + 0.6108{*}\text{X13} \\ & + 0.0838{*}\text{X19} \\ & + 0.0725{*}\text{X21}\end{aligned} \tag{23.1}$$

$$\begin{aligned}&\text{Prob (RESPONSE=1} \mid \text{CCA)} = \\ &\text{PROB\_COMPLETES} = \\ &\exp(\text{Logit (RESPONSE=1} \mid \text{CCA)}/(1+ \exp(\text{Logit (RESPONSE=1} \mid \text{CCA))}\end{aligned} \tag{23.2}$$

The performance of the CCA RESPONSE model uses a bootstrapped CCA dataset to remove noise in the original CCA. The decile analysis of the CCA RESPONSE model is in Table 23.4. A formal treatment of the bootstrapping is in Chapter 29.

TABLE 23.3
CCA Maximum Likelihood Estimates

Parameter	DF	Estimate	Standard Error	Wald Chi-Square	Pr > ChiSq
Intercept	1	−0.8387	0.0716	137.0543	<0.0001
X11	1	−1.59E−6	1.818E−7	76.0588	<0.0001
X12	1	−0.1920	0.0391	24.1679	<0.0001
X13	1	0.6108	0.0242	637.8281	<0.0001
X19	1	0.0838	0.0268	9.7610	0.0018
X21	1	0.0725	0.0248	8.5473	0.0035

TABLE 23.4
Decile Analysis of CCA RESPONSE Model

Decile	Number of Individuals	Number of Responses	RESPONSE RATE (%)	CUM RESPONSE RATE (%)	CUM LIFT (%)
top	1,600	1,118	69.9	69.9	314
2	1,600	637	39.8	54.8	247
3	1,601	332	20.7	43.5	195
4	1,600	318	19.9	37.6	169
5	1,600	165	10.3	32.1	144
6	1,601	165	10.3	28.5	128
7	1,600	158	9.88	25.8	116
8	1,601	256	16.0	24.6	111
9	1,600	211	13.2	23.3	105
bottom	1,600	199	12.4	22.2	100
Total	16,003	3,559			

23.4.1 CCA RESPONSE Model Results

The decile analysis of the CCA RESPONSE model is the final determinant of model performance. I discuss in detail the decile analysis of the CCA RESPONSE.* The DECILE column, as a noncalculated vertical identifier, renders the five arithmetically derived columns, Column #2 through Column #6. The individuals are ranked from high to low based on the logit (or PROB_COMPLETES). Ten equal-sized groups or deciles are created based on the ranked file. The five columns implicatively labeled show:

1. The sample size is 16,003, and there are 3,559 individuals with RESPONSE = 1. Thus, the mean RESPONSE is 22.2%, in Column #4, bottom decile.
2. The third column, RESPONSE RATE (%), shows the means at the decile level. The top decile mean (69.9%) and the bottom decile mean (12.4%) show a top-to-bottom ratio of 5.64. This ratio value indicates the model *very significantly discriminates* among the individuals.
3. The last column, CUM LIFT (%), presents the performance of the model. The top decile, CUM LIFT 314, means the model identifies the top 10% individuals whose

* Chapter 26 details thoroughly the construction and interpretation of the decile analysis. Indeed, the reader can take a quick detour to Chapter 26 and then return to this section. Or, the reader can go through the model results presented here, and after reading Chapter 26, the reader can revisit this section.

average response is 3.14 times the average RESPONSE RATE (22.2%) or 214% greater than the average RESPONSE RATE (22.2%).

4. The CUM LIFT(%) for the top two deciles (247) indicates the model identifies the top 20% (top and second deciles) individuals whose mean RESPONSE is 2.47 times (147% greater than) the average RESPONSE (22.2%).
5. For the remaining deciles, the interpretation of CUM LIFT(%) is similar.

In sum, the CCA RESPONSE model has very significant discriminatory power and identifies the best customers for effective target marketing campaigns.

23.5 Building the RESPONSE Model with Incomplete (ICA) Dataset

I build the RESPONSE model with incomplete (ICA) dataset based on the output of a PCA of ICA data. Chapter 7 thoroughly details the topic of PCA. The reader may need to review Chapter 7 and then return to this section.

PCA on ICA data requires a binary conversion of the ICA dataset, referred to as *BICA*. Using the subroutine in Appendix 23.D I convert the original ICA variables into one-zero values by replacing the missing values with ones and the nonmissing values with zeros. A listing of 10 random records showing the replacement of original variables Xs in ICA to the corresponding binary variables XXs is in Table 23.5.

TABLE 23.5

Converting Incompletes into Binary Variables for PCA

ID	X14	X15	X18	X20	X21	X22	X23
36	120,000	−1	.	.	0	0	.
74	.	1	2	2	0	0	2
83	.	2	0	0	2	.	2
27	60,000	1	−2	−1	−1	−1	−1
18	50,000	1	−1	.	−2	−2	−2
72	150,000	0	0	−1	0	0	−2
73	10,000	2	.	2	0	0	0
70	360,000	.	−1	2	0	−1	−1
76	200,000	2	2	.	.	2	2
48	.	2	2	2	3	.	2
ID	**XX14**	**XX15**	**XX18**	**XX20**	**XX21**	**XX22**	**XX23**
36	0	0	1	1	0	0	1
74	1	0	0	0	0	0	0
83	1	0	0	0	0	1	0
27	0	0	0	0	0	0	0
18	0	0	1	1	0	0	0
72	0	1	0	0	0	0	0
73	0	0	0	1	1	0	0
70	1	0	0	0	0	1	0
76	0	0	0	1	1	0	0
48	1	0	0	0	0	1	0

TABLE 23.6
Eigenvalues of PCA on BICA Data

	Eigenvalue	Difference	Proportion	Cumulative
1	1.16773489	0.02000119	0.1668	0.1668
2	1.14773370	0.00349223	0.1640	0.3308
3	1.14424148	0.01405275	0.1635	0.4942
4	1.13018873	0.13116876	0.1615	0.6557
5	0.99901997	0.00270757	0.1427	0.7984
6	0.99631240	0.58154357	0.1423	0.9407
7	0.41476883		0.0593	1.0000

TABLE 23.7
Eigenvectors of PCA on BICA Data

	NMISS_ PC1	NMISS_ PC2	NMISS_ PC3	NMISS_ PC4	NMISS_ PC5	NMISS_ PC6	NMISS_ PC7
XX14	−0.474478	−0.454708	−0.567826	0.186920	−0.012499	0.081685	0.451572
XX15	−0.301693	0.828850	−0.088169	−0.091948	0.041448	−0.027284	0.450886
XX18	0.524448	−0.144022	−0.190945	−0.684256	−0.043368	0.021479	0.444076
XX20	0.560151	0.039704	0.059096	0.685185	0.039075	−0.120934	0.442191
XX21	−0.289736	−0.272485	0.793519	−0.093766	−0.005879	0.058078	0.447138
XX22	−0.007323	−0.059711	−0.010477	−0.061341	0.989663	−0.114094	−0.007571
XX23	0.105058	0.078200	0.008103	0.079796	0.123491	0.980360	−0.007630

23.5.1 PCA on BICA Data

PCA as a datamining technique is one of the best for analyzing binary data. In this case, the binary data are BICA. PCA reveals the structure of the patterns of the binary version of incompletes. I use the *all possible subsets* variable selection approach, which results in conducting PCA on variables XX14, XX15, XX18, XX20, XX21, XX22, and XX23.

The two pieces of PCA output are the eigenvalues and the eigenvectors in Tables 23.6 and 23.7, respectively. Briefly, a PCA rule-of-thumb is eigenvalues greater than one have potential predictive power. There are four such eigenvalues, which account for 65.57% of the information in the patterns of the incompletes. The eigenvectors define the PCs, labeled NMISS_PC1 – NMISS_PC7.

23.6 Building the RESPONSE Model on PCA-BICA Data

I build a logistic regression of RESPONSE based on only two PCA-BICA variables, NMISS_PC2 and NMISS_PC3. The EDA approach discussed in Chapter 10 was the variable selection approach taken. The essential outputs of the logistic regression on NMISS_PC2 and NMISS_PC3 are the maximum likelihood estimates in Table 23.8.

TABLE 23.8

Maximum Likelihood Estimates of PCA on BICA Data

Parameter	DF	Estimate	Standard Error	Wald Chi-Square	Pr > ChiSq
Intercept	1	−1.2774	0.0206	3849.6342	<0.0001
NMISS_PC2	1	−0.1555	0.0213	53.0215	<0.0001
NMISS_PC3	1	−0.0722	0.0204	12.5473	0.0004

The PCA-BICA RESPONSE model is defined directly by logits in Equation 23.3 and indirectly by transforming logits into probabilities in Equation 23.4:

$$\text{Logit(RESPONSE=1 | PCA-BICA)} = -1.2774 - 0.1555\text{*NMISS\_PC2} - 0.0722\text{*NMISS\_PC3} \quad (23.3)$$

$$\text{Prob(RESPONSE=1 | PCA\_BICA)} = \text{PROB\_COMPLETES} =$$

$$\exp(\text{Logit(RESPONSE=1 | PCA-BICA)})/(1+\exp(\text{Logit(RESPONSE=1 | PCA-BICA)})) \quad (23.4)$$

Similar to the CCA RESPONSE model, the performance of the PCA-BICA RESPONSE model uses a bootstrapped PCA-BICA dataset to remove noise in the original PCA-BICA dataset. The decile analysis of the PCA-BICA RESPONSE model is in Table 23.9.

23.6.1 PCA-BICA RESPONSE Model Results

Similar to the discussion of the decile analysis of the CCA RESPONSE model, I detail the same points for the PCA-BICA RESPONSE model.

1. The sample size is 13,997, and there are 3,077 individuals with RESPONSE = 1. Thus, the mean RESPONSE is 22.0%, in Column #5, bottom decile.
2. In the fourth column, RESPONSE RATE (%) is the mean at the decile level. The top decile mean, 27.7%, and the bottom decile mean, 19.9%, show a top-to-bottom ratio of 1.39. This ratio value indicates the model *slightly significantly discriminates* among the individuals.
3. The last column, CUM LIFT (%), presents the performance of the model. The top decile, CUM LIFT 126, means the model identifies the top 10% individuals whose average response is 1.26 times the average RESPONSE 22.0%.
4. The CUM LIFT (%) for the top two deciles, 124, says the model identifies the top 20% (top and second deciles) of individuals whose mean RESPONSE is 1.24 times greater (24% greater than) than the average RESPONSE of 22.0%.
5. For the remaining deciles, the interpretation of CUM LIFT (%) is similar.

TABLE 23.9
Decile Analysis of PCA-BICA RESPONSE Model

Decile	Number of Individuals	Number of Individuals	RESPONSE RATE (%)	CUM RESPONSE RATE (%)	CUM LIFT (%)
top	1,399	388	27.7	27.7	126
2	1,400	373	26.6	27.2	124
3	1,400	285	20.4	24.9	113
4	1,400	304	21.7	24.1	110
5	1,399	404	28.9	25.1	114
6	1,400	258	18.4	24.0	109
7	1,400	274	19.6	23.3	106
8	1,400	306	21.9	23.1	105
9	1,400	206	14.7	22.2	101
bottom	1,399	279	19.9	22.0	100
Total	13,997	3,077			

On the surface, the PCA-BICA RESPONSE Model has slight significant discriminatory power and identifies the best customers (among all clients in BICA) for marginally effective target marketing campaigns. However, the marginal effectiveness of this model can be substantially upgraded if the database marketer develops a campaign to stimulate the best customers into becoming better than they are. In other words, the PCA-BICA RESPONSE Model, with the appropriate promotion, discounts, gift cards, and so on, can harvest these somewhat poor performers to be all that they can be to the marketer's benefit.

23.6.2 Combined CCA and PCA-BICA RESPONSE Model Results

I combine the CCA and PCA-BICA RESPONSE Models to obtain a measure of performance of modeling RESPONSE with all information in the original GENMERCH dataset. The decile analysis of the Combined CCA and PCA-BICA RESPONSE Model is in Table 23.10.

Similar to the discussions of the CCA and PCA-BICA Decile Analyses, I detail the same points for the Combined CCA and PCA-BICA RESPONSE Model in Table 23.10.

1. The sample size is 30,000, and there are 6,636 individuals with RESPONSE = 1. Thus, the mean RESPONSE is 22.1%, in Column #5, bottom decile.
2. In the fourth column, RESPONSE RATE (%), is the mean at the decile level. The top decile mean, 57.0%, and the bottom decile mean, 12.7%, show a top-to-bottom ratio of 4.49. This ratio value indicates the model *very significantly discriminates* among the individuals.
3. The last column, CUM LIFT (%), presents the performance of the model. The top decile, CUM LIFT 258, means the model identifies the top 10% individuals whose average RESPONSE is 2.58 times (158% greater than) the average RESPONSE of 22.1%.
4. The CUM LIFT (%) for the top two deciles, 190, says the model identifies the top 20% (top and second deciles) individuals whose mean RESPONSE is 1.90 times (90% greater than) the average RESPONSE of 22.1%.
5. For the remaining deciles, the interpretation of CUM LIFT (%) is similar.

TABLE 23.10

Decile Analysis of Combined CCA and PCA-BICA RESPONSE Model

Decile	Number of Individuals	Number of Responses	RESPONSE RATE (%)	CUM RESPONSE RATE (%)	CUM LIFT (%)
top	3,000	1,710	57.0	57.0	258
2	3,000	811	27.0	42.0	190
3	3,000	591	19.7	34.6	156
4	3,000	631	21.0	31.2	141
5	3,000	653	21.8	29.3	132
6	3,000	618	20.6	27.9	126
7	3,000	372	12.4	25.6	116
8	3,000	424	14.1	24.2	109
9	3,000	446	14.9	23.2	105
bottom	3,000	380	12.7	22.1	100
Total	30,000	6,636			

In sum, the Combined CCA and PCA-BICA RESPONSE Model has very significant discriminatory power and identifies the best customers for effective target marketing campaigns. Of course, the combined model has (slightly) lower performance than the CCA model, but the combined model provides a full-strength way to assess the predictive power of the full GENMERCH dataset, complete and incomplete together.

23.7 Summary

I provide a method for model building with big complete and incomplete data by responding to missing data's declaration, "You can't win unless you learn how to accept me." I exercise the EDA tenets. Tenet #3, Innovation: I split the original dataset complete and incomplete. I build a logistic regression on the complete, yielding very good performance. Tenet #1, Flexibility: I employ PCA on the incomplete, after converting the values to ones and zeros. I build a logistic regression on the incomplete, yielding okay performance. Tenet #4, Universality: I blend two diverse practices of long standing by upgrading the maligned traditional CCA and the force of data-mighty PCA. Lastly, Tenet #2, Practicality: I display the complete and incomplete decile analyses, separately and combined.

The complete model is very good and directly target the most responsive individuals. The incomplete model is marginal but its value can be substantial if the marketer develops a campaign to stimulate the customers into becoming better than they are. The incomplete model, with appropriate incentives, can harvest these somewhat poor performers to be all that they can be to the marketer's benefit. In sum, the proposed method provides a full-strength way to assess the predictive power of a full dataset, complete and incomplete together. I provide SAS subroutines for the proposed method, which should become a utile technique for the statistical model builder.

Appendix 23.A NMISS

```
libname ca 'c:\0-CCA_PCA';

PROC MEANS data=ca.CCA_PCA nmiss;
var X2-X24;
run;
```

Appendix 23.B Testing CCA Samsizes

```
data ca.CCA_PCA;
set ca.CCA_PCA;
/* trial and error for different subsets of variables with small NMISS */
CCA_SAMSIZE_ X11_X13X19X21=NMISS (of X11, X12, X13, X19, X21);
CCA=.;
if CCA_SAMSIZE_ X11_X13X19X21 eq 0 then CCA=1;
if CCA_SAMSIZE_ X11_X13X19X21 ne 0 then CCA=0;
run;

PROC FREQ ca.CCA_PCA;
table CCA_SAMSIZE_ X11_X13X19X21 CCA;
run;
```

Appendix 23.C CCA-CIA Datasets

```
data ca.COMPLETES ca.INCOMPLETES;
set ca.CCA_PCA;
if CCA=1 then output ca.COMPLETES;
if CCA=0 then output ca.INCOMPLETES;
run;
```

Appendix 23.D Ones and Zeros

```
data ca.INCOMPLETES;
set ca.INCOMPLETES;

array X(7) X14 X15 X18 X20 X21 X22 X23;
array XX(7) XX14 XX15 XX18 XX20 XX21 XX22 XX23;
```

```
do i = 1 to 7;
if X(i)=. then XX(i)=1; else XX(i)=0;
drop i;
end;
run;
```

Reference

1. Schafer, J. L., and Graham, J. W., Missing data: Our view of the state of the art, *Psychological Methods*, 7 (2) 147–177, 2002.

24

Art, Science, Numbers, and Poetry

24.1 Introduction

From the ancient Egyptian civilization circa 3200 BC, there is limestone concrete proof that art and science coexist. The Egyptian pyramids are a feat of scientific engineering and creation of real art. Quickly moving up the time line, Leonardo da Vinci (1452–1519) said, "Art is the Queen of all sciences communicating knowledge to all the generations of the world" (http://art-artificial-evolution.dei.uc.pt/preface.htm).

Advancing further, in the late nineteenth century, the period of Impressionism and Postimpressionism was born with the focus on the effects of color and light on canvas (or wood panels). Vincent van Gogh (1853–1890) practiced his painting in the open air with pure, high-keyed color to capture a fleeting impression, "Conveying a sense of trembling as the light and color of the landscape shift and time passes" (http://www.artic.edu/aic/education/sciarttech/lecturers.html).

From limestone to color and light, in 1905 light takes on a new meaning. Einstein's *Special Theory of Relativity* showed the world that light travels at a constant, finite speed of 186,000 miles per second.* Einstein, a Jew, as a pre-teenager practiced his religion with an orthodox adherence. In his twenties, Einstein changed his mind, heart, or soul and declared he was an atheist. Ironically, after discovering one of the keys to our universe, he was asked by the press whether he believed in G-d. Einstein's famous reply, although misunderstood by the media, was "G-d does not play with dice" [1]. What Einstein meant was "If G-d exists, He has not much to do, but watch the world" [1], even during the first 7 days.

(Stephen Hawking, who did not discover black holes or the big bang, and is arguably the predecessor to Einstein, is famous for the theoretical prediction that black holes emit radiation. Like Einstein, when asked by the media whether Hawking believes in G-d, Hawking, always a devout atheist, said no.)

When Einstein aged into his iconic image of a wiry, white-haired physicist and elder world statesman, he was quoted saying, "The more I study science, the more I believe in G-d" [1]. Moreover, Einstein is also quoted saying, "All religions, arts, and sciences are branches of the same tree" [1].

I present the aforementioned narratives to provide the framework for two objects of word art and one object of noesis. First, the word *art* is an energy of output experienced by the creation of the dataset of zeros and ones, illustrated in the bottom panel of Table 23.5 of Chapter 23, Section 23.2, re-displayed here in Table 24.1. Second, its origin, not recalled,

* An interesting and incredible consequence of the speed of light: If you throw a 12-inch ruler fast—very, very, very, very fast—the ruler starts getting shorter. If you windup and throw the ruler at the speed of light, the 12-inch ruler will disappear.

TABLE 24.1

Zeros and Ones

ID	XX14	XX15	XX18	XX20	XX21	XX22	XX23
36	0	0	1	1	0	0	1
74	1	0	0	0	0	0	0
83	1	0	0	0	0	1	0
27	0	0	0	0	0	0	0
18	0	0	0	1	0	0	0
72	0	0	0	0	0	0	0
73	0	0	1	0	0	0	0
70	0	1	0	0	0	0	0
76	0	0	0	1	1	0	0
48	1	0	0	0	0	1	0

other than being like silky leaves blowing in the wind, the word art can proudly drape on the three branches of Einstein's tree. Third, the art of the *statistical golden rule* is inspired by da Vinci's Queen.

24.2 Zeros and Ones

Zeros and Ones

In a world of zeros and ones
I ride daily waves of both.
One day swings up right
Another slides down left.
In my life I have loved them all.
For all the days of my life
My ticket is neither right nor left.
After all I am a lucky guy.

Bruce Ratner

24.3 Power of Thought

Power of Thought

In the beginning
There was human thought.

Thought begets form and fitness
Spreads light over the surface of

Yesterday today tomorrow.
Thought pours out knowledge
Knowledge waters our world
Our world's abloom.

Consider numbers:
Count numbers ticked ages ago.
William Jones in 1706 entangled
An absorbing faceless number
He labelled pi (=3.14159 ...).
Humans with accidental smarts
Were using pi forever, least 2700 BC.
The noisy string (...) of counts with
No face value has material earth
No pi no pyramids, to cite but one.

Rational beings distant of 3.14
Considered pi an irrational number!
Irrational numbers and their
Existent kin the rational numbers
Frame the set of real numbers.
Included are 0 and 1.

Real numbers have ghostly kins
Imaginary numbers with a ghastly i.
The unit of the imaginaries
Was coined in early 1700s.
(i unit is defined as i*i = –1)

Another faceless number
2.71828 ... was discovered by
Jacob Bernoulli in 1690s.
This stringy face was called e
In honor of Leonhard Euler, 1740s.
Euler painted an abstract mosaic
With five dabs of constants
e, i, pi, 1, and 0
(e^(i*pi) + 1 = 0).

The power of human thought
Unfolding any canvas of constants
Is itself almost beyond thought.

The interloper in the beginning
Has imbued us with the power of
Thought in our world universe or

Extra-universe, if you dare
With reason, with order, with time
(the 4th dimension of our 3D world).

Consider another something.
Thought was is will be.

BRUCE RATNER

24.4 The Statistical Golden Rule: Measuring the Art and Science of Statistical Practice

I propose the *statistical golden rule*—an application of the well-known golden ratio (circa 490–430 BC) in the context of statistical practice. The proposed rule promises to service statisticians as the golden ratio has guided artists and architects of the past and in the present. "Many artists and architects have proportioned their works to approximate the golden ratio ... believing this proportion to be aesthetically pleasing" (http://en.wikipedia.org/wiki/Golden_ratio). The end product of a statistics project, say, an estimated logistic regression model—performed by a statistician who uses a good blend of art and science—is statistically complete in form (compact equation) and fitness (accurate equation).

24.4.1 Background

Two quantities, **a** and **b**, are in the golden ratio if the ratio **a/b** is equal to the ratio **(a+b)/a**, where **a** > **b**. When the quantities are *golden*, these ratios have the unique value of 1.6180339887..., denoted by the Greek letter phi (φ). See Figure 24.1 (http://en.wikipedia.org/wiki/Golden_ratio).

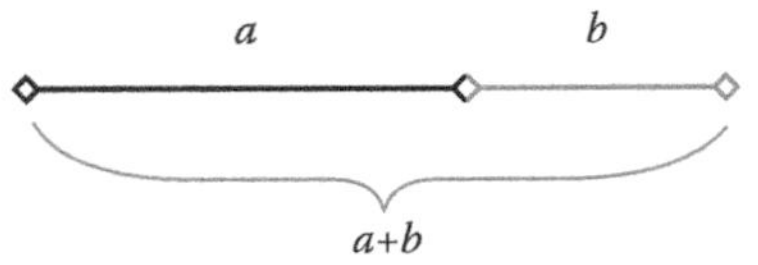

$a+b$ is to a as a is to b

Line segments in the golden ratio

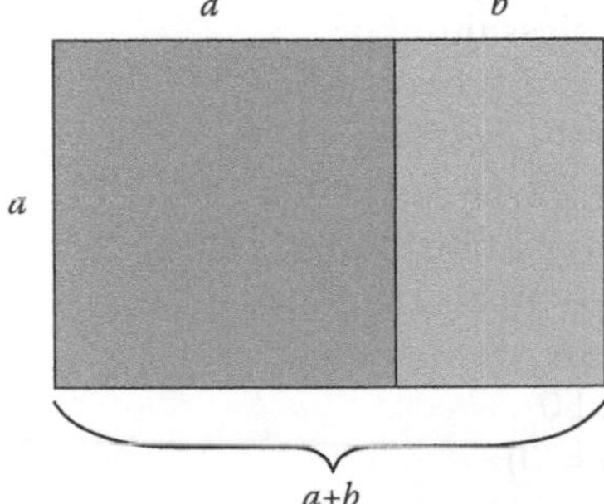

A golden rectangle with longer side a and shorter side b, when placed adjacent to a square with sides of length a, will produce a similar golden rectangle with longer side $a+b$ and shorter side a. This illustrates the relationship $\frac{a+b}{a} = \frac{a}{b} \equiv \varphi$.

FIGURE 24.1
Golden rule.

24.4.1.1 The Statistical Golden Rule

The statistical golden rule (SGR) is the average of the two golden ratios, in which the quantities **a** and **b** are, say, science units (e.g., measured in talent, time, mental strength, etc.) and art units (corresponding to the science units) employed in a statistical undertaking. The assignment of the units can be self-reported by the engaged statistician, or an independent overseer can make the allocation. Note, the quantities **a** and **b** do not have to correspond to science and art, respectively; they can just as easily be reversed with art and science, respectively. The interchangeability property of SGR renders it a symmetric relationship. A detailing of the symmetric property of SGR is in view as follows.

I define SGR1 = **(a+b)/a** and SGR2 = **a/b**. SGR is the average of SRG1 and SGR2, specifically, the harmonic mean of the two ratios: SGR = (2*SGR1*SGR2)/(SGR1+SGR2).*

To determine the closeness of SGR to 1.6180, I compare the log(SGR) to log(1.6180).† I define the ratio *quality of SGR* (QSGR) as:

QSGR = max (log(SGR), log(1.6180))/min (log(SGR), log(1.6180))

I establish a medal award criterion for an art-science blend:

- If QSGR is 1.00, then the statistician's art-science blend is *gold*.
- If QSGR lies within (1.00, 1.15], then the statistician's art-science blend is *white gold*.
- If QSGR lies within (1.15, 1.30], then the statistician's art-science blend is *silver*.
- If QSGR lies within (1.30, 1.40], then the statistician's art-science blend is *bronze*.
- If QSGR is greater than 1.40, then the statistician's art-science blend is *lead*.

Illustration #1: To see how SGR works, consider one of my recent building of a logistic regression model, in which my statistics art skill is running at 90 mpg and my statistics science adroitness at 70 mpg. The SGR statistics are:

- SGR1 = (90+70)/90 = 1.77778
- SGR2 = 90/70 = 1.2857
- SGR = 1.49223, which is somewhat off from the golden ratio 1.6180

For this illustration, log(SGR) = 0.40027 and log(1.6180) = 0.48243. The resultant QSGR = 1.20525 (=0.48243/0.40027). Thus, my art-science statistics blend in the building of the logistic regression model receives a silver medal. The estimated model has a fulfilling equation in form and fitness.

Illustration #2: The units of the quantities **a** and **b** can be of any scale. Consider another of my modeling projects, in which I use percentages, such that for art and science they total 100%. I use 20% of art skill and 80% of science craft. The SGR statistics are:

- SGR1 = (80+20)/80 = 1.2500
- SGR2 = 80/20 = 4.0000
- SGR = 1.90476

* The harmonic mean is the appropriate average for ratios. The weakness of the mean is apparent when the ratios are disparate. See Illustration #3.

† It is best to use logs when comparing ratios.

- Log(SGR) = 0.64436 and log(1.6180) = 0.48243
- QSGR = 1.33566

For this statistics project, I use a bronze blend of art-science statistics. Thus, the estimated model is of sound form and good fitness.

Illustration #3: This is an extreme illustration. Let **a** and **b** be 99 and 1, respectively. The SGR statistics are:

- SGR1 = (99+1)/99 = 1.01010
- SGR2 = 99/1 = 99.00000
- SGR = 1.99980
- Log(SGR) = 0.69305 and log(1.6180) = 0.48243
- QSGR = 1.43659

The statistical task related to the quantities 99 and 1 receives, not surprisingly, a lead medal. The implication of this scenario is in the summary that follows.

24.5 Summary

The statistical golden rule encourages the statistician to be mindful of blending art and science in statistical practice to ensure a winning analysis and modeling effort. The concept of balancing art and science should not imply a liberal view of art being more important than science. In practical terms, the demonstration of balancing art and science is a yin and yang relationship, easily understood. (Mathematically, SGR is a symmetric relationship represented by the notation *art*SGR*science* = *science*SGR*art*.)

Revisiting the first illustration in which my art skill is at 90 mpg and my science adroitness is 70 mpg, I reverse the quantities such that 70 and 90 mpg are for science and art, respectively. QSGR is still 1.20525. The implication is the "proportional parts" of art and science units are the key to satisfying the statistical golden rule. So, if I am running slowly on science, I must speed up on art to maintain the golden ratio expressions.

Illustration #3 is a demonstration of the proportional parts issue raised in the first example. The implication is that too much of a good thing (i.e., 99 for art) is not necessarily a good thing. There must be a gilt-edged mixture of art and science in statistical practice.

The statistical golden rule brings an indispensable self-monitoring check for statisticians and provides an invaluable aid to supervisors of statisticians and workers of data.

Reference

1. *Quotable Einstein: An A to Z Glossary of Quotations*, Ayres, A., Ed., Quotable Wisdom Books, 2015.

25

*Identifying Your Best Customers: Descriptive, Predictive, and Look-Alike Profiling**

25.1 Introduction

Marketers typically attempt to improve the effectiveness of their campaigns by targeting their best customers. Unfortunately, many marketers are unaware that typical target methods develop a descriptive profile of their target customer—an approach that often results in less-than-successful campaigns. The purpose of this chapter is to illustrate the inadequacy of the *descriptive* approach and to demonstrate the benefits of the correct *predictive profiling* approach. I explain the predictive profiling approach and then expand the approach to look-alike profiling.

25.2 Some Definitions

It is helpful to have a general definition of each of the three concepts discussed in this chapter. Descriptive profiles report the characteristics of a group of individuals. These profiles *do not allow* for drawing inferences about the group. The value of a descriptive profile lies in its definition of the target group's salient characteristics, which are used to develop an effective marketing strategy.

Predictive profiles report the characteristics of a group of individuals. These profiles *do allow* for drawing inferences about a specific behavior, such as response. The value of a predictive profile lies in its predictions of the behavior of individuals in a target group. Generating a list of likely responders to a marketing campaign is not possible without the predictive profile predictions.

A look-alike profile is a predictive profile based on a group of individuals who look like the individuals in a target group. When resources do not allow for gathering information on a target group, a predictive profile built on a surrogate or *look-alike* group provides a viable approach for predicting the behavior of the individuals in the target group.

* This chapter is based on an article with the same title in *Journal of Targeting, Measurement and Analysis for Marketing*, 10, 1, 2001. Used with permission.

25.3 Illustration of a Flawed Targeting Effort

Consider a hypothetical test mailing to a sample of 1,000 individuals conducted by Cell-Talk, a cellular phone carrier promoting a new bundle of phone features. Three hundred individuals responded, yielding a 30% response rate. (The offer also included the purchase of a cellular phone for individuals who do not have one but now want one because of the attractive offer.) Cell-Talk analyzes the responders and profiles them in Tables 25.1 and 25.2 by using variables GENDER and OWN_CELL (current cellular phone ownership), respectively. Ninety percent of the 300 responders are males, and 55% already own a cellular phone. Cell-Talk concludes the typical responder is a male and owns a cellular phone.

Cell-Talk plans to target the next "features" campaign to males and owners of cellular phones. The effort is sure to fail. The reason for the poor prediction is the profile of their best customers (responders) is descriptive not predictive. That is, the descriptive responder profile describes responders without regard to responsiveness. Therefore, the profile does not imply the best customers are responsive.*

Using a descriptive profile for predictive targeting draws a false implication of the descriptive profile. In our example, the descriptive profile of "90% of the responders are males" does not imply 90% of males are responders or even that males are more likely to respond.† Also, "55% of the responders who own cellular phones" does not imply 55% of cellular phone owners are responders or even that cellular phone owners are more likely to respond.

The value of a descriptive profile lies in its definition of the best customers' salient characteristics, which are used to develop an effective marketing strategy. In the illustration,

TABLE 25.1

Responder and Nonresponder Profile Response Rates by GENDER

	Responders		Nonresponders		
GENDER	**Count**	**%**	**Count**	**%**	**Response Rate %**
Female	30	10	70	10	30
Male	270	90	630	90	30
Total	300	100	700	100	

TABLE 25.2

Responder and Nonresponder Profile Response Rates by OWN_CELL

	Responders		Nonresponders		
OWN_CELL	**Count**	**%**	**Count**	**%**	**Response Rate %**
Yes	165	55	385	55	30
No	135	45	315	45	30
Total	300	100	700	100	

* A descriptive responder profile may also describe a typical nonresponder. In fact, this is the situation in Tables 25.1 and 25.2.

† More likely to respond than a random selection of individuals.

knowing the target customer is a male and owns a cellular phone, I would instruct Cell-Talk to position the campaign offer with a man wearing a cellular phone on his belt instead of a woman reaching for a cellular phone in her purse. Accordingly, a descriptive profile tells how to talk to the target audience. As discussed in the next section, a predictive profile helps find the target audience.

25.4 Well-Defined Targeting Effort

A predictive profile describes responders *concerning* responsiveness—that is, regarding variables that discriminate between responders and nonresponders. Effectively, the discriminating or predictive variables produce *varied* response rates and imply an expectation of responsiveness. To clarify this, consider the response rates for GENDER in Table 25.1. The response rates for both males and females are 30%. Accordingly, GENDER does not discriminate between responders and nonresponders (regarding responsiveness). Similar results for OWN_CELL are in Table 25.2.

Hence, GENDER and OWN_CELL have no value as predictive profiles. Targeting of males and current cellular phone owners by Cell-Talk is expected to generate the average or sample response rate of 30%. In other words, this profile in a targeting effort will not produce more responders than will a random sample.

I now introduce a new variable, CHILDREN with hopefully predictive value. CHILDREN equals "yes" if an individual belongs to a household with children and equals "no" if an individual does not belong to a household with children. Instead of discussing CHILDREN using a tabular display (such as in Tables 25.1 and 25.2), I prefer the user-friendly visual display of chi-squared automatic interaction detection (CHAID) trees.

The CHAID tree provides an excellent display of response rates. I review the GENDER and OWN_CELL variables in the tree displays in Figures 25.1 and 25.2, respectively. From this point, I refer only to the tree in this discussion, underscoring the utility of a tree as a profiler and reducing the details of tree building to nontechnical summaries.

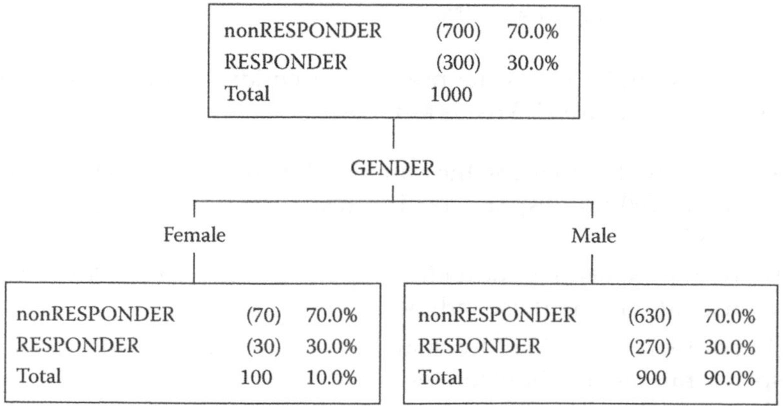

FIGURE 25.1
GENDER tree.

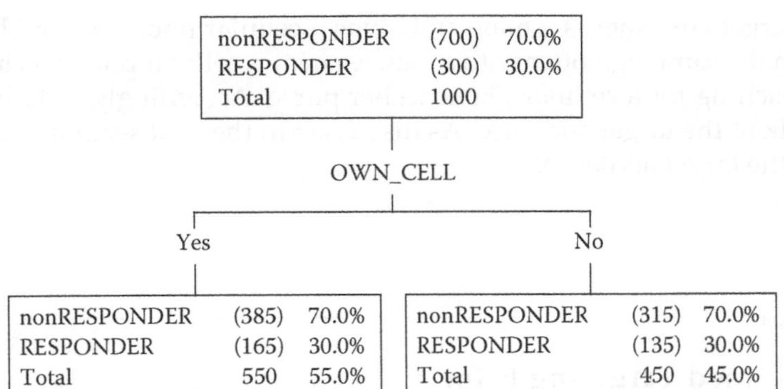

FIGURE 25.2
OWN_CELL tree.

The GENDER tree in Figure 25.1 reads as follows:

1. The top box indicates that for the sample of 1,000 individuals, there are 300 responders and 700 nonresponders. The response rate is 30%, and the nonresponse rate is 70%.
2. The left box represents 100 *females,* consisting of 30 responders and 70 nonresponders. The response rate among the 100 females is 30%.
3. The right box represents 900 *males,* consisting of 270 responders and 630 nonresponders. The response rate among the 900 males is 30%.

The OWN_CELL tree in Figure 25.2 reads as follows:

1. The top box indicates that for the sample of 1,000 individuals, there are 300 responders and 700 nonresponders. The response rate is 30%, and the nonresponse rate is 70%.
2. The left box represents 550 individuals who *own* a cell phone. The response rate among these individuals is 30%.
3. The right box represents 450 individuals who *do not own* a cell phone. The response rate among these individuals is 30%.

The new variable CHILDREN is the presence of children in the household (yes/no). The CHILDREN tree in Figure 25.3 reads as follows:

1. The top box indicates that for the sample of 1,000 individuals, there are 300 responders and 700 nonresponders. The response rate is 30%, and the nonresponse rate is 70%.
2. The left box represents 545 individuals belonging to households *with children.* The response rate among these individuals is 45.9%.
3. The right box represents 455 individuals belonging to households *with no children.* The response rate among these individuals is 11.0%.

CHILDREN has value as a predictive profile because it produces varied response rates of 45.9% and 11.0% for CHILDREN equal to yes and no, respectively. If Cell-Talk targets

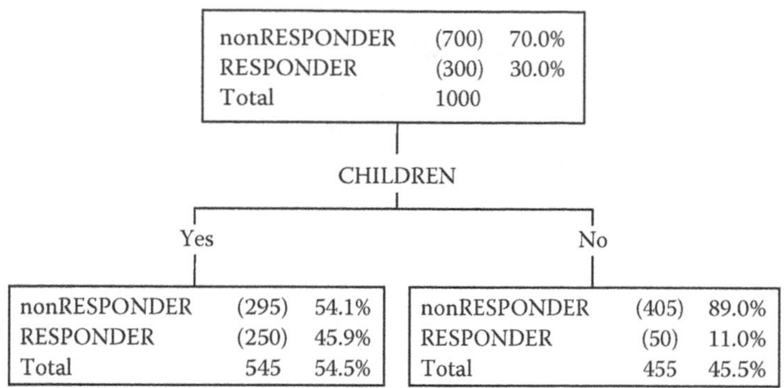

FIGURE 25.3
CHILDREN tree.

individuals belonging to households with children, the expected response rate is 45.9%, which represents a *profile lift* of 153. (*Profile lift* is defined as profile response rate of 45.9% divided by sample response rate of 30% multiplied by 100.) Thus, a targeting effort to the predictive profile is expected to produce 1.53 times more responses than expected from a random solicitation.

25.5 Predictive Profiles

Using additional variables, I can grow a single-variable tree into a full tree with many interesting and complex predictive profiles. Although the actual building of a full tree is beyond the scope of this chapter, it suffices to say that *a tree is grown to create end-node profiles (segments) with the greatest variation in response rates across all segments.* A tree has value as a *set of predictive profiles* to the extent (1) the number of segments with response rates greater than the sample response rate is large and (2) the corresponding profile (segment) lifts are large.*

Consider the full tree defined by GENDER, OWN_CELL, and CHILDREN in Figure 25.4. The tree reads as follows:

1. The top box indicates that for the sample of 1,000 individuals, there are 300 responders and 700 nonresponders. The response rate is 30%, and the nonresponse rate is 70%.
2. I reference the end-node segments from left to right, 1 through 7.
3. Segment 1 represents 30 *females* who *own* a cellular phone and belong to households *with children*. The response rate among these individuals is 50.0%.
4. Segment 2 represents 15 *females* who *do not own* a cellular phone and belong to households *with children*. The response rate among these individuals is 100.0%.
5. Segment 3 represents 300 *males* who *own* a cellular phone and belong to households *with children*. The response rate among these individuals is 40.0%.

* The term *large* is subjective. Accordingly, the tree-building process is subjective, which is an inherent weakness in CHAID trees.

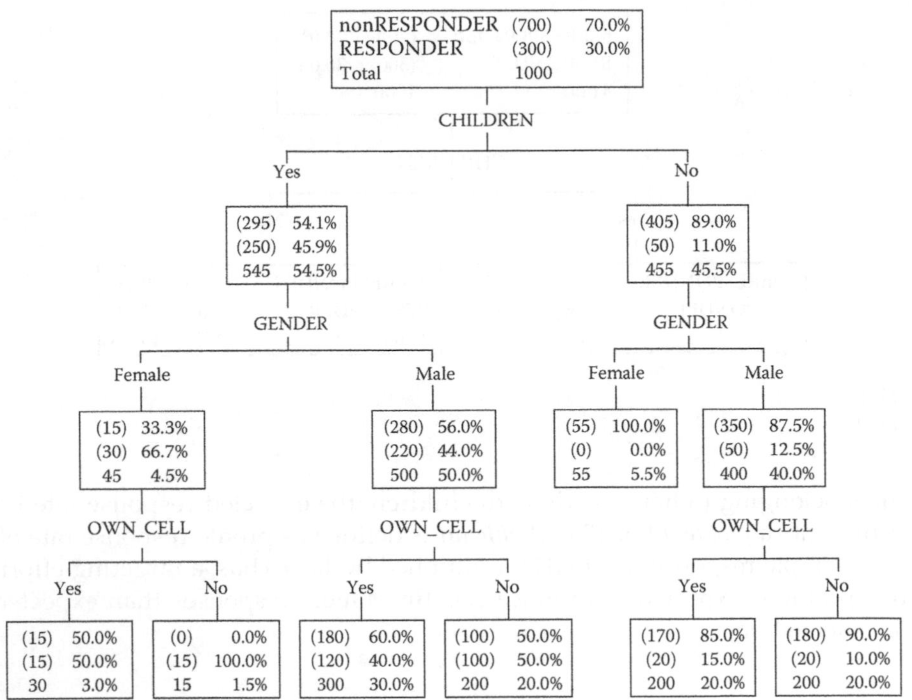

FIGURE 25.4
Full tree defined by GENDER, OWN_CELL, and CHILDREN.

6. Segment 4 represents 200 *males* who *do not own* a cellular phone and belong to households *with children*. The response rate among these individuals is 50.0%.
7. Segment 5 represents 55 *females* who belong to households *with no children*. The response rate among these individuals is 0.0%.
8. Segment 6 represents 200 *males* who *own* a cellular phone and belong to households *with no children*. The response rate among these individuals is 15.0%.
9. Segment 7 represents 200 *males* who *do not own* a cellular phone and belong to households *with no children*. The response rate among these individuals is 10.0%.

I provide a summary of the segments' response rates in the *gains chart* in Table 25.3. The construction and interpretation of the gains chart are as follows:

1. Segments are ranked descendingly by segment response rate.
2. In addition to the descriptive statistics (size of the segment, the number of responses, segment response rate), various calculated statistics are in the chart. They include the self-explanatory cumulative responses, segment response rate, and cumulative response rate.
3. The statistic posted in the rightmost column is the cumulative lift. *Cumulative lift* is the cumulative response rate divided by the sample response rate multiplied by 100. It measures the incremental gain in responses by targeting various levels of aggregated segments over chance. Cumulative lift is discussed in detail further in this section.

TABLE 25.3

Gains Chart for Tree Defined by GENDER, OWN CELL, and CHILDREN

Segment[a]		Size of Segment	Number of Responses	Cumulative Responses	Segment Response Rate (%)	Cumulative Rate (%)	Cumulative Lift
2	OWN_CELL, no GENDER, female CHILDREN, yes	15	15	15	100.0	100.0	333
1	OWN_CELL, yes GENDER, female CHILDREN, yes	30	15	30	50.0	66.7	222
3	OWN_CELL, no GENDER, male CHILDREN, yes	200	100	130	50.0	53.1	177
4	OWN_CELL, yes GENDER, male CHILDREN, yes	300	120	250	40.0	45.9	153
6	OWN_CELL, yes GENDER, male CHILDREN, no	200	30	280	15.0	37.6	125
7	OWN_CELL, no GENDER, male CHILDREN, no	200	20	300	10.0	31.7	106
5	GENDER, female CHILDREN, no	55	0	300	0.0	30.0	100
		1,000	300		30.0		

[a] Segments are ranked by response rates.

4. CHAID trees unvaryingly identify *sweet spots*—segments with above-average response rates* that account for small percentages of the sample. Under the working assumption that the sample is random and accurately reflects the population under study, sweet spots account for small percentages of the population. There are two sweet spots: Segment 2 has a response rate of 100% and accounts for only 1.5% (= 15/1000) of the sample/population; Segment 1 has a response rate of 50% and accounts for only 3.0% (= 30/1000) of the sample/population.
5. A targeting strategy to a single sweet spot is limited because it is effective only for solicitations of widely used products to large populations. Consider Sweet Spot 2 with a population of 1.5 million. Targeting this segment produces a campaign of size of 22,500 with an expected yield of 22,500 responses. Large campaigns of widely used products have low break-even points, which make sweet-spot targeting profitable.
6. Reconsider Sweet Spot 2 in a moderate size population of 100,000. Targeting this segment produces a campaign size of 1,500 with an expected yield of 1,500 responses. However, small campaigns for mass products have high break-even

* Extreme small segment response rates (close to both 0% and 100%) reflect another inherent weakness of CHAID trees.

points, which render such campaigns neither practical nor profitable. In contrast, upscale product small campaigns have low break-even points, which make sweet-spot targeting (e.g., to potential Rolls-Royce owners) both practical and profitable.

7. For moderate-sized populations, the targeting strategy is to solicit an aggregate of several top consecutive responding segments to yield a campaign of a cost-efficient size to ensure a profit. Here, I recommend a solicitation consisting of the top three segments, which would account for 24.5% (= (15 + 30 + 200)/1000) of the population with an expected yield of 53.1%. (See the cumulative response rate for the Segment 3 row in Table 25.3.) Consider a population of 100,000. Targeting the aggregate of Segments 2, 1, and 3 yields a campaign size of 24,500 with an expected 11,246 (= 53.1%*24,500) responses. A company benefits from the aggregate targeting approach when it has a product offering with a good profit margin.
8. On the cumulative lift, Cell-Talk can expect the following:
    a. Cumulative lift of 333 by targeting the top segment, which accounts for only 1.5% of the population. The top segment is sweet-spot targeting as previously discussed.
    b. Cumulative lift of 222 by targeting the top two segments, which account for only 4.5% (= (15 + 30)/1,000) of the population. The top two segments are effectively an enhanced sweet spot targeting because the percentage of the aggregated segments is small.
    c. Cumulative lift of 177 by targeting the top three segments, which account for 24.5% of the population. The top three segments are the recommended targeting strategy as previously discussed.
    d. Cumulative lift of 153 by targeting the top four segments, which account for 54.5% ((15 + 30 + 200 + 300)/1,000) of the population. Unless the population is not too large, a campaign targeted to the top four segments may be cost prohibitive.

25.6 Continuous Trees

So far, the profiling uses only categorical variables, that is, variables that assume two or more discrete values. Fortunately, trees can accommodate continuous variables or variables that assume many numerical values, which allows for developing expansive profiles.

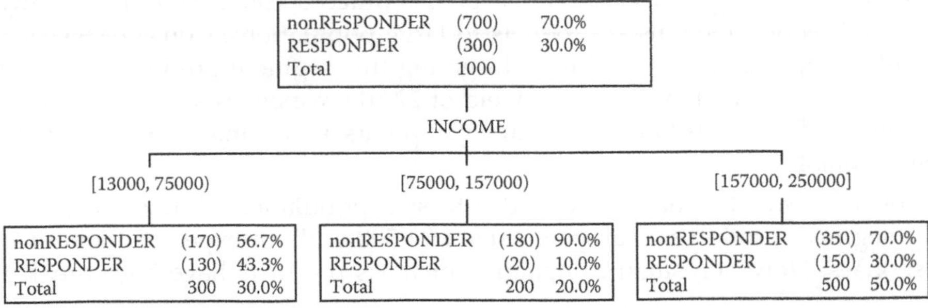

FIGURE 25.5
INCOME tree.

Consider a new variable INCOME. The INCOME tree in Figure 25.5 reads as follows:

1. Tree notation: Trees for a continuous variable denote the continuous values in ranges—a closed interval or a left-closed right-open interval. The former is denoted by [x, y], indicating all values between and including x and y. The latter is denoted by [x, y), indicating all values greater than or equal to x and less than y.
2. The top box indicates that for the sample of 1,000 individuals, there are 300 responders and 700 nonresponders. The response rate is 30%, and the nonresponse rate is 70%.
3. Segment 1 represents 300 individuals with income in the interval [$13,000, $75,000). The response rate among these individuals is 43.3%.
4. Segment 2 represents 200 individuals with income in the interval [$75,000, $157,000). The response rate among these individuals is 10.0%.
5. Segment 3 represents 500 individuals with income in the interval [$157,000, $250,000]. The response rate among these individuals is 30.0%.

A computer-intensive heuristic iterative algorithm determines the number of nodes and the range of an interval. Specifically, a tree is grown to create nodes/segments with the greatest variation in response rates across all segments.

The full tree with the variables GENDER, OWN_CELL, CHILDREN, and INCOME is in Figure 25.6. The gains chart of this tree is in Table 25.4. Cell-Talk can expect a cumulative lift of 177, which accounts for 24.5% (= (15+200+30/1,000) of the population, by targeting the top three segments.

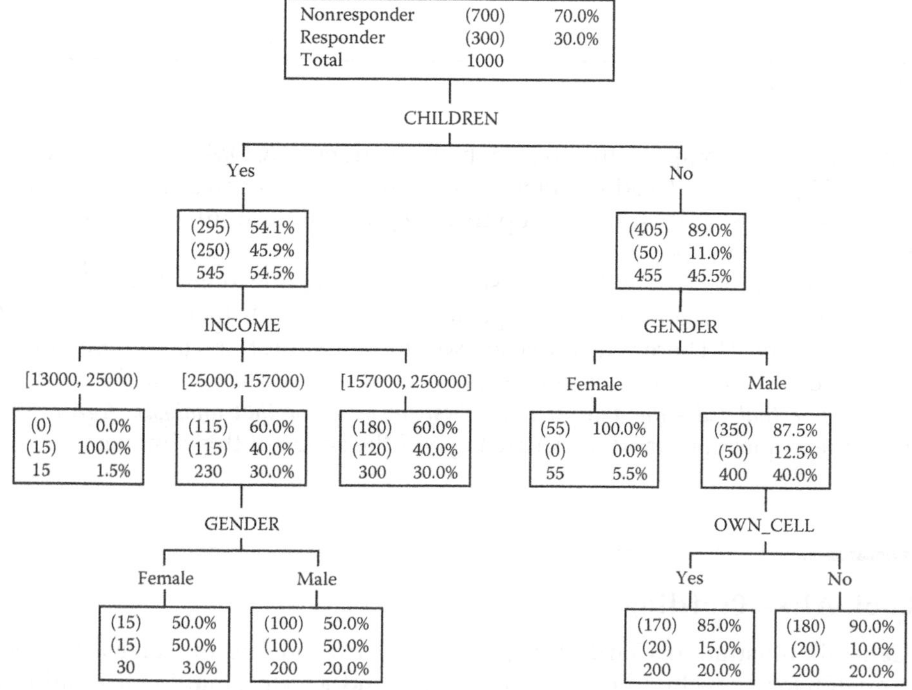

FIGURE 25.6
Full tree defined by GENDER, OWN_CELL, CHILDREN, and INCOME.

TABLE 25.4

Gains Chart for Tree Defined by GENDER, OWN CELL, CHILDREN, and INCOME

Segment[a]		Size of Segment	Number of Responses	Cumulative Responses	Segment Response Rate (%)	Cumulative Response Rate (%)	Cumulative Lift
1	INCOME, [13000, 25000) CHILDREN, yes	15	15	15	100.0	100.0	333
3	GENDER, male INCOME, [25000, 157000) CHILDREN, yes	200	100	115	50.0	53.5	178
2	GENDER, female INCOME, [25000, 157000) CHILDREN, yes	30	15	130	50.0	53.1	177
4	INCOME [157000, 250000] CHILDREN, yes	300	120	250	40.0	45.9	153
6	OWN_CELL, yes GENDER, male CHILDREN, no	200	30	280	15.0	37.6	125
7	OWN_CELL, no GENDER, male CHILDREN, no	200	20	300	10.0	31.7	106
5	GENDER, female CHILDREN, no	55	0	300	0.0	30.0	100
		1,000	300		30.0		

[a] Segments are ranked by response rates.

It is interesting to compare this tree, which includes INCOME, to the tree without INCOME in Figure 25.4. Based on Tables 25.3 and 25.4, these two trees have the same performance statistics, at least for the top three segments, a cumulative lift of 177, which accounts for 24.5% of the population.

The issue of including INCOME raises some interesting questions. Does INCOME add any noticeable predictive power? How important is INCOME? Which tree is better? Which set of variables is best? There is a simple answer to these questions (and much more tree-related questions): *An analyst can grow many equivalent trees to explain the same response behavior. The tree that suits the analyst is the best (at least for that analyst).* Again, detailed answers to these questions (and more) are beyond the scope of this chapter.

25.7 Look-Alike Profiling

In the Cell-Talk illustration, Cell-Talk requires predictive profiles to increase the response to its campaign for a new bundle of features. They conduct a test mailing to obtain a group of their best customers—responders of the new bundle offer—on which to develop the profiles.

Now, consider that Cell-Talk wants predictive profiles to target a solicitation based on a rental list of names, for which only demographic information is available and the ownership of a cellular phone is not known. The offer is a discounted rate plan, which should be attractive to cellular phone owners with high monthly usage (around 500 min of use per month).

Even though Cell-Talk does not have the time or money to conduct another test mailing to obtain their target group of responders with high monthly usage, they can still develop profiles to help in their targeting efforts as long as they have a notion of what their target group looks like. Cell-Talk can use a look-alike group—individuals who look like individuals in the target group—as a substitute for the target group. This substitution allows Cell-Talk to develop look-alike profiles: profiles that identify individuals (in this case, persons on the rental list) who are most likely to look like individuals in the target group.

The construction of the look-alike group is important. The greater the similarity between the look-alike group and target group, the greater the reliability of the resultant profiles will be.

Accordingly, the definition of the look-alike group should be as precise as possible to ensure that the look-alikes are good substitutes for the target individuals. The definition of the look-alike group can include as many variables as needed to describe pertinent characteristics of the target group. Note, the definition always involves at least one variable that is not available on the solicitation file or, in this case, the rental list. If all the variables are available, then there is no need for look-alike profiles.

Cell-Talk believes the target group looks like their current upscale cellular phone subscribers. Because cellular conversation is not inexpensive, Cell-Talk assumes that heavy users must have a high income to afford the cost of cellular use. Accordingly, Cell-Talk defines the look-alike group as individuals with a cellular phone (OWN_CELL = yes) and INCOME greater than $175,000.

The assumption of look-alike profiles is: Individuals who look like individuals in a target group have levels of responsiveness similar to the group. Thus, the look-alike individuals serve as surrogate or would-be responders.

Keep in mind that individuals identified by look-alike profiles are expected probabilistically to look like the target group but not expected necessarily to respond. In practice, the look-alike assumption is tenable as solicitations based on look-alike profiles produce noticeable response rates.

Look-alike profiling via tree analysis identifies variables that discriminate between look-alike individuals and nonlook-alike individuals (the balance of the population without the look-alike individuals). Effectively, the discriminating variables produce varied look-alike rates. I use the original sample data in Table 25.1 along with INCOME to create the LOOK-ALIKE variable required for the tree analysis. LOOK-ALIKE equals 1 if an individual has OWN_CELL = yes and INCOME greater than $175,000; otherwise, LOOK-ALIKE equals 0. There are 300 look-alikes and 700 nonlook-alikes, resulting in a look-alike sample rate of 30.0%.* These figures are in the top box of the look-alike tree in Figure 25.7. (Note, the look-alike sample rate and the original sample response rate are equal; this is purely coincidental.)

The gains chart for the look-alike tree is in Table 25.5. Targeting the top segment (#4) yields a cumulative lift of 333 (= (100%/30%)*100), with a 30% depth of population. The top segment lift means the predictive look-alike profile is expected to identify 3.33 times more

* The sample look-alike rate sometimes needs to be adjusted to equal the incidence of the target group in the population. This incidence is rarely known and must be estimated.

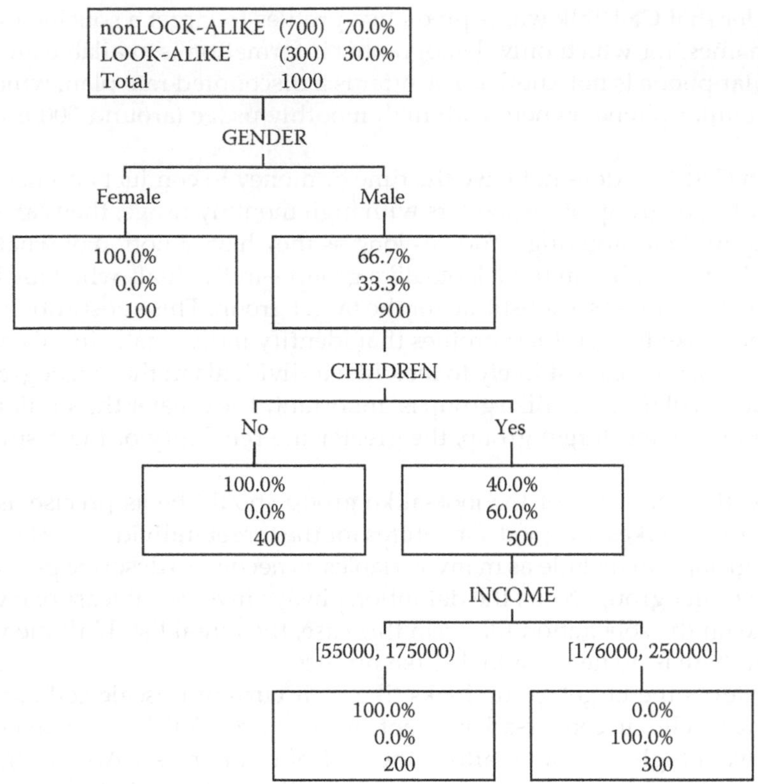

FIGURE 25.7
Look-alike tree.

TABLE 25.5
Gains Chart for Look-Alike Tree Defined by GENDER, CHILDREN, and INCOME

Segment[a]		Size of Segment	Number of Responses	Cumulative Responses	Segment Response Rate (%)	Cumulative Response Rate (%)	Cumulative Lift
4	INCOME, [176000, 250000) CHILDREN, yes GENDER, male	300	300	300	100.0	100.0	333
2	CHILDREN, no GENDER, male	400	0	300	0.0	42.9	143
1	GENDER, female	100	0	300	0.0	37.5	125
3	INCOME [55000, 175000) CHILDREN, yes GENDER, male	200	0	300	0.0	30.0	100
		1,000	300		30.0		

[a] Segments are ranked by look-alike rates.

individuals who look like the target group than expected from a random selection of 30% of rented names.

A closer look at the look-alike tree raises a question. INCOME is in the defining of both the LOOK-ALIKE variable and the profiles. Does this indicate that the tree is poorly defined? No. For this particular example, INCOME is a required variable. Without INCOME, this tree could not guarantee that the identified *males with children* have high incomes, a requirement for being a look-alike.

25.8 Look-Alike Tree Characteristics

It is instructive to discuss a noticeable characteristic of look-alike trees. In general, upper segment rates in a look-alike tree are quite large and often reach 100%. Similarly, lower segment rates are quite small and often fall to 0%. I observe these patterns in Table 25.5. There is one segment with a 100% look-alike rate and three segments with a 0% look-alike rate.

The implication is as follows:

1. It is easier to identify an individual who looks like someone with predefined characteristics (e.g., gender and children) than someone who behaves in a particular manner (e.g., responds to a solicitation).
2. The resultant look-alike rates are biased estimates of target response rates to the extent the defined look-alike group differs from the target group. Care is in order when defining the look-alike group because it is easy to include individuals inadvertently unwanted.
3. The success of a solicitation based on look-alike profiles, regarding the actual responses obtained, depends on the disparity of the defined look-alike group and target group and the tenability of the look-alike assumption.

25.9 Summary

Marketers typically attempt to improve the effectiveness of their campaigns by targeting their best customers. However, targeting only the best customers (responders) based on their characteristics is sure to fail. The descriptive profile, which represents responders without reference to nonresponders, is a nonpredictive profile. Therefore, targeting with a descriptive profile does not provide any assurance that the best customers will respond to a new campaign. The value of the descriptive profile lies in its definition of the salient characteristics of the target group, which are used to develop an effective marketing strategy.

I contrast descriptive and predictive profiles. A predictive profile describes responders concerning responsiveness—that is, regarding variables that discriminate between responders and nonresponders. The predictive profile is essential for finding responders to a new campaign, after which the descriptive profile is used to communicate effectively with those customers.

Then, I introduce the tree analysis method of developing a set of complex and interesting predictive profiles. With an illustration, I present the gains chart as the standard report of the predictive power of the tree-based predictive profiles. The gains chart indicates the expected response rates on implementing the profiles in a solicitation.

Last, I expand the predictive profiling approach to look-alike profiling, a reliable method when actual response information is not available. A look-alike profile is a predictive profile based on a group of individuals who look like the individuals in a target group, thus serving as surrogate responders. I address a warning that the look-alike rates are biased estimates of target response rates because the look-alike profiles are about surrogate responders not actual responders.

26

Assessment of Marketing Models[*]

26.1 Introduction

Marketers use decile analysis to assess their models regarding classification or prediction accuracy. The uninformed marketers do not know that additional information from the decile analysis can be extracted to supplement the model assessment. The purpose of this chapter is to present two additional concepts of model assessment—precision and separability—and to illustrate these concepts by further use of the decile analysis.

I begin the discussion with the traditional concepts of accuracy for response and profit models and illustrate the basic measures of accuracy. Then, I introduce the accuracy measure used in marketing, known as Cum Lift. The discussion of Cum Lift is in the context of a decile analysis, which is the usual approach marketers utilize to evaluate the performance of response and profit models. I provide a systematic procedure for conducting a decile analysis with an illustration.

I continue with the illustration to present the new concepts, precision, and separability. Last, I provide guidelines for using all three measures in assessing marketing models.

26.2 Accuracy for Response Model

How well does a response model correctly classify individuals as responders and nonresponders? The traditional measure of accuracy is the proportion of total correct classifications (PTCC), calculated from a simple cross tabulation.

Consider the classification results in Table 26.1 of a response model based on the validation sample consisting of 100 individuals with a 15% response rate. The "Total" column indicates there are 85 actual nonresponders and 15 actual responders in the sample. The "Total" row indicates the model predicts 76 nonresponders and 24 responders. The model correctly classifies 74 nonresponders and 13 responders. Accordingly, the PTCC is 87% (= (74 + 13)/100).

Although PTCC is frequently used, it may not be appropriate for the given situation. For example, if the assessment criterion imposes a penalty for misclassifications,[†] then PTCC must be either modified or discarded for a more relevant measure.

[*] This chapter is based on an article with the same title in *Journal of Targeting, Measurement and Analysis for Marketing*, 7, 3, 1998. Used with permission.

[†] For example, there is a $2 loss if a responder is classified as a nonresponder and a $4 loss if a nonresponder is classified as a responder.

TABLE 26.1

Classification Results of Response Model

		Predicted		
		Nonresponder	**Responder**	**Total**
Actual	Nonresponder	74	11	85
	Responder	2	13	15
Total		76	24	100

Marketers have defined their measure of accuracy for response models: Cum Lift. They use response models to identify those individuals most likely to respond to a solicitation. They create a solicitation list of the most likely individuals to obtain an advantage over a random selection of individuals. The Cum Lift indicates the incremental gain in responses over chance for various depths-of-file. Specifically, the Cum Lift is an index (ratio) of the expected response rate based on a response model to the expected response rate based on a random selection (no model, more accurately, a chance model). Before I illustrate the calculation of the Cum Lift, I provide a companion exposition of the accuracy for profit model.

26.3 Accuracy for Profit Model

How well does a profit model correctly predict an individual's profit value? There are several measures of prediction accuracy, all of which use the concept of *error*, namely, actual profit minus predicted profit. The mean squared error (MSE) is by far the most popular measure, but it is flawed, thus necessitating three alternative measures. I briefly review the four error measures.

1. MSE is the mean of individual squared errors. It gives greater importance to larger errors and tends to underestimate the predictive accuracy of the model. This point is in the discussion that follows.
2. MPE is the mean of individual percentage errors. It measures the bias in the estimates. Percentage error for an individual is the error divided by the actual profit multiplied by 100.
3. MAPE is the mean of individual absolute percentage errors. It disregards the sign of the error.
4. MAD is the mean of absolute individual deviations (errors). It disregards the sign of the error.

Consider the prediction results in Table 26.2 of the profit model (not shown). The validation sample consists of 30 individuals with a mean profit of $6.76. The first two columns after the ID number column are the *actual* profit and *predicted* profit produced by the profit model, respectively. The remaining columns in order from left to right are "Error," "Squared Error," "Percentage Error," "Absolute Error," and "Absolute Percentage Error." The bottom row "MEAN" consists of the means, based on the 30 individuals, for the last four columns on the right. The mean values are 23.22, 52.32%, 2.99, and 87.25%, for MSE, MPE, MAD, and MAPE, respectively. These measures are only indicators of a good model: Smaller values tend to correspond to better models.

TABLE 26.2

Profit Model: Four Measures of Errors

ID#	Actual PROFIT	Predicted PROFIT	Error	Squared Error	Percentage Error (%)	Absolute Error	Absolute Percentage Error (%)
1	0.60	0.26	0.34	0.12	132.15	0.34	132.15
2	1.60	0.26	1.34	1.80	519.08	1.34	519.08
3	0.50	0.26	0.24	0.06	93.46	0.24	93.46
4	1.60	0.26	1.34	1.80	519.08	1.34	519.08
5	0.50	0.26	0.24	0.06	93.46	0.24	93.46
6	1.20	0.26	0.94	0.89	364.31	0.94	364.31
7	2.00	1.80	0.20	0.04	11.42	0.20	11.42
8	1.30	1.80	–0.50	0.25	27.58	0.50	27.58
9	2.50	1.80	0.70	0.50	39.27	0.70	39.27
10	2.20	3.33	–1.13	1.28	33.97	1.13	33.97
11	2.40	3.33	–0.93	0.87	27.96	0.93	27.96
12	1.20	3.33	–2.13	4.54	63.98	2.13	63.98
13	3.50	4.87	–1.37	1.87	28.10	1.37	28.10
14	4.10	4.87	–0.77	0.59	15.78	0.77	15.78
15	5.10	4.87	0.23	0.05	4.76	0.23	4.76
16	5.70	6.40	–0.70	0.50	11.00	0.70	11.00
17	3.40	7.94	–4.54	20.62	57.19	4.54	57.19
18	9.70	7.94	1.76	3.09	22.14	1.76	22.14
19	8.60	7.94	0.66	0.43	8.29	0.66	8.29
20	4.00	9.48	–5.48	30.01	57.80	5.48	57.80
21	5.50	9.48	–3.98	15.82	41.97	3.98	41.97
22	10.50	9.48	1.02	1.04	10.78	1.02	10.78
23	17.50	11.01	6.49	42.06	58.88	6.49	58.88
24	13.40	11.01	2.39	5.69	21.66	2.39	21.66
25	4.50	11.01	–6.51	42.44	59.14	6.51	59.14
26	30.40	12.55	17.85	318.58	142.21	17.85	142.21
27	12.40	15.62	–3.22	10.40	–20.64	3.22	20.64
28	13.40	17.16	–3.76	14.14	21.92	3.76	21.92
29	26.20	17.16	9.04	81.71	52.67	9.04	52.67
30	7.40	17.16	–9.76	95.28	56.88	9.76	56.88
			MEAN	23.22	52.32	2.99	87.25
			MEAN without ID 26	13.03	49.20	2.47	85.30

To highlight the sensitivity of MSE to far-out values, I calculate MSE, MPE, MAD, and MAPE based on the sample without the large error of 318.58 corresponding to individual 26. The adjusted mean values for MSE, MPE, MAD, and MAPE are 13.03, 49.20%, 2.47, and 85.30%, respectively. The sensitivity of MSE is clear as it is dramatically reduced by almost 50%, whereas MPE, MAD, and MAPE remain relatively stable.

Except for the occasional need for an individual-level profit accuracy assessment (requiring one of the four error measures), marketers use their measure of accuracy, the Cum Lift. Cum Lift for a profit model—Cum Lift (profit)—is similar to the Cum Lift for a response model, except for slight changes in assessment and interpretation. Marketers use profit models to identify individuals contributing maximum profit from a solicitation and create a solicitation list of those individuals to obtain an advantage over a random selection. The Cum Lift (profit) indicates the incremental gain in profit over chance for various depths-of-file. Specifically, the Cum Lift (profit) is an index (ratio) of the expected profit based on a profit model to the expected profit based on a random selection, equivalently a chance model.

26.4 Decile Analysis and Cum Lift for Response Model

The decile analysis is a tabular display of model performance. I illustrate the construction and interpretation of the decile analysis for a response model (not shown) in Table 26.3.

1. Score the sample (i.e., calibration or validation file) using the response model under consideration. Every individual receives a model score, Prob_est, the estimated probability of response of the model.
2. Rank the scored file in descending order by Prob_est.
3. Divide the ranked and scored file into 10 equal groups. The *decile* variable is created, which takes on 10 ordered labels: top (1), 2, 3, 4, 5, 6, 7, 8, 9, and bottom (10). The top decile consists of the best 10% of individuals most likely to respond; Decile 2 consists of the next 10% of individuals most likely to respond and so on for the remaining deciles. Accordingly, the decile separates and orders the individuals on an ordinal scale ranging from most to least likely to respond.
4. *Number of individuals* is the number of individuals in each decile—10% of the total size of the file.
5. *Number of responses (actual)* is the actual—not predicted—number of responses in each decile. The model identifies 911 actual responders in the top decile. In Decile 2, the model identifies 544 actual responders. For the remaining deciles, the model identifies the actual responders similarly.

TABLE 26.3

Response Decile Analysis

Decile	Number of Individuals	Number of Responders	Decile Response Rate (%)	Cumulative Response Rate (%)	Cumulative Lift
top	7,410	911	12.3	12.3	294
2	7,410	544	7.3	9.8	235
3	7,410	437	5.9	8.5	203
4	7,410	322	4.3	7.5	178
5	7,410	258	3.5	6.7	159
6	7,410	188	2.5	6.0	143
7	7,410	130	1.8	5.4	129
8	7,410	163	2.2	5.0	119
9	7,410	124	1.7	4.6	110
bottom	7,410	24	0.3	4.2	100
Total	74,100	3,101	4.2		

6. *Decile response rate* is the actual response rate for each decile group. It is *number of responses* divided by *number of individuals* for each decile group. For the top decile, the response rate is 12.3% (= 911/7410). For the second decile, the response rate is 7.3% (= 544/7410) and so on for the remaining deciles.
7. *Cumulative response rate* for a given depth of file (the aggregated or cumulative deciles) is the response rate among the individuals in the cumulative deciles. For example, the cumulative response rate for the top decile (10% depth of file) is 12.3% (= 911/7410). For the top two deciles (20% depth of file), the cumulative response rate is 9.8% (= (911 + 544)/(7410 + 7410)). For the remaining deciles, the calculation of cumulative response rates is similar.
8. *Cum Lift* for a given depth of file is the *cumulative response rate* divided by the overall response rate of the file multiplied by 100. It measures how many more responders one can expect using a model over not using a model. For example, a Cum Lift of 294 for the top decile means that when soliciting to the top 10% of the file based on the model, one expects 2.94 times the total number of responders obtained without a model. The Cum Lift of 235 for the top two deciles means that when soliciting to 20% of the file based on the model, one can expect 2.35 times the total number of responders obtained without a model. For the remaining depths of file, the interpretations of the Cum Lifts are similar.

Rule: The larger the Cum Lift value is, the better the accuracy for a given depth of file will be.

26.5 Decile Analysis and Cum Lift for Profit Model

Calculation of the decile analysis for a profit model (not shown) is similar to that of the decile analysis for a response model with response and response rates replaced by profit and mean profit, respectively. I illustrate the construction and interpretation of the profit decile analysis in Table 26.4.

1. Score the sample (i.e., calibration or validation file) using the profit model under consideration. Every individual receives a model score, Pred_est, the predicted profit of the model.
2. Rank the scored file in descending order by Pred_est.
3. Divide the ranked and scored file into 10 equal groups, producing the *decile* variable. The top decile consists of the best 10% of individuals contributing maximum profit. Decile 2 consists of the next 10% of individuals contributing maximum profit and so on for the remaining deciles. Accordingly, the decile separates and orders the individuals on an ordinal scale ranging from maximum to minimum contribution of profit.
4. *Number of individuals* is the number of individuals in each decile—10% of the total size of the file.
5. *Total profit (actual)* is the actual—not predicted—total profit in each decile. The model identifies individuals contributing \$47 profit in the top decile. In Decile 2, the model identifies individuals contributing \$60.30 profit and so on for the remaining deciles.

TABLE 26.4
Profit Decile Analysis

Decile	Number of Individuals	Total Profit	Decile Mean Profit	Cumulative Mean Profit	Cumulative Lift
top	3	$47.00	$15.67	$15.67	232
2	3	$60.30	$20.10	$17.88	264
3	3	$21.90	$7.30	$14.36	212
4	3	$19.40	$6.47	$12.38	183
5	3	$24.00	$8.00	$11.51	170
6	3	$12.70	$4.23	$10.29	152
7	3	$5.80	$1.93	$9.10	135
8	3	$5.80	$1.93	$8.20	121
9	3	$2.70	$0.90	$7.39	109
bottom	3	$3.30	$1.10	$6.76	100
Total	30	$202.90	$6.76		

6. *Decile mean profit* is the actual mean profit for each decile group. It is *total profit* divided by *number of individuals* for each decile group. For the top decile, the actual mean profit is $15.67 (= $47/3). For the second decile, the value is $20.10 (= $60.30/3) and so on for the remaining deciles.
7. *Cumulative mean profit* for a given depth of file (the aggregated or cumulative deciles) is the mean profit among the individuals in the cumulative deciles. For example, the cumulative mean profit for the top decile (10% of the file) is $15.67 (= $47/3). For the top two deciles (20% depth-of-file), the cumulative mean profit is $17.88 (= ($47 + $60.30)/(3 + 3)) and so on for the remaining deciles.
8. *Cum Lift* for a given depth of file is the *cumulative mean profit* divided by the overall profit of the file multiplied by 100. It measures how much more profit one can expect using a model over not using a model. For example, a Cum Lift of 232 for the top decile means that when soliciting to the top 10% of the file based on a profit model, one expects 2.32 times the total profit obtained without a model. The Cum Lift of 264 for the top two deciles means that when soliciting to 20% of the file based on a profit model, one expects 2.64 times the total profit obtained without a model. For the remaining depths-of-file, the interpretations of the Cum Lifts are similar. Note, the nondecreasing profit values throughout the decile suggest that something is wrong with this model (e.g., the exclusion of an important predictor variable or a needed reexpression of the predictor variable).

Rule: The larger the Cum Lift value is, the better the accuracy for a given depth of file will be.

26.6 Precision for Response Model

How close are the predicted probabilities of response to the true probabilities of response? Closeness or precision of response cannot directly be determined because an individual's true probability is not known—if it were, then there is no need for a model. I recommend and illustrate the method of *smoothing* to provide estimates of the true probabilities.

Then, I present the Hosmer–Lemeshow goodness-of-fit measure (or HL index) as the measure of precision for response models.

Smoothing is the averaging of values within neighborhoods. In this application of smoothing, I average actual responses within *decile* neighborhoods formed by the model. Continuing with the response model illustration, the actual response rate for a decile, Column 4 in Table 26.3, is the estimate of the true probability of response for the group of individuals in that decile.

Next, I calculate the mean predicted probabilities of response based on the model response scores (Prob_est) among the individuals in each decile. I insert these predicted means in Column 4 in Table 26.5. Also, I insert the actual response rate (fourth column from the left in Table 26.3) in Column 3 in Table 26.5. I can now determine model response precision.*

Comparing Columns 3 and 4 (in Table 26.5) is informative. I see that for the top decile, the model underestimates the probability of a response: 12.3% actual versus 9.6% predicted. Similarly, the model underestimates for Deciles 2 through 4. Decile 5 is perfect. Going down from Decile 6 to the bottom decile, one can see clearly that the model is overestimating. This type of evaluation for precision is perhaps too subjective. An objective summary measure of precision is needed.

I present the HL index as the measure of precision. The calculations for the HL index are in Table 26.5 for the response model illustration:

1. Columns 1, 2, and 3 are available from the decile analysis for a response model.
2. Calculate the mean predicted probability of response for each decile from the model scores, Prob_est (Column 4).

TABLE 26.5

Response Model: HL and CV Indices

	Column 1	Column 2	Column 3	Column 4	Column 5	Column 6	Column 7
Decile	**Number of Individuals**	**Number of Responders**	**Decile Response Rate (Actual) (%)**	**Prob_est (Predicted) (%)**	**Square of (Column 3 – Column 4) Times Column 1**	**Column 4 Times (1–Column 4)**	**Column 5 Divided by Column 6**
top	7,410	911	12.3	9.6	5.40	0.086	62.25
2	7,410	544	7.3	4.7	5.01	0.044	111.83
3	7,410	437	5.9	4.0	2.68	0.038	69.66
4	7,410	322	4.3	3.7	0.27	0.035	7.49
5	7,410	258	3.5	3.5	0.00	0.033	0.00
6	7,410	188	2.5	3.4	0.60	0.032	18.27
7	7,410	130	1.8	3.3	1.67	0.031	52.25
8	7,410	163	2.2	3.2	0.74	0.031	23.92
9	7,410	124	1.7	3.1	1.45	0.030	48.35
bottom	7,410	24	0.3	3.1	5.81	0.030	193.40
Total	74,100	3,101	4.2				
	Separability CV		80.23		Precision HL		587.40

* This assessment is considered at a 10% level of smooth. A ventile-level analysis with the scored, ranked file divided into 20 groups provides an assessment of model precision at a 5% level of smooth. There is no agreement on a reliable level of smooth among statisticians.

3. Calculate Column 5: Take the difference between Column 3 and Column 4. Square the results. Then, multiply by Column 1.
4. Column 6: Column 4 times the quantity 1 minus Column 4.
5. Column 7: Column 5 divided by Column 6.
6. HL index: The sum of the 10 elements of Column 7.

Rule: The smaller the HL index value is, the better the precision will be.

26.7 Precision for Profit Model

How close are the predicted profits to the true profits? As in the case of the response model, determination of closeness is not directly possible because an individual's true profit is not known, and I recommend and illustrate the method of smoothing to provide estimates of the true profit values. Then, I present the smooth weighted mean of the absolute deviation (SWMAD) index as the measure of precision for a profit model.

To obtain the estimates of the true profit values, I average actual profit within decile neighborhoods formed by the profit model. Continuing with the profit model illustration, the mean actual profit for a decile, the fourth column from the left in Table 26.4, is the estimate of true mean profit for the group of individuals in that decile. Next, I calculate the mean predicted profit based on the model scores (Pred_est) among the individuals in each decile. I insert these predicted means in Column 2 in Table 26.6. Also, I insert the mean actual profit (fourth column from the left in Table 26.4) in Column 1 in Table 26.6. I can now determine the model precision.

TABLE 26.6

Profit Model: SWMAD and CV Indices

	Column 1	Column 2	Column 3	Column 4	Column 5	Column 6	Column 7	Column 8
Decile	**Decile Mean Profit (Actual)**	**Decile Mean Pred_est (Predicted Profit)**	**Absolute Error**	**Rank of Decile Actual Profit**	**Rank of Decile Predicted Profit**	**Absolute Difference between Ranks**	**Wt**	**Weighted Error**
top	$15.67	$17.16	$1.49	2	1	1.0	1.10	1.64
2	$20.10	$13.06	$7.04	1	2	1.0	1.10	7.74
3	$7.30	$10.50	$3.20	4	3	1.0	1.10	3.52
4	$6.47	$8.97	$2.50	5	4	1.0	1.10	2.75
5	$8.00	$7.43	$0.57	3	5	2.0	1.20	0.68
6	$4.23	$4.87	$0.63	6	6	0.0	1.00	0.63
7	$1.93	$3.33	$1.40	7.5	7	0.5	1.05	1.47
8	$1.93	$1.80	$0.14	7.5	8	0.5	1.05	0.15
9	$0.90	$0.26	$0.64	10	9.5	0.5	1.05	0.67
bottom	$1.10	$0.26	$0.84	9	9.5	0.5	1.05	0.88
Separability CV	95.86					SUM	10.8	20.15
						Precision	SWMAD	1.87

Comparing Columns 1 and 2 (in Table 26.6) is informative. The ranking of the deciles based on the mean actual profit values is not strictly descending—not a desirable indicator of a good model. The mean profit values for the top and second deciles are reversed, $15.67 and $20.10, respectively. Moreover, the third-largest decile mean profit value ($8.00) is in the fifth decile. This type of evaluation is interesting, but a quantitative measure for non-descending rankings is preferred.

26.7.1 Construction of SWMAD

I present the measure SWMAD for the precision of a profit model: a weighted mean of the absolute deviation between smooth decile actual and predicted profit values; the weights reflect discordance between the rankings of the smooth decile actual and predicted values. The steps for the calculation of SWMAD for the profit model illustration are next, and the results are in Table 26.6:

1. Column 1 is available from the decile analysis for a profit model.
2. Calculate the mean predicted profit for each decile from the model scores, Pred_est (Column 2).
3. Calculate Column 3: Take the absolute difference between Column 1 and Column 2.
4. Column 4: Rank the deciles based on actual profit (Column 1); assign the lowest rank value to the highest decile mean actual profit value. Tied ranks are assigned the mean of the corresponding ranks.
5. Column 5: Rank the deciles based on predicted profit (Column 2); assign the lowest rank value to the highest decile mean predicted profit value. Tied ranks are assigned the mean of the corresponding ranks.
6. Column 6: Take the absolute difference between Column 4 and Column 5.
7. Column 7: The weight variable (Wt) is Column 6 divided by 10 plus 1.
8. Column 8 is Column 3 times Column 7.
9. Calculate SUMWGT, the sum of the 10 values of Column 7.
10. Calculate SUMWDEV, the sum of the 10 values of Column 8.
11. Calculate SWMAD: SUMWDEV/SUMWGT.

Rule: The smaller the SWMAD value is, the better the precision will be.

26.8 Separability for Response and Profit Models

How different are the individuals across the deciles regarding likelihood to respond or contribution to profit? Is there a real variation or separation of individuals as identified by the model? I can measure the variability across the decile groups by calculating the traditional coefficient of variation (CV) among the decile estimates of the true probability of response for the response model and the decile estimates of true profit for the profit model.

I illustrate the calculation of CV with the response and profit model illustrations. CV (response) is the standard deviation of the 10 smooth values of Column 3 in Table 26.5

divided by the mean of the 10 smooth values and multiplied by 100. CV (response) is 80.23 in Table 26.5. CV (profit) is the standard deviation of the 10 smooth values of Column 1 in Table 26.6 divided by the mean 10 smooth values and multiplied by 100. CV (profit) is 95.86 in Table 26.6.

Rule: The larger the CV value is, the better the separability will be.

26.9 Guidelines for Using Cum Lift, HL/SWMAD, and CV

The following are guidelines for selecting the best model based on the three assessment measures Cum Lift, HL/SWMAD, and CV:

1. In general, a good model has large HL/SWMAD and CV values.
2. If maximizing response rate/mean profit is not the objective of the model, then the best model is among those with the smallest HL/SWMAD values and largest CV values. Because small HL/SWMAD values do not necessarily correspond with large CV values, the data analyst must decide on the best balance of small HL/SWMAD and large CV values for declaring the best model.
3. If maximizing response rate/mean profit is the objective, then the best model has the largest Cum Lift. If there are several models with comparable largest Cum Lift values, then the model with the best HL/SWMAD–CV combination is declared the best model.
4. If decile-level response/profit prediction is the objective of the model, then the best model has the smallest HL/SWMAD value. If there are several models with comparable smallest HL/SWMAD values, then the model with the largest CV value is declared the best.
5. The measure of separability of CV itself has no practical value. A model that is selected solely on the largest CV value will not necessarily have good accuracy or precision. Separability should be used in conjunction with the other two measures of model assessment as discussed.

26.10 Summary

The traditional measures of model accuracy are a PTCC and MSE or a variant of mean error for response and profit models, respectively. These measures have limited value in marketing. Marketers have their measure of model accuracy—Cum Lift—that takes into account the way they implement the model. They use a model to identify individuals most likely to respond or contribute profit and create a solicitation list of those individuals to obtain an advantage over a random selection. The Cum Lift is an index of the expected response/profit with a selection based on a model compared with the expected response/profit with a random selection (no model). A maxim of Cum Lift is the larger the Cum Lift value, the better the accuracy.

I discuss the Cum Lift by illustrating the construction of the decile analysis for both response and profit models. Then, using the decile analysis as a backdrop, I present two additional measures of model assessment—HL/SWMAD for response/profit precision and the traditional CV for separability for response and profit models. Because the true response/profit values are unknown, I estimate the true values by smoothing at the decile level. With these estimates, I illustrate the calculations for HL and SWMAD. A maxim of HL/SWMAD rule is the smaller the HL/SWMAD values are, the better the precision will be.

Separability addresses the question of how different the individuals are regarding their likelihoods to respond or to contribute profit across the deciles. I use traditional CV as the measure of separability among the estimates of true response/profit values. Thus, a maxim of CV is that the larger the CV value is, the better the separability will be.

Last, I provide guidelines for using all three measures together in selecting the best model.

27

Decile Analysis: Perspective and Performance

27.1 Introduction

Marketers use decile analysis to assess predictive incremental gains of their response models over responses obtained by chance (i.e., random selection of individuals for a solicitation). Underlying decile analysis is the traditional two-by-two classification table, commonly called the confusion matrix. The interpretation of the confusion matrix depends on the goal of the analysis. There are many perspectives, which account for the confusion matrix branding. The obvious statistic—percent of total correctly classified rate—is not necessarily important for all studies. In marketing, the perspective precision, unbeknownst to marketers and apparently not in the literature, is the underpinning of the decile analysis. The purpose of this chapter is to provide a comprehensive study of the decile analysis beyond the basics covered in Chapter 26. I provide an illustration in a pedagogical manner to a fuller appreciation of the decile analysis so users of the decile analysis can obtain a finer assessment of their response models. I provide the SAS© subroutines for constructing two new metrics and the proposed procedure, which will be a trusty tool for marketing statisticians. The subroutines are also available for downloading from my website: http://www.geniq.net/articles.html#section9.

27.2 Background

Framing the discussion, I build a logistic regression model for predicting RESPONSE (=yes/1 vs. =no/0), say, from a marketing solicitation. The RESPONSE model produces estimates of the probability of RESPONSE, labeled, PROB_est. PROB_est and a cut-off value (CV) define the variable PREDICTED. If PROB_est is greater than CV, then PREDICTED=1, otherwise PREDICTED=0. CV is usually the response rate of the marketing solicitation.

I discuss the basics of the classification table of binary variables, RESPONSE (yes=1, no=0) and PREDICTED (1=RESPONSE of yes, 0=RESPONSE of no) in Table 27.1.

The four cells of the table represent the following counts:

1. The top row of RESPONSE (=0) represents
    a. **a** individuals who are predicted to respond no and actually respond no. The model correctly predicts these individuals.
    b. **b** individuals who are predicted to respond yes but actually respond no. The model incorrectly predicts these individuals.

TABLE 27.1
Two-by-Two Classification

RESPONSE	PREDICTED		
Frequency	**0**	**1**	**Total**
0	a	b	a+b
1	c	d	c+d
Total	a+c	b+d	a+b+c+d (N)

2. The bottom row of RESPONSE (=1) represents
   a. **c** individuals who are predicted to respond no but actually respond yes. The model incorrectly predicts these individuals.
   b. **d** individuals who are predicted to respond yes and actually respond yes. The model correctly predicts these individuals.
   c. **a+b** represents the total number of individuals who actually respond no.
   d. **a+b+c+d (=N)** is the total number of individuals in the sample, the sample size.
   e. **a+b/N** is the percent of individuals who actually respond no.
   f. **c+d** represents the total number of individuals who actually respond yes.
   g. **c+d/N** is the percent of individuals who actually respond yes.
   h. **a+b/N** and **c+d/N** represent the RESPONSE distribution (%). The sum of the two percents equals 100%.
3. The first column of PREDICTED (=0) represents
   a. **a+c** represents the total number of individuals who are predicted to respond no. The model predicts **a+c** individuals who respond no.
4. The second column of PREDICTED (=1) represents
   a. **b+d** represents the total number of individuals who are predicted to respond yes. The model predicts **b+d** individuals who respond yes.

There are many ways of examining a classification table [1]. I present four widely used perspectives of model performance but limit special attention to two, which are relevant to the focal marketing applications of this book.

ACCURACY

1. **a+d/N** is the *accuracy* of the RESPONSE model, the percent of total correctly classified (PTCC).
2. Accuracy is misleading when **a** is disproportionately large. I discuss this issue in the illustration.

SPECIFICITY

1. **a/(a+b)** is *specificity*, the model's estimate of the probability of predicting an individual responding no given the individual's actual response is no.

2. The specificity metric addresses the question: When the model predicts an individual's response as no, how often is it correct?
3. If the model is highly specific and the model predicts a response of no, the individual is nearly certain to have a response of no.
4. Biology and medical diagnosis, not marketing, often use specificity. I present it here for the sake of completeness.

SENSITIVITY

1. **d/(c+d)** is *sensitivity*, the model's estimate of the probability of predicting an individual responding yes given the individual's actual response is yes.
2. The sensitivity metric addresses the question: When the model predicts an individual's response as yes, how often is it correct?
3. If the model is highly sensitive and the model predicts yes, the individual is nearly sure to have a response of yes.
4. Biology and medical diagnosis, not marketing, often use sensitivity. I present it here for the sake of completeness.

PRECISION

1. **d/(b+d)** is *precision*, the model's estimate of the probability of predicting an individual responding yes among all the individual's predicted to respond yes.
2. The precision metric addresses the question: When the model predicts an individual's response as yes, how often are the "yes" predictions correct?
3. A perfect precision score of 1.0 means every individual predicted as belonging to cell defined by RESPONSE=1 and PREDICTED=1, denoted by cell-d, does indeed belong to cell-d. But, a perfect score provides no insight as to the incorrectly predicted individuals from RESPONSE=1 (i.e., **c** individuals).
4. Precision is typically the focal point in marketing, in which situation **d** is small, especially compared to **a**. Small **d** values render a challenge for the statistician predicting individuals who truly belong to cell d.

27.2.1 Illustration

For this illustration, I use the RESPONSE model presented in Chapter 23, Section 23.4. Recall, I build a logistic regression model for predicting RESPONSE (=yes/1 vs. =no/0) based on a random sample (size = 16,003) from the marketing solicitation file. The number of individuals who respond to the solicitation is 3,559. The solicitation yields a response rate 22.24%, which I use for the CV to define the PREDICTED variable. Clearly, PREDICTED is required to create a two-by-two (2 × 2) classification table given in Table 27.2, which in turn is needed to assess the performance of the RESPONSE model.

TABLE 27.2

Classification Table of RESPONSE Model

RESPONSE	PREDICTED		
Frequency Percent	**0**	**1**	**Total**
0	9,134 57.08	3,310 20.68	12,444 77.76
1	1,349 8.43	2,210 13.81	3,559 22.24
Total	10,483 65.51	5,520 34.49	16,003 100.00

27.2.1.1 Discussion of Classification Table of RESPONSE Model

The 2 × 2 table indicates the following:

1. The distributions of RESPONSE (77.76%, 22.24%) and PREDICTED (65.51%, 34.49%) are roughly equal. That is, RESPONSE 77.76% and PREDICTED 65.51% are relatively close in their values. Similarly, RESPONSE 22.24% and PREDICTED 34.49% are relatively close in their values. The approximate equivalence of the distributions is a necessary indicator of a good model. So, the RESPONSE model has the promise of a good model. If the distributions are too disparate, then misclassification rates for RESPONSE levels (yes and no) are unacceptably large. Keep in mind: The CV determines the distribution of PREDICTED, which in turn, shapes the construction of the 2 × 2 table.
2. If the distributions are not reasonably equal, then fine-tuning CV may produce close-enough distributions but not without a cost. Fine-tuning is a euphemism for spurious data mining. The selected fine-turned CV value is likely a fortuitous finding (i.e., not reliable). Testing the reliability of CV, the statistician must construct multiple classification tables using many fresh samples of the solicitation file or resample via bootstrapping of the full solicitation file.
   a. If the CV holds up with the reliability testing, then it defines PREDICTED and consequentially construction of the classification table proceeds.
   b. If a reliable CV is not determined, then the original classification table yields suboptimal or biased model performance findings. In other words, the original table is an indicator that the model is not good, just acceptable.
3. The accuracy of the RESPONSE model is 70.89% (= (9134+2210)/16003).
4. The specificity of the RESPONSE model is 73.40% (= 9134/(9134+3310)).
5. The sensitivity of the RESPONSE model is 62.10% (= 2210/(1349+2210)).
6. The precision of the RESPONSE model is 40.04% (= 2210/(3310+2210)).

In sum, the RESPONSE model appears to have good performance indicators, accuracy and precision with values 70.89% and 40.04%, respectively. I do not cite the specificity and sensitivity value here and hereafter as they are not relevant to marketing applications.

27.3 Assessing Performance: RESPONSE Model versus Chance Model

Accuracy and precision values without corresponding baseline values (i.e., chance model performance measures) are subjectively qualified as excellent, good, poor, or whatever hyperbolic adjective the statistician chooses. The actual (RESPONSE) not predicted (PREDICTED) data are in the calculations of the baselines of accuracy and precision. Once the baselines are known, the statistician can compare the observed metric values and determine the performance of the model quantitatively.

The RESPONSE distribution completely determines the accuracy of the chance model. There is a simple formula* for determining the accuracy of the chance model for a 2 × 2 table. The formula is the sum of the squared actual percents, ((**a**+**b**)/**N**)*((**a**+**b**)/**N**) + ((**c**+**d**)/**N**)*((**c**+**d**)/**N**).

For the RESPONSE data, the accuracy of the chance model is 65.41% (=(77.76%*76.76%) + (22.24%*22.24%)). Hence, the *incremental* gain in RESPONSE model accuracy over the chance model accuracy is 8.37% (=(70.89% – 65.41%)/65.41%). The incremental gain of 8.37% is relatively good, in that accuracy is not the focal performance measure (see Table 27.3). The subroutine to reproduce Table 27.3 is in Appendix 27.A.

Worthy of note, accuracy as traditionally defined here is not an adequate measure when **a** is greatly larger than **d**. In this instance, **a** is greater than **d** by a factor of 4.13 (=9134/2210). In such situations, the geometric mean of **d**/(**c**+**d**) and **d**/(**b**+**d**), among other accuracy measures, is appropriate [2].

The formula for determining the precision of the chance model is somewhat complex. I obtain the chance model precision of 22.24% by running the subroutine in Appendix 27.B, Incremental Gain in Precision: Model versus Chance. The incremental gain in RESPONSE model precision over the chance model precision is 17.80% (=(40.04% – 22.24%)/22.24%). The incremental gain 17.80% is very good (see Table 27.4). The subroutine to reproduce Table 27.4 is in Appendix 27.B.

TABLE 27.3

Incremental Gain in Accuracy: Model versus Chance

MODEL_ ACCURACY	CHANCE_ ACCURACY	ACCURACY_ INCREMENTAL_ GAIN
70.89%	65.41%	8.37%

TABLE 27.4

Incremental Gain in Precision: Model versus Chance

MODEL_ PRECISION	CHANCE_ PRECISION	PRECISION_ INCREMENTAL_ GAIN
40.04%	22.24%	17.80%

* This simple formula is a good approximation when RESPONSE and PREDICTED distributions are relatively close. When the distributions are equal the formula produces exact results.

27.4 Assessing Performance: The Decile Analysis

Marketers use the decile analysis; the presentation of its details of construction and interpretation are in Section 26.4. In all likeliness, marketers are not aware the decile analysis is a horizontally sliced version of the 2 × 2 table. The 2 × 2 table has inherent horizontal slices, namely, RESPONSE=0 and RESPONSE=1. The decile analysis reslices the original two RESPONSE slices into 10 equal slices, commonly called deciles, which account for 10% of the sample.

Recall that PROB_est ranks the individuals in the sample descendingly to define the deciles. The first (top) decile consists of individuals with the 10% largest PROB_est values. Accordingly, the RESPONSE model, based on a current solicitation, identifies these top decile individuals as the most likely to respond to a solicitation akin to the current one. The second decile consists of individuals with the next largest 10% of PROB_est values. And so on for the remaining deciles, the third through the bottom (tenth) deciles. After the decile construction, the decile analysis adds five columns, unlike the two PREDICTED columns of the 2 × 2 table. Valuable performance information within the new five columns comes from the PREDICTED columns.

27.4.1 The RESPONSE Decile Analysis

Recall, the traditional discussion of the RESPONSE model decile analysis, redisplayed in Table 27.5, is in Section 23.4. Here, I focus on the proposed performance metrics, *model decile-analysis precision* and *chance decile-analysis precision*. These metrics allow for singular decile analysis assessment of a given model. In this case, the comparison between RESPONSE decile-analysis precision and the chance decile-analysis precision quantifies the performance of the RESPONSE model. The model and chance precision metrics, defined and illustrated in the 2 × 2 table, are implanted in the decile analysis. Decile-analysis precision is the underpinning of the Cum Lift, which is the chief factor in the success of the decile analysis.

I provide an illustration, using the RESPONSE model, in a pedagogical manner to a fuller appreciation of the decile analysis, so users of the decile analysis can obtain a finer

TABLE 27.5

RESPONSE Model Decile Analysis

Decile	Number of Individuals	Number of Responses	RESPONSE RATE (%)	CUM RESPONSE RATE (%)	CUM LIFT (%)
top	1,600	1,118	69.9	69.9	314
2	1,600	637	39.8	54.8	247
3	1,601	332	20.7	43.5	195
4	1,600	318	19.9	37.6	169
5	1,600	165	10.3	32.1	144
6	1,601	165	10.3	28.5	128
7	1,600	158	9.88	25.8	116
8	1,601	256	16.0	24.6	111
9	1,600	211	13.2	23.3	105
bottom	1,600	199	12.4	22.2	100
	16,003	3,559			

assessment of their response models. I reconfigure the RESPONSE decile analysis to bring out a new method of assessing the traditional, old-style decile analysis with a new, unique delivery of output performance metrics.

The steps to reconfigure the old into the new enhanced decile analysis are as follows:

1. The decile analysis has two variables, ACTUAL (RESPONSE) and PREDICTED response, denoted PPROB_est, for a pre-selected CV value. Accordingly, I determine the minimum and maximum PROB_est values, for each of the 10 deciles from a tabulation underlying the decile analysis itself. See the Decile PROB_est values in Table 27.6. The subroutine to reproduce Table 27.6 is in Appendix 27.C.
2. The Decile PROB_est table has all but the last two Cum columns of the RESPONSE decile analysis. The table does not include Cum RESPONSE RATE and CUM LIFT. Thus, the table is not a cumulative table. So to speak, the Cum items are replaced by the PROB_est minimum and maximum values. The Decile PROB_est table is strictly a reporting of RESPONSE model information at the decile level.

   An evaluation of the minimum and maximum values is an exercise beyond the intended purpose. Observations such as: Why are the minimum values frequently equal to the maximum values of the immediate lower decile? What does it mean for Deciles 6 and 7, where this observation does not occur? Answers to these questions are in my link http://www.geniq.net/res/SmartDecileAnalysis.html.
3. The utility of the PROB_est is to construct the PREDICTED variable for ten 2 × 2 tables, one for each decile. For example, for decile top, the minimum value is 0.4602. Thus, if PROB_est is greater than or equal to 0.4602, then PREDICTED=1, otherwise PREDICTED=0. For Decile 2, the minimum value is 0.2654. Thus, if PROB_est is greater than or equal to 0.2654, then PREDICTED=1, otherwise PREDICTED=0. And so on, for the remaining deciles.
4. The 2 × 2 tables for the first five deciles, 0 (=top) through Decile 4 (=5), are in Table 27.7. The 2 × 2 tables for the bottom five deciles, 6 (=5) through Decile 9 (=bottom), are in Table 27.8. The subroutines for reproducing Tables 27.7 and 27.8 are in Appendix 27.D. Note, the corresponding PROB_est values for each decile are in the table titles.

TABLE 27.6

RESPONSE Model Decile PROB_est Values

		RESPONSE		PROB_est	
Decile	**N**	**Mean**	**Sum**	**Min**	**Max**
top	1,600	0.6988	1,118	0.4602	0.9922
2	1,600	0.3981	637	0.2654	0.4602
3	1,601	0.2074	332	0.2354	0.2654
4	1,600	0.1988	318	0.2138	0.2354
5	1,600	0.1031	165	0.2008	0.2138
6	1,601	0.1031	165	0.1782	0.2008
7	1,600	0.0988	158	0.1427	0.1776
8	1,601	0.1599	256	0.1060	0.1427
9	1,600	0.1319	211	0.0682	0.1060
bottom	1,600	0.1244	199	0.0228	0.0682
All	16,003	0.2224	3,559	0.0228	0.9922

TABLE 27.7

2 × 2 Tables for First Five Deciles

Decile = 0 PROB_est = 0.4602

ACTUAL Frequency Percent	PREDICTED 0	 1	 Total
0	0 0.00	482 30.13	482 30.13
1	12 0.75	1,106 69.13	1,118 69.88
Total	12 0.75	1,588 99.25	1,600 100.00

Decile = 1 PROB_est = 0.2654

ACTUAL Frequency Percent	PREDICTED 0	 1	 Total
0	0 0.00	963 60.19	963 60.19
1	0 0.00	637 39.81	637 39.81
Total	0 0.00	1,600 100.00	1,600 100.00

Decile = 2 PROB_est = 0.2354

ACTUAL Frequency Percent	PREDICTED 0	 1	 Total
0	9 0.56	1,260 78.70	1,269 79.26
1	1 0.06	331 20.67	332 20.74
Total	10 0.62	1,591 99.38	1,601 100.00

Decile = 3 PROB_est = 0.2138

ACTUAL Frequency Percent	PREDICTED 0	 1	 Total
0	0 0.00	1,282 80.13	1,282 80.13
1	0 0.00	318 19.88	318 19.88
Total	0 0.00	1,600 100.00	1,600 100.00

Decile = 4 PROB_est = 0.2008

ACTUAL Frequency Percent	PREDICTED 0	 1	 Total
0	0 0.00	1,435 89.69	1,435 89.69
1	0 0.00	165 10.31	165 10.31
Total	0 0.00	1,600 100.00	1,600 100.00

TABLE 27.8

2 × 2 Tables for Bottom Five Deciles

Decile = 5 PROB_est = 0.1782

ACTUAL Frequency Percent	PREDICTED 0	 1	Total
0	0 0.00	1,436 89.69	1,436 89.69
1	0 0.00	165 10.31	165 10.31
Total	0 0.00	1,601 100.00	1,601 100.00

Decile = 6 PROB_est = 0.1427

ACTUAL Frequency Percent	PREDICTED 0	 1	Total
0	41 2.56	1,401 87.56	1,442 90.13
1	2 0.13	156 9.75	158 9.88
Total	43 2.69	1,557 97.31	1,600 100.00

Decile = 7 PROB_est = 0.1060

ACTUAL Frequency Percent	PREDICTED 0	 1	Total
0	0 0.00	1,345 84.01	1,345 84.01
1	4 0.25	252 15.74	256 15.99
Total	4 0.25	1,597 99.75	1,601 100.00

Decile = 8 PROB_est = 0.0682

ACTUAL Frequency Percent	PREDICTED 0	 1	Total
0	0 0.00	1,389 86.81	1,389 86.81
1	0 0.00	211 13.19	211 13.19
Total	0 0.00	1,600 100.00	1,600 100.00

Decile = 9 PROB_est = 0.02228

ACTUAL Frequency Percent	PREDICTED 0	 1	Total
0	0 0.00	1,401 87.56	1,401 87.56
1	0 0.00	199 12.44	199 12.44
Total	0 0.00	1,600 100.00	1,600 100.00

5. The 2 × 2 table cell entries **a, b, c, d,** and **N** by decile are in Table 27.9. Decile-precision requires the sum of each cell column (**a, b, c,** and **d**) across all deciles. Applying the formula for precision (**d/(b+d)**), I obtain RESPONSE model decile-precision of 22.22% in Table 27.10. The subroutines for reproducing Tables 27.9 and 27.10 are in Appendix 27.D.

   The RESPONSE model decile-precision of 22.22% is the supplemental statistic, which offers a compact performance measure instead of Cum Lifts at various depths of the file. I declare a thorough assessment of the performance of a response model should include the model decile-precision stamped on the original decile analysis.
6. I obtain the cell entries needed to apply the formula for chance decile-precision. Using the cell entries, I calculate chance decile-precision in Table 27.11.
7. The desired chance model decile-precision is the geometric mean of the decile-precision values in Table 27.11. The geometric mean is the tenth root of the product of the 10 decile-precision values. The chance model decile-precision is 17.78% in Table 27.12.
8. In the final analysis, the singular triplet of decile analysis metrics to tell the full performance of the RESPONSE model is in Table 27.13. The RESPONSE model precision is a significant 24.94% gain over the chance model precision. Thus, the performance of RESPONSE model is excellent. The subroutines for reproducing Tables 27.11, 27.12, and 27.13 are in Appendix 27.D.

TABLE 27.9

2 × 2 Table Cell Entries by Decile

Decile	a	b	c	d	N
1	0	482	12	1,106	1,600
2	0	963	0	637	1,600
3	9	1,260	1	331	1,601
4	0	1,282	0	318	1,600
5	0	1,435	0	165	1,600
6	0	1,436	0	165	1,601
7	41	1,401	2	156	1,600
8	0	1,345	4	252	1,601
9	0	1,389	0	211	1,600
10	0	1,401	0	199	1,600

TABLE 27.10

Model Decile-Precision from the Sums of 2 × 2 Cell Entries

MODEL_ Decile_ Precision	a	b	c	d	N
0.2222	50	12,394	19	3,540	16,003

TABLE 27.11

2 × 2 Table Cell Entries for Chance Decile-Precision

Decile	CHANCE_ Decile_ Precision	a	b	c	d
top	0.6965	0.00	30.13	0.75	69.13
2	0.3981	0.00	60.19	0.00	39.81
3	0.2080	0.56	78.70	0.06	20.67
4	0.1988	0.00	80.13	0.00	19.88
5	0.1031	0.00	89.69	0.00	10.31
6	0.1031	0.00	89.69	0.00	10.31
7	0.1002	2.56	87.56	0.13	9.75
8	0.1578	0.00	84.01	0.25	15.74
9	0.1319	0.00	86.81	0.00	13.19
bottom	0.1244	0.00	87.56	0.00	12.44

TABLE 27.12

Chance Model Decile-Precision

CHANCE_Decile_Precision
0.1778

TABLE 27.13

Response Model Decile-Precision

MODEL_ decile_ Precision	CHANCE_ decile_ Precision	DECILE_ PRECISION_ INCREMENT_GAIN
22.22%	17.78%	24.94%

27.5 Summary

Marketers use decile analysis to assess predictive incremental gains of their response models over responses obtained by chance. I argue, unbeknownst to marketers and apparently not in the literature, the correct evaluation of the decile analysis does not lie in the Cum Lift but in the metric precision as traditionally used in the 2 × 2 classification table. I define two new metrics, response model decile-analysis precision and chance model decile-analysis precision. I provide an illustration in a pedagogical manner to a fuller appreciation of the decile analysis, so users of the decile analysis can obtain a finer assessment of their response models. I provide the SAS subroutines for constructing the two new metrics and the proposed procedure, which will be a trusty tool for marketing statisticians.

Appendix 27.A Incremental Gain in Accuracy: Model versus Chance

```
libname da 'c://0-da';
data dec;
  set da.score;
  PREDICTED=0;
  if prob_hat > 0.222 then PREDICTED=1;
run;

data dec_;
  set dec end=last;
  wght=1; output;
  if last then do;
    predicted=0; wght=0; output;
    predicted=1; wght=0; output;
  end;
run;

PROC FREQ data=dec_;
  table RESPONSE*PREDICTED / norow nocol sparse out=D;
  weight wght / zeros;
run;

PROC TRANSPOSE data=D out=transp;
run;

data COUNT;
set transp;
if _NAME_="COUNT";
array col(4) col1-col4;
array cell(4) a b c d;
do i=1 to 4;
cell(i)=col(i);
drop i col1-col4;
end;
a_d=a+d;
N=a+b+c+d;
if _NAME_="COUNT" then MODEL_ACCURACY= (a+d)/(a+b+c+d);
if _NAME_="COUNT" then MODEL_GEOM_ACCURACY= SQRT((d/(c+d)) * (d/(b+d)));

drop a--d _NAME_ _LABEL_;
m=1;
run;

data PERCENT;
set transp;
if _NAME_="PERCENT";
array col(4) col1-col4;
array cell(4) a b c d;
```

```
do i=1 to 4;
cell(i)=col(i)/100;
drop i col1-col4;
end;
ab_sq=(a+b)**2;
cd_sq=(c+d)**2;
if _NAME_="PERCENT" then CHANCE_ACCURACY=( ((a+b)**2)+((c+d)**2));
drop a--d _NAME_ _LABEL_;
m=1;
run;

PROC SORT data=COUNT; by m;
PROC SORT data=PERCENT; by m;
run;

data PERCOUNT;
merge PERCENT COUNT; by m;
drop m;
keep MODEL_ACCURACY CHANCE_ACCURACY MODEL_GEOM_ACCURACY;
run;

data ACCURACY;
set PERCOUNT;
ACCURACY_INCREMENTAL_GAIN=
  ((MODEL_ACCURACY - CHANCE_ACCURACY)/CHANCE_ACCURACY);
run;

PROC PRINT data=ACCURACY;
var MODEL_ACCURACY CHANCE_ACCURACY
      ACCURACY_INCREMENTAL_GAIN;
format CHANCE_ACCURACY MODEL_ACCURACY
      ACCURACY_INCREMENTAL_GAIN percent8.2;
run;
```

Appendix 27.B Incremental Gain in Precision: Model versus Chance

```
libname da 'c://0-da';
options pageno=1;

data dec;
set da.score;
PREDICTED=0;
if prob_hat > 0.222 then PREDICTED=1;
run;

data dec_;
set dec end=last;
```

```
wght=1; output;
if last then do;
predicted=0; wght=0; output;
predicted=1; wght=0; output;
end;
run;

PROC FREQ data=dec_;
table RESPONSE*PREDICTED / norow nocol sparse out=D;
weight wght / zeros;
run;

PROC TRANSPOSE data=D out=transp;
run;

data Precision_IMPROV;
retain d;
set transp;
array col(4) col1-col4;
array cell(4) a b c d;
do i=1 to 4;
cell(i)=col(i);
drop i col1-col4;
end;

b_d=b+d;
if _NAME_="COUNT" ;
if _NAME_="COUNT" then MODEL_Precision= d/(b+d);
c_d=c+d;
N=a+b+c+d;
if _NAME_="COUNT" then CHANCE_Precision= (c+d)/N;
PRECISION_INCREMENTAL_GAIN=MODEL_Precision - CHANCE_Precision ;
drop a--c _NAME_ _LABEL_;
run;

PROC PRINT data=Precision_IMPROV;
var MODEL_Precision CHANCE_Precision PRECISION_INCREMENTAL_GAIN;
format MODEL_Precision CHANCE_Precision PRECISION_INCREMENTAL_GAIN
percent8.2;
run;
```

Appendix 27.C RESPONSE Model Decile PROB_est Values

```
libname ca 'c:\0-PCA_CCA';
options pageno=1;
title' completes X10-X14';

PROC LOGISTIC data=ca.completes nosimple des noprint outest=coef;
```

```
model RESPONSE = X10-X14;
run;

PROC SCORE data=ca.completes predict type=parms score=coef out=score;
var X10-X14;
run;

data score;
set score;
estimate=response2;
label estimate='estimate';
run;

data notdot;
set score;
if estimate ne .;

PROC MEANS data=notdot noprint sum; var wt;
output out=samsize (keep=samsize) sum=samsize;
run;

data scoresam (drop=samsize);
set samsize score;
retain n;
if _n_=1 then n=samsize;
if _n_=1 then delete;
run;

PROC SORT data=scoresam; by descending estimate;
run;

PROC FORMAT;
value Decile
0='top '
1='2'
2='3'
3='4'
4='5'
5='6'
6='7'
7='8'
8='9'
9='bottom';

data complete_probs;
set scoresam;
if estimate ne . then cum_n+wt;
if estimate = . then Decile=.;
else Decile=floor(cum_n*10/(n+1));
```

```
PROB_est=exp(estimate)/(1+ exp(estimate) );
keep PROB_est response wt Decile;
run;

PROC TABULATE data=complete_probs missing;
class Decile;
var response PROB_est;
table Decile all, (n*f=comma8.0 response *(mean*f=8.4 sum*f=comma8.0)
 (PROB_est) *(( min max)*f=8.4));
format Decile Decile.;
run;
```

Appendix 27.D 2 × 2 Tables by Decile

```
libname da 'c://0-da';
%let problist=%str(0.4602, 0.2654, 0.2354, 0.2138, 0.2008, 0.1782, 0.1427, 0.1060, 0.0682, 0.02228);
%macro dochance;
  %do k=0 %to 9;
    %let count=%eval(&k+1);
    %let prob=%qscan(&problist,&count,%str(','));
    %put k=&k prob=&prob;

  title2 "Decile=&k PROB_est=&prob ";
  data dec&k;
    set da.score;
    if dec=&k;
    ACTUAL=RESPONSE;
    PREDICTED=0;
    if prob_hat> &prob then PREDICTED=1;
  run;

  data dec_&k;
    set dec&k end=last;
    wght=1; output;
    if last then do;
      predicted=0; wght=0; output;
      predicted=1; wght=0; output;
    end;
  run;

 PROC FREQ data=dec_&k ;
 table ACTUAL*PREDICTED / norow nocol sparse out=D&k;
 weight wght / zeros;
 run;
```

```
PROC TRANSPOSE data=D&k out=transp&k;
run;
 %end;
%mend;

%doChance

title' ';
data PERCENT;
retain Decile;
set transp0 transp1 transp2 transp3 transp4
transp5 transp6 transp7 transp8 transp9;
if _NAME_="PERCENT" ;
if _NAME_="PERCENT" then CHANCE_Decile_Precision= (col4/(col2+col4));
drop _LABEL_ _NAME_;
Decile+1;
array col(4) col1-col4;
array cell(4) a b c d;
do i=1 to 4;
cell(i)=col(i);
drop i ;
end;
run;

PROC PRINT data=PERCENT noobs;
var Decile CHANCE_Decile_Precision a b c d;
title'Chance Decile Precision';
format a b c d 5.2;
format CHANCE_Decile_Precision 6.4;
run;

data CHANCE_Decile_Precision;
set PERCENT;
keep CHANCE_Decile_Precision;
PROC TRANSPOSE out=transCHANCE_Decile_Precision;
var CHANCE_Decile_Precision;
run;

PROC PRINT data=transCHANCE_Decile_Precision;
title 'data=transCHANCE_Decile_Precision';
run;

data CHANCE_Decile_Precision;
set transCHANCE_Decile_Precision;
CHANCE_Decile_Precision=
        geomean(col1, col2, col3, col4, col5, col6, col7, col8, col9, col10);
m=1;
keep CHANCE_Decile_Precision;
run;
```

```
PROC PRINT data=CHANCE_Decile_Precision noobs;
title' geometric mean - CHANCE_Decile_Precision ';
format CHANCE_Decile_Precision 6.4;
run;

title' ';
data COUNT;
retain Decile;
set
transp0 transp1 transp2 transp3 transp4
transp5 transp6 transp7 transp8 transp9;
if _NAME_="COUNT" ;
drop _LABEL_ _NAME_;
Decile+1;
array col(4) col1-col4;
array cell(4) a b c d;
do i=1 to 4;
cell(i)=col(i);
drop i ;
end;

N=a+b+c+d;
run;

PROC PRINT data=COUNT;
var Decile a b c d N;
title'COUNT';
run;

PROC SUMMARY data=COUNT;
var col1 col2 col3 col4;
output out=sum_counts sum=;
run;

data sum_counts;
set sum_counts;
_NAME_="COUNT";
drop _TYPE_ _FREQ_;
sum_counts=sum(of col1-col4);
N=sum(of col1-col4);
MODEL_Decile_Precision=col4 /( col2 + col4);
array col(4) col1-col4;
array cell(4) a b c d;
do i=1 to 4;
cell(i)=col(i);
drop i;
end;
m=1;
run;
```

```
PROC PRINT data=sum_counts;var MODEL_Decile_Precision a b c d N ;
format MODEL_Decile_Precision 6.4;
title 'sum_counts';
run;

PROC SORT data=CHANCE_Decile_Precision; by m;
PROC SORT data=sum_counts; by m;
run;

data PERCOUNTS;
merge CHANCE_Decile_Precision sum_counts;
run;

PROC PRINT data=PERCOUNTS;
title 'PERCOUNTS';
run;

data Decile_Precision_Gain;
set PERCOUNTS;
DECILE_PRECISION_INCREMENT_GAIN=
((MODEL_Decile_Precision - CHANCE_Decile_Precision)/CHANCE_Decile_Precision);
run;

title ' ';
PROC PRINT data=Decile_Precision_Gain noobs;
var MODEL_Decile_Precision CHANCE_Decile_Precision
        DECILE_PRECISION_INCREMENT_GAIN;
format MODEL_Decile_Precision CHANCE_Decile_Precision
        DECILE_PRECISION_INCREMENT_GAIN percent8.2;
title 'Decile_Precision_Gain';
run;
```

References

1. Landis, J.R., and Koch, G. G., The measurement of observer agreement for categorical data, *Biometrics*, 33(1), 159–174, 1977.
2. Kubat, M., Holte, R., and Matwin, S., Machine learning for the detection of oil spills in satellite radar images, *Machine Learning*, 30, 195–215, 1998.

28

Net T-C Lift Model: Assessing the Net Effects of Test and Control Campaigns

28.1 Introduction

Direct marketing statisticians assuredly have the task of building a response model based on a campaign, current or past. The response model provides a way of predicting responsive customers to select for the next campaign. Model builders use the customary decile analysis to assess the performance of the response model. Unwittingly, the modeler is assessing the predictive power of the response model against the chance model (i.e., random selection of individuals for the campaign). This assessment is wanting because it does not account for the impact of a nonchance control model on an individual's responsiveness to the campaign.

Using the basic principle of hypothesis testing, the proper evaluation of a test campaign involves conducting the test campaign along with a control campaign and appropriately analyzing the net difference between test and control outcomes within the framework of a decile analysis. There is large literature, albeit confusing and conflicting, on the methodologies of the net difference between test versus control campaigns. The purpose of this chapter is to propose an another approach, the *Net T-C Lift* Model, to moderate the incompatible literature on this topic, by offering a simple, straightforward, reliable model, easy to implement and understand. I provide the SAS© subroutines for the Net T-C Lift Model to enable statisticians to conduct net lift modeling without purchasing proprietary software. The subroutines are also available for downloading from my website: http://www.geniq.net/articles.html#section9.

28.2 Background

Two decades ago, the originative concept of *uplift* modeling to determine the net difference between test and control campaigns began. The first article published on this topic was by Decisionhouse [1]. The initial interest in the uplift model was not overwhelming. Disinterest in the new model not only baffled the developer of the model but the author* as well (personal communication with Quadstone salesperson, 2000). Nonrecognition of the new model merit reflects a short circuit between statistics education and the business world. All college graduates, including business majors, take the required basic Statistics 101 course, which covers the fundamentals of significance testing of differences. During the past 5 years, the model has received its well-deserved appreciation and has gained acceptance [2–5].

* I have been using net-effects modeling coincidentally since Decisionhouse's article was published.

There is large literature, albeit confusing, on the methodologies of testing the net difference between the probabilities of responding yes both to the test and control campaigns, denoted by *Diff_Probs*. The net difference approaches go by many names and use different algorithms. The names do not suggest which approach uses which algorithm. In addition, the types of approaches range from simple to complex. Critics of the simple approaches claim the results are always inferior to the results of the complex methods. The complex approaches have unfounded consistent superiority over their simpler counterparts. Theoretically based methods claim optimal results but produce nonunique results (e.g., a set of several solutions). In sum, theoretical net difference models produce good, suboptimal solutions comparable to heuristic net difference models. Empirical studies indicate the intricacy of the complex models is superfluous because its results are neither stable or without meaningful improvement over the simpler models.

There are many approaches of net effects models—uplift alternative model [6,7], decision trees [8,9], naïve Bayes [10,11], net lift model [12], differential response model [13], and association rules [14], to mention a few. Irrespective of the approach, the model is variously referred to as an uplift model, a differential response model, incremental impact model, and a true lift model. To lessen the literature muddle, I brand the proposed approach *net-effects* and use the suggestive algorithm name, *Net T-C Lift Model*. The Net T-C Lift Model is true to the original definition of the net incremental (lift) improvement of a test campaign over a control campaign within the framework of the decile analysis.

All net-effects models either directly or indirectly use the two-model paradigm as their underpinning. The list of weaknesses of the two-method conceptual framework is quite a barrelful [15]. The disadvantages are:

1. The set of predictor variables must be the same for both the TEST and CONTROL models.
2. The set of predictor variables of the TEST model must be different than the set of predictor variables of the CONTROL model.
3. The set of predictor variables of the TEST model must include the set of predictor variables of the CONTROL plus have another set of predictor variables specific to the TEST model.
4. Variable selection is especially important.
5. Nonlinearity is particularly important.
6. TEST and CONTROL group sizes must be equal.
7. The net-effects model is unsatisfactory with *small* net-effects.
8. The fitness function Diff_Probs is not explicitly optimized; that is, the Diff_Probs is a surrogate for the true Diff_Probs, which is a censored variable.
    a. True. However, most applications of statistical modeling use surrogates and yield excellent results.
    b. Lest one forgets, consider logistic regression (LR), the workhorse for response modeling. The LR fitness function is the log-likelihood function, which serves as a surrogate for the classical confusion matrix or its derivative the decile analysis.
    c. There are an uncountable number of success stories of decile analysis assessment of LR, for response modeling as well as for nonmarketing applications.
    d. Well-chosen surrogates perform well—not necessarily optimal, but near optimal—and serve as prime performers.

The proposed Net T-C Lift Model, plainly a two-model method, has the following features:

1. The Net T-C Lift Model works: Its foundation is Fisher's formalized set of methods that evolved into the practice of hypothesis testing.
2. Net T-C Lift Model's pathway to its reliability is randomization testing and real study feedback-loop tweaking, producing the production version.
3. The set of predictor variables for either TEST or CONTROL can be any mixture of variables.
4. Variable selection and new variable construction are no more important than for building any model.
5. TEST and CONTROL group sizes can be unequal or equal.
6. Net-effects can be small.
7. The surrogacy of Diff_Probs is tightened to get close to the true fitness function by the net T-C lift algorithm.
8. The only net-effects model that implicitly uses the true Diff_Probs fitness function is the GenIQ TAC Model, based on the GenIQ Model of Chapter 40. GenIQ TAC is proprietary and requires a long learning curve.

The benefits of the net-effects model are:

1. Statistically, targeting customers whose responsiveness has the highest positive impact from a solicitation, by maximizing the net incremental gain of test campaign over control campaign
2. Balancing the expected net incremental gains within budgeted spending levels
3. Selection of a statistically gainful list of responsive customers to target, rendering an operational decrease in some unfavorable responses
4. Construction of a rich list of customers, who are likely to respond without solicitation
5. Generation of do-not-disturb customers as they become inactive customers

28.3 Building TEST and CONTROL Response Models

Direct marketers are always implementing test campaigns as part of their company's marketing strategy to increase response and profit. Implementing a test campaign, which offers, say, the latest smart thing at a discounted price or maybe a coupon to get a good deal on the product, starts by defining a target population that is relevant to the offering. The initial target population, a subset drawn from the marketer's database (population), consists of customers with a hearty blend of product preferences for the offering. The initial target population is fine-tuned based on a set of offering-related pre selects (e.g., males between 18 and 35 years old and college graduates). The initial target population, although part of the population, is still assuredly large. Thus, the test target sample is randomly drawn from the initial target population such that its size, the test group size, is typically from 15,000 to 35,000 customers. The test target group is the test campaign sample.

Similar to the test sample, a control sample comes from a random sample of the initial target population. The control, as compared with the test, has two modifications—one minor and one major. The minor change is in the size of the control group, i.e., smaller than the test group, typically from 10,000 to 20,000 customers. The major change is that the control campaign does *not* offer anything, no smart things or the like, no coupons, and so on. The control campaign is effectively a communication element of the company's general product line.

The test and control campaigns are "dropped" and, say, 6 weeks later, a collection of most of the responses is complete. The intermediary marketing effort is measured by the number of positive responses. The direct marketer assesses test responses by comparing test and control response rates.

Assuming the test campaign outperforms the control campaign, the real value of the marketing effort lies in building a response model on the test campaign data (i.e., a test response model). The test response model serves as an instrument for achieving future well-performing campaigns. Specifically, the building of the test response model is the vehicle for statistically smart selection of customers, who have a high propensity to respond favorably to the next campaign. The working assumption of the implementation of the test response model is that the target population has not considerably changed, and the offering is the same or a similar smart thing.

28.3.1 Building TEST Response Model

I have the data from the test and control smart-thing campaigns. The test campaign group consists of 30,000 plus customers, and the control campaign group consists of 20,000 customers. I randomly split the two datasets into training and validation datasets for both test and control. The candidate predictor variables include the standard recency–frequency–monetary value variables, along with purchase history, and usual demographic, socio-economic, and lifestyle variables. I rename the original variable names to Xs to avoid the unnecessary distraction of checking face validity of the model. It suffices to say that I build the best possible model, adhering to the topics covered in

1. Chapter 10, Section 10.14, Visual Indicators of Goodness of Model Predictions
2. Chapter 26, Assessment of Marketing Models, via performance measures based on the decile analysis

I present the test response modeling process, step by step.

1. I define the dependent variable TEST for the test campaign:
   TEST = 1, if customer responds "yes," i.e., made a purchase
   TEST = 0, if customer responds "no," i.e., made no purchase
2. Using the test training dataset, I perform variable selection to determine the best set of predictor variables and use Proc Logistic and the decile analysis subroutine in Appendix 28.A.
   a. The analysis of maximum likelihood estimators (MLEs) and the TEST decile analysis, based on the validation dataset, are in Tables 28.1 and 28.2, respectively. The estimates indicate all statistically significant variables define the TEST model. The popular measures of goodness-of-fit (not shown) also indicate the model is well defined. The popular measures of goodness-of-fit (not shown) are also significant.

TABLE 28.1

TEST Maximum Likelihood Estimators

Parameter	DF	Estimate	Standard Error	Wald Chi-Square	Pr > ChiSq
Intercept	1	−1.9390	0.0283	4699.7650	<0.0001
X14	1	0.4936	0.1378	12.8309	0.0003
X23	1	2.7311	0.8352	10.6926	0.0011
X25	1	1.6004	0.5495	8.4840	0.0036
X26	1	0.3541	0.1058	11.2046	0.0008
X36	1	−0.1368	0.0374	13.3525	0.0003
X41	1	−0.1247	0.0372	11.2192	0.0008
X42	1	0.1976	0.0569	12.0531	0.0005
X49	1	1.6366	0.5313	9.4896	0.0021

TABLE 28.2

TEST Decile Analysis

Decile	TEST Group (n)	TEST Count (n)	TEST RATE (%)	CUM TEST RATE (%)	CUM LIFT (%)
top	1,545	255	16.50	16.50	136
2	1,546	197	12.74	14.62	120
3	1,545	193	12.49	13.91	115
4	1,546	193	12.48	13.56	112
5	1,545	193	12.49	13.34	110
6	1,546	199	12.87	13.26	109
7	1,546	186	12.03	13.09	108
8	1,545	172	11.13	12.84	106
9	1,546	153	10.28	12.56	103
bottom	1,545	129	8.35	12.14	100
Total	15,455	1,876			

b. The TEST decile analysis, the true arbitrator of model performance,* indicates the model is a little problematic. The model produces almost decreasing monotonicity in its counts (in the third column), except for Decile 6, where there is a slight jump up to 199.

c. The TEST top-bottom decile ratio of 1.98 (=16.50%/8.35%) indicates a good model with satisfactory discrimination of the customers across the deciles.

The construction and reading of a decile analysis are in Chapter 26. Here, I review the interpretation of a decile analysis with a focus on the objective of the proposed Net T-C Lift Model. The direct marketer observes the validation dataset consists of 15,455 customers, of whom 1,876 favorably respond to the campaign, in Table 28.2. The overall TEST response rate is 12.14% (penultimate column, last row). The top Cum Lift of 136 indicates the top 10% of the file has the most responsive customers, whose response rate is 1.36 times (or 36% greater than) the overall TEST response rate. The second Cum Lift of 120 indicates the top 20% of the most responsive customers have a response rate 1.20 times (or 20% greater than) the overall TEST response rate. For the remaining deciles, the interpretation of the Cum Lift is similar.

* This assertion is the opinion of the author, who has present reasons and gained a long list of statisticians who agree.

TABLE 28.3
No-Model CONTROL of TEST Decile Analysis

Decile	TEST Group (n)	TEST Count (n)	TEST RATE (%)	No-Model CUM TEST RATE (%)	No-Model CUM LIFT (%)
top	1,545	192	12.43	12.43	102
2	1,546	189	12.23	12.33	102
3	1,545	191	12.36	12.34	102
4	1,546	203	13.13	12.54	103
5	1,545	160	10.36	12.10	100
6	1,546	190	12.29	12.13	100
7	1,546	199	12.87	12.24	101
8	1,545	200	12.94	12.33	102
9	1,546	181	11.71	12.26	101
bottom	1,545	171	11.07	12.14	100
Total	15,455	1,876			

I digress with two points relevant to the net-effects approach. First, from the fundamentals of hypotheses testing, the statistician uses a two independent sample method to measure the effectiveness of treatment (TEST) over CONTROL groups. (The well-known t-test determines the significance of the difference between the means of two independent samples.) For the task at hand, the net-effects approach is appropriate to test the difference between two independent decile analyses. Accordingly, Cum *Net* Lift is the focus of assessing the net difference between TEST and CONTROL campaigns within the framework of the decile analysis.

Second, to set the stage for generating net-effects duo-deciles, I find it instructive to construct a CONTROL decile analysis based on *no model* (chance) for the TEST campaign itself in Table 28.2. The No-Model CONTROL decile analysis, Table 28.3, by definition randomly distributes the 1,876 responses across the deciles. (The Cum Lifts would be exactly 100% for each decile if the TEST count sizes in Column 3 were divisible by 10.) It is clear the No-Model CONTROL of TEST decile analysis, Table 28.3, is randomly generated by chance. The No-Model CUM TEST RATE (%) per decile is principally 12.14% (randomly varying from 12.10% to 12.54%). Thus, assessing the TEST campaign itself is tantamount to comparing the TEST and No-Model CONTROL decile analyses. For each decile, dividing the CUM TEST RATE (Column 5) in Table 28.2 by the No-Model CUM TEST RATE (Column 5) in Table 28.3 yields the equivalent results (values at or close to 100; namely, 100–103) in TEST CUM LIFT. The equivalent results are due to the chance variation of the No-Model CONTROL decile analysis.

28.3.2 Building CONTROL Response Model

I present the step-by-step CONTROL response modeling process before I introduce the NET T-C Lift Model.

1. I define the dependent variable CONTROL for the test campaign:

   CONTROL = 1 if customer responded "yes," i.e., made a purchase

   CONTROL = 0 if customer responded "no," i.e., made no purchase

2. Using the control training dataset, I perform variable selection to determine the best set of predictor variables and use Proc Logistic and the decile analysis in subroutine Appendix 28.B. The LR estimates are in Table 28.4, and the CONTROL decile analysis on the validation dataset is in Table 28.5. The estimates indicate all statistically significant variables define the CONTROL model. The popular necessary-not-sufficient goodness-of-fit measures (not shown) are also significant. And, the decile analysis confirms the model is good, in that the model produces reasonably monotonic decreasing counts.

For the sake of completeness, I review the interpretation of the CONTROL decile analysis with attention to the objective of the NET T-C Lift Model. The validation CONTROL dataset consists of 10,000 customers, of whom 919 favorably responded to the campaign. The CONTROL response rate is 9.19%. The top Cum Lift indicates the top 10% of the file has the most responsive customers with a response rate 1.27 times (or 27% greater than) the CONTROL response rate of 9.19%. The second Cum Lift indicates the top 20% of the file has the most responsive customers with a response rate 1.19 times (or 19% greater than) the CONTROL response rate of 9.19%. For the remaining deciles, the interpretation of the Cum Lift is similar.

TABLE 28.4

CONTROL Maximum Likelihood Estimators

Parameter	DF	Estimate	Standard Error	Wald Chi-Square	Pr > ChiSq
Intercept	1	−2.2602	0.0397	3241.3831	<0.0001
X17	1	1.2055	0.4287	7.9059	0.0049
X23	1	2.0529	0.8849	5.3826	0.0203
X29	1	1.3911	0.4730	8.6513	0.0033
X37	1	−0.0482	0.0188	6.5955	0.0102
X42	1	0.2146	0.0772	7.7227	0.0055

TABLE 28.5

CONTROL Decile Analysis

Decile	Number of Customers	Number of Responses	CONTROL RATE (%)	CUM CONTROL RATE (%)	CUM LIFT (%)
top	1,000	117	11.70	11.70	127
2	1,000	102	10.20	10.95	119
3	1,000	98	9.80	10.57	115
4	1,000	93	9.30	10.25	112
5	1,000	90	9.00	10.00	109
6	1,000	92	9.20	9.87	107
7	1,000	87	8.70	9.70	106
8	1,000	91	9.10	9.63	105
9	1,000	78	7.80	9.42	103
bottom	1,000	71	7.10	9.19	100
Total	10,000	919			

28.4 Net T-C Lift Model

Two decades ago, the issue of properly assessing the net effects of TEST versus CONTROL campaigns finally came to the fore. The central concept of the difference between TEST and CONTROL at the individual level is in equivalent equations (Equations 28.1 and 28.2, respectively).

$$\text{Diff\_Probs} = \text{Prob (Response} = 1 \mid \text{TEST=1)} - \text{Prob (Response} = 1 \mid \text{CONTROL=1)} \quad (28.1)$$

$$\text{Diff\_Probs} = \text{Prob (Response} = 1 \mid \text{Offer)} - \text{Prob (Response} = 1 \mid \text{No Offer)} \quad (28.2)$$

As simple as the individual-level equations appear, a closer look reveals they require the impossibility of the same individual to receive both an offer and no offer. Surmounting the inability to operationalize a mathematical calculation to proceed toward a solution, the method of measuring Diff_Probs, actually mean of Diff_Probs, is based on individuals who are in the same neighborhoods (i.e., deciles) for both the TEST and CONTROL groups. For example, consider customers BR and AR, the only two customers in the top decile. Their probabilities of response to the TEST and CONTROL campaigns are:

- BR has Prob (Response = 1 | TEST=1) = 0.78. BR buys from the offer campaign.
- AR has Prob (Response = 1 | CONTROL=1) = 0.48. AR buys from no-offer campaign.

Thus, the mean of Diff_Probs for top decile is =0.30 (= 0.78 – 0.48).

There are many net-effects models that use different operationalizations of Diff_Probs. These models go by various names, such as the uplift model, incremental impact models, true lift model, and now the NET T-C Lift Model. Although the models use the focal equations (Equations 28.1 and 28.2), software developers of their models put their proprietary code within their algorithms. Some models are too simple and provide poor results. The complex models have nice extensions of the net-effects concept, but they also perform poorly due to the untenable assumptions, which are never satisfied in real studies. Each net-effects model is in its developer's proprietary software package. Commercial models are for sale (with high price tags). Other net-effects models use free source software. I discuss the problems of free source software (in Section 43.2.1). Briefly, there is no vetting of free source, no minimal-level assurance of reliability, and no means of direct support. In other words, *caveat utilitor* (let the user beware).

The Net T-C Lift Model is free to users who have BASE SAS and SAS/STAT. Net T-C Lift is accurate and reliable, and accommodates virtually all analytic scenarios, which the other net-effects models cannot address. For example, the other net-effects models require equal-sized control and test groups. This requirement is not always practical because marketers do not like to incur costs for control groups, which by definition offer no expectation of yielding profit. Lastly, Net T-C Lift is demonstratively faithful in its output because its report format is the decile analysis with the addition of columns related to net-effects. Diff_Probs are not in the standard Net T-C Lift decile analysis output, but the actual response counts are in the decile analysis. The decile analysis can include Diff_Probs by easily modifying the code.

28.4.1 Building the Net T-C Lift Model

The topline results of the TEST and CONTROL response models are in Tables 28.2 and 28.5, respectively. Now, I present the two models in detail, step by step, required for the process of building the Net T-C Lift Model:

1. Build a TEST response model on a training dataset and validate it on a hold-out (validation) dataset. (This task is discussed in Section 28.3.1.)
   a. The TEST model is defined directly in logits, and equivalently probabilities, in Equations 28.3 and 28.4, respectively:

$$\text{logit(TEST=1)} = -1.93 + 0.49^*\text{X14} + 2.73^*\text{X23} + 1.60^*\text{X25} + 0.35^*\text{X26} - 0.13^*\text{X36} - 0.12^*\text{X41} + 0.19^*\text{X42} + 1.63^*\text{X49} \tag{28.3}$$

$$\text{prob\_TEST} = \exp(\text{logit(TEST)})/(1 + \exp(\text{logit(TEST)})) \tag{28.4}$$

2. Build a CONTROL response model on a training dataset and validate it on a hold-out dataset. (This task is outlined in Section 28.3.2.)
   a. The CONTROL model is defined directly in logits, and equivalently probabilities, as:

$$\text{logit(CONTROL=1)} = -2.26 + 1.20^*\text{X17} + 2.05^*\text{X23} + 1.39^*\text{X29} - 0.048^*\text{X37} + 0.21^*\text{X42} \tag{28.5}$$

$$\text{prob\_CONTROL} = \exp(\text{logit(CONTROL)})/(1 + \exp(\text{logit(CONTROL)})) \tag{28.6}$$

3. Score the hold-out datasets for the final Net T-C Lift Model, using the prob_TEST and prob_CONTROL in Equations 28.4 and 28.6.
4. Calculate Diff_Probs = prob_TEST – prob_CONTROL, by running the subroutine in Appendix 28.C.
5. Run the Net T-C Lift Model subroutine in Appendix 28.D. The Diff_Probs is the metric of the NET T-C Decile Analysis. The *net* results of the Net T-C Lift Model are in Table 28.6 and are discussed in the next section.

28.4.1.1 Discussion of the Net T-C Lift Model

I use hold-out datasets for validating the NET T-C Lift Model.

1. The datasets consist of TEST and CONTROL group of sizes 15,500 and 10,006, respectively.
2. TEST and CONTROL group sizes for the top three deciles are glaringly unequal. Clearly, the poorly balanced sizes are because the total group sizes are unequal. There is no way to evenly squeeze 10,006 individuals across 15,500 individuals. The model dictates how the distribution of the CONTROL individuals falls within the deciles.

TABLE 28.6

Net T-C Lift Model (Unequal Group Sizes)

Decile	TEST Group (N)	CONTROL Group (N)	TEST Count (N)	CONTROL Count (N)	Net T-C Count (N)	Net T-C Improv (%)	Cum Net T-C Count (N)	Cum Net T-C Lift (%)
top	1,550	131	178	10	168	10.8	168	17.3
2	1,550	354	210	33	177	11.4	345	35.6
3	1,550	705	210	50	160	10.3	505	52.1
4	1,550	1,218	208	110	98	6.3	603	62.2
5	1,550	1,350	194	117	77	5.0	680	70.1
6	1,550	1,550	195	129	66	4.3	746	76.9
7	1,550	1,349	184	146	38	2.5	784	80.8
8	1,550	1,137	174	81	93	6.0	877	90.4
9	1,550	1,050	182	115	67	4.3	944	97.3
bottom	1,550	1,162	150	124	26	1.7	970	100.0
	15,500	**10,006**	**1,885**	**915**				
			Overall NET T-C: 3.0%					

3. "Overall NET T-C: 3.0%" is stamped below the column sums. Of course, the overall statistic is TEST response rate of 12.1% (=1885/15,500) minus CONTROL response rate of 9.1% (=915/10,006).
4. The focal point for model assessment is NET T-C Count, which indicates good performance of response improvement of TEST over CONTROL by decile. Similar to a regular decile analysis, the decreasing monotonicity of the relevant column, which in this case is NET T-C Count, is an indicator a good performance.
   a. There are only two moderate jumps in NET T-C Count, decreases in counts (i.e., second-decile 177 jumps up from top-decile 168 and eighth-decile 93 jumps up from seventh-decile 38).
   b. The spread of NET T-C Count from the top-two decile average 172.5 (= (168+177)/2) to the bottom-two decile average 46.5 (= (67+26)/2) produces an impressive top-two to bottom-two ratio of 3.7 (=172.5/46.5). This ratio indicates the model's good discriminating power of net-effects throughout the deciles.
   c. NET T-C Lift Model is declared *contingently* good, as will be discussed at the end of this section.
   d. NET T-C IMPROV equals NET T-C Count(n)/TEST Group decile size. For example, for top decile, NET T-C IMPROV is 10.8% (=168/1550).
5. CUM NET T-C Count is an instrumental measure in which its value shows where the determinant CUM NET T-C Lift indicates how net-effects improvement cumulates throughout various depths-of-file.
   a. For example, the top decile captures 17.3% (=168/970) of total net improvement predicted by the model. Similarly, the top two deciles capture 35.67% (=345/970) of total net improvement based on the model. For the remaining deciles, the interpretation of CUM NET T-C Lift is similar.

The NET T-C Lift Model performance is contingently good, as previously stated. Depending on the cost–benefit analysis, discussed here, implementation of the model for future solicitations is applied at a cost-effective depth-of-file. For example, if the cost–benefit analysis indicates the top four deciles are significantly above the breakeven point, then the expected net improvement of TEST over CONTROL is 62.2%.

NET T-C Lift Model performance is considered contingently good because of the performance of the model, and value to the company lies in the cost–benefit analysis. The final cost–benefit analysis, which incorporates any one or more bottom-line values such as profit, gross and net revenue, and return on investment (ROI), is the ultimate determinant. The NET T-C Lift Model Decile Analysis shows the model has better-than-average quantitative functioning, based on standard statistical modeling practices. The NET T-C Lift Model predicts, in part, response counts. However, depending on the profit-related value of a single response, the decile-aggregated counts yield the all-important performance of the NET T-C Lift Model itself. The latter is the foundation of the final analysis. Given the NET T-C Lift Model foundation is solid, the model will be an absolute good performer.

28.4.1.2 Discussion of Equal-Group Sizes Decile of the Net T-C Lift Model

The discussion in this section essentially reflects the debate among developers of the various approaches of net-effects models. Most net-effects software requires equal sizes of the TEST and CONTROL groups. This position is at odds with direct-marketing campaign designs. It is virtually standard practice to set the control group size as small as possible for the simple reason that control campaigns do not yield any significant gains; it is a money loser—only a necessity of having a traditional benchmark against which to measure a test campaign.

Generating an equal group net-effects decile analysis, I rerun the NET T-C Lift Model, deleting the extra individuals of the test group randomly. The resultant decile analysis is in Table 28.7.

I compare the decile analyses with unequal and equal group sizes in Tables 28.6 and 28.7, respectively. Again, the focal column is NET T-C Count. The equal group size decile has two jumps in the fifth and eighth deciles with values of 40 (from 31 in the sixth decile) and 36 (from 5 in the seventh decile). These jumps can be considered a bit bumpier than the two jumps in the unequal group size decile. (See the aforementioned Step #4a.) The comparative findings have no real diagnostic value.

CUM NET T-C LIFT of the unequal size and equal size groups do have comparative value. CUM NET T-C LIFT of the equal group has larger values than those of the unequal group.

a. For the equal group decile analysis, CUM NET T-C LIFT ranges from 27.9% (top decile) to 82.3% (the actionable fifth decile).
b. For the unequal group decile analysis, CUM NET T-C LIFT ranges from 17.3% (top decile) to 70.1% (the actionable fifth decile).

The implication is that the equal group model captures 17.4% (=(82.3%–70.1%)/70.1%) increase net improvement—of actual response counts. Of further comparative value, the equal group NET T-C Lift Model identifies negative net-effects, albeit, nominal counts of 1 and 12 in the bottom two deciles, respectively. The implication is that implementing the NET T-C Lift Model with equal size groups indicates excluding the bottom 20% of the file.

Arithmetic note: CUM NET T-C LIFT values in the bottom two deciles have awkward values greater than 100%, 104.3% and 103.4%, respectively. These values are an artifact of the negative net-effects.

28.5 Summary

Net-effects modeling, after its introduction 20 years ago, has finally gained the recognition it deserves because it is a statistically logical approach of modeling net improvement of a test campaign, accounting for the proper integration of a control campaign. The net-effects modeling has an important body of literature, albeit confusing and conflicting. I hope this chapter provides the reader with the current span of research done on this worthy topic as well as sorting out the unclear and mixed messages of the research findings. I propose an another approach, the *Net T-C Lift Model,* to moderate the incompatible literature on this topic, by offering a simple, straightforward, reliable model, easy to implement and understand. I provide the SAS subroutines for the Net T-C Lift Model to enable statisticians to conduct net lift modeling without purchasing proprietary software.

TABLE 28.7

Net T-C Lift Model (Equal Group Sizes)

Decile	TEST Group (N)	CONTROL Group (N)	TEST Count (N)	CONTROL Count (N)	Net T-C Count (N)	Net T-C Improv (%)	Cum Net T-C Count (N)	Cum Net T-C Lift (%)
top	1,000	1,000	166	81	85	8.5	85	27.9
2	1,001	1,001	133	76	57	5.7	142	46.6
3	1,001	1,001	136	98	38	3.8	180	59.0
4	1,000	1,000	116	85	31	3.1	211	69.2
5	1,001	1,001	125	85	40	4.0	251	82.3
6	1,001	1,001	134	108	26	2.6	277	90.8
7	1,000	1,000	92	87	5	0.5	282	92.5
8	1,001	1,001	114	78	36	3.6	318	104.3
9	1,001	1,001	104	105	−1	−0.1	317	103.9
bottom	1,000	1,000	100	112	−12	−1.2	305	100.0
	10,006	10,006	1,220	915				
				Overall NET T-C: 3.0%				

Appendix 28.A TEST Logistic with Xs

```
libname upl 'c://0-upl';
options pageno=1;

%let depvar=TEST;
%let indvars= X14 X23 X25 X26 X36 X41 X42 X49;
PROC LOGISTIC data= upl.upl_datanumkpX nosimple des outest=coef;
model &depvar = &indvars;
run;

PROC SCORE data=upl.upl_datanumkpX predict type=parms score=coef out=score;
var &indvars;
run;

data score;
set score;
logit=&depvar.2;
prob_TEST=exp(logit)/(1+ exp(logit));

data score;
set score;
estimate=&depvar.2;
run;

data notdot;
set score;
if estimate ne .;
run;

PROC MEANS data=notdot sum noprint; var wt;
output out=samsize (keep=samsize) sum=samsize;
run;

data scoresam (drop=samsize);
set samsize score;
retain n;
if _n_=1 then n=samsize;
if  _n_=1  then  delete;
run;

PROC SORT data=scoresam; by descending estimate;
run;

data score;
set scoresam;
if estimate ne . then cum_n+wt;
if estimate = . then dec=.;
```

```
else dec=floor(cum_n*10/(n+1));
run;

PROC SUMMARY data=score missing;
class dec;
var &depvar wt;
output out=sum_dec sum=sum_can sum_wt;

data sum_dec;
set sum_dec;
avg_can=sum_can/sum_wt;
run;

data avg_rr;
set sum_dec;
if dec=.;
keep avg_can;
run;

data sum_dec1;
set sum_dec;
if dec=. or dec=10 then delete;
cum_n +sum_wt;
r =sum_can;
cum_r +sum_can;
cum_rr=(cum_r/cum_n)*100;
avg_cann=avg_can*100;
run;

data avg_rr;
set sum_dec1;
if dec=9;
keep avg_can;
avg_can=cum_rr/100;
run;

data scoresam;
set avg_rr sum_dec1;
retain n;
if _n_=1 then n=avg_can;
if _n_=1 then delete;
lift=(cum_rr/n);
if dec=0 then decc=' top ';
if dec=1 then decc=' 2  ';
if dec=2 then decc=' 3 ';
if dec=3 then decc=' 4 ';
if dec=4 then decc=' 5 ';
```

```
if dec=5 then decc=' 6 ';
if dec=6 then decc=' 7  ';
if dec=7 then decc=' 8 ';
if dec=8 then decc=' 9  ';
if dec=9 then decc='bottom';
if dec ne .;
run;

title2' Decile Analysis based on ';
title3" &depvar Regressed on &indvars ";

PROC PRINT data=scoresam d split='*' noobs;
var decc sum_wt r avg_cann cum_rr lift;
label decc='DECILE'
sum_wt ='NUMBER OF*CUSTOMERS'
r ='NUMBER OF*RESPONSES'
cum_r ='CUM No. CUSTOMERS w/* RESPONSES'
avg_cann ='TEST*RATE (%)'
cum_rr ='CUM TEST* RATE (%)'
lift =' C U M *LIFT (%)';
sum sum_wt r;
format sum_wt r cum_n cum_r comma10.;
format avg_cann cum_rr 5.2;
format lift 3.0;
run;

Data upl.score_RESP;
set scoresam;
run;
```

Appendix 28.B CONTROL Logistic with Xs

```
libname upl 'c://0-upl';
options pageno=1 ;
title ' ';
title2 'CONTROL';
%let depvar=CONTROL;
%let indvars= X17 X23 X29 X37 X42 ;
title3  "adjust_n  =  &Control_n";
title4 "adjust_dot = &Contrl_dot";

PROC LOGISTIC data= upl.CONTROL nosimple des outest=coef;
model &depvar = &indvars;
run;

PROC SCORE data=upl.CONTROL predict type=parms score=coef out=score;
var &indvars;
run;
```

```
data score;
set score;
logit=&depvar.2;
prob_CONTROL=exp(logit)/(1+ exp(logit));
data score;
set score;
estimate=&depvar.2;
run;

data notdot;
set score;
if estimate ne .;

PROC MEANS data=notdot sum noprint; var wt;
output out=samsize (keep=samsize) sum=samsize;
run;

data scoresam (drop=samsize);
set samsize score;
retain n;
if _n_=1 then n=samsize;
if  _n_=1  then  delete;
run;

PROC SORT data=scoresam; by descending estimate;
run;

data score;
set scoresam;
if estimate ne . then cum_n+wt;
if estimate = . then  dec=.;
else dec=floor(cum_n*10/(n+1));
run;

PROC SUMMARY data=score missing;
class dec;
var &depvar wt;
output out=sum_dec sum=sum_can sum_wt;

data sum_dec;
set sum_dec;
avg_can=sum_can/sum_wt;
run;

data avg_rr;
set sum_dec;
```

```
if dec=.;
keep avg_can;
run;

data sum_dec1;
set sum_dec;
if dec=. or dec=10 then delete;
cum_n +sum_wt;
r =sum_can;
cum_r +sum_can;
cum_rr=(cum_r/cum_n)*100;
avg_cann=avg_can*100;
run;

data avg_rr;
set sum_dec1;
if dec=9;
keep avg_can;
avg_can=cum_rr/100;
run;

data scoresam;
set avg_rr sum_dec1;
retain n;
if _n_=1 then n=avg_can;
if _n_=1 then delete;
lift=(cum_rr/n);
if dec=0 then decc=' top ';
if dec=1 then decc=' 2  ';
if dec=2 then decc=' 3 ';
if dec=3 then decc=' 4 ';
if dec=4 then decc=' 5 ';
if dec=5 then decc=' 6 ';
if dec=6 then decc=' 7 ';
if dec=7 then decc=' 8 ';
if dec=8 then decc=' 9  ';
if dec=9 then decc='bottom';
if dec ne .;
run;

title5" &depvar Regressed on &indvars ";
PROC PRINT data=scoresam d split='*' noobs;
var decc sum_wt r avg_cann cum_rr lift;
label decc='DECILE'
sum_wt ='NUMBER OF*CUSTOMERS'
r ='NUMBER OF*RESPONSES'
cum_r ='CUM No. CUSTOMERS w/* CONTROLS'
```

```
avg_cann ='CONTROL *RATE (%)'
cum_rr ='CUM CONTROL * RATE (%)'
lift =' C U M *LIFT (%)';
sum sum_wt r;
format sum_wt r cum_n cum_r comma10.;
format avg_cann cum_rr 5.2;
format lift 3.0;
run;
footnote;

Data upl.score_CNTRL;
set scoresam;
run;
```

Appendix 28.C Merge Score

```
libname upl 'c://0-upl';

data score_RESP_uni;
set upl.score_RESP;
uni=uniform(12345);

data score_CNTRL_uni;
set upl.score_CNTRL;
uni=uniform(12345);

PROC SORT data=score_RESP_uni; by uni;
PROC SORT data=score_CNTRL_uni; by uni;
run;

data RESP_CNTRL_scores_uni;
merge
score_RESP_uni (in=r)
score_CNTRL_uni (in=c); by uni;
if r=1 then wtT=1; else wtT=0;
if c=1 then wtC=1; else
wtC=0;
run;

data upl.diff_probs_uni;
set RESP_CNTRL_scores_uni;
diff_probs=prob_RESPONSE-prob_CONTROL;
run;

PROC MEANS data=upl.diff_probs_uni n nmiss min max mean;
var diff_probs prob_RESPONSE prob_CONTROL;
run;
```

Appendix 28.D NET T-C Decile Analysis

```
libname upl 'c://0-upl';
options pageno=1 ps=33;

data score;
set upl.diff_probs_uniBS;
estimate=diff_probs;
do until (-0.435 < uni < 0.12345);
uni=uniform(12345);
end;
if estimate=. then estimate=uni;
TEST=RESPONSE;
keep _n_ wt wtC TEST CONTROL estimate;
run;

data notdot;
set score;
PROC MEANS data=notdot sum noprint; var wt;
output out=samsize (keep=samsize) sum=samsize;
run;

data scoresam (drop=samsize);
set samsize score;
retain n;
if _n_=1 then n=samsize;
if _n_=1 then delete;
run;

PROC SORT data=scoresam; by descending estimate;
run;

data score;
set scoresam;
if estimate ne . then cum_n+wt;
if estimate = . then dec=.;
else dec=floor(cum_n*10/(n+1));
if dec=. then delete;
run;

PROC SUMMARY data=score missing;
class dec;
var TEST wt;
output out=sum_decT sum=sum_canT sum_wtT;
run;

data sum_decT;
set sum_decT;
```

```
avg_canT=sum_canT/sum_wtT;
run;

data avg_rrT;
set  sum_decT;
if dec=.;
keep avg_canT;
run;

data sum_dec1T;
set  sum_decT;
if dec=. or dec=10 then delete;
cum_nT +sum_wtT;
rT     =sum_canT;
cum_rT +sum_canT;
cum_rrT=(cum_rT/cum_nT)*100;
avg_cannT=avg_canT*100;
run;

data scoresamT;
set  avg_rrT sum_dec1T;
retain n;
if _n_=1 then n=avg_canT;
if _n_=1 then delete;
liftT=(cum_rrT/n);
run;

PROC SUMMARY data=score missing;
class dec;
var  CONTROL wtC;
output out=sum_decC sum=sum_canC sum_wtC;
run;

data sum_decC;
set  sum_decC;
avg_canC=sum_canC/sum_wtC;
run;

data avg_rrC;
set  sum_decC;
if dec=.;
keep avg_canC;
run;

data sum_dec1C;
set  sum_decC;
```

```
if dec=. or dec=10 then delete;
cum_nC +sum_wtC;
rC    =sum_canC;
cum_rC +sum_canC;
cum_rrC=(cum_rC/cum_nC)*100;
avg_cannC=avg_canC*100;
run;

data scoresamC ;
set  avg_rrC sum_dec1C;
retain n;
if _n_=1 then n=avg_canC;
if _n_=1 then delete;
liftC=(cum_rrC/n);
run;

PROC SORT data=scoresamC (drop= _FREQ_ _type_ n liftC); by  dec;
PROC SORT data=scoresamT (drop= _FREQ_ _type_ n liftT); by  dec;

data scoresam_TAC;
merge  scoresamC  scoresamT; by dec;
run;

data scoresam_TAC;
set  scoresam_TAC;
CNTRL_SIZE =sum_wtC;
TEST_SIZE  =sum_wtT;
TEST_CUM   =cum_rrT;
CNTRL_CUM  =cum_rrC;
CNTRL_RESP =rC;
TEST_RESP  =rT;
TEST_MEAN  =avg_cannT;
CNTRL_MEAN =avg_cannC;
NET_TEST_MEAN = (TEST_MEAN-CNTRL_MEAN);
CUM_TAC+NET_TEST_MEAN;
NET_TAC=TEST_RESP-CNTRL_RESP;
CUM_TAC1+TEST_RESP-CNTRL_RESP;
m=1;
run;

data lift_base;
set  scoresam_TAC;
if dec=9 ;
lift_base=CUM_TAC1;
m=1;
keep lift_base  m;
```

```
PROC SORT data=scoresam_TAC; by m;
PROC SORT data= lift_base; by m;
data scoresam_TAC_LIFT;
merge scoresam_TAC lift_base; by m;
drop m;
CUM_LIFT=(CUM_TAC/lift_base)*100;
CUM_LIFT1=(CUM_TAC1/lift_base)*100;

data LIFT;
set scoresam_TAC_LIFT;
if dec=0 then decc=' top ';
if dec=1 then decc=' 2 ';
if dec=2 then decc=' 3 ';
if dec=3 then decc=' 4 ';
if dec=4 then decc=' 5 ';
if dec=5 then decc=' 6 ';
if dec=6 then decc=' 7 ';
if dec=7 then decc=' 8 ';
if dec=8 then decc=' 9 ';
if dec=9 then decc='bottom';
if dec ne .;
run;

PROC SORT data=scoresam_TAC_LIFT; by dec;
PROC SORT data=LIFT; by dec;

data upl.final_NET_TAC;
merge LIFT scoresam_TAC_LIFT; by dec;
overall_net=(test_cum-cntrl_cum)/100;
call symputx('overall',put(overall_net,percent8.1));
NET_IMPROV=(NET_TAC/TEST_SIZE)*100;
run;
footnote "    Overall NET T-C: &overall";

PROC PRINT data=upl.final_NET_TAC d split='*' noobs;
var decc
TEST_SIZE   CNTRL_SIZE
TEST_RESP   CNTRL_RESP
NET_TAC NET_IMPROV CUM_TAC1 CUM_LIFT1 ;

label
decc='DECILE'
TEST_SIZE ='TEST*Group*(n)'
CNTRL_SIZE ='CONTROL*Group*(n)'

TEST_RESP =' TEST*Count*(n)'
CNTRL_RESP='CONTROL*Count*(n)'
```

```
TEST_MEAN ='TEST*Rate*(%)'
CNTRL_MEAN='CNTRL*Rate*(%)'

TEST_CUM ='CUM*TEST*Rate*(%)'
CNTRL_CUM ='CUM*CONTROL*Rate*(%)'

NET_TEST_MEAN ='NET T-C*Rate*(%)'
NET_TAC       ='NET T-C*Count*(n)'
NET_IMPROV    ='NET T-C*IMPROV*(%)'
CUM_TAC1 ='CUM*NET T-C*Count*(n)'
CUM_LIFT1 ='CUM*NET T-C*LIFT*(%)';
sum
TEST_SIZE CNTRL_SIZE TEST_RESP CNTRL_RESP;
format TEST_SIZE CNTRL_SIZE TEST_RESP CNTRL_RESP comma6.;
format NET_TEST_MEAN 6.1;
format TEST_CUM  CNTRL_CUM  4.1;
format TEST_MEAN CNTRL_MEAN  4.1;
format NET_TAC 3.0;
format CUM_TAC1 comma6.0;
format CUM_LIFT1 5.1;
format NET_IMPROV 4.1;
run;
footnote;
```

References

1. The Decisionhouse Uplift Model software produced by Quadstone Limited, 1996.
2. Lee, T., Zhang, R., Meng, X., and Ryan, L., *Incremental Response Modeling Using SAS Enterprise Miner*, Paper 096-2013, SAS Global Forum, San Francisco, CA, 2013.
3. Surry, P.D., and Radcliffe, N.J., 2011. *Quality Measures for Uplift Models*, Submitted to KDD2011.
4. Rzepakowski, P., and Jaroszewicz, S., Uplift modeling for clinical trial data, *ICML Workshop on Machine Learning for Clinical Data Analysis*, 2012.
5. Rzepakowski, P., and Jaroszewicz, S., Decision trees for uplift modeling, in *IEEE Conference on Data Mining*, pp. 441–450, 2010.
6. Zaniewicz, L., and Jaroszewicz, S., Support vector machines for uplift modeling, *IEEE ICDM Workshop on Causal Discovery*, 2013.
7. Radcliffe, N.J., Using control groups to target on predicted lift: Building and assessing uplift model, *Direct Marketing Analytics Journal, An Annual Publication from the Direct Marketing Association Analytics Council*, 14–21, 2012.
8. Radcliffe, N.J., and Surry, P.D., *Real-World Uplift Modelling with Significance-Based Trees*, Portrait Technical Report TR-2011-1, Stochastic Solutions, p. 14, 2011.
9. Jaroskowski, M., and Jaroszewicz, S., Uplift modeling for clinical trial data, *ICML Workshop on Machine Learning for Clinical Data Analysis*, Edinburgh, Scotland, UK, 2012.
10. Larsen, K., Generalized naïve Bayes classifiers, *SIGKDD Explorations*, 7(1), 76–81, 2005.
11. Hand, D.J., and Keming, Y., Idiot's Bayes—Not so stupid after all? *International Statistical Review*, 69(3), 385–398, 2001.

12. Larsen, K., Net lift models, Verified email at www.cs.aau.dk, 2010.
13. Radcliffe, N.J., and Surry, P.D., Differential response analysis: Modeling true response by isolating the effect of a single action, *Proceedings of Credit Scoring and Credit Control VI*, Credit Research Centre, University of Edinburgh Management School, SIAM, Philadelphia, PA, 1999.
14. Piatetsky-Shapiro, G., Discovery, analysis, and presentation of strong rules, in Piatetsky-Shapiro, G., and Frawley, W.J., Eds., *Knowledge Discovery in Databases*, AAAI/MIT Press, Cambridge, MA, 1991.
15. Lo, V.S., The true lift model, *ACM SIGKDD Explorations Newsletter*, 4(2), 78–86, 2002.

29

*Bootstrapping in Marketing: A New Approach for Validating Models**

29.1 Introduction

Traditional validation of a marketing model uses a holdout sample consisting of individuals who are not part of the sample used in building the model itself. Using resampling methods helps ensure validation results are unbiased and complete. This chapter points to the weaknesses of the traditional validation and then presents a bootstrap approach for validating response and profit models as well as measuring the efficiency of the models.

I provide SAS© subroutines for performing bootstrapped decile analysis in Chapter 44.

29.2 Traditional Model Validation

The data analyst's first step in building a marketing model is to split randomly the original data file into two mutually exclusive parts: a calibration sample for developing the model and validation or holdout sample for assessing the reliability of the model. If the analyst is lucky to split the file to yield a holdout sample with *favorable* characteristics, then a better-than-true biased validation is obtained. If unlucky and the sample has *unfavorable* characteristics, then a worse-than-true biased validation is obtained. Lucky or not, or even if the validation sample is a true reflection of the population under study, a single sample cannot provide a measure of variability that would otherwise allow the analyst to assert a level of confidence about the validation.

In sum, the traditional single-sample validation provides neither assurance that the results are not biased nor any measure of confidence in the results. These points are made clear with an illustration using a response model (RM), and all results and implications apply equally to profit models.

* This chapter is based on an article with the same title in *Journal of Targeting, Measurement and Analysis for Marketing*, 6, 2, 1997. Used with permission.

29.3 Illustration

As marketers use the Cum Lift measure from a decile analysis to assess the goodness of a model, the validation of the model* consists of comparing the Cum Lifts from the calibration and holdout decile analyses based on the model. Expected shrinkage in the Cum Lifts occurs: Cum Lifts from the holdout sample are typically smaller (less optimistic) than those from the calibration sample from which they originated. The Cum Lifts on a fresh holdout sample, which does not contain the calibration idiosyncrasies, provide a more realistic assessment of the quality of the model. The calibration Cum Lifts inherently capitalize on the idiosyncrasies of the calibration sample due to the modeling process, which favors large Cum Lifts. If both the Cum Lift shrinkage and the Cum Lift values themselves are acceptable, then the model is considered successfully validated and ready to use; otherwise, reworking the model is required until successfully validated.

Consider an RM that produces the decile analysis validation in Table 29.1 based on a sample of 181,100 customers with an overall response rate of 0.26%. (Recall from Chapter 26, the Cum Lift is a measure of predictive power; it indicates the expected gain from a solicitation implemented with a model over a solicitation implemented *without* a model.) The Cum Lift for the top decile is 186. This lift indicates that when soliciting to the top decile—the top 10% of the customer file identified by the RM—there is an expected 1.86 times the number of responders obtained without a model. Similar to that for the second decile, the Cum Lift of 154 indicates that when soliciting to the top two deciles—the top 20% of the customer file based on the RM—there is an expected 1.54 times the number of responders obtained without a model.

TABLE 29.1

Response Decile Analysis

Decile	Number of Individuals	Number of Responders	Decile Response Rate (%)	Cumulative Response Rate (%)	Cum Lift
top	18,110	88	0.49	0.49	186
2	18,110	58	0.32	0.40	154
3	18,110	50	0.28	0.36	138
4	18,110	63	0.35	0.36	137
5	18,110	44	0.24	0.33	128
6	18,110	48	0.27	0.32	123
7	18,110	39	0.22	0.31	118
8	18,110	34	0.19	0.29	112
9	18,110	23	0.13	0.27	105
bottom	18,110	27	0.15	0.26	100
Total	181,100	474	0.26		

* Validation of any response or profit model built from any modeling technique (e.g., discriminant analysis, logistic regression, neural network, genetic algorithms, or chi-squared automatic interaction detection [CHAID]).

TABLE 29.2

Cum Lifts for Three Validations

Decile	First Sample	Second Sample	Third Sample
top	186	197	182
2	154	153	148
3	138	136	129
4	137	129	129
5	128	122	122
6	123	118	119
7	118	114	115
8	112	109	110
9	105	104	105
bottom	100	100	100

As luck would have it, the data analyst finds two additional samples on which to perform two additional decile analysis validations. Not surprisingly, the Cum Lifts for a given decile across the three validations are somewhat different. The reason for this is the *expected* sample-to-sample variation, attribution of chance. There is a large variation in the top decile (range is 15 = 197 – 182) and a small variation in decile 2 (range is 6 = 154 – 148). These results in Table 29.2 raise obvious questions.

29.4 Three Questions

With many decile analysis validations, the expected sample-to-sample variation within each decile points to the uncertainty of the Cum Lift estimates. If there is an observed large variation for a given decile, then there is less confidence in the Cum Lift for that decile. If there is an observed small variation, then there is more confidence in the Cum Lift. Thus, there are the following questions:

1. With many decile analysis validations, how can an *average* Cum Lift (for a decile) be defined to serve as an *honest* estimate of the Cum Lift? Also, how many validations are needed?
2. With many decile analysis validations, how can the *variability* of an honest estimate of Cum Lift be assessed? That is, how can the standard error (a measure of the precision of an estimate) of an honest estimate of the Cum Lift be calculated?
3. With only a single validation dataset, can an honest Cum Lift estimate and its standard error be calculated?

The answers to these questions and more lie in the bootstrap methodology.

29.5 The Bootstrap Method

The bootstrap method is a computer-intensive approach to statistical inference [1]. It is the most popular resampling method, using the computer to resample the sample at hand* [2]. By random selection with replacement from the sample, some individuals occur more than once in a *bootstrap* sample, and some individuals occur not at all. Each same-size bootstrap sample will be slightly different from the others. This variation makes it possible to induce an empirical sampling distribution† of the desired statistic, which determines the estimates of bias and variability.

The bootstrap is a flexible technique for assessing the accuracy‡ of any statistic. For well-known statistics, such as the mean, the standard deviation, regression coefficients, and R-squared, the bootstrap provides an alternative to traditional parametric methods. For statistics with unknown properties, such as the median and Cum Lift, traditional parametric methods do not exist. Thus, the bootstrap provides a viable alternative to the inappropriate use of traditional methods, which yield questionable results.

The bootstrap also falls into the class of nonparametric procedures. It does not rely on unrealistic parametric assumptions. Consider testing the significance of a variable§ in a regression model built using ordinary least squares estimation. Say the error terms do not have a normal distribution, a clear violation of the ordinary least squares assumptions [3]. The significance testing may yield inaccurate results due to the model assumption not being met. In this situation, the bootstrap is a feasible approach in determining the significance of the coefficient without concern for any assumptions. As a nonparametric method, the bootstrap does not rely on theoretical derivations required in traditional parametric methods. I review the well-known parametric approach to the construction of confidence intervals (CIs) to demonstrate the utility of the bootstrap as an alternative technique.

29.5.1 Traditional Construction of Confidence Intervals

Consider the parametric construction of a confidence interval for the population mean. I draw a random sample, A, of five numbers from a population. Sample A consists of (23, 4, 46, 1, 29). The sample mean is 20.60, the sample median is 23, and the sample standard deviation is 18.58.

The parametric method follows from the central limit theorem, which states that the theoretical sampling distribution of the sample mean is normal with an analytically defined standard error [4]. Thus, the 100(1 – a)% CI for the mean is:

$$\text{Sample mean value} \pm |Z_{a/2}| {*}\text{Standard error}$$

where sample mean value is simply the mean of the five numbers in the sample.

$|Z_{a/2}|$ is the value from the standard normal distribution for a 100(1 – a)% CI. The $|Z_{a/2}|$ values are 1.96, 1.64, and 1.28 for 95%, 90%, and 80% CIs, respectively. Standard error (SE) of the sample mean has the analytic formula: SE = the sample standard deviation divided by the square root of the sample size.

---

* Other resampling methods include, for example, the jackknife, infinitesimal jackknife, delta method, influence function method, and random subsampling.

† A sampling distribution can be considered as the frequency curve of a sample statistic from an infinite number of samples.

‡ Accuracy includes bias, variance, and error.

§ That is, is the coefficient equal to zero?

An often used term is the *margin of error*, defined as $|Z_{a/2}|$*SE.

For Sample A, SE equals 8.31, and the 95% CI for the population mean is between 4.31 (= (20.60 – 1.96*8.31)) and 36.89 (= (20.60 + 1.96*8.31)). The usual, yet incorrect, interpretation of the CI interval is: There is 95% confidence that the population mean lies between 4.31 and 36.89. The statistically correct statement for the 95% CI for the unknown population mean is: If one repeatedly calculates such intervals from, say, 100 independent random samples, 95% of the constructed intervals would contain the true population mean. Do not be fooled. Given a calculated confidence interval, the true mean is either in the interval or not in the interval. Thus, 95% confidence refers to the procedure for constructing the interval not the observed interval itself. This parametric approach for the construction of confidence intervals for statistics, such as the median and the Cum Lift, does not exist because the theoretical sampling distributions (which provide the standard errors) of the median and Cum Lift are not known. If desired, a resampling methodology such as the bootstrap can produce the confidence intervals for the median and Cum Lift.

29.6 How to Bootstrap

The key assumption of the bootstrap[*] is that the sample is the best estimate[†] of the unknown population. Treating the sample as the population, the analyst repeatedly draws same-size random samples with replacement from the original sample. The analyst estimates the sampling distribution of the desired statistic from the many bootstrap samples and can calculate a bias-reduced bootstrap estimate of the statistic and a bootstrap estimate of the SE of the statistic.

The bootstrap procedure consists of 10 simple steps.

1. State desired statistic, say, Y.
2. Treat sample as population.
3. Calculate Y on the sample/population; call it SAM_EST.
4. Draw a bootstrap sample from the population, that is, a random selection with the replacement of size n, the size of the original sample.
5. Calculate Y on the bootstrap sample to produce a pseudo-value; call it BS_1.
6. Repeat Steps 4 and 5 "m" times.[‡]
7. From Steps 1 to 6, there are BS_1,BS_2, ..., BS_m.
8. Calculate the bootstrap estimate of the statistic:[§]

$$BS_{est}(Y) = 2*SAM\_EST - mean(BS_i).$$

* This bootstrap method is the normal approximation. Others are percentile, B-C percentile, and percentile-t.

† Actually, the sample distribution function is the nonparametric maximum likelihood estimate of the population distribution function.

‡ Studies showed the precision of the bootstrap does not significantly increase for m > 250.

§ This calculation arguably ensures a bias-reduced estimate. Some analysts question the use of the bias correction. I feel that this calculation adds precision to the decile analysis validation when conducting small solicitations and has no noticeable effect on large solicitations.

9. Calculate the bootstrap estimate of the standard error of the statistic:

$$SE_{BS}(Y) = \text{standard deviation of } (BS_i).$$

10. The 95% bootstrap confidence interval is

$$BS_{est}(Y) \pm |Z_{0.025}|{*}SE_{BS}(Y).$$

29.6.1 Simple Illustration

Consider a simple illustration.* I have a Sample B from a population (no reason to assume a normally distributed population) that produces the following 11 values:

Sample B: 0.1, 0.1, 0.1, 0.4, 0.5, 1.0, 1.1, 1.3, 1.9, 1.9, 4.7

I want a 95% confidence interval for the population standard deviation. If I knew the population was normal, I would use the parametric chi-square test and obtain the confidence interval:

$$0.93 < \text{population standard deviation} < 2.35$$

I apply the bootstrap procedure on Sample B:

1. The desired statistic is the standard deviation (StD).
2. Treat Sample B as the population.
3. I calculate StD on the original sample/population, SAM_EST = 1.3435.
4. I randomly select 11 observations with replacement from the population. This is the first bootstrap sample.
5. I calculate StD on this bootstrap (BS) sample to obtain a pseudovalue, $BS_1 = 1.3478$.
6. I repeat Steps 4 and 5 an additional 99 times.
7. I have $BS_1, BS_2, \ldots, BS_{100}$ in Table 29.3.
8. I calculate the bootstrap estimate of StD:

$$BS_{est}(StD) = 2{*}SAM\_EST - \text{mean}(BS_i) = 2{*}1.3435 - 1.2034 = 1.483$$

9. I calculate the bootstrap estimate of the standard error of StD:

$$SE_{BS}(StD) = \text{standard deviation } (BS_i) = 0.5008$$

10. The bootstrap 95% confidence interval for the population standard deviation is 0.50 < population standard deviation < 2.47.

* This illustration draws on Sample B and comes from Mosteller, F., and Tukey, J.W., Data Analysis and Regression, Addison-Wesley, Reading, MA, 139–143, 1977.

TABLE 29.3
100 Bootstrapped StDs

1.3431	0.60332	1.7076	0.6603	1.4614	1.43
1.4312	1.2965	1.9242	0.71063	1.3841	1.7656
0.73151	0.70404	0.643	1.6366	1.8288	1.6313
0.61427	0.76485	1.3417	0.69474	2.2153	1.2581
1.4533	1.353	0.479	0.62902	2.1982	0.73666
0.66341	1.4098	1.8892	2.0633	0.73756	0.69121
1.2893	0.67032	1.7316	0.60083	1.4493	1.437
1.3671	0.49677	0.70309	0.51897	0.65701	0.59898
1.3784	0.7181	1.3802	1.3985	1.4356	1.2972
0.47854	2.0658	1.7825	0.63281	1.8755	0.39384
0.69194	0.6343	1.31	1.3491	0.70079	
0.29609	1.5522	0.62048	1.8657	1.3919	
1.3726	2.0877	1.6659	1.4372	0.72111	
0.83327	1.4056	1.7404	1.796	1.7957	
1.399	1.3653	1.8172	1.2665	1.2874	
1.322	0.56569	0.74863	1.4085	1.6363	
0.60194	1.9938	0.57937	0.74117	1.6596	
1.3476	0.6345	1.7188	1.3994	1.3754	

NP 0.93xxxxxxxxxxxxxxx2.35
BS 0.50xxxxxxxxxxxxxxxxxxxx2.47

FIGURE 29.1
Bootstrap versus normal estimates.

As you may have suspected, the sample is from a normal population (NP). Thus, it is instructive to compare the performance of the bootstrap with the theoretically correct parametric chi-square test. The bootstrap confidence interval is somewhat wider than the chi-square/NP interval in Figure 29.1. The BS confidence interval covers values between 0.50 and 2.47, which also includes the values within the NP confidence interval (0.93, 2.35).

These comparative performance results are typical. Performance studies indicate the bootstrap methodology provides results that are consistent with the outcomes of the parametric techniques. Thus, the bootstrap is a reliable approach to inferential statistics in most situations.

Note that the bootstrap estimate offers a more honest* and bias-reduced† point estimate of the standard deviation. The original sample estimate is 1.3435, and the bootstrap estimate is 1.483. There is a 10.4% (1.483/1.3435) bias reduction in the estimate.

29.7 Bootstrap Decile Analysis Validation

Continuing with the RM illustration, I execute the 10-step bootstrap procedure to perform a bootstrap decile analysis validation. I use 50 bootstrap samples,‡ each of a size

* Due to the many samples used in the calculation.
† Attributable, in part, to sample size.
‡ I experience high precision in bootstrap decile validation with just 50 bootstrap samples.

TABLE 29.4

Bootstrap Response Decile Validation (Bootstrap Sample Size n = 181,000)

Decile	Bootstrap Cum Lift	Bootstrap SE	95% Bootstrap CI
top	183	10	163, 203
2	151	7	137, 165
3	135	4	127, 143
4	133	3	127, 139
5	125	2	121, 129
6	121	1	119, 123
7	115	1	113, 117
8	110	1	108, 112
9	105	1	103, 107
bottom	100	0	100, 100

equal to the original sample size of 181,100. The bootstrap Cum Lift for the top decile is 183 and has a bootstrap standard error of 10, in Table 29.4. Accordingly, for the top decile, the 95% bootstrap confidence interval is 163 to 203. The second decile has a bootstrap Cum Lift of 151, and a bootstrap 95% confidence interval is between 137 and 165. Specifically, this bootstrap validation indicates that the expected Cum Lift is 135, using the RM to select the top 30% of the most responsive individuals from a randomly drawn sample of size 181,100 from the target population or database. Moreover, the Cum Lift is expected to lie between 127 and 143 with 95% confidence. The remaining deciles have a similar reading.

Bootstrap estimates and confidence intervals for decile Cum Lifts easily convert into bootstrap estimates and confidence intervals for decile response rates. The conversion formula involves the following: Bootstrap decile response rate equals bootstrap decile Cum Lift divided by 100 then multiplied by overall response rate. For example, for the third decile, the bootstrap response rate is 0.351% (= (135/100)*0.26%). The lower and upper confidence interval end points are 0.330% (= (127/100)*0.26%) and 0.372% (= (143/100)*0.26%), respectively.

29.8 Another Question

Quantifying the predictive certainty, that is, constructing prediction confidence intervals, is likely to be more informative to data analysts and their management than obtaining a *point estimate* alone. A single calculated value of a statistic such as Cum Lift can serve as a point estimate that provides the best guess of the true value of the statistic. However, there is an obvious need to quantify the certainty associated with such a point estimate. The decision-maker wants the margin of error of the estimate. The margin of error value is added and subtracted to the estimated value to yield an interval in which there is a reasonable confidence that the true (Cum Lift) value lies. If the confidence interval (equivalently, the standard error or margin of error) is too large for the business objective at hand, then is there a procedure to increase the confidence level?

TABLE 29.5

Bootstrap Response Decile Validation (Bootstrap Sample Size n = 225,000)

Decile	Bootstrap Cum Lift	Bootstrap SE	95% Bootstrap CI
top	185	5	163, 203
2	149	3	137, 165
3	136	3	127, 143
4	133	2	127, 139
5	122	1	121, 129
6	120	1	119, 123
7	116	1	113, 117
8	110	0.5	108, 112
9	105	0.5	103, 107
bottom	100	0	100, 100

The answer rests on the well-known, fundamental relationship between sample size and confidence interval length: increasing (decreasing) sample size increases (decreases) confidence in the estimate. Equivalently, increasing (decreasing) sample size decreases (increases) the standard error [5]. The sample size and confidence length relationship can serve as a guide to increase confidence in the bootstrap Cum Lift estimates in two ways:

1. If ample additional customers are available, add them to the original validation dataset until the enhanced validation dataset size produces the desired standard error and confidence interval length.
2. Simulate or bootstrap the original validation dataset by increasing the bootstrap sample size until the enhanced bootstrap dataset size produces the desired standard error and confidence interval length.

Returning to the RM illustration, I increase the bootstrap sample size to 225,000 from the original sample size of 181,100 to produce a slight decrease in the standard error from 4 to 3 for the cumulative top three deciles. This simulated validation in Table 29.5 indicates that an expected Cum Lift of 136 is centered in a slightly shorter 95% confidence interval (between 130 and 142) when using the RM to select the top 30% of the most responsive individuals from a randomly selected sample of size 225,000 from the database. Note, the bootstrap Cum Lift estimates also change (in this instance, from 135 to 136) because their calculations use new, larger samples. Bootstrap Cum Lift estimates rarely show big differences when the enhanced dataset size increases. In the next section, I continue the discussion on the sample size confidence and length relationship as it relates to a bootstrap assessment of model implementation performance.

29.9 Bootstrap Assessment of Model Implementation Performance

Statisticians are often asked, "How large a sample do I need to have confidence in my results?" The traditional answer, based on parametric theoretical formulations, depends on the statistic, such as response rate or mean profit, and on additional required input. These additions include the following:

1. The expected value of the desired statistic.
2. The preselected level of confidence is related to the probability the decision-maker is willing to wrongly reject a true null hypothesis (e.g., H0: no relationship). The level of confidence may involve incorrectly including that a relationship exists when, in fact, it does not.
3. The preselected level of *power to detect* a relationship. Power is one minus the probability of wrongly failing to reject a false null hypothesis. The level of power involves making the right decision, rejecting a false null hypothesis when it is false.
4. The assurance that sundry theoretical assumptions hold.

Regardless of who built the marketing model, once it is ready for implementation, the question is essentially the same: How large a sample do I need to implement a solicitation *based on the model* to obtain the desired performance quantity? The answer, which in this case does not require most of the traditional input, depends on one of two performance objectives. Objective 1 is to maximize the performance quantity, namely, the Cum Lift, for a specific depth-of-file. Determining how large the sample should be—the smallest sample necessary to obtain the Cum Lift value—involves the concepts discussed in the previous section that correspond to the relationship between confidence interval length and sample size. The following procedure answers the sample size question for Objective 1.

1. For a desired Cum Lift value, identify the decile and its confidence interval containing Cum Lift values closest to the desired value based on the decile analysis validation at hand. If the corresponding confidence interval length is acceptable, then the size of the validation dataset is the required sample size. Draw a random sample of that size from the database.
2. If the corresponding confidence interval is too large, then increase the validation sample size by adding individuals or bootstrapping a larger sample size until the confidence interval length is acceptable. Draw a random sample of that size from the database.
3. If the corresponding confidence interval is unnecessarily small, which indicates that a smaller sample can be used to save time and cost of data retrieval, then decrease the validation sample size by deleting individuals or bootstrapping a smaller sample size until the confidence interval length is acceptable. Draw a random sample of that size from the database.

Objective 2 adds a constraint to the performance quantity of Objective 1. Sometimes, it is desirable not only to reach a performance quantity but also to discriminate among the *quality* of individuals who are selected to contribute to the performance quantity. Reaching a performance quantity is particularly worth doing when a targeted solicitation is to a relatively homogeneous population of finely graded individuals regarding responsiveness or profitability.

The quality constraint imposes the restriction that the performance value of individuals within a decile is different across the deciles within the preselected depth-of-file range.

Achieving the quality constraint involves (1) determining the sample size that produces decile confidence intervals that do not overlap *and* (2) ensuring an overall confidence level that the individual decile confidence intervals are *jointly* valid. That is, the true Cum Lifts are in their confidence intervals. Achieving the former condition involves increasing the sample size. Accomplishing the latter condition involves employing the Bonferroni method, which allows the analyst to assert a confidence statement that multiple confidence intervals are *jointly* valid.

Briefly, the Bonferroni method is as follows: Assume the analyst wishes to combine k confidence intervals that individually have confidence levels $1 - a_1$, $1 - a_2$, ..., $1 - a_k$. The analyst wants to make a joint confidence statement with confidence level $1 - a_J$. The Bonferroni method states that the joint confidence level $1 - a_J$ is greater than or equal to $1 - a_1 - a_2 \ldots - a_k$. The joint confidence level is a conservative lower bound on the actual confidence level for a joint confidence statement. The Bonferroni method is conservative in the sense that it provides confidence intervals that have confidence levels larger than the actual level.

I apply the Bonferroni method for four common individual confidence intervals: 95%, 90%, 85%, and 80%.

1. For combining 95% confidence intervals, there is at least 90% confidence that the two true decile Cum Lifts lie between their respective confidence intervals. There is at least 85% confidence that the three true decile Cum Lifts lie between their respective confidence intervals. And, there is at least 80% confidence that the four true decile Cum Lifts lie between their respective confidence intervals.
2. For combining 90% confidence intervals, there is at least 80% confidence that the two true decile Cum Lifts lie between their respective confidence intervals. There is at least 70% confidence that the three true decile Cum Lifts lie between their respective confidence intervals. And, there is at least 60% confidence that the four true decile Cum Lifts lie between their respective confidence intervals.
3. For combining 85% confidence intervals, there is at least 70% confidence that the two true decile Cum Lifts lie between their respective confidence intervals. There is at least 55% confidence that the three true decile Cum Lifts lie between their respective confidence intervals. And, there is at least 40% confidence that the four true decile Cum Lifts lie between their respective confidence intervals.
4. For combining 80% confidence intervals, there is at least 60% confidence that the two true decile Cum Lifts lie between their respective confidence intervals. There is at least 40% confidence that the three true decile Cum Lifts lie between their respective confidence intervals. And, there is at least 20% confidence that the four true decile Cum Lifts lie between their respective confidence intervals.

The following procedure determines how large the sample should be. The smallest sample necessary to obtain the desired quantity and quality of performance involves the following:

1. For a desired Cum Lift value for a preselected depth-of-file range, identify the decile and its confidence interval containing Cum Lift values closest to the desired value based on the decile analysis validation at hand. If the decile confidence intervals within the preselected depth-of-file range do not overlap (or concededly

have acceptable minimal overlap), then the validation sample size at hand is the required sample size. Draw a random sample of that size from the database.

2. If the decile confidence intervals within the preselected depth-of-file range are too large or overlap, then increase the validation sample size by adding individuals or bootstrapping a larger sample size until the decile confidence interval lengths are acceptable and do not overlap. Draw a random sample of that size from the database.
3. If the decile confidence intervals are unnecessarily small and do not overlap, then decrease the validation sample size by deleting individuals or bootstrapping a smaller sample size until the decile confidence interval lengths are acceptable and do not overlap. Draw a random sample of that size from the database.

29.9.1 Illustration

Consider a three-variable RM predicting response from a population with an overall response rate of 4.72%. The decile analysis validation based on a sample size of 22,600, along with the bootstrap estimates, is in Table 29.6. The 95% margin of errors ($|Z_{a/2}|*SE_{BS}$) for the top four decile 95% confidence intervals is considered too large for using the model with confidence. Moreover, the decile confidence intervals severely overlap. The top decile has a 95% confidence interval of 160 to 119, with an expected bootstrap response rate of 6.61% (= (140/100)*4.72%; the bootstrap Cum Lift for the top decile is 140). The top two decile levels have a 95% confidence interval of 113 to 141, with an expected bootstrap response rate of 5.99% (= (127/100)*4.72%; the bootstrap Cum Lift for the top two decile levels is 127).

I create a new bootstrap validation sample of size 50,000. The 95% margins of error in Table 29.7 are still unacceptably large, and the decile confidence intervals still overlap. I further increase the bootstrap sample size to 75,000 in Table 29.8. There is no noticeable change in 95% margins of error, and the decile confidence intervals overlap.

I recalculate the bootstrap estimates using 80% margins of error on the bootstrap sample size of 75,000. The results in Table 29.9 show that the decile confidence intervals have

TABLE 29.6

Three-Variable Response Model 95% Bootstrap Decile Validation (Bootstrap Sample Size n = 22,600)

Decile	Number of Individuals	Number of Responses	Response Rate (%)	Cum Response Rate (%)	Cum Lift	Bootstrap Cum Lift	95% Margin of Error	95% Lower Bound	95% Upper Bound
top	2,260	150	6.64	6.64	141	140	20.8	119	160
2	2,260	120	5.31	5.97	127	127	13.8	113	141
3	2,260	112	4.96	5.64	119	119	11.9	107	131
4	2,260	99	4.38	5.32	113	113	10.1	103	123
5	2,260	113	5.00	5.26	111	111	9.5	102	121
6	2,260	114	5.05	5.22	111	111	8.2	102	119
7	2,260	94	4.16	5.07	107	107	7.7	100	115
8	2,260	97	4.29	4.97	105	105	6.9	98	112
9	2,260	93	4.12	4.88	103	103	6.4	97	110
bottom	2,260	75	3.32	4.72	100	100	6.3	93	106
Total	22,600	1,067							

TABLE 29.7

Three-Variable Response Model 95% Bootstrap Decile Validation (Bootstrap Sample Size n = 50,000)

Decile	Model Cum Lift	Bootstrap Cum Lift	95% Margin of Error	95% Lower Bound	95% Upper Bound
top	141	140	15.7	124	156
2	127	126	11.8	114	138
3	119	119	8.0	111	127
4	113	112	7.2	105	120
5	111	111	6.6	105	118
6	111	111	6.0	105	117
7	107	108	5.6	102	113
8	105	105	5.1	100	111
9	103	103	4.8	98	108
bottom	100	100	4.5	95	104

TABLE 29.8

Three-Variable Response Model 95% Bootstrap Decile Validation (Bootstrap Sample Size n = 75,000)

Decile	Model Cum Lift	Bootstrap Cum Lift	95% Margin of Error	95% Lower Bound	95% Upper Bound
top	141	140	12.1	128	152
2	127	127	7.7	119	135
3	119	120	5.5	114	125
4	113	113	4.4	109	117
5	111	112	4.5	107	116
6	111	111	4.5	106	116
7	107	108	4.2	104	112
8	105	105	3.9	101	109
9	103	103	3.6	100	107
bottom	100	100	3.6	96	104

acceptable lengths and are almost nonoverlapping. Unfortunately, the joint confidence levels are quite low: at least 60%, at least 40%, and at least 20% for the top two, top three, and top four decile levels, respectively.

After successively increasing the bootstrap sample size, I reach a bootstrap sample size (175,000) that produces acceptable and virtually nonoverlapping 95% confidence intervals in Table 29.10. The top four decile Cum Lift confidence intervals are (133, 147), (122, 132), (115, 124), and (109, 116), respectively. The joint confidence for the top three deciles and top four deciles is respectable levels of at least 85% and at least 80%, respectively. It is interesting to note that increasing the bootstrap sample size to 200,000, for either 95% or 90% margins of error, does not produce any noticeable improvement in the quality of performance. (Validations for bootstrap sample size 200,000 are not shown.)

TABLE 29.9

Three-Variable Response Model 80% Bootstrap Decile Validation (Bootstrap Sample Size n = 75,000)

Decile	Model Cum Lift	Bootstrap Cum Lift	80% Margin of Error	80% Lower Bound	80% Upper Bound
top	141	140	7.90	132	148
2	127	127	5.10	122	132
3	119	120	3.60	116	123
4	113	113	2.90	110	116
5	111	112	3.00	109	115
6	111	111	3.00	108	114
7	107	108	2.70	105	110
8	105	105	2.60	103	108
9	103	103	2.40	101	106
bottom	100	100	2.30	98	102

TABLE 29.10

Three-Variable Response Model 95% Bootstrap Decile Validation (Bootstrap Sample Size n = 175,000)

Decile	Model Cum Lift	Bootstrap Cum Lift	95% Margin of Error	95% Lower Bound	95% Upper Bound
top	141	140	7.4	133	147
2	127	127	5.1	122	132
3	119	120	4.2	115	124
4	113	113	3.6	109	116
5	111	111	2.8	108	114
6	111	111	2.5	108	113
7	107	108	2.4	105	110
8	105	105	2.3	103	108
9	103	103	2.0	101	105
bottom	100	100	1.9	98	102

29.10 Bootstrap Assessment of Model Efficiency

The bootstrap approach to decile analysis of marketing models can provide an assessment of model efficiency. Consider an alternative Model A to the best Model B (both models predict the same dependent variable). Model A is said to be *less efficient* than Model B if either condition is true:

1. Model A yields the same results as Model B: Cum Lift margins of error are equal and Model A sample size is larger than Model B sample size.
2. Model A yields worse results than Model B: Model A Cum Lift margins of error are greater than those of Model B and sample sizes for models A and B are equal.

TABLE 29.11

Bootstrap Model Efficiency (Bootstrap Sample Size n = 175,000)

	Eight-Variable Response Model A					Three-Variable Response Model B	
Decile	**Model Cum Lift**	**Bootstrap Cum Lift**	**95% Margin of Error**	**95% Lower Bound**	**95% Upper Bound**	**95% Margin of Error**	**Efficiency Ratio (%)**
top	139	138	8.6	129	146	7.4	86.0
2	128	128	5.3	123	133	5.1	96.2
3	122	122	4.3	117	126	4.2	97.7
4	119	119	3.7	115	122	3.6	97.3
5	115	115	2.9	112	117	2.8	96.6
6	112	112	2.6	109	114	2.5	96.2
7	109	109	2.6	107	112	2.4	92.3
8	105	105	2.2	103	107	2.3	104.5
9	103	103	2.1	101	105	2.0	95.2
bottom	100	100	1.9	98	102	1.9	100.0

The measure of efficiency is reported as the ratio of either of the following quantities: (1) the sample sizes necessary to achieve equal results or (2) the variability measures (Cum Lift margin of error) for equal sample size. The efficiency ratio is defined as follows: Model B (quantity) over Model A (quantity). Efficiency ratio values less (greater) than 100% indicate that Model A is less (more) efficient than Model B.

I illustrate how a model with unnecessary predictor variables is less efficient (has larger prediction error variance) than a model with the right number of predictor variables. Returning to the three-variable RM (considered the best model, Model B), I create an alternative model (Model A) by adding five unnecessary predictor variables to Model B. The extra variables include irrelevant variables (not affecting the response variable) and redundant variables (not adding anything to the predicting of response). Thus, the eight-variable Model A can be considered an overloaded and noisy model, which should produce unstable predictions with large error variance. The bootstrap decile analysis validation of Model A based on a bootstrap sample size of 175,000 is in Table 29.11. To facilitate the discussion, I added the Cum Lift margins of error for Model B (from Table 29.10) to Table 29.11. The efficiency ratios for the decile Cum Lifts are in the rightmost column of Table 29.11. Note, the decile confidence intervals overlap.

It is clear that Model A is less efficient than Model B. Efficiency ratios are less than 100%, ranging from a low of 86.0% for the top decile to 97.3% for the fourth decile, with one exception for the eighth decile, which has an anomalous ratio of 104.5%. The implication is that Model A predictions are less unstable than Model B predictions (i.e., Model A has larger prediction error variance relative to Model B).

The broader implication is a warning: Review a model with too many variables for justification of the contribution of each variable in the model. Otherwise, there can be an expectation that the model has unnecessarily large prediction error variance. In other words, apply the bootstrap methodology during the model-building stages. A bootstrap decile analysis of model calibration, similar to bootstrap decile analysis of model validation, can be another technique for variable selection and other assessments of model quality.

29.11 Summary

Traditional validation of marketing models involves comparing the Cum Lifts from the calibration and holdout decile analyses based on the model under consideration. If the expected shrinkage (difference in decile Cum Lifts between the two analyses) and the Cum Lift values themselves are acceptable, then the model is considered successfully validated and ready to use; otherwise, rework the model until it is successfully validated. I illustrate with an RM case study that the single-sample validation provides neither assurance that the results are not biased nor any measure of confidence in the Cum Lifts.

I propose the bootstrap method—a computer-intensive approach to statistical inference—as a methodology for assessing the bias and confidence in the Cum Lift estimates. I introduce the bootstrap briefly, along with a simple 10-step procedure for bootstrapping any statistic. I illustrate the procedure for a decile validation of the RM in the case study. I compare and contrast the single-sample and bootstrap decile validations for the case study. It is clear that the bootstrap provides necessary information for a complete validation: the biases and the margins of error of decile Cum Lift estimates.

I address the issue of margins of error (confidence levels) that are too large (low) for the business objective at hand. I demonstrate how to use the bootstrap to decrease the margins of error.

Then, I continue the discussion on the margin of error to a bootstrap assessment of model implementation performance. I speak to the issue of how large a sample is needed to implement a solicitation based on the model to obtain the desired performance quantity. Again, I provide a bootstrap procedure for determining the smallest sample necessary to obtain the desired quantity and quality of performance and illustrate the procedure with a three-variable RM.

Last, I show how the bootstrap decile analysis serves as an assessment of model efficiency. Continuing with the three-variable RM study, I show the efficiency of the three-variable RM relative to the eight-variable alternative model. The latter model has more unstable predictions than the former model. The implication is that review is necessary for a model with too many variables. Required is the justification of each variable's contribution in the model. Otherwise, the model presumably has unnecessarily large prediction error variance. The broader implication is that a bootstrap decile analysis of model calibration, similar to a bootstrap decile analysis of model validation, can be another technique for variable selection and other assessments of model quality.

I provide SAS subroutines for performing bootstrap decile analysis in Chapter 44.

References

1. Noreen, E.W., *Computer Intensive Methods for Testing Hypotheses*, Wiley, New York, 1989.
2. Efron, B., *The Jackknife, the Bootstrap and Other Resampling Plans*, SIAM, Philadelphia, PA, 1982.
3. Draper, N.R., and Smith, H., *Applied Regression Analysis*, Wiley, New York, 1966.
4. Neter, J., and Wasserman, W., *Applied Linear Statistical Models*, Irwin, Homewood, IL., 1974.
5. Hayes, W.L., *Statistics for the Social Sciences*, Holt, Rinehart and Winston, New York, 1973.

30

Validating the Logistic Regression Model: Try Bootstrapping

30.1 Introduction

The purpose of this how-to chapter is to introduce the principal features of a bootstrap validation method for the ever-popular logistic regression model (LRM).

30.2 Logistic Regression Model

Lest the model builder forgets, the LRM depends on the assumption of linearity in the logit. The logit of Y, a binary dependent variable (typically, assume 0 or 1) is a *linear* function of the predictor variables defining LRM (= $b_0 + b_1X_1 + b_1X_2 + \dots + b_nX_n$). Recall, the transformation to convert logit Y = 1 into the probability of Y = 1 is 1 divided by exp(–LRM), where exp is the exponentiation function e^x, and e is the number 2.718281828....

The standard use of the Hosmer–Lemeshow (HL) test, confirming a fitted LRM is the correct model, is not reliable because it does not distinguish between nonlinearity and noise in checking the model fit [1]. Also, the HL test is sensitive to multiple datasets.

30.3 The Bootstrap Validation Method

The *bootstrap validation method* provides (1) a measure of sensitivity of LRM predictions to changes (random perturbations) in the data and (2) a procedure for selecting the best LRM if it exists. Bootstrap samples are randomly different by way of sampling with replacement from the original (training) data. The model builder performs repetitions of logistic regression modeling based on various bootstrap samples.

1. If the bootstrapped LRMs are *stable* (in form and predictions), then the model builder picks one of the candidate (bootstrapped) LRMs as the final model.
2. If the bootstrapped LRMs are *not* stable, then the model builder can either (a) add more data to the original data or (b) take a new, larger sample to serve as the training data. Thus, the model builder starts rebuilding an LRM on the appended

training data or the new sample. These tasks prove quite effective in producing a stable bootstrapped LRM. If the tasks are not effective, then the data are not homogeneous and the model builder proceeds to Step 3.

3. One approach for rendering homogeneous data, discussed in Chapter 38, addresses *overfitting*. Delivering homogeneous data is tantamount to eliminating the components of an overfitted model. The model builder may have to stretch his or her thinking from issues of overfitting to issues of homogeneity, but the effort will be fruitful.

Lest the model builder gets the wrong impression about the proposed method, the bootstrap validation method is not an elixir. Due diligence dictates that a fresh dataset—perhaps one that includes ranges outside the original ranges of the predictor variables X_i—must be used on the final LRM. In sum, I use the bootstrap validation method to yield reliable and robust results along with the final LRM tested on a fresh dataset.

30.4 Summary

This how-to chapter introduces a bootstrap validation method for the ever-popular LRM. The *bootstrap validation method* provides (1) a measure of sensitivity of LRM predictions to changes (random perturbations) in the data and (2) a procedure for selecting the best LRM if it exists.

Reference

1. Hosmer, D.W., and Lemeshow, S., *Applied Logistic Regression*, Wiley, New York, 1989.

31

*Visualization of Marketing Models:** *Data Mining to Uncover Innards of a Model*

31.1 Introduction

Visual displays—commonly known as graphs—have been around for a long time but are currently at their peak of popularity. The popularity is due to the massive amounts of data flowing from the digital environment and the accompanying increase in data visualization software. Visual displays play an essential part in the exploratory phase of data analysis and model building. An area of untapped potential for visual displays is "what the final marketing model is doing" upon the implementation of its intended task of predicting response. Such displays would increase the confidence in the model builder while engendering confidence in the marketer, an end user of marketing models. The purpose of this chapter is to introduce two data mining graphical methods—star graphs and profile curves—for *uncovering the innards of a model:* a visualization of the characteristic and performance levels of the individuals predicted by the model. Also, I provide SAS© subroutines for generating star graphs and profile curves. The subroutines are also available for downloading from my website: http://www.geniq.net/articles.html#section9.

31.2 Brief History of the Graph

The first visual display had its pictorial debut about 2000 BC when the Egyptians created a real estate map describing data such as property outlines and owners. The Greek Ptolemy created the first world map circa AD 150 using latitudinal and longitudinal lines as coordinates to represent the earth on a flat surface. In the fifteenth century, Descartes realized that Ptolemy's geographic mapmaking could serve as a graphical method to identify the relationship between numbers and space, such as patterns [1]. Thus, the common graph was born: a horizontal line (X-axis) and a vertical line (Y-axis) intersecting perpendicularly to create a visual space that occupies numbers defined by an ordered pair of X–Y coordinates. The original Descartes graph, embellished with more than 500 years of knowledge and technology, is the genesis of the discipline of *data visualization,* which is experiencing an unprecedented growth due to the advances in microprocessor technology and a plethora of visualization software.

The scientific community was slow to embrace the Descartes graph, first on a limited basis from the seventeenth century to the mid-eighteenth century, then with much more

* This chapter is based on the following: Ratner, B., Profile curves: a method of multivariate comparison of groups, *The DMA Research Council Journal,* 28–45, 1999. Used with permission.

enthusiasm toward the end of the eighteenth century [2,3]. At the end of the eighteenth century, Playfield initiated work in the area of statistical graphics with the invention of the bar diagram (1786) and pie chart (1801).* Following Playfield's progress with graphical methods, Fourier presented the cumulative frequency polygon, and Quetelet created the frequency polygon and histogram [4]. In 1857, Florence Nightingale—who was a self-educated statistician—unknowingly reinvented the pie chart, which she used in her report to the royal commission to force the British army to maintain nursing and medical care for soldiers in the field [5].

In 1977, Tukey started a revolution of numbers and space with his seminal book, *Exploratory Data Analysis* (EDA) [6]. Tukey explained, by setting forth in careful and elaborate detail, the unappreciated value of numerical, counting, and graphical detective work performed by simple arithmetic and easy-to-draw pictures. Almost three decades after reinventing the concept of graph-making as a means of encoding numbers for strategic viewing, Tukey's *graphic* offspring are everywhere. They include the box-and-whiskers plot taught in early grade schools and easily generated computer-animated and interactive displays in three-dimensional space and with 64-bit color used as a staple in business presentations. Tukey, who has been called the "Picasso of statistics," has visibly left his imprint on the visual display methodology of today [6].

In the previous chapters, I have presented model-based graphical data mining methods, including smooth and fitted graphs and other Tukey-esque displays for the identification of structure of data and fitting of models. Geometry-based graphical methods, which show how the dependent variable varies over the pattern of internal model variables (variables defining the model), are in an area that has not enjoyed growth and can benefit from a Tukey-esque innovation [7]. In this chapter, I introduce two data mining methods to show what the final model is doing. Specifically, the methods provide visualization of the individuals identified by the model regarding *variables of interest*—internal model variables or external model variables (variables not defining the model)—and the levels of performance of those individuals. I illustrate the methods using a response model, but they equally apply to a profit model.

31.3 Star Graph Basics

A table of numbers neatly shows "important facts contained in a jungle of figures" [6]. However, more often than not, the table leaves room for further untangling of the numbers. Graphs can help the data analyst out of the "numerical" jungle with a visual display of where the analyst has been and what he or she has seen. The data analyst can go only so far into the thicket of numbers until the eye–brain connection needs a *data mining* graph to extract patterns and gather insight from within the table.

A star graph† is a visual display of multivariate data, for example, a table of many rows of many variables [8]. It is especially effective for a small table, such as the decile analysis table. The basics of star graph construction are as follows:

1. Identify the units of the star graphs: the *j observations* and the *k variables.* Consider a set of *j* observations. Each observation, which corresponds to a star graph, is defined by an array or row of *k* variables X.

* Michael Friendly (2008). "Milestones in the history of thematic cartography, statistical graphics, and data visualization". 13–14. Retrieved July 7, 2008.

† Star graphs are also known as star glyphs.

2. There are k equidistant rays emitting from the center of the star for each observation.
3. The lengths of the rays correspond to the row of X values. The variables are to be measured assuming relatively similar scales. If not, the data must be transformed to induce comparable scales. A preferred method to accomplish comparable scales is *standardization,* which transforms all the variables to have the same mean value and the same standard deviation value. The mean value used is essentially arbitrary but must satisfy the constraint that the transformed standardized values are positive. The standard deviation value is 1*. The standardized version of X, Z(X), is defined as follows: Z(X) = (X − mean(X))/standard deviation(X). If X assumes all negative values, there is a preliminary step of multiplying X by −1, producing −X. Then, the standardization is performed on −X.
4. The ends of the rays are connected to form a polygon or star for each observation.
5. Circumscribed around each star is a circle. The circumference provides a reference *line,* which aids in interpreting the star. The centers of the star and circle are the same point. The radius of the circle is equal to the length of the largest ray.
6. A star graph typically does not contain labels indicating the X values. If transformations are required, then the transformed values are virtually *meaningless.*
7. The assessment of the *relative differences in the shapes* of the star graphs untangles the numbers and brings out the true insights within the table.

SAS/Graph has a procedure to generate star graphs. I provide the SAS code at the end of the chapter for the illustrations presented.

31.3.1 Illustration

Marketers use models to identify potential customers. Specifically, a marketing model provides a way of identifying individuals into 10 equal-sized groups (deciles), ranging from the top 10% most likely to perform (i.e., respond or contribute profit) to the bottom 10% least likely to perform. The traditional approach of profiling the individuals identified by the model is to calculate the means of variables of interest and assess their values across the deciles.

Consider a marketing model for predicting response to a solicitation. The marketer is interested in what the final model is doing: what the individuals look like in the top decile, in the second decile, and so on. How do the individuals differ across the varying levels (deciles) of performance regarding the usual demographic variables of interest: AGE, INCOME, EDUCATION, and GENDER? Answers to this and related questions provide the marketer with strategic marketing intelligence to put together an effective targeted campaign.

The means of the four demographic variables across the RESPONSE model deciles are in Table 31.1. From the tabular display of means by deciles, I conclude the following:

1. AGE: Older individuals are more responsive than younger individuals.
2. INCOME: High-income individuals are more responsive than low-income individuals.

* The value of 1 is arbitrary but works well in practice.

TABLE 31.1

Response Decile Analysis: Demographic Means

Decile	AGE (years)	INCOME ($000)	EDUCATION (years of Schooling)	GENDER (1 = male, 0 = female)
top	63	155	18	0.05
2	51	120	16	0.10
3	49	110	14	0.20
4	46	111	13	0.25
5	42	105	13	0.40
6	41	95	12	0.55
7	39	88	12	0.70
8	37	91	12	0.80
9	25	70	12	1.00
bottom	25	55	12	1.00

3. EDUCATION: Individuals who attain greater education are more responsive than individuals who attain less education.
4. GENDER: Females are more responsive than males. Note, a mean GENDER of 0 and 1 implies all females and all males, respectively.

This interpretation is correct, albeit not exhaustive because it only considers one variable at a time, a topic covered further in the chapter. It does describe the individuals identified by the model regarding the four variables of interest and responsiveness. However, it does not beneficially stimulate the marketer into strategic thinking for insights. A graph—*an imprint of many insights*—is needed.

31.4 Star Graphs for Single Variables

The first step in constructing a star graph is to identify the units, which serve as the observations and the variables. For decile-based single-variable star graphs, the variables of interest are the j observations, and the 10 deciles (top, 2, 3, …, bot) are the k variables. For the RESPONSE model illustration, there are four star graphs in Figure 31.1, one for each demographic variable. Each star has 10 rays, which correspond to the 10 deciles.

I interpret the star graphs as follows:

1. For AGE, INCOME, and EDUCATION: There is a decreasing trend in the mean values of the variables as the individuals are successively assigned to the top decile down through the bottom decile. These star graphs display the following: Older individuals are more responsive than younger individuals. High-income earners are more responsive than low-income earners. And, individuals who attain greater education are more responsive than individuals who attain less education.
2. AGE and INCOME star graphs are virtually identical, except for the ninth decile, which has a slightly protruding vertex. The implication is that AGE and INCOME

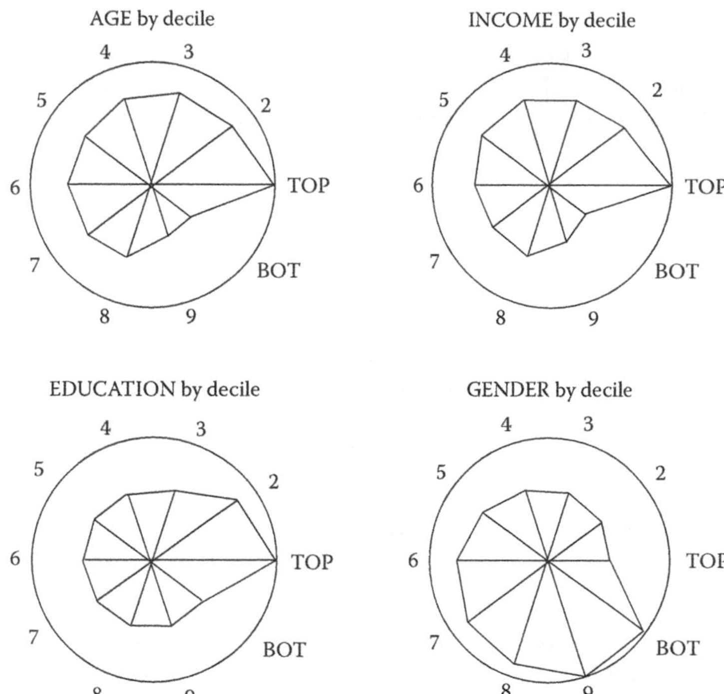

FIGURE 31.1
Star graphs for AGE, INCOME, EDUCATION, and GENDER.

have a similar effect on RESPONSE. Specifically, a standardized unit increase in AGE and INCOME produce a similar change in response.

3. For GENDER: There is an increasing trend in the incidence of males as the individuals are successively assigned to the top decile down through the bottom decile. Keep in mind GENDER is coded zero for females.

In sum, the star graphs provide a unique visual display of the conclusions in Section 31.3.1. However, they are one-dimensional portraits of the effect of each variable on RESPONSE. To obtain a deeper understanding of how the model works as it assigns individuals to the decile analysis table, a full profile of the individuals using all variables *considered jointly* is needed. In other words, data mining to uncover the innards of the decile analysis table is a special need that is required. A multiple-variable star graph provides an unexampled full profile.

31.5 Star Graphs for Many Variables Considered Jointly

As with the single-variable star graph, the first step in constructing the many variable star graphs is to identify the units. For decile-based many variable star graphs, the 10 deciles are the *j* observations, and the variables of interest are the *k* variables. For the RESPONSE

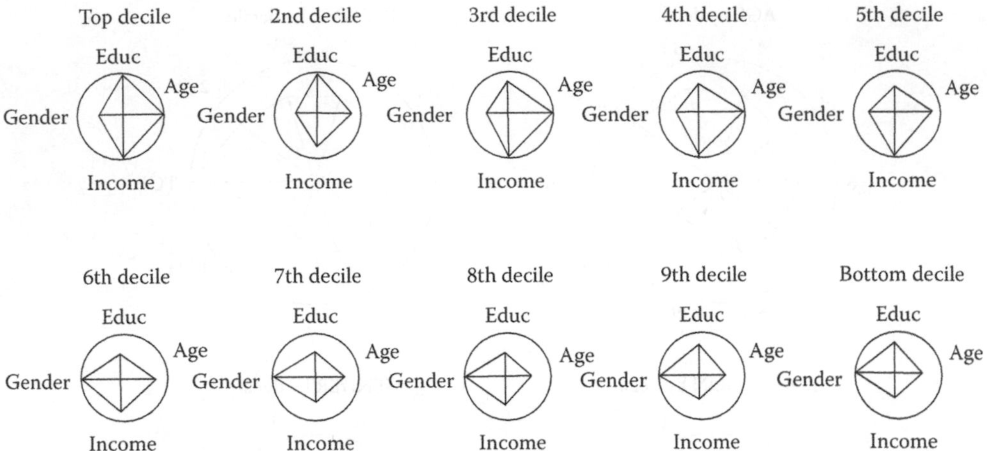

FIGURE 31.2
Star graphs for four demographic variables considered jointly.

model illustration, there are 10 star graphs, one for each decile, in Figure 31.2. Each decile star has four rays, which correspond to the four demographic variables.

I interpret a star graph for an array of variables in a *comparative* context. Because star graphs have no numerical labels, I assess the *shapes* of the stars by observing their *movement* within the reference circle as I go from the top to bottom deciles.

1. *Top decile star:* The rays of AGE, INCOME, and EDUCATION touch or nearly touch the circle circumference. These long rays indicate older, more educated individuals with higher income. The short ray of GENDER indicates these individuals are mostly females. The individuals in the top decile comprise the reference group for a comparative analysis of the other decile stars.
2. *Second-decile star:* Individuals in this decile are slightly younger, with less income than individuals in the top decile.
3. *Third- to fifth-decile stars:* Individuals in these deciles are less educated substantially than individuals in the top two deciles. Also, the education level decreases for individuals in Deciles 3 through 5.
4. *Sixth-decile star:* The shape of this star makes a significant departure from the top five decile stars. This star indicates individuals who are mostly males (because the GENDER ray touches the circle) and are younger, less-educated individuals with less income than the individuals in the upper deciles.
5. *Seventh- to bottom-decile stars:* These stars hardly move within the circle across the lower deciles (Decile 6 to bottom). The stability within the lower deciles indicate the individuals across the least responsive deciles are essentially the same.

In sum, the 10 star graphs provide a storyboard animation of how the four-dimensional profiles change as the model assigns the individuals into the 10 deciles. The first five deciles show a slight progressive decrease in EDUCATION and AGE means. Between the fifth and sixth deciles, there is a sudden change in full profile as the GENDER contribution to profile is now skewed to males. Across the bottom five deciles, there is a slight progressive decrease in INCOME and AGE means.

31.6 Profile Curves Method

I present the profile curves method as an alternative geometry-based data mining graphical method for the problem previously tackled by the star graphs but from a slightly different angle. The star graphs provide the marketer with strategic marketing intelligence for campaign development, as obtained from an ocular inspection and complete descriptive account of their customers regarding variables of interest. In contrast, the profile curves provide the marketer with strategic marketing intelligence for model implementation, specifically determining the number of reliable decile groups.

The profile curves are demanding to build and interpret conceptually, unlike the star graphs. Their construction is not intuitive because they use a series of unanticipated trigonometric functions, and their display is disturbingly abstract. However, the value of profile curves in a well-matched problem solution (as presented here) can offset the initial reaction and difficulty in their use. From the discussion, the profile curves method is clearly a unique data mining method, uncovering unsuspected patterns of what the final model is doing on the implementation of its intended task of predicting.

I discuss the basics of profile curves and the *profile analysis*, which serves as a useful preliminary step to the implementation of the profile curves method. In the next sections, I illustrate profile curves and profile analysis with a RESPONSE model decile analysis. The profile analysis involves simple pairwise scatterplots. The profile curves method requires a special computer program, which SAS/Graph has. I provide the SAS code for the profile curves for the illustration presented in the appendices at the end of the chapter.

31.6.1 Profile Curves* Basics

Consider the curve function f(t) defined in Equation 31.1 [9]:

$$f(t) = X_1/\sqrt{2} + X_2\sin(t) + X_3\cos(t) + X_4\sin(2t) + X_5\cos(2t) + \ldots \quad (31.1)$$

where $-\pi \le t \le \pi$.

The curve function f(t) is a *weighted* sum of *basic curves* for an observation X, represented by many variables, that is, a multivariate data array, $X = \{X_1, X_2, X_3, \ldots, X_k\}$. The *weights* are the values of the Xs. The *basic curves* are trigonometric functions sine and cosine. The plot of f(t) on the Y-axis and t on the X-axis for a set of multivariate data arrays (rows) of mean values for *groups* of individuals are called *profile curves*.

Like the star graphs, the profile curves are a visual display of multivariate data, especially effective for a small table, such as the decile analysis table. Unlike the star graphs, which provide visual displays for single and many variables jointly, the profile curves only provide a visual display of the *joint effects* of the X variables across several groups. A profile curve for a single group is an abstract mathematical representation of the row of mean values of the variables. As such, a single group curve imparts no usable information. Extracting usable information comes from a comparative evaluation of two or more group curves. Profile curves permit a qualitative assessment of the differences among the persons across the groups. In other words, profile curves serve as a method of multivariate comparison of groups.

* Profile curves are also known as curve plots.

31.6.2 Profile Analysis

Database marketers use models to classify customers into 10 deciles, ranging from the top 10% most likely to perform to the bottom 10% least likely to perform. To communicate effectively to the customers, database marketers combine the deciles into groups, typically three: top, middle, and bottom *decile groups*. The top group typifies high-valued/high-responsive customers to harvest by sending enticing solicitations for preserving their performance levels. The middle group represents medium-valued/medium-responsive customers to retain and grow by including them in marketing programs tailored to keep and further stimulate their performance levels. Last, the bottom group depicts a segment of minimal performers and former customers, whose performance levels can be rekindled and reactivated by creative new product and discounted offerings.

Profile analysis is used to create decile groups. Profile analysis consists of (1) calculating the means of the variables of interest and (2) plotting the means of several *pairs* of the variables. These *profile plots* suggest how the individual deciles may be combined. However, because the profiles are multidimensional (i.e., defined by many variables), the assessment of the many profile plots provides an incomplete view of the groups. If the profile analysis is fruitful, it serves as guidance for the profile curves method in determining the number of reliable decile groups.

31.7 Illustration

Returning to the RESPONSE model decile analysis (Table 31.1), I construct three profile plots in Figures 31.3 through 31.5: AGE with INCOME, AGE with GENDER, and EDUCATION with INCOME. The AGE–INCOME plot indicates that AGE and INCOME jointly decrease down through the deciles. That is, the most responsive customers are older with more income, and the least responsive customers are younger with less income.

The AGE–GENDER plot shows that the most responsive customers are older and prominently female, and the least responsive customers are younger and typically male. The EDUCATION–INCOME plot shows that better-educated, higher-income customers are most responsive.

Constructing similar plots for the remaining three pairs of variables is not burdensome, but the task of interpreting all the pairwise plots is formidable.*

The three profile plots indicate various candidate decile group compositions:

1. The AGE–INCOME plot suggests defining the groups as follows:
   a. Top group: top decile.
   b. Middle group: second through eighth decile.
   c. Bottom group: ninth and bottom decile.
2. The AGE–GENDER plot does not reveal any grouping.

* Plotting three variables at a time can be done by plotting a pair of variables for each decile value of the third variable, clearly a challenging effort.

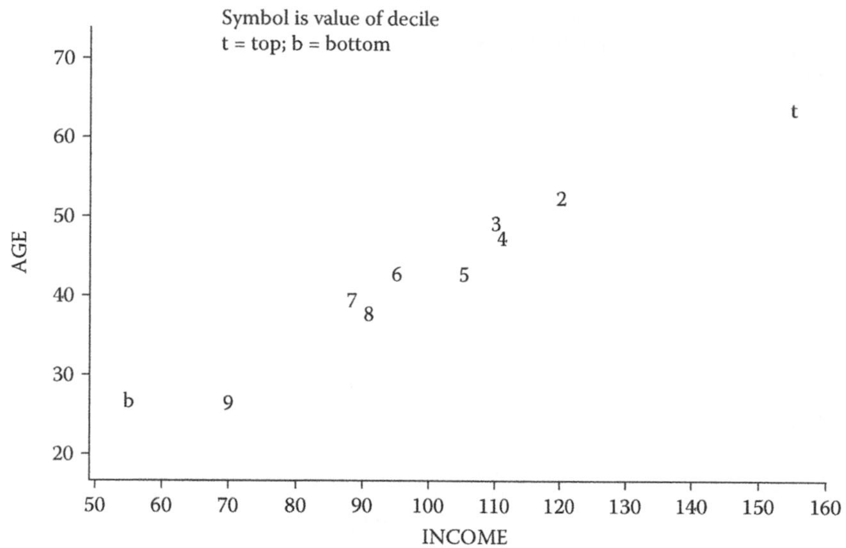

FIGURE 31.3
Plot of AGE and INCOME.

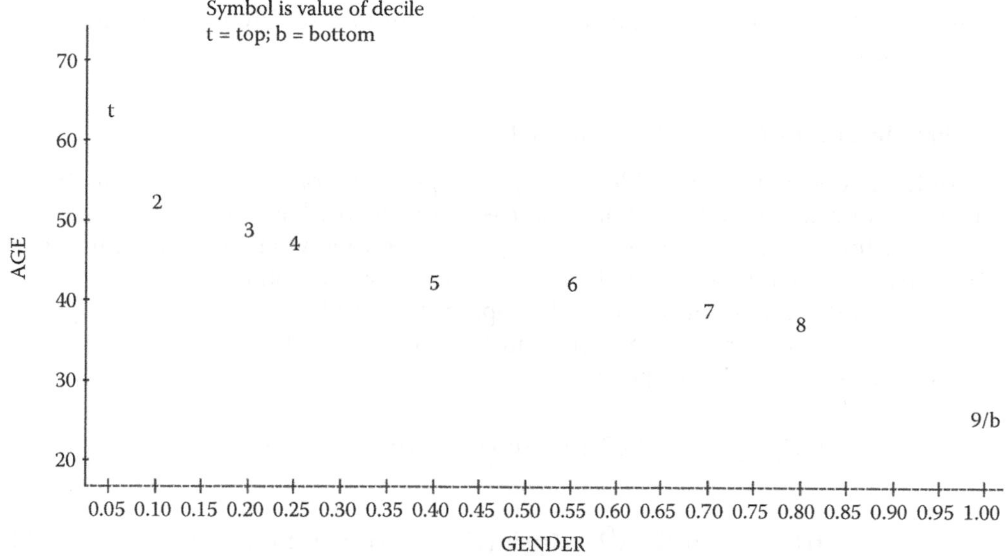

FIGURE 31.4
Plot of AGE and GENDER by decile.

3. The EDUCATION–INCOME plot indicates the following:
   a. Top group: top decile.
   b. Middle group: second decile.
   c. Bottom group: third through bottom deciles; the bottom group splits into two subgroups: Deciles 3 through 8 and Deciles 9 and bottom.

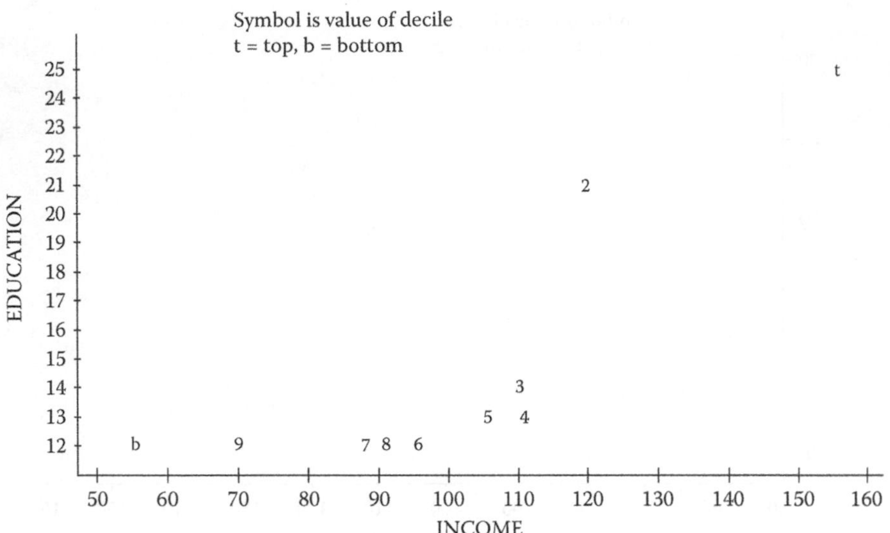

FIGURE 31.5
Plot of EDUCATION and INCOME by decile.

In this case, the profile analysis is not fruitful. It is not clear how to define the best decile grouping based on these findings. Additional plots would probably produce other inconsistent findings.

31.7.1 Profile Curves for RESPONSE Model

The profile curves method provides a graphical presentation (Figure 31.6) of the joint effects of the four demographic variables across all the deciles based on the decile analysis in Table 31.1. Interpretation of this graph is part of the following discussion, in which I illustrate the strategy for creating reliable decile groups with profile curves.

Under the working assumption that the top, middle, and bottom decile groups exist, I create profile curves for the top, fifth, and bottom deciles (Figure 31.7) as defined in Equations 31.2, 31.3, and 31.4, respectively.

$$f(t)_{top\_decile} = 63/\sqrt{2} + 155\sin(t) + 18\cos(t) + 0.05\sin(2t) \tag{31.2}$$

$$f(t)_{5th\_decile} = 42/\sqrt{2} + 105\sin(t) + 13\cos(t) + 0.40\sin(2t) \tag{31.3}$$

$$f(t)_{bottom\_decile} = 25/\sqrt{2} + 55\sin(t) + 12\cos(t) + 1.00\sin(2t) \tag{31.4}$$

The upper, middle, and lower profile curves correspond to the rows of means for the top, fifth, and bottom deciles, respectively, in Table 31.2. The three profile curves form two "hills." Based on a subjective assessment,* I declare that the profile curves are different

* I can test for statistical difference (see Andrew's article [9]); however, I am doing a visual assessment, which does not require any statistical rigor.

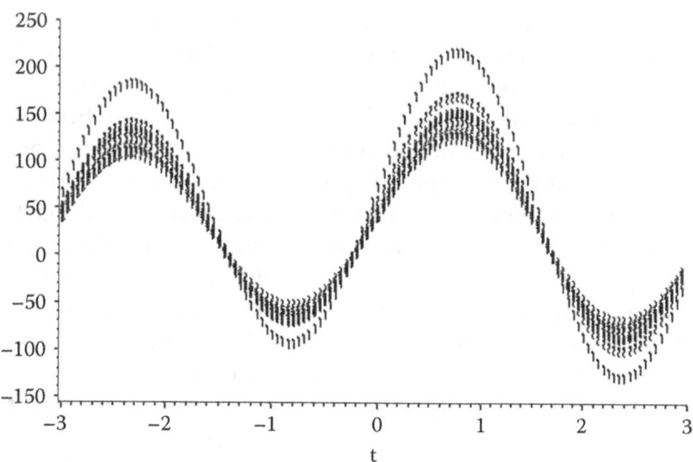

FIGURE 31.6
Profile curves: all deciles.

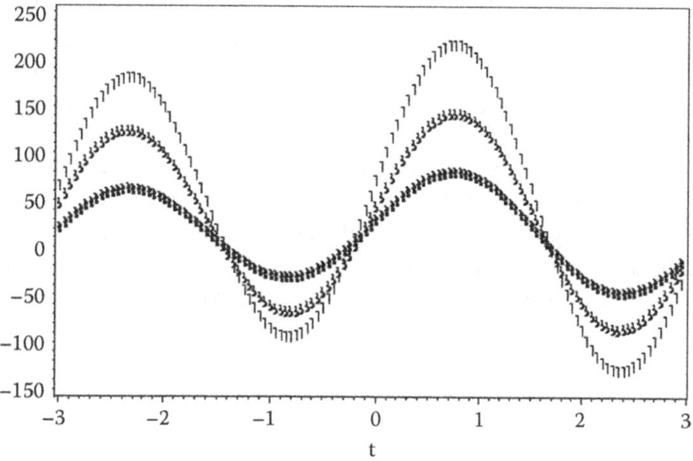

FIGURE 31.7
Profile curves for top, fifth, and bottom deciles.

TABLE 31.2

Response Decile Analysis: Demographic Means for Top, Fifth, and Bottom Deciles

Decile	AGE (years)	INCOME ($000)	EDUCATION (Years of Schooling)	GENDER (1 = male, 0 = female)
top	63	155	18	0.05
5	42	105	13	0.40
bottom	25	55	12	1.00

regarding the slope of the hill. The implication is that the individuals in each decile are different—on the four demographic variables considered jointly—from the individuals in each of the other two deciles; consequently, the deciles cannot be combined.

The graph of the three profile curves in Figure 31.7 exemplifies how the profile curves method works. Large variation among rows corresponds to a set of profile curves that greatly departs from a common shape. Disparate profile curves indicate that the rows should remain separate, not combined. Profile curves that slightly depart from a common shape indicate that the rows can be combined to form a more reliable row. Accordingly, I restate my preemptive implication: The graph of three profile curves indicates that individuals across the three deciles are diverse of the four demographic variables considered jointly, and the deciles cannot be aggregated to form a homogeneous group.

When the rows are obviously different, as in the case in Table 31.2, the profile curves method serves as a confirmatory method. When the variation in the rows is not apparent, as is likely with a large number of variables, noticeable row variation is harder to discern, in which case the profile curves method serves as an exploratory tool.

31.7.2 Decile Group Profile Curves

I have an initial set of decile groups: the top group is the top decile, the middle group is the fifth decile, and the bottom group is the bottom decile. I have to assign the remaining deciles to one of the three groups. Can I include the second decile in the top group? The answer lies in the graph for the top and second deciles in Figure 31.8, from which I observe that the top and second decile profile curves are different. Thus, the top group remains with only the top decile. I assign the second decile to the middle group. Discussion of this assignment follows later.

Can I add the ninth decile to the bottom group? From the graph for the ninth and bottom deciles in Figure 31.9, I observe that the two profile curves are not different. Thus, I add the ninth decile to the bottom group. Can the eighth decile also be added to the bottom group (now consisting of the ninth and bottom deciles)? I observe in the graph for the eighth to bottom deciles in Figure 31.10 that the eighth decile profile curve is somewhat different

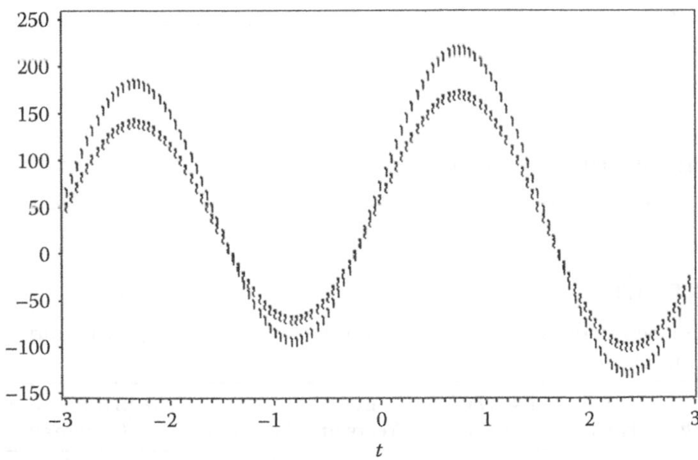

FIGURE 31.8
Profile curves for top and second deciles.

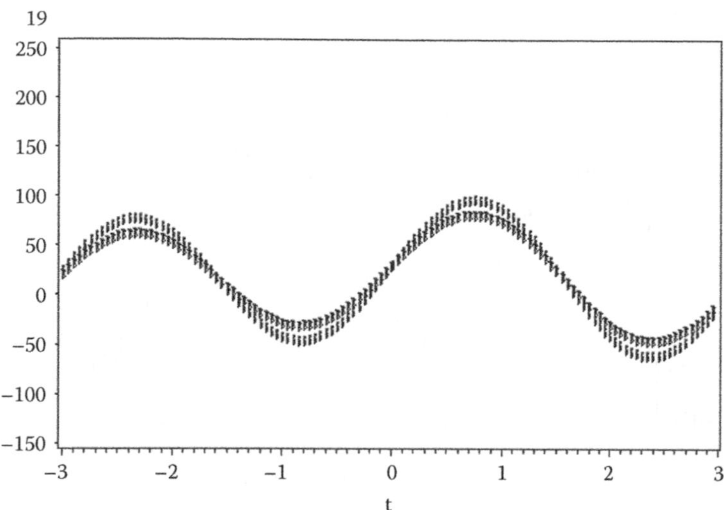

FIGURE 31.9
Profile curves for ninth and bottom deciles.

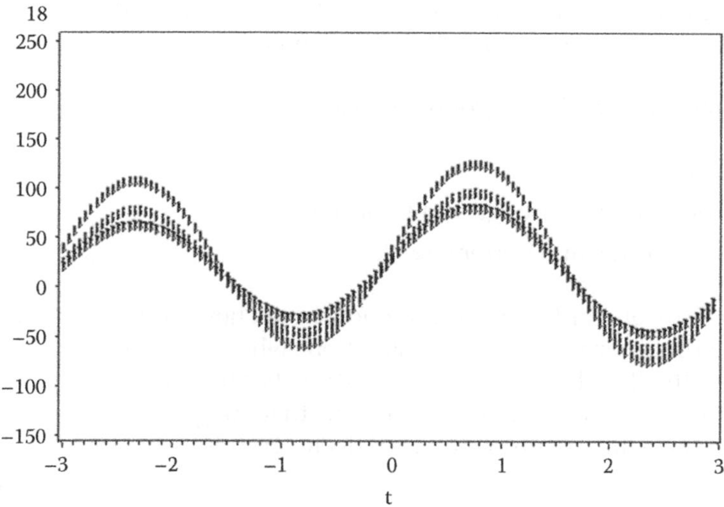

FIGURE 31.10
Profile curves for eighth to bottom deciles.

from the ninth and bottom decile profile curves. Thus, I do not include the eighth decile in the bottom group. I place the eighth decile in the middle group.

To ensure combining Deciles 2 through 8 correctly defines the middle group, I generate the corresponding graph in Figure 31.11. I observe a bold common curve formed by the seven profile curves tightly stacked together. The common curve suggests that the individuals across the deciles are similar. I conclude that the group comprised of Deciles 2 through 8 is homogeneous.

Two points are worth noting. First, the density of the common curve of the middle group is a measure of homogeneity among the individuals across these middle group deciles.

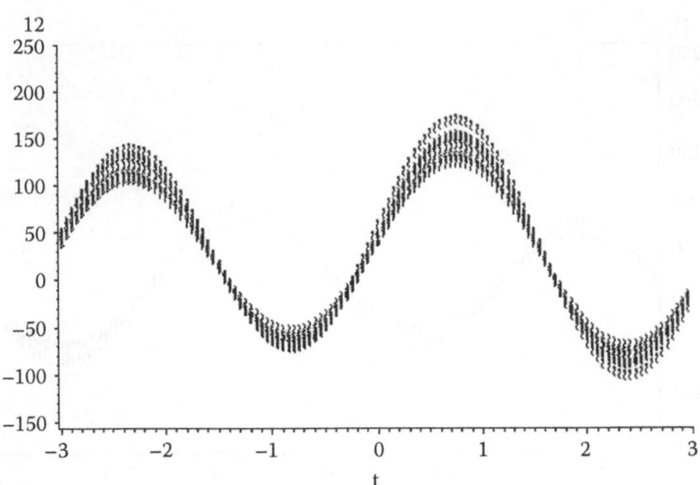

FIGURE 31.11
Profile curves for second through eighth deciles.

Except for a little *daylight* at the top of the right-side hill, the curve is solid. This observation indicates that nearly all the individuals within the deciles are alike. Second, if the curve had a *pattern* of daylight, then the middle group would be divided into subgroups defined by the pattern.

In sum, I define the final decile groups as follows:

1. Top group: top decile
2. Middle group: second through eighth deciles
3. Bottom group: ninth and bottom deciles

The sequential use of profile curves (as demonstrated here) serves as a general technique for multivariate profile analysis. In some situations when the profile data appear obvious, a single graph with all 10 decile curves may suffice. For the current illustration, the single graph in Figure 31.6 shows patterns of daylight that suggest the final decile groups. A closer look at the two hills confirms the final decile groups.

31.8 Summary

After taking a quick look at the history of the graph, from its debut as an Egyptian map to the visual displays of today in three dimensions with 64-bit color, I focus on two underused methods of displaying multivariate data. I propose the star graphs and profile curves as data mining methods of visualizing what the final model is doing on the implementation of its intended task of predicting.

I present the basics of star graph construction along with an illustration of a response model with four demographic variables. The utility of the star graph is apparent when I compare its results with those of traditional profiling, such as examining the means of variables of interest across the deciles. Traditional profiling provides good information, but it is neither compelling nor complete in its examination of how the response model works. It is incomplete in two respects: (1) It only considers one variable at a time without any display and (2) it does not beneficially stimulate the marketer into strategic thinking for insights. The star graph with its unique visual display does stimulate strategic thinking. However, like traditional profiling, the star graph only considers one variable at a time.

Accordingly, I extend the single-variable star graph application to the many-variable star graph application. Continuing with the response illustration, I generate many-variable star graphs, which clearly provide the beneficial stimulation desired. The many-variable star graphs display a full profile using all four demographic variables considered jointly for a complete understanding of how the response model works as it assigns individuals to the response deciles.

Last, I present the unique data mining profile curves method as an alternative method for the problem previously tackled by the star graphs, from a slightly different concern. The star graph provides the marketer with strategic intelligence for campaign development. In contrast, the profile curves provide the marketer with strategic information for model implementation, specifically determining the number of reliable decile groups.

A maiden encounter with profile curves can be unremittingly severe on the data miner. The curves are demanding not only in their construction, presenting an initial affront to the analytical senses, but also in their interpretation requiring visual dexterity. Hopefully, the demonstration of the profile curves reveals its utility.

In the appendices, I provide SAS subroutines for generating star graphs and profile curves.

Appendix 31.A Star Graphs for Each Demographic Variable about the Deciles

```
title1 'table';
data table;
input decile age income educ gender;
cards;
1 63 155 18 0.05
2 51 120 16 0.10
3 49 110 14 0.20
4 46 111 13 0.25
5 42 105 13 0.40
6 41 095 12 0.55
7 39 088 12 0.70
8 37 091 12 0.80
9 25 070 12 1.00
10 25 055 12 1.00
;
run;

PROC PRINT;
run;
```

```
PROC STANDARD data = table out = tablez mean = 4 std = 1;
var age income educ gender;
run;

title1 'table stdz';
PROC PRINT data = tablez;
run;

PROC FORMAT; value dec_fmt
1. = 'top' 2 = ' 2 ' 3 = ' 3 ' 4 = ' 4 ' 5 = ' 5 '
6. = ' 6 ' 7 = ' 7 ' 8 = ' 8 ' 9 = ' 9 ' 10 = 'bot';
run;

PROC GREPLAY nofs igout = work.gseg;
delete all;
run;
quit;

goptions reset = all htext = 1.05 device = win
targetdevice = winprtg ftext = swissb lfactor = 3
hsize = 2 vsize = 8;

PROC GREPLAY nofs igout = work.gseg;
delete all;
run;

goptions reset = all device = win
targetdevice = winprtg ftext = swissb lfactor = 3;
title1 'AGE by Decile';
PROC GCHART data = tablez;
format decile dec_fmt. ;
star decile/fill = empty discrete sumvar = age
slice = outside value = none noheading ;
run;
quit;

title1 'EDUCATON by Decile';
PROC GCHART data = tablez;
format decile dec_fmt. ;
star decile/fill = empty discrete sumvar = educ
slice = outside value = none noheading;
run;
quit;

title1 'INCOME by Decile';
PROC GCHART data = tablez;
format decile dec_fmt. ;
star decile/fill = empty discrete sumvar = income
slice = outside value = none noheading;
run;
quit;
```

```
title1 'GENDER by Decile';
PROC GCHART data = tablez;
format decile dec_fmt.;
star decile/fill = empty discrete sumvar = gender
slice = outside value = none noheading;
run;
quit;

PROC GREPLAY nofs igout = work.gseg tc = sashelp.templt template = l2r2s;
treplay 1:1 2:2 3:3 4:4;
run;
quit;
```

Appendix 31.B Star Graphs for Each Decile about the Demographic Variables

```
data table;
input decile age income educ gender;
cards;
1 63 155 18 0.05
2 51 120 16 0.10
3 49 110 14 0.20
4 46 111 13 0.25
5 42 105 13 0.40
6 41 095 12 0.55
7 39 088 12 0.70
8 37 091 12 0.80
9 25 070 12 1.00
10 25 055 12 1.00
;
run;

PROC STANDARD data = table out = tablez mean = 4 std = 1;
var age income educ gender;
title2 'table stdz';
PROC PRINT data = tablez;
run;

PROC TRANSPOSE data = tablez out = tablezt prefix = dec_;
var age income educ gender;
run;

PROC PRINT data = tablezt;
run;

PROC STANDARD data = tablezt out = tableztz mean = 4 std = 1;
var dec_1 - dec_10;
title2'tablezt stdz';
```

```
PROC PRINT data = tableztz;
run;

PROC TRANSPOSE data = tablez out = tablezt prefix = dec_;
var age income educ gender;
run;

PROC PRINT data = tablezt;
run;

PROC GREPLAY nofs igout = work.gseg;
delete all;
run;
quit;

goptions reset = all htext = 1.05 device = win
target = winprtg ftext = swissb lfactor = 3
hsize = 4 vsize = 8;
title1 'top decile';
PROC GCHART data = tableztz;
star name/fill = empty sumvar = dec_1
slice = outside value = none noheading;
run;
quit;

title1 '2nd decile';
PROC GCHART data = tableztz;
star name/fill = empty sumvar = dec_2
slice = outside value = none noheading;
run;
quit;

title1 '3rd decile';
PROC GCHART data = tableztz;
star name/fill = empty sumvar = dec_3
slice = outside value = none noheading;
run;
quit;

title1 '4th decile';
PROC GCHART data = tableztz;
star name/fill = empty sumvar = dec_4
slice = outside value = none noheading;
run;
quit;

title1 '5th decile';
proc gchart data = tableztz;
```

```
star name/fill = empty sumvar = dec_5
slice = outside value = none noheading;
run;
quit;

title1 '6th decile';
PROC GCHART data = tableztz;
star name/fill = empty sumvar = dec_6
slice = outside value = none noheading;
run;
quit;

title1 '7th decile';
PROC GCHART data = tableztz;
star name/fill = empty sumvar = dec_7
slice = outside value = none noheading;
run;
quit;

title1 '8th decile';
PROC GCHART data = tableztz;
star name/fill = empty sumvar = dec_8
slice = outside value = none noheading;
run;
quit;

title1 '9th decile';
PROC GCHART data = tableztz;
star name/fill = empty sumvar = dec_9
slice = outside value = none noheading;
run;
quit;

title1 'bottom decile';
PROC GCHART data = tableztz;
star name/fill = empty sumvar = dec_10
slice = outside value = none noheading;
run;
quit;

goptions hsize = 0 vsize = 0;
PROC GREPLAY Nofs TC = Sasuser.Templt;
Tdef L2R5 Des = 'Ten graphs: five across, two down'
1/llx = 0 lly = 51
ulx = 0 uly = 100
urx = 19 ury = 100
lrx = 19 lry = 51
2/llx = 20 lly = 51
ulx = 20 uly = 100
```

```
urx = 39 ury = 100
lrx = 39 lry = 51
3/llx = 40 lly = 51
ulx = 40 uly = 100
urx = 59 ury = 100
lrx = 59 lry = 51
4/llx = 60 lly = 51
ulx = 60 uly = 100
urx = 79 ury = 100
lrx = 79 lry = 51
5/llx = 80 lly = 51
ulx = 80 uly = 100
urx = 100 ury = 100
lrx = 100 lry = 51
6/llx = 0 lly = 0
ulx = 0 uly = 50
urx = 19 ury = 50
lrx = 19 lry = 0
7/llx = 20 lly = 0
ulx = 20 uly = 50
urx = 39 ury = 50
lrx = 39 lry = 0
8/llx = 40 lly = 0
ulx = 40 uly = 50
urx = 59 ury = 50
lrx = 59 lry = 0
9/llx = 60 lly = 0
ulx = 60 uly = 50
urx = 79 ury = 50
lrx = 79 lry = 0
10/llx = 80 lly = 0
ulx = 80 uly = 50
urx = 100 ury = 50
lrx = 100 lry = 0;
run;
quit;

PROC GREPLAY Nofs Igout = Work.Gseg
TC = Sasuser.Templt Template = L2R5;
Treplay 1:1 2:2 3:3 4:4 5:5 6:6 7:7 8:8 9:9 10:10;
run;
quit;
```

Appendix 31.C Profile Curves: All Deciles

```
title1'table';
data table;
```

```
input decile age income educ gender;
cards;
1 63 155 18 0.05
2 51 120 16 0.10
3 49 110 14 0.20
4 46 111 13 0.25
5 42 105 13 0.40
6 41 095 12 0.55
7 39 088 12 0.70
8 37 091 12 0.80
9 25 070 12 1.00
10 25 055 12 1.00
;
run;

data table;
set table;
x1 = age; x2 = income; x3 = educ; x4 = gender;

PROC PRINT;
run;

data table10;
sqrt2 = sqrt(2);
array f {10};
do t = -3.14 to 3.14 by.05;
do i = 1 to 10;
set table point = i;
f(i) = x1/sqrt2 + x4*sin(t) + x3*cos(t) + x2*sin(2*t);
end;
output;
label f1 = '00'x;
end;
stop;
run;

goptions reset = all device = win target = winprtg ftext = swissb lfactor = 3;
title1 'Figure 31.6 Profile Curves: All Deciles';
PROC GPLOT data = table10; plot
f1*t = 'T'
f2*t = '2'
f3*t = '3'
f4*t = '4'
f5*t = '5'
f6*t = '6'
f7*t = '7'
f8*t = '8'
f9*t = '9'
f10*t = 'B'
```

```
/overlay haxis = -3 -2 -1 0 1 2 3
nolegend vaxis = -150 to 250 by 50;
run;
quit;
```

References

1. Descartes, R., *The Geometry of Rene Descartes*, Dover, New York, 1954.
2. Costigan-Eaves, P., Data graphics in the 20th century: A comparative and analytical survey, PhD thesis, Rutgers University, Rutgers, NJ, 1984.
3. Funkhouser, H.G., Historical development of the graphical representation of statistical data, *Osiris*, 3, 269–404, 1937.
4. Du Toit, S.H.C., Steyn, A.G.W., and Stumpf, R.H., *Graphical Exploratory Data Analysis*, Springer-Verlag, New York, 1986, p. 2.
5. Salsburg, D., *The Lady Tasting Tea*, Freeman, New York, 2001.
6. Tukey, J.W., *The Exploratory Data Analysis*, Addison-Wesley, Reading, MA, 1977.
7. Snee, R.D., Hare, L.B., and Trout, J.R., *Experiments in Industry. Design, Analysis and Interpretation of Results*, American Society for Quality, Milwaukee, WI, 1985.
8. Friedman, H.P., Farrell, E.S., Goldwyn, R.M., Miller, M., and Siegel, J., A graphic way of describing changing multivariate patterns, in *Proceedings of the Sixth Interface Symposium on Computer Science and Statistics*, University of California Press, Berkeley, CA, 1972.
9. Andrews, D.F., Plots of high-dimensional data, *Biometrics*, 28, 125–136, 1972.

32

The Predictive Contribution Coefficient: A Measure of Predictive Importance

32.1 Introduction

Determining the most important predictor variables in a regression model is a vital element in the interpretation of the model. A general rule is to view the predictor variable with the largest standardized coefficient (SRC) as the most important variable, the predictor variable with the next largest SRC as the next important variable, and so on. This rule is intuitive, easy to apply, and provides practical information for understanding how the model works. Unknown to many, however, is that the rule is theoretically problematic. The purpose of this chapter is twofold: First, to discuss why the decision rule is theoretically amiss yet works well in practice, and second, to present an alternative measure—the *predictive contribution coefficient*—that offers greater utile information than the SRC as it is an assumption-free measure founded in the data mining paradigm.

32.2 Background

Let Y be a continuous dependent variable, and the array $X_1, X_2, \ldots, X_n$ is the predictor variables. The linear regression model is in Equation 32.1:

$$Y = b_0 + b_1{*}X_1 + b_2{*}X_2 + \ldots + b_n{*}X_n \tag{32.1}$$

The b's are the raw regression coefficients, estimated by the method of ordinary least squares. Once the coefficients are estimated, the calculation of an individual's predicted Y value is plugging in the values of the predictor variables for that individual in Equation 32.1.

The interpretation of the raw regression coefficient points to the following question: How does X_i affect the prediction of Y? The answer is the predicted Y experiences an average change of b_i with a unit increase in X_i when the other Xs are held constant. A common misinterpretation of the raw regression coefficient is that the predictor variable with the largest absolute value (coefficient sign ignored) has the greatest effect on the predicted Y. Unless the predictor variables are in the same units, the raw regression coefficient values can be so unequal that any comparison of the coefficients is nonmeaningful. The raw regression coefficients must be standardized to import the different units of measurement of the predictor variables to allow a fair comparison. This discussion is slightly modified for the situation of a binary dependent variable: The linear regression model is the logistic regression model, and the method of maximum likelihood is used to estimate the regression coefficients, which are linear in the logit of Y.

The standardized regression coefficients (SRCs) are just plain numbers, unitless values, allowing material comparisons among the predictor variables. The calculation of SRC for X_i is multiplying the raw regression coefficient for X_i by a conversion factor, a ratio of a unit measure of X_i variation to a unit measure of Y variation. The SRC calculation for an ordinary regression model is in Equation 32.2, where StdDevX_i and StdDevY are the standard deviations for X_i and Y, respectively.

$$\text{SRC for } X_i = (\text{StdDev}X_i/\text{StdDevY})^{*}\text{Raw reg. coeff. for } X_i \tag{32.2}$$

For the logistic regression model, the problem of calculating the standard deviation of the dependent variable, which is logit Y not Y, is complicated. The problem has received attention in the literature, with solutions that provide inconsistent results [1]. The simplest is the one used in SAS©, although it is not without its problems. The StdDev for the logit Y is 1.8138 (the value of the standard deviation of the standard logistic distribution). Thus, the SRC in the logistic regression model is in Equation 32.3.

$$\text{SRC for } X_i = (\text{StdDev}X_i/1.8138)^{*}\text{Raw reg. coeff. for } X_i \tag{32.3}$$

The SRC can also be obtained directly by performing the regression analysis on standardized data. Recall that standardizing the dependent variable Y and the predictor variables X_i's creates new variables, zY and zX_i, such that their means and standard deviations are equal to zero and one, respectively. The coefficients obtained by regressing zY on the zX_i's are, by definition, the SRCs.

The question of which variables in the regression model are its most important predictors in rank order requires annotation before being answered. The importance of ranking is traditionally regarding the statistical characteristic of reduction in prediction error. The usual answer invokes the decision rule: the variable with the largest SRC is the most important variable, the variable with the next largest SRC is the next important variable, and so on. This decision rule is correct with the *unnoted caveat that the predictor variables are uncorrelated.* There is a rank order correspondence between the SRC (coefficient sign ignored) and the reduction in prediction error in a regression model with only uncorrelated predictor variables. There is one other unnoted caveat for the proper use of the decision rule: The SRC for a dummy predictor variable (defined by only two values) is not reliable as the standard deviation of a dummy variable is not meaningful.

Regression models in virtually all applications have correlated predictor variables, which challenge the utility of the decision rule. The rule provides continuously useful information for understanding how the model works without raising sophistic findings. The reason for its utility is there is an unknown working assumption at play: The reliability of the ranking based on the SRC increases as the average correlation among the predictor variables decreases (see Chapter 16). Thus, for well-built models, which necessarily have a minimal correlation among the predictor variables, the decision rule remains viable in virtually all regression applications. However, there are caveats: Dummy variables cannot be ranked; composite variables and their elemental component variables (defining the composite variables) are highly correlated inherently and thus cannot be ranked reliably.

32.3 Illustration of Decision Rule

Consider RESPONSE (0 = no, 1 = yes), PROFIT (in dollars), AGE (in years), GENDER (1 = female, 0 = male), and INCOME (in thousand dollars) for 10 individuals in the small data in Table 32.1. I standardized the data to produce the standardized variables with the notation used previously: zRESPONSE, zPROFIT, zAGE, zGENDER, and zINCOME (data not shown).

I perform two ordinary regression analyses based on both the raw data and the standardized data. Specifically, I regress PROFIT on INCOME, AGE, and GENDER and regress zPROFIT on zINCOME, zAGE, and zGENDER. The raw regression coefficients and SRCs based on the raw data are in the "Parameter Estimate" and "Standardized Estimate" columns in Table 32.2, respectively. The raw regression coefficients for INCOME, AGE, and GENDER are 0.3743, 1.3444, and –11.1060, respectively. The SRCs for INCOME, AGE, and GENDER are 0.3516, 0.5181, and –0.1998, respectively.

The raw regression coefficients and SRCs based on the standardized data are in the "Parameter Estimate" and "Standardized Estimate" columns in Table 32.3, respectively. As expected, the raw regression coefficients—which are now the SRCs—are equal to the values in the "Standardized Estimate" column; for zINCOME, zAGE, and zGENDER, the values are 0.3516, 0.5181, and –0.1998, respectively.

I calculate the average correlation among the three predictor variables, which is a gross 0.71. Thus, the SRC provides a questionable ranking of the predictor variables for PROFIT. AGE is the most important predictor variable, followed by INCOME; the ranked position of GENDER is undetermined.

TABLE 32.1

Small Data

ID	RESPONSE	PROFIT	AGE	GENDER	INCOME
1	1	185	65	0	165
2	1	174	56	0	167
3	1	154	57	0	115
4	0	155	48	0	115
5	0	150	49	0	110
6	0	119	40	0	99
7	0	117	41	1	96
8	0	112	32	1	105
9	0	107	33	1	100
10	0	110	37	1	95

TABLE 32.2

PROFIT Regression Output Based on Raw Small Data

Variable	DF	Parameter Estimate	Standard Error	*t* Value	Pr > *t*	Standardized Estimate
Intercept	1	37.4870	16.4616	2.28	0.0630	0.0000
INCOME	1	0.3743	0.1357	2.76	0.0329	0.3516
AGE	1	1.3444	0.4376	3.07	0.0219	0.5181
GENDER	1	–11.1060	6.7221	–1.65	0.1496	–0.1998

TABLE 32.3

PROFIT Regression Output Based on Standardized Small Data

Variable	DF	Parameter Estimate	Standard Error	*t* Value	Pr 1+1	Standardized Estimate
Intercept	1	–4.76E–16	0.0704	0.0000	1.0000	0.0000
zINCOME	1	0.3516	0.1275	2.7600	0.0329	0.3516
zAGE	1	0.5181	0.1687	3.0700	0.0219	0.5181
zGENDER	1	–0.1998	0.1209	–1.6500	0.1496	–0.1998

TABLE 32.4

RESPONSE Regression Output Based on Raw Small Data

Variable	DF	Parameter Estimate	Standard Error	Wald Chi-Square	Pr > Chi-Square	Standardized Estimate
Intercept	1	–99.5240	308.0000	0.1044	0.7466	
INCOME	1	0.0680	1.7332	0.0015	0.9687	1.0111
AGE	1	1.7336	5.9286	0.0855	0.7700	10.5745
GENDER	1	14.3294	82.4640	0.0302	0.8620	4.0797

TABLE 32.5

RESPONSE Regression Output Based on Standardized Small Data

Variable	DF	Parameter Estimate	Standard Error	Wald Chi-Square	Pr > Chi-Square	Standardized Estimate
Intercept	1	–6.4539	31.8018	0.0412	0.8392	
zINCOME	1	1.8339	46.7291	0.0015	0.9687	1.0111
zAGE	1	19.1800	65.5911	0.0855	0.7700	10.5745
zGENDER	1	7.3997	42.5842	0.0302	0.8620	4.0797

I perform two logistic regression analyses* based on both the raw data and the standardized data. Specifically, I regress RESPONSE on INCOME, AGE, and GENDER and regress RESPONSE† on zINCOME, zAGE, and zGENDER. The raw and standardized logistic regression coefficients based on the raw data are in the "Parameter Estimate" and "Standardized Estimate" columns in Table 32.4, respectively. The raw logistic regression coefficients for INCOME, AGE, and GENDER are 0.0680, 1.7336, and 14.3294, respectively. The standardized logistic regression coefficients for zINCOME, zAGE, and zGENDER are 1.8339, 19.1800, and 7.3997, respectively (Table 32.5). Although this is a trivial example, it still serves valid for *predictive contribution coefficient* (PCC) calculations.

The raw and standardized logistic regression coefficients based on the standardized data are in the "Parameter Estimate" and "Standardized Estimate" columns in Table 32.5, respectively. Unexpectedly, the raw logistic regression coefficients—which are now the standardized logistic regression coefficients—do not equal the values in the "Standardized Estimate" column. Regarding ranking, the inequality of raw and standardized logistic

* I acknowledge that the dataset for a logistic regression model has complete separation of data. However, this condition does not affect the illustration of the PCC approach.

† I choose not to use zRESPONSE. Why?

regression coefficients presents no problem as the raw and standardized values produce the same rank order. If information about the expected increase in the predicted Y is required, I prefer the SRCs in the "Parameter Estimate" column as it follows the definition of SRC.

As previously determined, the average correlation among the three predictor variables is a gross 0.71. Thus, the SRC provides a questionable ranking of the predictor variables for RESPONSE. AGE is the most important predictor variable, followed by INCOME; the ranked position of GENDER is undetermined.

32.4 Predictive Contribution Coefficient

The proposed PCC is a development in the data mining paradigm. The PCC is flexible because it is an assumption-free measure that works equally well with ordinary and logistic regression models. The PCC has practicality and innovation because it offers greater utile information than the SRC, and above all, the PCC has simplicity because it is easy to understand and calculate, as is evident from the following discussion.

Consider the linear regression model built on standardized data and defined in Equation 32.4:

$$zY = b_0 + b_1{*}zX_1 + b_2{*}zX_2 + \ldots + b_i{*}zX_i + \ldots + b_n{*}zX_n \tag{32.4}$$

The PCC for zX_i, PCC(zX_i), is a measure of the contribution of zX_i relative to the contribution of the other variables to the predictive scores of the model. PCC(zX_i) is the average absolute ratio of the zX_i score-point contribution ($zX_i{*}b_i$) to the score-point contribution (total predictive score minus zX_i's score point) of the other variables. Briefly, the PCC reads as follows: The larger the PCC(zX_i) value is, the more significant the part zX_i has in the predictions of the model, and in turn, the greater importance zX_i has as a predictor variable. Exactly how the PCC works and what its benefits are over the SRC are discussed in the next section. Now, I provide justification for the PCC.

Justification of the trustworthiness of the PCC is required as it depends on the SRC, which itself, as discussed, is not a perfect measure to rank predictor variables. The effects of any impurities (biases) carried by the SRC on the PCC are presumably negligible due to the *wash cycle* calculations of the PCC. The six-step cycle, described in the next section, crunches the actual values of the SRC such that any original bias effects are washed out.

I now revisit the question of which variables in a regression model are the most important predictors. The predictive contribution decision rule is: The variable with the largest PCC is the most important variable, the variable with the next largest PCC is the next important variable, and so on. The predictor variables can be ranked from most to least important based on the descending order of the PCC. Unlike the decision rule for a reduction in prediction error, there are no presumable caveats for the predictive contribution decision rule. Correlated predictor variables, including composite and dummy variables, can be thus ranked.

32.5 Calculation of Predictive Contribution Coefficient

Consider the logistic regression model based on the standardized data in Table 32.5. Iillustrate in detail the calculation of PCC(zAGE) with the necessary data in Table 32.6.

1. Calculate the total predicted (logit) score for individuals in the data. For individual ID 1, the values of the standardized predictor variable in Table 32.6 multiplied by the corresponding SRC in Table 32.5 produce the total predicted score of 24.3854 (Table 32.6).
2. Calculate the zAGE score-point contribution for individuals in the data. For individual ID 1, the zAGE score-point contribution is 33.2858 (= 1.7354*19.1800).*
3. Calculate the other variables score-point contribution for individuals in the data. For individual ID 1, the other variables score-point contribution is –8.9004 (= 24.3854–33.2858).†
4. Calculate the zAGE_OTHVARS for individuals in the data; zAGE_OTHVARS is the absolute ratio of zAGE score-point contribution to the other variables' score-point contribution. For ID 1, zAGE_OTHVARS is 3.7398 (= absolute value of 33.2858/–8.9004).
5. Calculate PCC(zAGE), the average (median) of the zAGE_OTHVARS values: 2.8787. The zAGE_OTHVARS distribution has a typical skewness, which suggests that the median is more appropriate than the mean for the average.

I summarize the results of the PCC calculations after applying the five-step process for zINCOME and zGENDER (not shown).

1. AGE ranks top and is the most important predictor variable. GENDER is next important, and INCOME is last. Their PCC values are 2.8787, 0.3810, and 0.0627, respectively.

TABLE 32.6

Necessary Data

ID	zAGE	zINCOME	zGENDER	Total Predicted Score	zAGE Score-Point Contribution	OTHERVARS Score-Point Contribution	zAGE_ OTHVARS
1	1.7354	1.7915	–0.7746	24.3854	33.2858	–8.9004	3.7398
2	0.9220	1.8657	–0.7746	8.9187	17.6831	–8.7643	2.0176
3	1.0123	–0.0631	–0.7746	7.1154	19.4167	–12.3013	1.5784
4	0.1989	–0.0631	–0.7746	–8.4874	3.8140	–12.3013	0.3100
5	0.2892	–0.2485	–0.7746	–7.0938	5.5476	–12.6414	0.4388
6	–0.5243	–0.6565	–0.7746	–23.4447	–10.0551	–13.3897	0.7510
7	–0.4339	–0.7678	1.1619	–7.5857	–8.3214	0.7357	11.3105
8	–1.2474	–0.4340	1.1619	–22.5762	–23.9241	1.3479	17.7492
9	–1.1570	–0.6194	1.1619	–21.1827	–22.1905	1.0078	22.0187
10	–0.7954	–0.8049	1.1619	–14.5883	–15.2560	0.6677	22.8483

* For ID 1, zAGE = 1.7354; 19.1800 = SRC for zAGE.

† For ID 1, Total Predicted = 24.3854; zAGE score-point contribution = 33.2858.

2. AGE is the most important predictor variable with the largest and "large" PCC value of 2.8787. The implication is that AGE is clearly driving the predictions of the model. When a predictor variable has a large PCC value, it is known as a *key driver* of the model.

Note, I do not use the standardized variable names (e.g., zAGE instead of AGE) in the summary of findings. The standardized variable names reflect the issue of determining the importance of predictor variables in the content of the variable, clearly conveyed by the original name, not the technical name, which reflects a mathematical necessity of the calculation process.

32.6 Extra-Illustration of Predictive Contribution Coefficient

This section assumes an understanding of the decile analysis, discussed in full detail in Chapter 26. Readers who are not familiar with the decile analysis may still be able to glean the key points of the following discussion without reading Chapter 26.

The PCC offers greater utile information than the SRC, notwithstanding the working assumption and caveats of the SRC. Both coefficients provide an overall ranking of the predictor variables in a model. However, the PCC can extend beyond an overall ranking by providing a ranking at various levels of model performance. Moreover, the PCC allows for the identification of key drivers—salient features—which the SRC metric cannot legitimately yield. By way of continuing with the illustration, I discuss these two benefits of the PCC.

Consider the decile performance of the logistic regression model based on the standardized data in Table 32.5. The decile analysis in Table 32.7 indicates the model works well as it identifies the three responders in the top three deciles.

The PCC as presented so far provides an overall model ranking of the predictor variables. In contrast, the admissible calculations of the PCC at the decile level provide a decile ranking of the predictor variables with a response modulation, ranging from most

TABLE 32.7

Decile Analysis for RESPONSE Logistic Regression

Decile	Number of Individuals	Number of Responses	RESPONSE RATE (%)	CUM RESPONSE RATE (%)	CUM LIFT
top	1	1	100	100.0	333
2	1	1	100	100.0	333
3	1	1	100	100.0	333
4	1	0	0	75.0	250
5	1	0	0	60.0	200
6	1	0	0	50.0	167
7	1	0	0	42.9	143
8	1	0	0	37.5	125
9	1	0	0	33.3	111
bottom	1	0	0	30.0	100
Total	**10**	**3**			

to least likely to respond. To affect a decile-based calculation of the PCC, I rewrite Step 5 in Section 32.5 on the calculation of the PCC:

Step 5: Calculate PCC(zAGE)—the median of the zAGE_OTHVARS values for each decile.

The PCC decile-based small data calculations, which are obvious and trivial, are presented to make the PCC concept and procedure clear and to generate interest in its application. Each of the 10 individuals is itself a decile in which the median value is the value of the individual. However, this point is also instructional. The reliability of the PCC value is sample size dependent. In real applications, in which the decile sizes are large to ensure the reliability of the median, the PCC decile analysis is quite informational regarding how the predictor variables' rankings interact across the response modulation produced by the model. The decile PCCs for the RESPONSE model in Table 32.8 are clearly daunting. I present two approaches to analyze and draw implications from the seemingly scattered arrays of PCCs left by the decile-based calculations. The first approach ranks the predictor variables by the decile PCCs for each decile. The rank values, which descend from 1 to 3, from most to least important predictor variable, respectively, are in Table 32.9. Next, I will discuss the comparison between decile PCC rankings and the Overall PCC ranking.

The analysis and implications of Table 32.9 are as follows:

1. The Overall PCC importance ranking is AGE, GENDER, and INCOME, in descending order. The implication is that an inclusive marketing strategy is defined by a primary focus on AGE, a secondary emphasis on GENDER, and incidentally calling attention to INCOME.
2. The decile PCC importance rankings are in agreement with the Overall PCC importance ranking for all but Deciles 2, 4, and 6.
3. For Decile 2, AGE remains most important, whereas GENDER and INCOME are reversed in their importance about the overall PCC ranking.
4. For Deciles 4 and 6, INCOME remains the least important, whereas AGE and GENDER are reversed in their importance about the overall PCC ranking.

TABLE 32.8

Decile PCC: Actual Values

Decile	PCC(zAGE)	PCC(zGENDER)	PCC(zINCOME)
top	3.7398	0.1903	0.1557
2	2.0176	0.3912	0.6224
3	1.5784	0.4462	0.0160
4	0.4388	4.2082	0.0687
5	11.3105	0.5313	0.2279
6	0.3100	2.0801	0.0138
7	22.8483	0.3708	0.1126
8	22.0187	0.2887	0.0567
9	17.7492	0.2758	0.0365
bottom	0.7510	0.3236	0.0541
Overall	2.8787	0.3810	0.0627

TABLE 32.9

Decile PCC: Rank Values

Decile	PCC(zAGE)	PCC(zGENDER)	PCC(zINCOME)
top	1	2	3
2	1	3	2
3	1	2	3
4	2	1	3
5	1	2	3
6	2	1	3
7	1	2	3
8	1	2	3
9	1	2	3
bottom	1	2	3
Overall	1	2	3

5. The implication is that two decile tactics are necessary, beyond the inclusive marketing strategy. For individuals in Decile 2, careful planning includes particular prominence given to AGE, secondarily mentions INCOME, and incidentally addresses GENDER. For individuals in Deciles 4 and 6, careful planning includes particular prominence given to GENDER, secondarily mentions AGE, and incidentally addresses INCOME.

The second approach determines decile-specific key drivers by focusing on the actual values of the obsequious decile PCCs in Table 32.8. To put a methodology in place, a measured value of "large proportion of the combined predictive contribution" is needed. Remember that AGE is informally declared a key driver of the model because of its large PCC value, which indicates that the predictive contribution of zAGE represents a large proportion of the combined predictive contribution of zINCOME and zGENDER. Accordingly, I define predictor variable X_i as a key driver as follows: *X_i is a key driver if PCC(X_i) is greater than 1/(k – 1), where k is the number of other variables in the model; otherwise, X_i is not a key driver.* The value 1/(k – 1) is, of course, user-defined, but I presuppose that if the score-point contribution of a single predictor variable is greater than the rough average score-point contribution of the other variables, it can safely be declared a key driver of model predictions.

The key driver definition is used to recode the actual PCC values into 0–1 values, which represent non-key drivers/key drivers. The result is a key driver table, which serves as a means of formally declaring the decile-specific key drivers and overall model key drivers. In particular, the table reveals key driver patterns across the decile about the overall model key drivers. I use the key driver definition to recode the actual PCC values in Table 32.9 into key drivers in Table 32.10.

The analysis and implications of Table 32.10 are as follows:

1. There is a single overall key driver of the model, AGE.
2. AGE is also the sole key driver for Deciles top, 3, 7, 8, 9, and bottom.
3. AGE and INCOME are the drivers of Decile 2.
4. GENDER is the only driver for both Deciles 4 and 6.
5. AGE and GENDER are the drivers of Decile 5.

TABLE 32.10
Decile PCC: Key Drivers

Decile	AGE	GENDER	INCOME
top	1	0	0
2	1	0	1
3	1	0	0
4	0	1	0
5	1	1	0
6	0	1	0
7	1	0	0
8	1	0	0
9	1	0	0
bottom	1	0	0
Overall	1	0	0

6. The implication is that the salient feature of an inclusive marketing strategy is AGE. Thus, the "AGE" message must be tactically adjusted to fit the individuals in Deciles 1 through 6 (the typical range for model implementation). Specifically, for Decile 2, the marketing strategist must add INCOME to the AGE message; for Deciles 4 and 6, the marketing strategist must center attention on GENDER with an undertone of AGE; and for Decile 5, the marketing strategist must add a GENDER component to the AGE message.

32.7 Summary

I briefly review the traditional approach of using the magnitude of the raw regression coefficient or SRCs in determining which variables in a model are its most important predictors. I explain why neither coefficient yields a perfect importance ranking of predictor variables. The raw regression coefficient does not account for inequality in the units of the variables. Furthermore, the SRC is theoretically problematic when the variables are correlated as is virtually the situation in database applications. With a small dataset, I illustrate for ordinary and logistic regression models the often-misused raw regression coefficient and the dubious use of the SRC in determining the rank importance of the predictor variables.

I point out that the ranking based on the SRC provides useful information without raising sophistic findings. An unknown working assumption is that the reliability of the ranking based on SRC increases as the average correlation among the predictor variables decreases. Thus, for well-built models, which necessarily have a minimal correlation among the predictor variables, the resultant ranking is an accepted practice.

Then, I present the PCC, which offers greater utile information than the SRC. I illustrate how it works, as well as its benefits over the SRC using the small dataset.

Last, I provide an extra-illustration of the benefits of the new coefficient over the SRC. First, it can rank predictor variables at a decile level of model performance. Second, it can identify a newly defined predictor variable type, namely, the key driver. The new coefficient allows the data analyst to determine the overall model level and the individual decile levels at which predictor variables are the predominant or key driver variables of the predictions of the model.

Reference

1. Menard, S., *Applied Logistic Regression Analysis,* Quantitative Applications in the Social Sciences Series, Sage, Thousand Oaks, CA, 1995.

33

Regression Modeling Involves Art, Science, and Poetry, Too

33.1 Introduction

The statistician's utterance "regression modeling involves art and science" implies a mixture of *skill acquired by experience* (art) and a *technique that reflects a precise application of fact or principle* (science). The purpose of this chapter is to put forth my assertion that regression modeling involves the trilogy of art, science, and concrete poetry. With concrete poetry, the poet's intent is conveyed by graphic patterns of symbols (e.g., a regression equation) rather than by the conventional arrangement of words. As an example of the regression trilogy, I make use of a metrical "modelogue" to introduce the machine-learning technique GenIQ, an alternative to the statistical regression models. Interpreting the modelogue requires an understanding of the GenIQ Model. I provide the interpretation after presenting the Shakespearean modelogue "To Fit or Not to Fit Data to a Model."

33.2 Shakespearean Modelogue

To Fit or Not to Fit Data to a Model

To fit or not to fit data to a model—that is the question:
Whether 'tis nobler in the mind to suffer
The slings and arrows of outrageously using
The statistical regression paradigm of
Fitting data to a prespecified model, conceived and tested
Within the *small-data setting* of the day, 200 plus years ago,
Or to take arms against a sea of troubles
And, by opposing, move aside fitting data to a model.
Today's big data necessitates—*Let the data define the model.*
Fitting big data to a prespecified *small-framed* model
Produces a *skewed* model with
Doubtful interpretability and questionable results.
When we have shuffled off the expected coil,

There's the respect of the GenIQ Model,
A machine-learning alternative regression model
To the statistical regression model.
GenIQ is an assumption-free, free-form model that
Maximizes the Cum Lift statistic, equally, the decile table.

BRUCE "SHAKESPEARE" RATNER

33.3 Interpretation of the Shakespearean Modelogue

At the beginning of every day, model builders whose tasks are to predict continuous and binary outcomes are likely to be put to use the ordinary least squares (OLS) regression model and the logistic regression model (LRM), respectively. These regression techniques give the promise of another workday of successful models. The essence of any prediction model is the fitness function, which quantifies the optimality (goodness or accuracy) of a solution (predictions). The fitness function of the OLS regression model is mean squared error (MSE), which is minimized by calculus.

Historians date calculus to the time of the ancient Greeks, circa 400 BC. Calculus started making great strides in Europe toward the end of the eighteenth century. Leibniz and Newton pulled these ideas together. They are both credited with the independent invention of calculus. The OLS regression model is celebrating 200 plus years of popularity as the invention of the method of least squares was on March 6, 1805. The backstory of the OLS regression model follows [1].

In 1795, Carl Friedrich Gauss at the age of 18 developed the fundamentals of least squares analysis. Gauss did not publish the method until 1809.* In 1822, Gauss was able to state that the least squares approach to regression analysis is optimal, in the sense that the best linear unbiased estimator of the coefficients is the least squares estimator. This result is known as the Gauss–Markov theorem. However, the term *method of least squares* was coined by Adrien Marie Legendre (1752–1833). Legendre published his method in the appendix *Sur la Méthode des moindres quarrés* ("*On the Method of Least Squares*").† The appendix is dated March 6, 1805 [2].

The fitness function of the LRM is the likelihood function, which is maximized by calculus (i.e., the method of maximum likelihood). The logistic function has its roots spread back to the nineteenth century, when the Belgian mathematician Verhulst invented the function, which he named logistic, to describe population growth. The rediscovery of the function in 1920 is due to Pearl and Reed. The survival of the term *logistic* is due to Yule, and the introduction of the function in statistics is due to Berkson. Berkson used the logistic function in his regression model as an alternative to the normal probability probit model, usually credited to Bliss in 1934 and sometimes to Gaddum in 1933. (The probit can be first traced to Fechner in 1860.) As of 1944, the statistics community did not view Berkson's logistic as a viable alternative to Bliss's probit. After the ideological debate about

* Gauss did not publish the method until 1809, when it appeared in Volume 2 of his work on celestial mechanics, *Theoria Motus Corporum Coelestium in sectionibus conicis solem ambientium*. Gauss, C.F., *Theoria Motus Corporum Coelestium*, 1809.

† "On the Method of Least Squares" is the title of an appendix to *Nouvelles méthodes pour la détermination des orbites des comètes*.

the logistic and probit had abated in the 1960s, Berkson's logistic gained wide acceptance. Coining the term *logit* by analogy to the probit of Bliss, who coined the term *probit* for "probability unit," Berkson was much derided [3,4].

The not-yet-popular model is the GenIQ Model, a machine-learning alternative model to the statistical OLS and LRM. I conceived and developed this model in 1994. The modeling paradigm of GenIQ is to "let the data define the model," which is the complete antithesis of the statistical modeling paradigm, "fit the data to a model." GenIQ automatically (1) data mines for new variables, (2) performs the variable selection, and (3) specifies the model—to optimize the decile table* (i.e., to fill the upper deciles with as much profit or as many responses as possible). The fitness function of GenIQ is the decile table, maximized by the Darwinian-inspired machine-learning paradigm of genetic programming (GP). Operationally, optimizing the decile table creates the best possible descending ranking of the dependent variable (outcome) values. Thus, the prediction of GenIQ is that of identifying individuals who are most to least likely to respond (for a binary outcome) or who contribute large-to-small profits (for a continuous outcome).

In 1882, Galton was the first to use the term *decile* [5]. Historians traced the first use of the decile table, originally called a gains chart, to the direct mail business of the early 1950s [6]. The direct mail business has its roots inside the covers of matchbooks. More recently, the decile table has transcended the origin of the gains chart toward a generalized measure of model performance. The decile table is the model performance display used in direct and database marketing and telemarketing, marketing mix optimization programs, customer relationship management (CRM) campaigns, social media platforms and e-mail broadcasts, and the like. Historiographers cite the first experiments using GP by Stephen F. Smith (1980) and Nichael L. Cramer (1985) [7]. The seminal book, *Genetic Programming: On the Programming of Computers by Means of Natural Selection,* is by John Koza (1992), who is considered the inventor of GP [7–9].

Despite the easy implementation of GenIQ (simply insert the GenIQ equation into the scoring database), it is not yet the everyday regression model because of the following two checks:

1. Unsuspected equation: GenIQ output is the visual display called a parse tree, depicting the GenIQ Model and the GenIQ Model "equation" (which is a computer program/code). Regression modelers are taken aback when they see, say, the Pythagorean Theorem in GenIQ computer code (Figure 33.1).
2. Ungainly interpretation: The GenIQ parse tree and computer code can be a bit much to grasp (for unknown models/solutions). The visual display provides the modeler with an ocular sense of comfort and confidence in understanding and using the GenIQ Model. The GenIQ tree for the Pythagorean Theorem (Figure 33.2) is not so ungraspable because the solution is well known. (It is the sixth most famous equation [10].)

The GenIQ tree, although not a black box like most other machine-learning methods, gives the modeler a graphic, albeit Picasso-like, to interpret. GenIQ, for the everyday regression model, produces a GenIQ tree defined by branches formed by predictor variables attached to various functions (like that of the Pythagorean GenIQ tree).

* Optimizing the decile table is equivalent to maximizing the Cum Lift statistic.

```
x1 = height;
x2 = x1*x1;
x3 = length;
x4 = x3*x3;
x5 = x2 + x3;
x6 = SQRT(x5);
diagonal = x6;
```

FIGURE 33.1
GenIQ computer code.

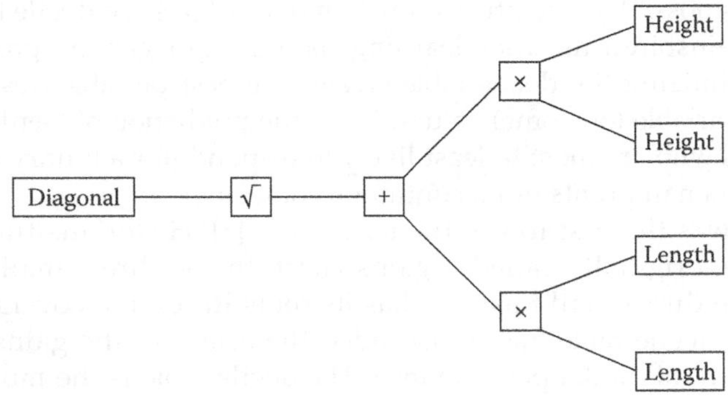

FIGURE 33.2
GenIQ tree for the Pythagorean theorem.

The GenIQ tree and computer code represent *feature 3* of GenIQ: The GenIQ Model serves as a nonstatistical, machine-learning regression method that automatically *specifies the model* for the problem at hand. As well, the GenIQ tree represents *feature 1* of GenIQ: The GenIQ Model automatically *data mines for new variables*. Continuing with the Pythagorean illustration, there are four new variables (branches): new_var1 = (height × height); new_var2 = (length × length); new_var3 = (new_var1 + new_var2); and last, new_var4 = SQRT(new_var3), which is the model itself. Thus, the GenIQ Model serves as a unique data mining method that creates new variables—which cannot be intuited by the model builder—via the GP process, which evolves structure (new variables) "without explicit programming" (Adams, 1959) [11]. Moreover, appending the new variables to the dataset with the original variables for building a statistical regression model produces a *hybrid statistics–machine-learning model*, along with the regression coefficients that provide the regression modeler the necessary comfort level for model acceptance.

For the regression modeler, interpretability is all about the regression coefficients. The regression coefficients provide the key to how the model works: Which predictor variables are most important, in rank order? What effect does each predictor variable have on the dependent variable? It is not well known that the standard method of interpreting regression coefficients often leads to an incorrect interpretation of the regression model and, specifically, incorrect answers to the two questions stated in this paragraph. GenIQ is a nonstatistical machine-learning method. Thus, it has no coefficients like a statistical method. But, GenIQ provides the answer to the first question by way of *feature 2* of GenIQ: The GenIQ Model provides a unique *variable selection* of important predictor variables by ranking* the relationship between each predictor variable with the dependent

* The ranking is based on the mean frequency of a variable across the top 18 models in the latest generation.

variable—accounting for the presence of the other predictor variables considered jointly. As for the second question, GenIQ provides the answer by analyzing the decile table (see Chapters 40 and 41).

With two checks against it: Why use GenIQ? How will GenIQ ever become popular? GenIQ is the appropriate model when the decile table is the unquestionable measure of model performance. For all other instances, a trade-off is necessary—the trade-off between the performance and no coefficients of GenIQ versus the interpretability and use of fitness functions of statistical regression, which often serve as surrogates for optimizing the decile table. Separation anxiety for something used for between 60 plus and 200 plus years is a condition that takes time to treat. Until the checks become ocularly palatable, which comes about with retraining statisticians to think out of the box, OLS and LRM will continue to be in use. With the eventual recognition that valuable information comes in unsuspected forms (e.g., a parse tree), GenIQ will be popular.

GenIQ is the model for today's data. It can accommodate big (and small) data as it is a flexible, assumption-free, nonparametric model whose engine lets the data define the model. In stark contrast, OLS and LRM conceived, tested, and experimented on the small data setting of their day. These models are suboptimal and problematic with today's big data [12]. The paradigm of OLS and LRM is "fit the data to" an unknown, prespecified, assumption-full, parametric model, which is best for small data settings. GenIQ will become popular as today's data grow, necessitating that the data define the model rather than fitting "square data in a round model."

33.4 Summary

The statistician's utterance that regression modeling involves art and science implies a mixture of skill acquired by experience and a technique that reflects a precise application of fact or principle. I assert the regression trilogy of art, science, and concrete poetry by bringing forth a modelogue to introduce the machine-learning technique of GenIQ, an alternative to the statistical OLS and LRM. I provide the interpretation of the poetry as best as I can, being the poet of "To Fit or Not to Fit Data to a Model" and the inventor of the GenIQ method.

References

1. Newton vs. Leibniz; The Calculus Controversy, 2010. http://www.angelfire.com/md/byme/mathsample.html.
2. Earliest Known Uses of Some of the Words of Mathematics, 2009. http://jeff560.tripod.com/m.html.
3. Finney, D., *Probit Analysis*, Cambridge University, Cambridge, UK, 1947.
4. Cramer, J.S., *Logit Models from Economics and Other Fields*, Cambridge University Press, Cambridge, UK, 2003.

5. Ratner, B., 2007, p. 107. http://www.geniq.net/res/Statistical-Terms-Who-Coined-Them-and-When.pdf.
6. Ratner, B., *Decile Analysis Primer*, 2008, http://www.geniq.net/DecileAnalysis Primer_2.html.
7. Smith, S.F., A learning system based on genetic adaptive algorithms, PhD Thesis, Computer Science Department, University of Pittsburgh, 1980.
8. Cramer, N.L., A representation for the adaptive generation of simple sequential programs, in *International Conference on Genetic Algorithms and their Applications (ICGA85)*, CMU, Pittsburgh, 1985.
9. Koza, J.R, *Genetic Programming: On the Programming of Computers by Means of Natural Selection*, MIT Press, Cambridge, MA, 1992.
10. Alfeld, P., University of Utah, Alfred's homepage, 2010.
11. Samuel, A.L., Some studies in machine learning using the game of checkers, *IBM Journal of Research and Development*, 3(3), 210–229, 1959.
12. Harlow, L.L., Mulaik, S.A., and Steiger, J.H., Eds., *What If There Were No Significance Tests?* Erlbaum, Mahwah, NJ, 1997.

34

Opening the Dataset: A Twelve-Step Program for Dataholics

34.1 Introduction

My name is Bruce Ratner, and I am a dataholic. I am also an artist* and poet† in the world of statistical data. I always await getting my hands on a new dataset to crack open and paint the untouched, untapped numbers into swirling equations and the pencil data gatherings into beautiful verse. The purpose of this eclectic chapter is to provide a staircase of twelve steps to ascend upon cracking open a dataset regardless of the application the datawork may entail. I provide SAS© subroutines of the twelve-step program in case the reader wants to take a nip. The subroutines are also available for downloading from my website: http://www.geniq.net/articles.html#section9.

34.2 Background

I see numbers as the prime coat for a bedazzled visual percept. Is not the nums pic grand in Figure 34.1 [1]? I also see mathematical devices as the elements in poetic expressions that allow truths to lay bare. A poet's rendition of love gives a thinkable pause in Figure 34.2 [2]. For certain, the irresistible equations are the poetry of numerical letters. The most powerful and famous equation is E = m*(c^2).‡ The fairest of them all is e^(i*pi) + 1 = 0.§ The aforementioned citations of the trilogy of art, poetry, and data, which makes an intensely imaginative interpretation of beauty, explain why I am a dataholic.

34.3 Stepping

Before painting the numbers by the numbers, penciling dataiku verses, and formulating equation poems, I brush my tabular canvas with four essentials markings

* See Chapter 33.

† See Chapter 24.

‡ Of course, Einstein.

§ Leonhard Euler (1707–1783) was a Swiss mathematician who made enormous contributions to mathematics and physics.

for the just-out dataset. The markings, first encountered upon rolling around the rim of the dataset, are:

- Step/Marking #1. Determine sample size, an indicator of data depth.
- Step/Marking #2. Count the number of numeric and character variables, an indicator of data breadth.
- Step/Marking #3. Air the listing of all variables in a Post-it© format. This unveiling permits copying and pasting of variables into a computer program editor. A copy-pasteable list forwards all statistical tasks.
- Step/Marking #4. Calculate the percentage of missing data for each numeric variable. This arithmetical reckoning provides an indicator of havoc on the assemblage of the variables due to the missingness. Character variables never have missing values: We can get something from nothing.

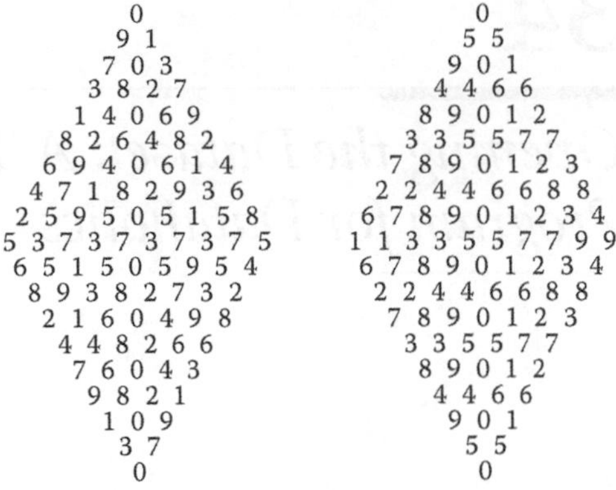

FIGURE 34.1
Parallel intervals.

$$\text{Love} = \lim_{\text{Ego}\to 0} \frac{1}{\text{Ego}}$$

FIGURE 34.2
Love equation.

The following eight steps complete my twelve-step program for dataholics, at least for cracking open a fresh dataset.

- Step #5. Follow the contour of each variable. The topographical feature of the variable offers a map of the variable's meaning through patterns of peaks, valleys, gatherings, and partings across all or part of the variable's plain.
- Step #6. Start a searching wind for the unexpected of each variable: improbable values, say, a boy named Sue; impossible values, say, age is 120 years; and undefined values due to irresponsibilities such as X/0.
- Step #7. Probe the underside of the pristine cover of the dataset. This poke uncovers the meanings of misinformative values, such as NA, the blank, the number 0, the letters o and O, the varied string of 9s, the dash, the dot, and many QWERTY expletives. Decoding the misinformation always yields unscrambled data wisdom.
- Step #8. Know the nature of numeric variables. Declare the formats of the numeric: decimal, integer, or date.
- Step #9. Check the reach of numeric variables. This task seeks values *far from* or *outside* the fences of the data.*

* Of course, this refers to J.W. Tukey (1905–2000), an American mathematician best known for his brilliant development of the fast Fourier transform (FFT) in 1965. Ironically, his simple yet powerful visual display of data—the boxplot—brought him greater fame than did his infinitely complex FFT. The boxplot, which is taught in elementary schools, is also taught in higher education, just in case the kiddies were sleeping that day. In the world of one-upmanship in statistics, Tukey considered himself a *data analyst*.

- Step #10. Check the angles of logic within the dataset. This biased way of looking allows for weighing contradictory values with conflict resolution rules.
- Step #11. Stomp on the lurking typos. These lazy and sneaky characters earn their keep by ambushing the integrity of data.
- Step #12. Find and be rid of noise within thy dataset. Noise, the idiosyncrasies of the data, the nooks and crannies, the particulars, are not part of the sought-after essence of the data. Ergo, the data particulars are lonely, not-really-belonging-to pieces of information that happen to be both in the population and the drawn data. Paradoxically, as the anamodel (word marriage of analysis and model) includes more and more of the prickly particulars, the anamodel build becomes better and better. Inconsistent with reason, the anamodel validation becomes worse and worse.

  Eliminating noise from the data is actioned by (1) identifying the idiosyncrasies and (2) deleting the records that define the idiosyncrasies of the data. After the data are noise-free, the anamodel reliably represents the sought-after essence of the data.

34.4 Brush Marking

I request that the reader allow the use of my poetic license. I illustrate the reveal of four data markings by using not only a minikin dataset but also by identifying the variable list itself. At the onset of a big data project, the sample size is perhaps the only knowable. The variable list is often not known; if so, it is rarely copy-pasteable. And for sure, the percentages of missing data are never in showy splendor.

- *Markings 1 and 2:* Determine sample size and count the number of numeric and character variables. Consider the otherwise unknown dataset IN in Appendix 34.A, I run the subroutine in Appendix 34.B to obtain the sample size and number of variables by numeric-character type (see Table 34.1).
- *Marking 3:* Air the listing of all variables in a Post-it format. I run the subroutine in Appendix 34.C to obtain the copy-pasteable variable list, which is in the Log window. The text shows the variable list for the copy-paste: VARLIST_IS_HERE in Figure 34.3.
- *Marking 4:* Calculate the percentage of missing data for each numeric variable. I run the subroutine in Appendix 34.D to obtain the percentages of missing data for variables X1, X2, X3, and X4 (see Table 34.2).

TABLE 34.1
Sample Size and Number of Variables by Type

Type	Sample_Size	Number_of_Variables
Numeric	5	4
Character	5	1

FIGURE 34.3
Copy-pasteable variable list.

TABLE 34.2
Percent Missing per Variables

PCT_MISSING_X1	PCT_MISSING_X2	PCT_MISSING_X3	PCT_MISSING_X4
20.0%	40.0%	60.0%	80.0%

34.5 Summary

My name is neither the significant statistic here nor is it that I am a functioning dataholic. I need to write dataiku verses and paint swirling equations to live before I die. As I did well in kindergarten ("he shares and plays well with others"), I hope my twelve-step program helps others, like me, who love data. I provide SAS subroutines of the twelve-step program in case the reader wants to take a nip.

Appendix 34.A Dataset IN

```
data IN;
input ID $1. X1 X2 X3 X4;
cards;
1 1 2 3 4
2 1 2 3 .
3 1 2 . .
4 1 . . .
5 . . . .
;
run;
```

Appendix 34.B Samsize Plus

```
PROC CONTENTS data=IN noprint
out=out1(keep=libname memname nobs type);
run;

PROC FORMAT;
value typefmt 1='Numeric' 2='Character';
run;

PROC SUMMARY data=out1 nway;
class libname memname type;
id nobs;
output out=out2
(drop=_type_ LIBNAME MEMNAME
    rename=(_freq_=NUMBER_of_VARIABLES NOBS=SAMPLE_SIZE));
format type typefmt.;
run;

PROC PRINT data=out2 noobs;
run;
```

Appendix 34.C Copy-Pasteable

```
PROC CONTENTS data=IN
out = vars (keep = name type)
noprint;
run;

PROC SQL noprint;
select name into :varlist_is_here separated by ''
from vars;
quit;
%put _global_ ;
```

Appendix 34.D Missings

```
PROC SUMMARY data=in;
var x1 x2 x3 x4;
output out=out3(drop=_type_ rename=(_freq_=sam_size)) nmiss=n_miss1-n_miss4;
run;
```

```
data out4;
set out3;
array nmiss n_miss1-n_miss4;
array pct_miss PCT_MISSING_X1-PCT_MISSING_X4;
do over nmiss;
pct_miss= nmiss/sam_size;
end;
keep PCT_MISSING_X1-PCT_MISSING_X4;
run;

PROC PRINT data=out4;
format PCT_MISSING_X1-PCT_MISSING_X4 PERCENT8.1;
run;
```

References

1. Mathematicalpoetry.blogspot.com
2. Hispirits.com

35

Genetic and Statistic Regression Models: A Comparison

35.1 Introduction

Statistical ordinary least squares (OLS) regression and logistic regression (LR) models are workhorse techniques for prediction and classification, respectively. In 1805, Legendre published the ordinary regression method. In 1944, Berkson developed the LR model. Something old is not necessarily useless today, and something new is not necessarily better than something old is. Lest one forgets, consider the wheel and the written word. With respect for statistical lineage, I maintain the statistical regression paradigm, which dictates "fitting the data to the model," is old because it was developed and tested within the small data setting of yesterday and has been shown untenable with big data of today. I further maintain the new machine-learning genetic paradigm, which "lets the data define the model," is especially effective with big data. Consequently, genetic models outperform the originative statistical regression models. The purpose of this chapter is to present a comparison of genetic and statistic LR in support of my assertions.

35.2 Background

Statistical OLS regression and LR models are workhorse techniques for prediction (of a continuous dependent variable) and classification (of a binary dependent variable), respectively. On March 6, 1805, Legendre published the OLS regression method. In 1944, Berkson developed the LR model. Something old (say, between 60 plus and 200 plus years ago) is not necessarily useless today, and something new is not necessarily better than something old is. Lest one forgets, consider the wheel and the written word. With respect for statistical lineage, I maintain the statistical regression paradigm, which dictates "fitting the data to the model" is old because it was developed and tested within the small data* setting of yesterday and has been shown untenable with big data† of today. The linear nature of the regression model renders the model insupportable in that it cannot capture the underlying structure of big data. I further maintain the new machine-learning genetic paradigm,

* Hand, D.J., Daly, F., Lunn, A.D., McConway, K.J., and Ostrowski, E. (Eds.), *A Handbook of Small Data Sets*, Chapman & Hall, London, 1994. This book includes many of the actual data from the early-day efforts of developing the OLS regression method of yesterday.

† See Chapters 8 and 9.

which "lets the data define the model," is especially effective with big data. Consequently, genetic models outperform the originative statistical regression models.

The engine behind any predictive model, statistical or machine learning, linear or nonlinear, is the fitness function, also known as the objective function. The OLS regression fitness function of mean squared error is easily understood. The fitness function is the average of the squares of the error, actual observation minus predicted observation. The LR model fitness function is the likelihood function,* an abstract mathematical expression for individuals not well-versed in mathematical statistics that is best understood from the classification table, indicating the primary measure of a total number of correctly classified individuals.† The estimation of the equations of the statistical regression models universally and routinely uses calculus.

35.3 Objective

The objective of this chapter is to present the GenIQ Model as the genetic LR alternative to the statistical LR model, in support of my assertions in Section 35.2. (The companion presentation of a comparison of genetic and OLS regression model, not provided, is straightforward with the continuous dependent variable replacing the binary dependent variable.)

Adding to my assertions, I bring attention to the continuing decades-long debate, *What If There Were No Significant Testing?* [1]. The debate raises doubts over the validity of the statistical regression paradigm: The regression modeler is dictated to "fit the data to" a prespecified parametric model, linear "in its coefficients," $Y = b_0 + b_1X_1 + b_2X_2 + \ldots + b_nX_n$. Specifically, the matter in dispute is the modeler must prespecify the *true unknowable* model under a null hypothesis. The issue of prespecifying an unknowable model is undoubtedly an inappropriate way to build a model. If the hypothesis testing yields rejection of the model, then the modeler tries repeatedly until an acceptable model is finally dug up. The try again task is a back-to-the-future data mining procedure that statisticians deemed, over three decades ago, a work effort of finding spurious relations among variables. In light of the debated issue, perhaps something new is better: The GenIQ Model, its definition, and an illustration follow.

35.4 The GenIQ Model, the Genetic Logistic Regression

The GenIQ Model is a flexible, any-size data method guided by the paradigm *let the data define the model*. The GenIQ Model

1. *Data mines* for new variables among the original variables.
2. *Performs* variable selection, which yields the best subset of the new and original variables.

* Also known as the joint probability function: P(X)*P(X)*¼P(X)*{(1 – P(X)}*{(1 – P(X)} … *{(1 – P(X)}.

† Actually, there are many other measures of classification accuracy. Perhaps even more important than total classified correctly is the number of "actual" individuals classified correctly.

3. *Specifies* the model to optimize the decile table. *The decile table is the fitness function.* The optimization of the fitness function uses the method of *genetic programming* (GP), which executes a computer—without explicit programming—to fill up the upper deciles with as many responses as possible, equivalently, to maximize the Cum Lift. The decile table fitness function is directly appropriate for model building in industry sectors such as direct and database marketing and in telemarketing; marketing mix optimization programs; customer relationship management campaigns; social media platforms and e-mail broadcasts; and the like. However, the decile table is becoming the universal statistic for virtually all model assessment of predictive performance.

Note, I use interchangeably the terms *optimize* and *maximize* as they are equivalent in the content of the decile table.

35.4.1 Illustration of "Filling Up the Upper Deciles"

I illustrate the concept of filling up the upper deciles. First, the decile table, the display that indicates model performance, is constructed. Construction consists of five steps: (1) apply, (2) rank, (3) divide, (4) calculate, and (5) assess (see Figure 35.1).

I present the *RESPONSE model criterion*, for a *perfect* RESPONSE model, along with the instructive decile table to show the visible representation of the concepts to which the GenIQ Model directs its efforts. Consider the RESPONSE model working with 40 responses among 100 individuals. In Figure 35.2, all 40 responses are in the upper deciles, indicating a perfect model.

In Figure 35.3, the decile table indicates that perfect models are hard to come by. Accordingly, I replace criterion with a goal. Thus, one seeks a *RESPONSE model goal* for which the desired decile table is one that *maximizes* the responses in the upper deciles.

The next two decile tables are exemplary in that they show how to read the predictive performance of a RESPONSE model, stemming from any predictive modeling approach. The decile tables reflect the result from a model validated with a sample of 46,170 individuals; among these, there are 2,104 responders. In Figure 35.4, the decile table shows the decile table along with the annotation of how to read the Cum Lift of the top decile. The Cum Lift for the top decile is the top decile Cum Response Rate (18.7%) divided by Total-Decile Response Rate (4.6%, the average response rate of the sample) multiplied by 100, yielding 411.

The decile table in Figure 35.5 shows the collapsed decile table on the top and second deciles combined, along with the annotation of how to read the Cum Lift of the top 20% of the sample. The Cum Lift for the top two deciles is the "top and 2 deciles" Cum Response Rate (13.5%) divided by Total-Decile Response Rate (4.6%, the average response rate

Model Performance Criterion

... is the DECILE ANALYSIS.

1. **Apply** model to the file (score the file).
2. **Rank** the scored file, in descending order.
3. **Divide** the ranked file into 10 equal groups.
4. **Calculate** Cum Lift.
5. **Assess** model performance. The **best** model identifies the **most** response in the **upper deciles.**

FIGURE 35.1
Construction of the decile analysis.

RESPONSE Model Criterion

How well the model **correctly classifies** response in the **upper** deciles.
Perfect response model among 100 individuals
- 40 responders
- 60 nonresponders

Decile	Number of Individuals	Total Response
top	10	10
2	10	10
3	10	10
4	10	10
5	10	0
6	10	0
7	10	0
8	10	0
9	10	0
bottom	10	0
Total	100	40

FIGURE 35.2
RESPONSE model criterion.

RESPONSE Model Goal

- **One seeks a model** that identifies the **maximum** responses in the **upper** deciles.

Decile	Total Response
top	Max
2	Max
3	Max
4	
5	
6	
7	
8	
9	
bottom	

FIGURE 35.3
RESPONSE model goal.

of the sample) multiplied by 100, yielding 296. The calculation of the remaining Cum Lifts follows the pattern of arithmetic tasks described.

35.5 A Pithy Summary of the Development of Genetic Programming

Unlike the statistical regression models that use calculus as their number cruncher to yield the model equation, the GenIQ Model uses GP as its number cruncher. As GP is relatively new, I provide a pithy summary of the development of GP.*

* Genetic programming, http://en.wikipedia.org/wiki/Genetic_programming. Accordingly, I replace criterion with goal. Thus, one seeks a *RESPONSE model goal* for which the desired decile table is one that *maximizes* the responses in the upper deciles.

One can expect 4.11 times the responders obtained using no model, targeting the "top" decile.

Response Decile Analysis

Decile	Number of Customers	Number of Responses	Decile Response Rate(%)	Cum Response Rate(%)	Cum Response Lift
top	4,617	865	18.7	18.7	411
2	4,617	382	8.3	13.5	296
3	4,617	290	6.3	11.1	244
4	4,617	128	2.8	9.0	198
5	4,617	97	2.1	7.6	167
6	4,617	81	1.8	6.7	146
7	4,617	79	1.7	5.9	130
8	4,617	72	1.6	5.4	118
9	4,617	67	1.5	5.0	109
bottom	4,617	43	0.9	4.6	100
Total	46,170	2,104	4.6		

FIGURE 35.4
The decile table.

One can expect 2.96 times the responders obtained using no model, targeting the "top two" deciles.

Response Decile Analysis

Decile	Number of Customers	Number of Responses	Decile Response Rate(%)	Cum Response Rate(%)	Cum Response Lift
top	9,234	1,247	18.7	18.7	411
2			8.3	13.5	296
3	4,617	290	6.3	11.1	244
4	4,617	128	2.8	9.0	198
5	4,617	97	2.1	7.6	167
6	4,617	81	1.8	6.7	146
7	4,617	79	1.7	5.9	130
8	4,617	72	1.6	5.4	118
9	4,617	67	1.5	5.0	109
bottom	4,617	43	0.9	4.6	100
Total	46,170	2,104	4.6		

FIGURE 35.5
The decile table collapsed with respect to the top two deciles.

In 1954, GP began with the evolutionary algorithms first utilized by Nils Aall Barricelli and applied to evolutionary simulations. In the 1960s and early 1970s, evolutionary algorithms became widely recognized as viable optimization approaches. John Holland was a highly influential force behind GP during the 1970s.

The first statement of tree-based GP (i.e., computer languages organized in tree-based structures and operated on by natural genetic operators, such as reproduction, mating, and mutation) was given by Nichael L. Cramer (1985). John R. Koza (1992), the leading exponent of GP who has pioneered the application of GP, greatly expanded this work to various complex optimization and search problems [2].

In the 1990s, GP was used mainly to solve relatively simple problems because the computational demand of GP could not be met effectively by the CPU power of the day. Recently, GP has produced many outstanding results due to improvements in GP technology and the exponential growth in CPU power. In 1994, the GenIQ Model, a Koza-GP machine-learning alternative model to the statistical OLS and LR models, was introduced.

Statisticians do not have a scintilla of doubt about the mathematical statistics solutions of the OLS and LR models. Statisticians may be rusty with calculus to the extent they may have to spend some time bringing back to mind how to derive the estimated regression models. Regardless, the estimated models are in constant use without question (notwithstanding the effects of annoying modeling issues, e.g., multicollinearity). There is no burden of proof required for the acceptance of statistical model results.

Unfortunately, statisticians are not so kind with GP because I dare say they have no formal exposure to GP principles or procedures. So, I acknowledge the current presentation may not sit well with the statistical modeler. The GenIQ Model specification is set forth for the attention of statisticians, along with their acceptance of the GenIQ Model *without my obligation* to provide a primer on GP. GP methodology itself is not difficult to grasp but requires much space for discourse not available here. I hope this acknowledgment allows the statistical modeler *to accept the trustworthiness of GP and the presentation of comparing genetic and statistic regression models assumedly*. Note, I do present GP methodology proper in Chapter 40.

35.6 The GenIQ Model: A Brief Review of Its Objective and Salient Features

The operational objective of the GenIQ Model is to find a set of *functions* (e.g., arithmetic, trigonometric) and *variables* such that the equation (the GenIQ Model), represented symbolically as GenIQ = functions + variables, maximizes response in the upper deciles. GP determines the functions and variables (see Figure 35.6).

35.6.1 The GenIQ Model Requires Selection of Variables and Function: An Extra Burden?

The GenIQ Model requires selection of variables and function, whereas the statistic regression model requires presumably only selection of variables. The obvious question

GenIQ model

- OBJECTIVE
  - To find a set of **functions** and **variables** such that the equation (GenIQ Model)
    - ▶ GenIQ = **functions** + **variables**
    - ▶ **Maximizes response in upper deciles.**
- GP determines the functions, variables.

FIGURE 35.6
The GenIQ Model—objective and form.

is whether the GenIQ Model places an extra burden on the modeler who uses GenIQ over the statistical model.

The selection of functions is neither a burden for the modeler nor a weakness in the GP paradigm. Function selection is a nonissue as indicated by the success of GP and the GenIQ Model. More to the point, as a true burden often unheeded, the statistical regression model *also requires* the selection of functions as the pre-modeling, mandatory exploratory data analysis (EDA) seeks the best transformations, such as log X, 1/X, or $(5 - x)^2$. Due to the evolutionary process of GP, there is no longer the need for EDA. The elimination of EDA is not hyperbole. It is a salient feature of GP and consequently of GenIQ. The only GP data preparatory work called for is to eliminate impossible or improbable values (e.g., age is 120 years old, and a boy named Sue, respectively).

35.7 The GenIQ Model: How It Works

I acknowledge my assertion of the machine-learning genetic paradigm is *especially effective with big data,* yet I illustrate the new technique with the smallest of data. I take poetic license in using a small dataset for the benefit of presenting the GenIQ Model effectively and thereby making it easy for the reader to follow the innards of GenIQ. That GenIQ works especially well with big data does not preclude the predictive prowess of GenIQ on small data.

I offer to view how the GenIQ Model works with a simple illustration that stresses the predictive power of GenIQ. Consider 10 customers (five responders and five nonresponders) and predictor variables X1 and X2. The dependent variable is RESPONSE. The objective is to build a model that maximizes the *upper four deciles* (see Figure 35.7). Note that the dataset of size 10 is used to simulate a decile table in which each decile is of size 1.

I build two RESPONSE models, a statistical LR model and a genetic LR model (GenIQ). GenIQ identifies three of four responders, a 75% response rate for the upper four deciles. LRM identifies two of four responders, a 50% response rate for the upper four deciles (see Figure 35.8).

GenIQ: How it works

Consider 10 customers:

- Their **RESPONSE** and two predictors **X1 and X2**
- Objective: to build a model that maximizes **upper four deciles**

i	**RESPONSE**	X1	X2
1	R	45	5
2	R	35	21
3	R	31	38
4	R	30	30
5	R	6	10
6	N	45	37
7	N	30	10
8	N	23	30
9	N	16	13
10	N	12	30

FIGURE 35.7
The GenIQ Model—how it works.

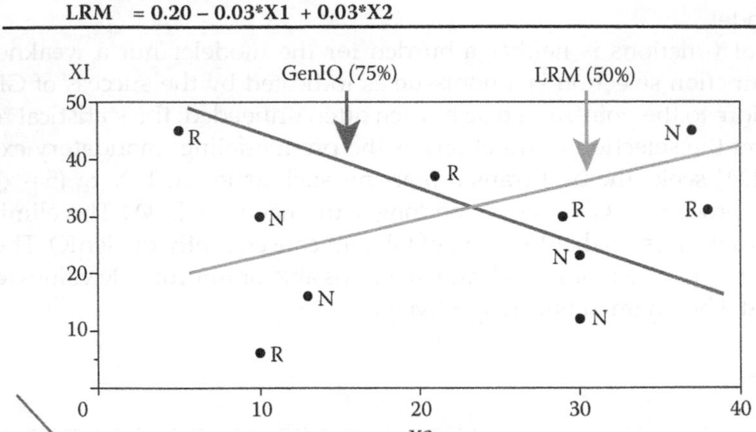

FIGURE 35.8
Genetic versus statistic logistic regression models (LRM).

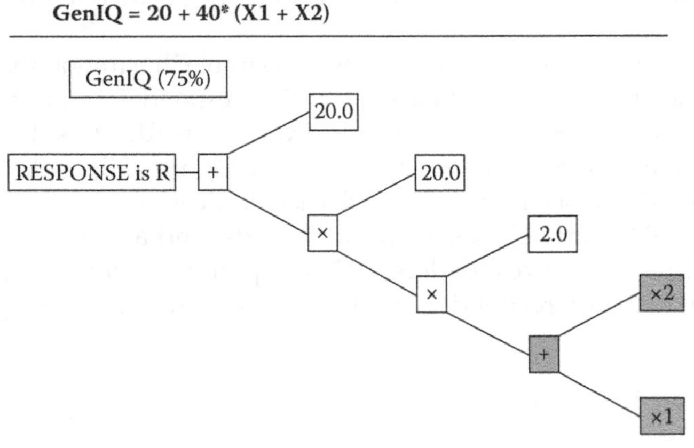

FIGURE 35.9
GenIQ parse tree.

The GenIQ Model, a by-product of being a GP-based model, provides a unique visual display of its equation in the form of a *parse tree*. The parse tree for the presented GenIQ Model is in Figure 35.9. From the tree, I point out the following:

1. GenIQ has data mined a new variable: X1 + X2 as indicated in the tree by a gray-shaded branch.
2. X1 + X2 is a result of the variable selection of GenIQ, albeit it is a singleton best subset.
3. The tree itself specifies the model.

In sum, I provide an exemplary illustration of GenIQ outperforming a LR model (It may seen the illustration should show validation results. However, if GenIQ does not outperform the LR model, then validation is obviously unnecessary). Consequently, I show the features of the GenIQ Model as described in Section 35.4: It (1) data mines for new variables, (2) performs the variable selection, and (3) specifies the model to optimize the decile table (actually, the upper four deciles in this illustration).

35.7.1 The GenIQ Model Maximizes the Decile Table

I reconsider the dataset in Figure 35.7. I seek a GenIQ Model that maximizes the *full* decile table. From the modeling session of the first illustration, I obtain such a model in Figure 35.10. Before discussing the model, I explain the process of generating the model. GenIQ modeling produces, in a given session, about 5–10 *equivalent* models. Some models perform better, say, in the top four deciles (like the previous GenIQ Model). Some models perform better all the way from the top to the bottom deciles, the full decile table. Also, some models have the desired discreteness of predicted scores. And, some models have undesired clumping of, or gaps among, the GenIQ predicted scores. The GenIQer selects the model that meets the objective at hand along with desired characteristics. It is important to note that the offering by GenIQ of equivalent models is *not* spurious data mining: There is no rerunning the data through various variable selections until the model satisfies the given objective.

GenIQ Model Tree

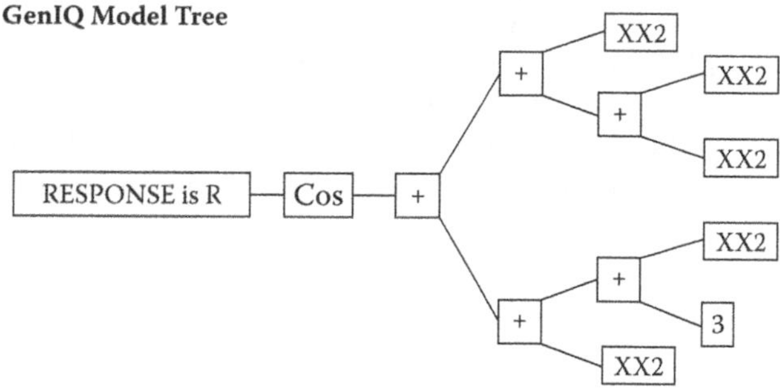

GenIQ Model—Computer Code

```
x1 = XX2;
  x2 = 3;
    x3 = XX1;
  x2 = x2 + x3;
If x1 NE 0 Then x1 = x2/x1; Else x1 = 1;
  x2 = XX2;
    x3 = XX2;
  x2 = x3 – x2;
      x3 = XX2;
    x2 = x2 + x3;
  If x1 NE 0 Then x1 = x2/x1; Else x1 = 1;
  x1 = Cos(x1);
 GenIQvar = x1;
```

Table 1.
Scored and ranked data by GenIQvar

ID	XX1	XX2	RESPONSE	GenIQvar
1	45	5	R	0.86740
2	35	21	R	0.57261
5	6	10	R	0.11528
3	31	38	R	0.05905
4	30	30	R	−0.53895
9	16	13	N	−0.86279
10	12	30	N	−0.95241
6	45	37	N	−0.96977
7	30	10	N	−0.99381
8	23	30	N	−0.99833

FIGURE 35.10
The GenIQ Model maximizes the decile table.

The GenIQ Model that maximizes the full decile table is in Figure 35.10. There are three elements of the GenIQ Model: (1) parse tree; (2) original dataset (with XXs replacing Xs, respectively*) with the appended GenIQ Model score GenIQvar; and (3) the GenIQ Model equation, which is actually a computer code defining the model. From the parse tree, the GenIQ Model selects the following functions: addition, subtraction, division, and the cosine. Also, GenIQ selects the two original variables XX1 and XX2 and the number 3, which in the world of GP is considered a variable.

The scored and ranked, by GenIQvar, dataset indicates a perfect ranking of the RESPONSE variable, 5 Rs followed by 5 Ns. This perfect ranking is tantamount to a maximized decile table. Hence, GenIQvar achieves the sought-after objective. Note, this illustration implies the weakness of statistical regression modeling: Statistical modeling offers only one model, unless there is a spurious rerunning of the data through various variable selections, versus the several models of varying decile performances offered by genetic modeling.

35.8 Summary

With respect for statistical lineage, I maintain the statistical regression paradigm, which dictates fitting the data to the model, is old because it was developed and tested within the small data setting of yesterday and has been shown untenable with big data of today. I further maintain the new machine-learning genetic paradigm, which lets the data define the model, is especially effective with big data. Consequently, genetic models outperform the originative statistical regression models. I present a comparison of genetic versus statistical LR, showing great care and completeness. My comparison shows the promise of the GenIQ Model as a flexible, any-size data alternative method to the statistical regression models.

References

1. Harlow, L.L., Mulaik, S.A., & Steiger, J.H., Eds., *What If There Were No Significance Testing?* Erlbaum, Mahwah, NJ, 1997.
2. Koza, J.R, *Genetic Programming: On the Programming of Computers by Means of Natural Selection*, MIT Press, Cambridge, MA, 1992.

* The renaming of the X variables to XX is a necessary task as GenIQ uses X-named variables as an intermediate variable for defining the computer code of the model.

36

Data Reuse: A Powerful Data Mining Effect of the GenIQ Model

36.1 Introduction

The purpose of this chapter is to introduce the concept of *data reuse*, a powerful data mining effect of the GenIQ Model. Data reuse is appending new variables, found when building a GenIQ Model, to the original dataset. As the new variables are reexpressions of the original variables, the correlations among the original variables and the GenIQ data-mined variables are expectedly high. In the context of statistical modeling, the occurrence of highly correlated predictor variables is a condition known as *multicollinearity*. One effect of multicollinearity is unstable regression coefficients, an unacceptable result. In contrast, multicollinearity is a nonissue for the GenIQ Model because it has no coefficients. The benefit of data reuse is apparent: The original dataset has the addition of new, predictive-full, GenIQ data-mined variables. I provide two illustrations of data reuse as a powerful data mining technique.

36.2 Data Reuse

Data reuse is appending new variables, found when building a GenIQ Model, to the original dataset. As the new variables are reexpressions of the original variables, the correlations among the original variables and the GenIQ data-mined variables are expectedly high. In the context of statistical modeling, the occurrence of highly correlated predictor variables is a condition known as *multicollinearity*. The effects of multicollinearity are inflated standard errors of the regression coefficients, unstable regression coefficients, lack of valid declaration of the importance of the predictor variables, and an indeterminate regression equation when the multicollinearity is severe. The simplest solution of guess-n-check, although inefficient, to the multicollinearity problem is deleting suspect variables from the regression model.

Multicollinearity is a nonissue for the GenIQ Model because it has no coefficients. The benefit of data reuse is apparent: The original dataset has the addition of new, predictive-full, GenIQ data-mined variables. Illustration best explains the concept and benefit of data reuse. I provide two such illustrations.

36.3 Illustration of Data Reuse

To illustrate data reuse, I build an ordinary least squares (OLS) regression model with a dependent variable PROFIT, predictor variables XX1 and XX2, and the data in Table 36.1.

The OLS Profit_est model in Equation (36.1) is

$$\text{Profit\_est} = 2.42925 + 0.16972*\text{XX1} - 0.06331*\text{XX2} \tag{36.1}$$

The assessment of the Profit_est model regards decile table performance. The individuals in Table 36.1 are scored and ranked by Profit_est. The resultant performance is in Table 36.2. The PROFIT ranking *is not perfect,* although three individuals are correctly placed in the proper positions. Individual ID 1 is correctly placed in the top position (top decile, decile of size 1); individuals ID 9 and ID 10 are correctly placed in the bottom two positions (ninth and bottom deciles, each decile of size 1).

36.3.1 The GenIQ Profit Model

The GenIQ Profit Model is in Figure 36.1. The assessment of the GenIQ Profit Model regards decile table performance. The individuals in Table 36.1 are scored and ranked by GenIQvar.

TABLE 36.1

Profit Dataset

ID	XX1	XX2	PROFIT
1	45	5	10
2	32	33	9
3	33	38	8
4	32	23	7
5	10	6	6
6	46	38	5
7	25	12	4
8	23	30	3
9	5	5	2
10	12	30	1

TABLE 36.2

Scored and Ranked Data by Profit_est

ID	XX1	XX2	PROFIT	Profit_est
1	45	5	10	9.74991
6	46	38	5	7.83047
4	32	23	7	6.40407
7	25	12	4	5.91245
2	32	33	9	5.77099
3	33	38	8	5.62417
8	23	30	3	4.43348
5	10	6	6	3.74656
9	5	5	2	2.96129
10	12	30	1	2.56660

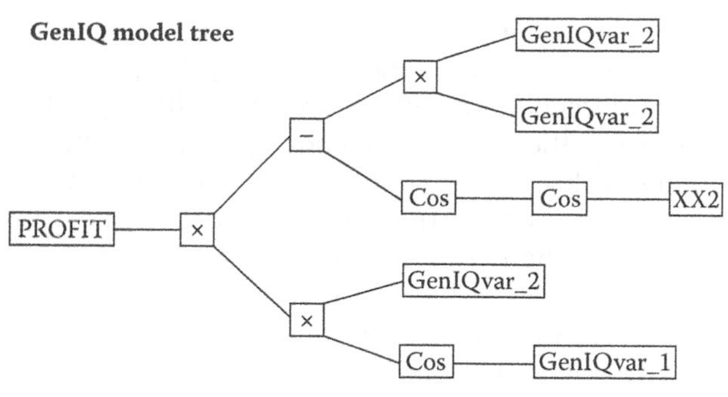

GenIQ model—computer code

```
x1 = GenIQvar_1;
   x1 = Cos(x1);
      x2 = GenIQvar_2;
   x1 = x1 * x2;
      x2 = XX2;
      x2 = Cos(x2);
      x2 = Cos(x2);
         x3 = GenIQvar_2;
            x4 = GenIQvar_2;
         x3 = x3 * x4;
      x2 = x3 – x2;
   x1 = x1 * x2;
GenIQvar = x1;
```

Table 1.
Scored and ranked data by GenIQvar

ID	XX1	XX2	PROFIT	GenIQvar
1	45	5	10	9.27644
2	32	33	9	6.22359
3	33	38	8	0.60790
4	32	23	7	0.29207
5	10	6	6	0.09151
6	46	38	5	0.06930
7	25	12	4	−0.12350
8	23	30	3	−0.19258
9	5	5	2	−0.26727
10	12	30	1	−0.31133

FIGURE 36.1
The PROFIT GenIQ Model.

The resultant performance is in Table 1 of Figure 36.1. The PROFIT ranking *is perfect.* I hold off declaring GenIQ outperforms OLS.

The GenIQ Model computer code is the GenIQ Profit Model equation. The GenIQ Model tree indicates the GenIQ Profit Model has three predictor variables: the original XX2 variable and two data-reused variables, GenIQvar_1 and GenIQvar_2. Before I discuss GenIQvar_1 and GenIQvar_2, I provide a formal treatment of the fundamental characteristics of data-reused variables.

36.3.2 Data-Reused Variables

Data-reused variables, as found in GenIQ modeling, come about by the inherent mechanism of genetic programming (GP). GP starts out with an initial random set, a *genetic population,* of, say, 100 genetic models, defined by a set of predictor variables, numeric variables (numbers are considered variables within the GP paradigm), and a set of functions (e.g., arithmetic and trigonometric). The initial genetic population, generation 0, is subjected to GP evolutionary computation that includes mimicking the natural biological, genetic operators: (1) copying/reproduction, (2) mating/sexual combination, and (3) altering/mutation. Copying of some models occurs. Mating of most models occurs. Two parent models evolve two offspring models with probabilistically greater predictive power than the parent models. Altering a few models occurs by changing a model characteristic (e.g., replacing the addition function by the multiplication function). Altered versions of the models have probabilistically greater predictive power than the original models. The resultant genetic population of 100 models, generation 1, has probabilistically greater predictive power than the models in generation 0. The models in generation 1 are now selected for another round of copying–mating–altering (based on proportionate to total decile table fitness), resulting in a genetic population of 100 models, generation 2. Iteratively, the sequence of copying–mating–altering continues with jumps in the decile table fitness values until the fitness values start to flatten (i.e., there is no observed significant improvement in fitness). Equivalently, the decile table remains stable.

Including the data-reused variables, the iterative process is somewhat modified. When there is a jump in the decile table fitness values, the GenIQer stops the process to (1) take the corresponding GenIQ Model, which is a variable itself, conventionally labeled GenIQvar_i and (2) append the data-reused variable to the original dataset. The GenIQer captures and appends two or three GenIQvar variables. Then, the GenIQer builds the final GenIQ Model with the original variables and the appended GenIQvar variables. If there is an observed extreme jump in the final step of building the GenIQ Model, the GenIQer can, of course, append the corresponding GenIQvar variable to the original dataset and then restart the final GenIQ Model building.

36.3.3 Data-Reused Variables GenIQvar_1 and GenIQvar_2

GenIQvar_1 and GenIQvar_2 are displayed in Figures 36.2 and 36.3. GenIQvar_1 is defined by the original variables XX1 and XX2 and the cosine function, which is not surprising as the trigonometric functions are known to maximize the decile table fitness function. Note, the bottom branch of the GenIQvar_1 tree: XX1/XX1. The GP method periodically *edits* the tree for such branches (in this case, XX1/XX1 would be replaced by 1). Also, the GP method edits to eliminate redundant branches. GenIQvar_2 is defined by GenIQvar_1, –0.345, and –0.283 and the following functions: addition, and multiplication.

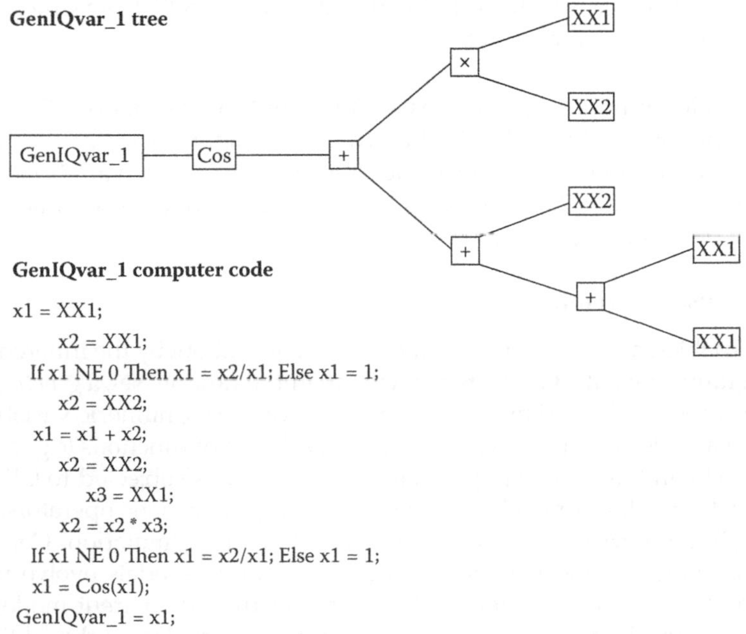
GenIQvar_1 tree
XX1
×
XX2
GenIQvar_1
Cos
+
XX2
+
XX1
+
XX1
GenIQvar_1 computer code
x1 = XX1;
x2 = XX1;
If x1 NE 0 Then x1 = x2/x1; Else x1 = 1;
x2 = XX2;
x1 = x1 + x2;
x2 = XX2;
x3 = XX1;
x2 = x2 * x3;
If x1 NE 0 Then x1 = x2/x1; Else x1 = 1;
x1 = Cos(x1);
GenIQvar_1 = x1;

FIGURE 36.2
Data-reused variable GenIQvar_1.

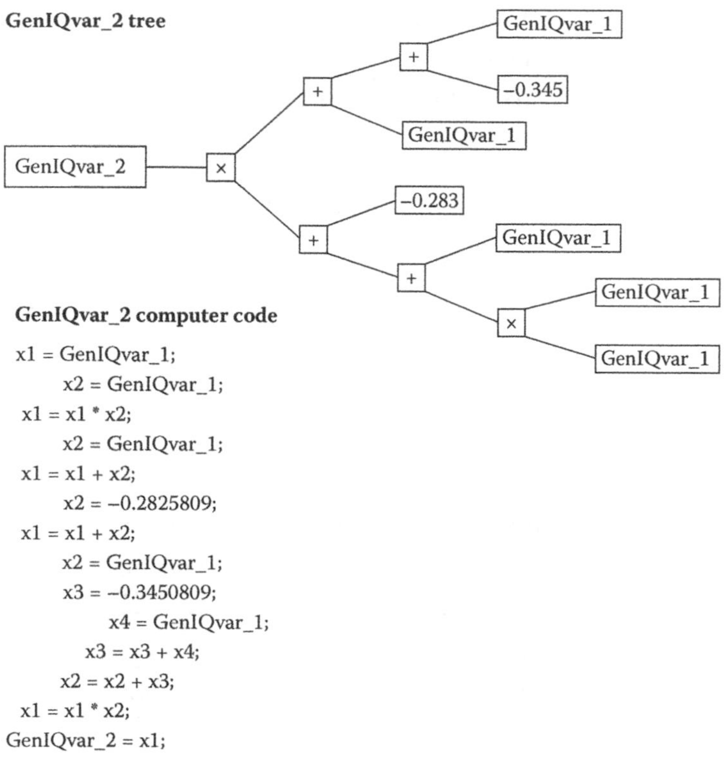
GenIQvar_2 tree
GenIQvar_2
×
+
+
GenIQvar_1
−0.345
GenIQvar_1
+
−0.283
+
GenIQvar_1
×
GenIQvar_1
GenIQvar_1
GenIQvar_2 computer code
x1 = GenIQvar_1;
x2 = GenIQvar_1;
x1 = x1 * x2;
x2 = GenIQvar_1;
x1 = x1 + x2;
x2 = −0.2825809;
x1 = x1 + x2;
x2 = GenIQvar_1;
x3 = −0.3450809;
x4 = GenIQvar_1;
x3 = x3 + x4;
x2 = x2 + x3;
x1 = x1 * x2;
GenIQvar_2 = x1;

FIGURE 36.3
Data-reused variable GenIQvar_2.

In sum, the seemingly simple PROFIT data in Table 36.1 is not so simple after all if one seeks to rank the PROFIT variable perfectly. Now, I declare GenIQ outperforms the logistic regression model (LRM). (Recall, the caveat at the Section 35.7). (Recall, the caveat at the Section 35.7). This exemplary illustration points to the superiority of GenIQ modeling over statistical regression modeling.

36.4 Modified Data Reuse: A GenIQ-Enhanced Regression Model

I modify the definition of data reuse to let the data mining prowess of GenIQ enhance the results of an already-built statistical regression model. I define data reuse as appending new variables, found when building a GenIQ or *any model*, to the original dataset. I build a GenIQ Model with only one predictor variable, the already-built model score. This approach of using GenIQ as a GenIQ enhancer of an existing regression model has been shown empirically to improve significantly the existent model 80% of the time.

36.4.1 Illustration of a GenIQ-Enhanced LRM

I build an LRM with a dependent variable RESPONSE, predictor variables XX1 and XX2, and the data in Table 36.3.

TABLE 36.3

Response Data

ID	XX1	XX2	RESPONSE
1	31	38	Yes
2	12	30	No
3	35	21	Yes
4	23	30	No
5	45	37	No
6	16	13	No
7	45	5	Yes
8	30	30	Yes
9	6	10	Yes
10	30	10	No

TABLE 36.4

Scored and Ranked Data by Prob_of_Response

ID	XX1	XX2	RESPONSE	Prob_of_ Response
7	45	5	Yes	0.75472
10	30	10	No	0.61728
3	35	21	Yes	0.57522
5	45	37	No	0.53452
6	16	13	No	0.48164
8	30	30	Yes	0.46556
9	6	10	Yes	0.42336
1	31	38	Yes	0.41299
4	23	30	No	0.40913
2	12	30	No	0.32557

The LRM is in Equation 36.2.

$$\text{Logit of RESPONSE } (= \text{Yes}) = 0.1978 - 0.0328{*}\text{XX1} + 0.0308{*}\text{XX2} \tag{36.2}$$

The assessment of the RESPONSE LRM regards decile table performance. The individuals in Table 36.2 are scored and ranked by Prob_of_Response. The resultant performance is in Table 36.4. The RESPONSE ranking *is not perfect*, although three individuals are correctly placed in the proper positions. Specifically, individual ID 7 is correctly placed in the top position (top decile, decile of size 1); individuals ID 4 and ID 2 are correctly placed in the bottom two positions (ninth and bottom deciles, respectively, each decile of size 1).

Now, I build a GenIQ Model with only one predictor variable—the LRM defined in Equation 36.2. The assessment of the resultant GenIQ-enhanced RESPONSE model (Figure 36.4) regards decile table performance. The individuals in Table 36.2 are scored and ranked GenIQvar. The ranking performance is in Table 36.5. The RESPONSE ranking *is perfect*.

The GenIQ-enhanced computer code is the GenIQ-enhanced RESPONSE model equation.

The GenIQ-enhanced tree indicates the GenIQ-enhanced RESPONSE model has the predictor variable Prob_of_Response and two numeric variables, 3 and 0.1, combined with the Sine (Sin), division, and multiplication functions.

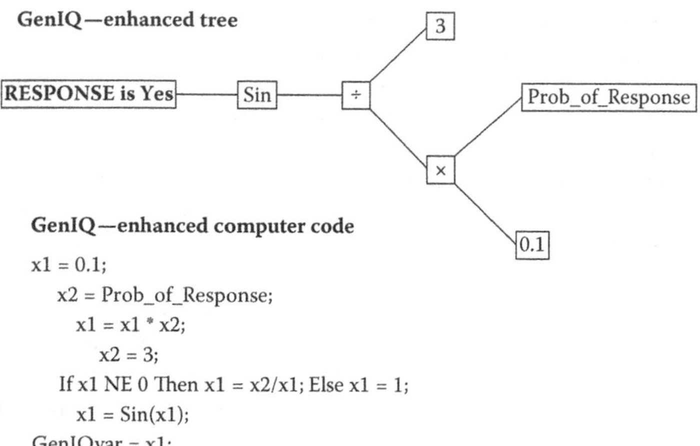
GenIQ—enhanced tree
3
RESPONSE is Yes
Sin
÷
Prob_of_Response
×
0.1
GenIQ—enhanced computer code
x1 = 0.1;
x2 = Prob_of_Response;
x1 = x1 * x2;
x2 = 3;
If x1 NE 0 Then x1 = x2/x1; Else x1 = 1;
x1 = Sin(x1);
GenIQvar = x1;

FIGURE 36.4
The GenIQ-enhanced RESPONSE.

TABLE 36.5
Scored and Ranked Data by GenIQvar

ID	XX1	XX2	Prob_of_ Response	RESPONSE	GenIQvar
8	30	30	0.46556	Yes	0.99934
9	6	10	0.42336	Yes	0.98444
3	35	21	0.57522	Yes	0.95002
7	45	5	0.75472	Yes	0.88701
1	31	38	0.41299	Yes	−0.37436
5	45	37	0.53452	No	−0.41121
6	16	13	0.48164	No	−0.51791
2	12	30	0.32557	No	−0.86183
4	23	30	0.40913	No	−0.87665
10	30	10	0.61728	No	−0.99553

In sum, the GenIQ-enhanced RESPONSE Model clearly improves the already-built RESPONSE LRM. As the GenIQ Model is inherently a nonlinear model, the GenIQ-enhanced model captures nonlinear structure in the data that the statistical regression model cannot capture because it is, after all, a linear model. This exemplary illustration implies the GenIQ-enhanced model is worthy of application.

36.5 Summary

I introduce the concept of data reuse, a powerful data mining effect of the GenIQ Model. Data reuse is the appending of new variables, found when building a GenIQ Model, to the original dataset. The benefit of data reuse is apparent: The original dataset has the addition of new GenIQ variables filled with predictive power.

I provide a formal treatment of the fundamental characteristics of data-reused variables. Then, I illustrate data reuse as a powerful data mining effect of the GenIQ Model. First, I build an OLS regression model. After discussing the results of the OLS modeling, I build a corresponding GenIQ Model. Comparing and contrasting the two models, I show the predictive power of the data reuse technique.

I modify the definition of data reuse to let the data mining prowess of GenIQ enhance the results of an already-built statistical regression model. I redefine data reuse as appending new variables, found when building a GenIQ or any model, to the original dataset. I build a GenIQ Model with only one predictor variable, the regression equation of an already-built LRM. This approach of using GenIQ as a GenIQ enhancer of an existing regression model shows significant improvement over the statistical model. Consequently, I illustrate the GenIQ enhancer is a powerful ful technique to extract the nonlinear structure in the data, which the statistical regression model cannot capture in its final regression equation because it is a linear model.

37

A Data Mining Method for Moderating Outliers Instead of Discarding Them

37.1 Introduction

In statistics, an outlier is an observation whose position falls outside the overall pattern of the data. Outliers are problematic: Statistical regression models are quite sensitive to outliers, which render an estimated regression model with questionable predictions. The common remedy for handling outliers is to determine and discard them. The purpose of this chapter is to present an alternative data mining method for moderating outliers instead of discarding them. I illustrate the data mining feature of the GenIQ Model as a method for moderating outliers with a simple, compelling presentation.

37.2 Background

There are numerous statistical methods for identifying outliers.* The most popular methods are the univariate tests.† There are many multivariate methods,‡ but they are not the first choice because of the advanced expertise required to understand their underpinnings (e.g., the Mahalanobis distance). The assumption of normally distributed data, an untenable condition to satisfy with either big or small data, is inherent in almost all the univariate and multivariate methods. If the test for the normality assumption fails, then the decision—there is an outlier—may be due to the nonnormality of the data rather than the presence of an outlier. There are tests for nonnormal data,§ but they are difficult to use and not as powerful as the tests for normal data.

* In statistics, an *outlier* is an observation whose position falls outside the overall pattern of the data. This is my definition of an outlier. There is no agreement for the definition of an outlier. There are many definitions, which in my opinion only reflect each author's writing style.

† Three popular, classical, univariate methods for normally distributed data are (a) z-score method, (b) modified z-score method, and (c) the Grubbs' test. One exploratory data analysis (EDA) method, which assumes no distribution of the data, is the boxplot method.

‡ The classical approach of identifying outliers is to calculate the Mahalanobis distance using robust estimators of the covariance matrix and the mean array. A popular class of robust estimators is that of the M estimators, first introduced by Huber [1].

§ Popular tests for nonnormal data are given in Barnett, V., and Lewis, T., *Outliers in Statistical Data*, Wiley, New York, 1984.

The statistical community[*] has not addressed uniting the outlier detection methodology and the reason for the existence of the outlier. Hence, I maintain the current approach of determining and discarding an outlier, based on the application of tests *with* the untenable assumption of normality and *without* accounting for the reason of existence, is wanting.

The qualified outlier is a serious matter in the context of statistical regression modeling. Statistical regression models are quite sensitive to outliers, which render an estimated regression model with questionable predictions. Without a workable robust outlier detection approach, statistical regression models go unwittingly in production with indeterminable predictions. In the next section, I introduce an alternative approach to the much-needed outlier detection methodology. The alternative approach uses the bivariate graphic outlier technique, the scatterplot, and the GenIQ Model.

37.3 Moderating Outliers Instead of Discarding Them

Comparing relationships between pairs of variables, in scatterplots, is a way of drawing closer to outliers. The scatterplot is an effective nonparametric (implication: flexible), assumption-free (no assumption of normality of data) technique. Using the scatterplot in tandem with the GenIQ Model provides a perfect pair of techniques for moderating outliers instead of discarding them. I illustrate this perfect duo with a seemingly simple dataset.

37.3.1 Illustration of Moderating Outliers Instead of Discarding Them

I put forth for the illustration the dataset[†] as described in Table 37.1.

The scatterplot is a valuable visual display to check for outliers.[‡] Checking for outliers is identifying data points off the observed linear relationship between two variables under consideration. In the illustration, the scatterplot of (XX, Y) (Figure 37.1) suggests the single circled point (1, 20), labeled A, is an outlier, among the linear relationship of the 100 points, consisting of the four mass points labeled Y. I assume, for a moment, the reason for the existence of the outlier is sound.

TABLE 37.1

Dataset of Paired Variables (XX, Y)

- Consider the dataset of 101 points (XX, Y).
  - There are four "mass" points; each has 25 observations:
    - (17, 1) has 25 observations
    - (18, 2) has 25 observations
    - (19, 4) has 25 observations
    - (20, 4) has 25 observations
  - There is one "single" point.
    - (1, 20) has 1 observation

* To the best of my knowledge.
† This dataset is in Huck, S.W., Perfect correlation ¼ if not for a single outlier, *STAT*, 49, 9, 2008.
‡ Outliers in two dimensions.

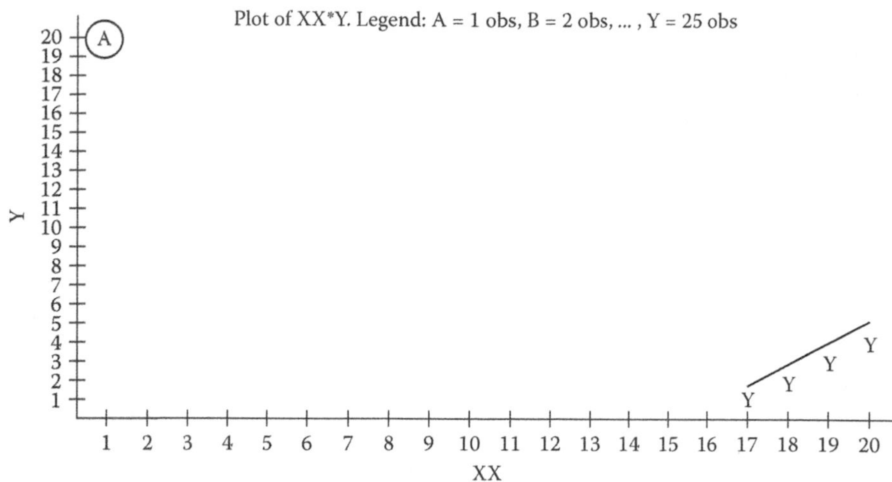

FIGURE 37.1
Scatterplot of (XX, Y).

TABLE 37.2

Correlation Coefficients: $N = 101$
Prob > r under H0: Rho=0

	XX	GenIQvar
Y	–0.41618	0.84156
	<0.0001	<0.0001

The correlation coefficient of (XX, Y) is –0.41618 (see Table 37.2). The correlation coefficient value implies the strength of the linear relationship between XX and Y *only if* the corresponding scatterplot indicates an underlying linear relationship between XX and Y. Under the momentary assumption, the relationship between XX and Y is not linear. Thus, the correlation coefficient value of –0.41618 is meaningless.

If no reason is put forward for the existence of outlier point (1, 20), then the relationship is assumed to be curvilinear, as depicted in the scatterplot in Figure 37.2. In this case, there is no outlier, and the model builder seeks to reexpress the paired variables (XX, Y) to straighten the curvilinear relationship and then to observe the resultant scatterplot for new outliers.

I have doubts about point (1, 20) as an outlier. Thus, I attempt to reexpress the paired variables (XX, Y) to straighten the curvilinear relationship. I apply the GenIQ Model to paired variables (XX, Y). The GenIQ reexpression of (XX, Y) is (GenIQvar, Y), where GenIQvar is the GenIQ data-mined transformation of XX. (Discussion of specification of the transformation is in Section 37.3.2) The scatterplot of (GenIQvar, Y) (Figure 37.3) shows no outliers. More important, the scatterplot indicates a linear relationship between GenIQvar and Y, and the meaningful correlation coefficient of (GenIQvar, Y), 0.84156 (see Table 37.2), quantifies the linear relationship as strong.

A post-transformation justification of point (1, 20) as a nonoutlier requires an explanation in light of the four mass points where GenIQ is –1.00, on the horizontal axis, labeled GenIQvar. The four points are in a vertical trend, viewed as a variation about

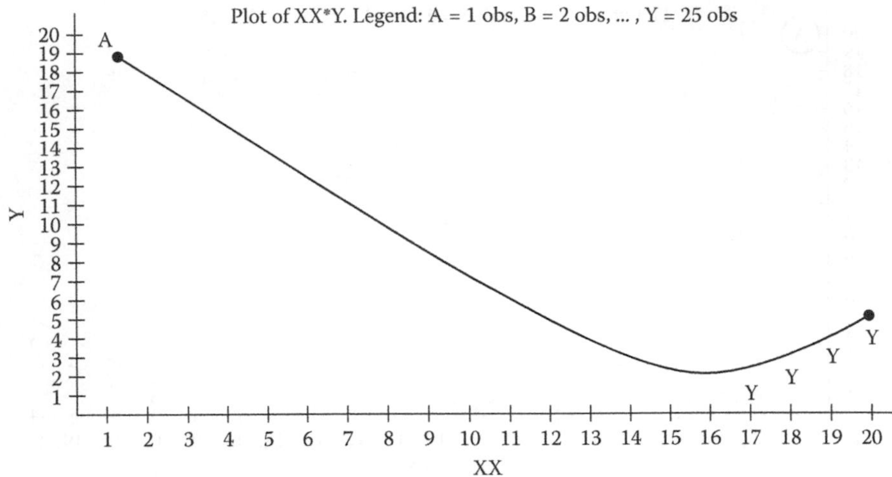

FIGURE 37.2
Scatterplot of (XX, Y), depicting a nonlinear relationship.

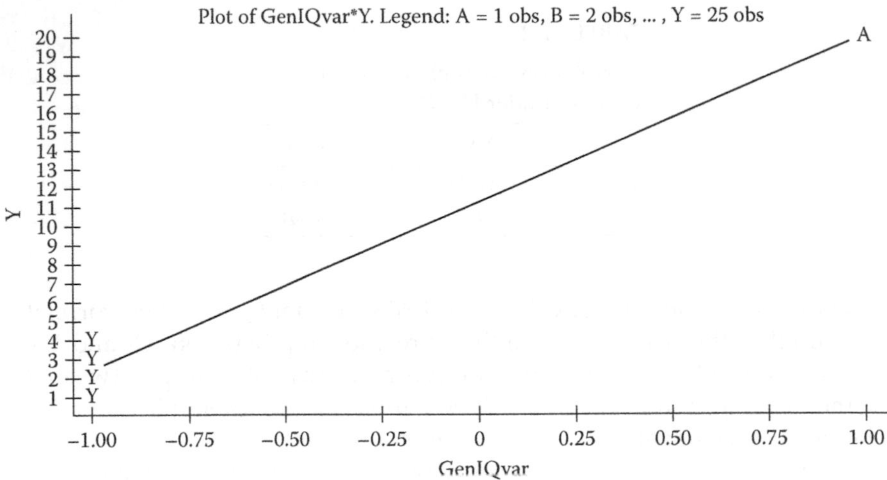

FIGURE 37.3
Scatterplot of (GenIQvar, Y).

a "true" transformed mass point. The scatterplot shows the quintessential straight line defined by two points: the true transformed point, positioned at the middle of the four Ys, (–1.00, 2.5) and the point (1.00, 20), labeled A.

37.3.2 The GenIQ Model for Moderating the Outlier

The GenIQ Model moderates the outlier point (1, 20) by reexpressing all 101 points. The outlier at position top left in Figure 37.1 goes to position top right in Figure 37.3. The remaining four mass points at position bottom right in Figure 37.1 go to position bottom right in Figure 37.3. The GenIQ Model transformation, defined by the GenIQvar tree and GenIQvar computer code, is in Figure 37.4.

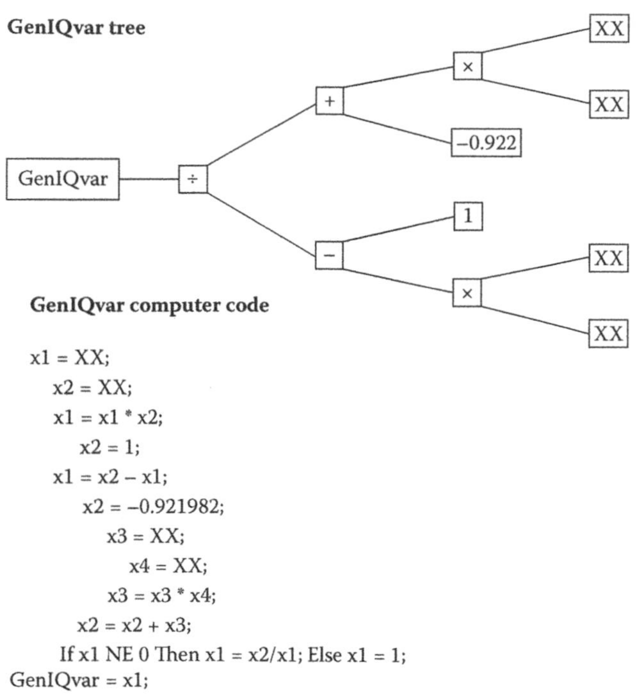

GenIQvar tree
GenIQvar
÷
+
−
×
−0.922
1
×
XX
XX
XX
XX
GenIQvar computer code
x1 = XX;
x2 = XX;
x1 = x1 * x2;
x2 = 1;
x1 = x2 – x1;
x2 = –0.921982;
x3 = XX;
x4 = XX;
x3 = x3 * x4;
x2 = x2 + x3;
If x1 NE 0 Then x1 = x2/x1; Else x1 = 1;
GenIQvar = x1;

FIGURE 37.4
GenIQ Model for moderating the outlier.

The GenIQ Model as illustrated to moderate a *single* outlier can handily serve as a multivariate method for moderating virtually *all* outliers in the data. Multivariate moderation is possible due to GenIQ's fitness function. A discussion of how optimizing the decile table fitness function brings forth a moderation of virtually all outliers is beyond the scope of this chapter. Suffice it to say that such optimization is equivalent to straightening the data by repositioning the outliers into a multilinear pattern, thus moderating the outliers instead of discarding them.

37.4 Summary

The common remedy for handling outliers is to determine and discard them. I present an alternative data mining method for moderating outliers instead of discarding them. I illustrate the data mining feature of the GenIQ Model as the alternative method for handling outliers.

Reference

1. Huber, P. J., Robust Estimation of a Location Parameter. *Annals of Mathematical Statistics,* 35:73–101, 1964.

38

Overfitting: Old Problem, New Solution

38.1 Introduction

Overfitting, a problem akin to model inaccuracy, is as old as model building itself, as it is part of the modeling process. The effect of overfitting is an inaccurate model. The purpose of this chapter is to introduce a new solution, based on the data mining feature of the GenIQ Model, to the old problem of overfitting. I illustrate how the GenIQ Model identifies the complexity of the idiosyncrasies and subsequently instructs for deletion of the individuals that contribute to the complexity of the data under consideration.

38.2 Background

Overfitting, a problem akin to model inaccuracy, is as old as model building itself as it is part of the modeling process. An overfitted model is one that *approaches reproducing* the training data on which the model is built—by capitalizing on the idiosyncrasies of the training data. The model reflects the complexity of the idiosyncrasies by including extra variables, interactions, and variable constructs. It follows that a key characteristic of an overfitted model is the model has too many variables: an overfitted model is too complex.

In another rendering, an overfitted model can be considered a *too perfect picture* of the predominant pattern in the data; the model memorizes the training data instead of capturing the desired pattern. Individuals of validation data (drawn from the population of the training data) are strangers who are unacquainted with the training data and cannot expect to fit into the model's perfect picture of the predominant pattern.

When the accuracy of a model based on the validation data is out of the neighborhood of the accuracy of the model based on the training data, the problem is one of overfitting. As the fit of the model *increases* by *including more information* (seemingly to be a good thing), the predictive performance of the model on the validation data *decreases*. This "good thing" is the *paradox of overfitting*.

Related to overfitted models is the concept of prediction error variance. Overfitted models have large predictive error variance: The confidence interval about the prediction error is large.

38.2.1 Idiomatic Definition of Overfitting to Help Remember the Concept

A model is built to *represent* training data, not to *reproduce* training data. Otherwise, a visitor (an individual's data point) from validation data will not feel at home with the model. The visitor encounters an uncomfortable fit in the model because he or she probabilistically *does not* look like a typical data point from the training data. The misfit visitor takes a poor prediction. The model is overfitted.

The underfitted model, a nonfrequenter model, has too few variables. The underfitted model is too simple. An underfitted model can be considered a *poorly rendered picture* of the predominant pattern. Without recollection of the training data, the model captures poorly the desired pattern. Individuals of validation data are strangers who have no familiarity with the training data and cannot expect to fit into the model's portraiture of the predominant pattern.

As overfitted models affect large (prediction) error variance, underfitted models affect error bias. *Bias* is the difference between the predicted score and the true score. Underfitted models have large error bias: Predicted scores are extremely far from the true scores. Figure 38.1 is a graphical depiction of overfitted and underfitted models.*

Consider the two models: the simple model *g(x)* in the left-hand graph and the *zigzag* model in the right-hand graph in Figure 38.1. Clearly, I want a model that best represents the predominant pattern of the parabola as depicted by the data points indicated by circles. I fit the points with the straight-line model *g(x)*, using only one variable (too few variables). The model is visibly too simple. It does not do a good job of fitting the data and would not do well in predicting for new data points. This model is underfitted.

For the *rough* zigzag model, I fit the data to hit every data point by using too many variables. The model does a perfect job of reproducing the data points but would not do well in predicting for new data points. This model is utterly overfitted. The model does not reflect the obvious *smooth* parabolic pattern. As is plainly evident, I want a model between the *g(x)* and zigzag models, a model that is powerful enough to represent the apparent pattern of a parabola. The discussion of the conceptual building of the desired model follows.

It is "fitting" to digress here for a discussion of model accuracy. A well-fitted model is one that *faithfully represents* the sought-after predominant pattern in the data, ignoring the idiosyncrasies in the training data. A well-fitted model is typically defined by a handful of variables. Individuals of validation data, the everyman and everywoman incognizant with the training data can expect to fit into the model's faithfully rendered picture of the

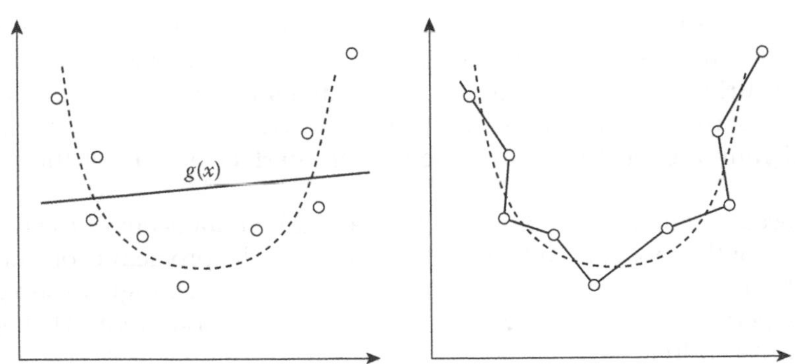

FIGURE 38.1
Over- and underfitted models.

* At the time of printing this edition, the original source is no longer available. http://www.willamette.edu/gorr/classes/cs449.html, 2010.

Model's accuracy (HOLDOUT) ≈ **Model's accuracy** (TRAINING)

FIGURE 38.2
Definition of well-fitted model.

Model's accuracy (HOLDOUT) ≤ **Model's accuracy** (TRAINING)

FIGURE 38.3
Definition of overfitted model.

predominant pattern. The accuracy of the well-fitted model on validation/holdout data is in the neighborhood of the model's accuracy based on the training data. Thus, the well-fitted model can be defined by the *approximately equal* equality, as depicted in Figure 38.2.

In contrast, the accuracy of the overfitted model on validation data will be outside the neighborhood of the model's accuracy based on the training data. Thus, the overfitted model can be defined by the *less than* inequality, as depicted in Figure 38.3.

38.3 The GenIQ Model Solution to Overfitting

I introduce a new solution, based on the data mining feature of the GenIQ Model, to the old problem of overfitting. The GenIQ Model solution to overfitting consists of the following steps:

1. Identify the complexity of the idiosyncrasies, the variables, and their constructs of the idiosyncrasies.
2. Delete the individuals that contribute to the complexity of the data from the dataset under consideration.
3. A model can now be built on *clean* data to represent the predominant pattern honestly, yielding a well-fitted model.

I illustrate how the GenIQ Model identifies the complexity of the idiosyncrasies and subsequently instructs for deletion of the individuals that contribute to the complexity of the data, from the dataset under consideration. Using the popular random-split validation, I create a variable RANDOM_SPLIT (R-S) that randomly divides the dataset into equal halves (50%–50%).* The SAS© code for the construction of RANDOM_SPLIT within a real case study dataset OVERFIT is

```
data OVERFIT;
set OVERFIT;
RANDOM_SPLIT = 0;
if uniform(12345) = le 0.5 then RANDOM_SPLIT = 1;
run;
```

* For unequal splits (e.g., 60%–40%), the approach is the same.

TABLE 38.1
Overfit Data: Variables and Type

No.	Variable	Type
1	RANDOM_SPLIT	Num
2	REQUESTE	Num
3	INCOME	Num
4	TERM	Num
5	APPTYPE	Char
6	ACCOMMOD	Num
7	CHILDREN	Num
8	MOVES5YR	Num
9	MARITAL	Num
10	EMPLOYEE	Num
11	DIRECTDE	Char
12	CONSOLID	Num
13	NETINCOM	Num
14	EMPLOY_1	Char
15	EMAIL	Num
16	AGE	Num
17	COAPP	Num
18	GENDER	Num
19	INCOMECO	Num
20	COSTOFLI	Num
21	PHCHKHL	Char
22	PWCHKHL	Char
23	PMCHKHL	Char
24	NOCITZHL	Char
25	EMPFLGHL	Char
26	PFSFLGHL	Char
27	NUMEMPLO	Num
28	BANKAFLG	Char
29	EMPFLGML	Char
30	PFSFLGML	Char
31	CIVILSML	Char
32	TAXHL	Num
33	TAXML	Num
34	NETINCML	Num
35	LIVLOANH	Num
36	LIVCOSTH	Num
37	CARLOAN	Num
38	CARCOST	Num
39	EDLOAN	Num
40	EDCOST	Num
41	OTLOAN	Num
42	OTCOST	Num
43	CCLOAN	Num
44	CCCOST	Num
45	EMLFLGHL	Char
46	PHONEH	Num
47	PHONEW	Num
48	PHONEC	Num
49	REQCONSR	Num
50	TIMEEMPL	Num
51	AGECOAPP	Num
52	APPLIEDY	Num
53	GBCODE	Num

The variables and their type (numeric or character) in the OVERFIT dataset are in Table 38.1. There are three possible modeling events on the condition of OVERFIT:

1. If OVERFIT has *no noise*, building a model with dependent variable RANDOM_SPLIT is *impossible*. The decile table has Cum Lifts equal to 100 throughout, from top to bottom deciles. OVERFIT is *clean* of idiosyncrasies. Accurate predictions result when building a model with OVERFIT data.
2. If OVERFIT has *negligible noise*, building a model is *most likely* possible. The decile table has Cum Lifts *within* [98, 102] in the upper deciles, say, top to third. OVERFIT is *almost* clean of idiosyncrasies. Highly probable accurate predictions result when building a model with OVERFIT data.
3. If OVERFIT has *unacceptable noise*, building a model is *possible*: The decile table has Cum Lifts *outside* [98, 102] in the upper deciles, say, top to third. OVERFIT has idiosyncrasies causing *substantial overfitting*. The cleaning of OVERFIT consists of deleting the individuals in the deciles with Cum Lifts outside [98, 102]. As a result, a well-fitted model uses the clean version of OVERFIT. Accurate predictions result when building a model with the clean OVERFIT data.

38.3.1 RANDOM_SPLIT GenIQ Model

The GenIQ Model consists of two components: a tree display and computer code. Using the OVERFIT dataset, I build a GenIQ Model with the dependent variable RANDOM_SPLIT. The RANDOM_SPLIT GenIQ Model tree display (Figure 38.4) identifies the complexity of the idiosyncrasies (noise) in OVERFIT. The RANDOM_SPLIT GenIQ Model computer code is in Figure 38.5.

38.3.2 RANDOM_SPLIT GenIQ Model Decile Analysis

The decile analysis from the RANDOM_SPLIT GenIQ Model* built on the OVERFIT data indicates OVERFIT has unacceptable noise.

1. The decile table, in Table 38.2, has Cum Lifts for the top and second deciles, 126 and 105, respectively, outside [98, 102]. Individuals in these deciles cause substantial overfitting. Individuals with Cum Lifts of 103 in deciles 4, 7, and 8 are perhaps iffy, as per the *quasi N-tile analysis* in Section 38.3.3.
2. The decile table is constructed by the brute (dumb) division of the data into 10 rows of *equal size regardless* of individual model scores. As model scores often spill over into lower deciles, the dumb decile table typically yields a biased assessment of model performance.
3. In Table 38.2, the top decile minimum score (2.09) spills over into the second decile. Not knowing how many model scores of 2.09 fall in the second deciles, I can only guess the percentage of data with 2.09 noise: The percentage falls in the fully open interval (0%, 20%).
4. A quasi N-tile analysis sheds light on the *spilling over of scores* (in not only the top and second deciles but also any number of consecutive deciles).

* Note: The logistic regression model could be used instead of the GenIQ Model, but the results would be the identification of only linear noise, consisting of a few variables with two functions, addition and subtraction.

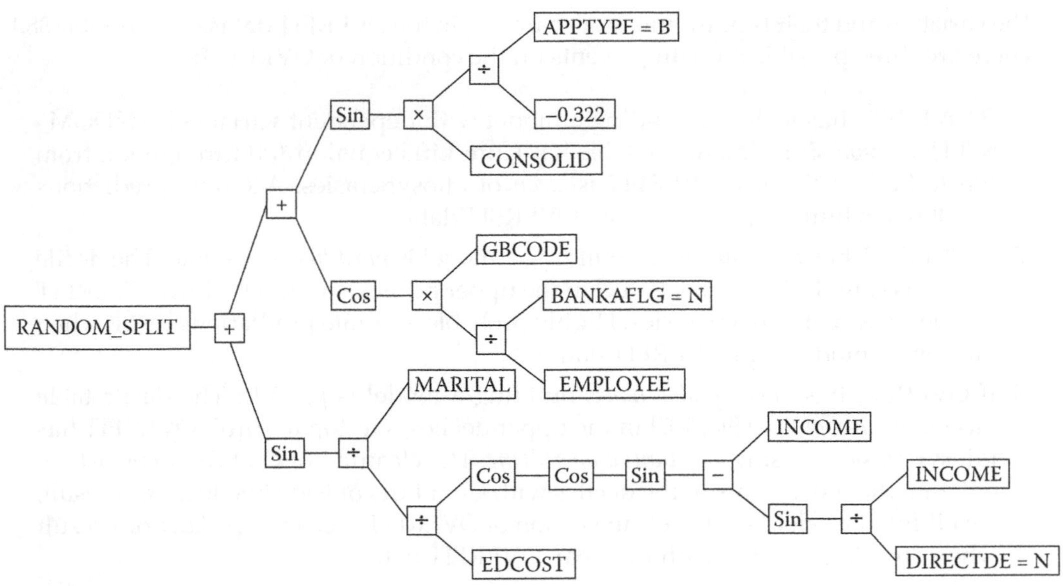

FIGURE 38.4
RANDOM_SPLIT GenIQ Model tree.

```
x1 = EDCOST;
    If DIRECTDE = "N" Then x2 = 1; Else x2 = 0;
        x3 = INCOME;
    If x2 NE 0 Then x2 = x3/x2; Else x2 = 1;
    x2 = Sin(x2);
        x3 = INCOME;
    x2 = x3 – x2; x2 = Sin(x2);
    x2 = Cos(x2); x2 = Cos(x2);
If x1 NE 0 Then x1 = x2/x1; Else x1 = 1;
    x2 = MARITAL;
If x1 NE 0 Then x1 = x2/x1; Else x1 = 1;
x1 = Sin(x1);
    x2 = EMPLOYEE;
        If BANKAFLG = "N" Then x3 = 1; Else x3 = 0;
    If x2 NE 0 Then x2 = x3/x2; Else x2 = 1;
        x3 = GBCODE;
x2 = x2*x3; x2 = Cos(x2);
        x3 = CONSOLID;
            x4 = –.3223163;
                If APPTYPE = "B" Then x5 = 1; Else x5 = 0;
            Lf x4 NE 0 Then x4 = x5/x4; Else x4 = 1;
        x3 = x3 * x4; x3 = Sin(x3);
    x2 = x2 + x3;
x1 = x1 + x2;
GenIQvar = x1;
```

FIGURE 38.5
RANDOM_SPLIT GenIQ Model computer code.

TABLE 38.2
RANDOM_SPLIT Decile Table

Decile	Predicted Random_Split	Random_ Split Rate (%)	Cum Random_ Split Rate (%)	Cum Lift	Min Score	Max Score
top	88	62.86	62.86	126	2.09	2.91
2	59	42.14	52.50	105	1.91	2.09
3	67	47.86	50.95	102	1.84	1.91
4	73	52.14	51.25	103	1.84	1.84
5	65	46.43	50.29	101	1.65	1.84
6	76	54.29	50.95	102	1.14	1.65
7	78	55.71	51.63	103	1.02	1.14
8	69	49.29	51.34	103	0.46	1.02
9	62	44.29	50.56	101	0.23	0.46
bottom	63	45.00	50.00	100	–0.96	0.23

TABLE 38.3
Quasi 20-tile Analysis

Select Number of Tiles	20					
N-Tile	**Number of Individuals**	**Number of Random_Split**	**Random_ Split Rate (%)**	**Cum Random_ Split Rate (%)**	**Cum% of Sample**	**Cum Lift (%)**
top	56	40	71.43	71.43	04.00	143
2	23	15	65.22	69.62	05.64	139
3	16	8	50.00	66.32	06.79	133
4	33	20	60.61	64.84	09.14	130
5	44	19	43.18	59.30	12.29	119
6	503	240	47.71	50.67	48.21	101

38.3.3 Quasi N-tile Analysis

A quasi N-tile analysis is needed to determine how many model scores fall about consecutive deciles. A quasi N-tile analysis divides model scores into *score groups* or *N-tiles,* which consist of a *distinct* score for individuals within an N-tile and *different* distinct scores across N-tiles. A quasi-*smart* analysis eliminates model scores spilling over to provide an unbiased assessment of model performance.

The quasi-*smart* analysis of the RANDOM_SPLIT decile table is shown in Table 38.3. Not knowing the distribution of the model scores, I instruct for a 20-tile analysis:

1. The smart analysis produces only six N-tiles as there are apparently only six distinct scores.
2. The top five N-tiles have Cum Lifts outside [98, 102].
3. Within these N-tiles, there are 172 (= 56 + 23 + 16 + 33 + 44) individuals.
4. The Cum Lift at the fifth score group is 119, after which the Cum Lifts of the RANDON_SPLIT Decile in Table 38.2 range between [100, 102/103].
5. Deleting the 172 individuals *removes the source of noise* in OVERFIT data.

TABLE 38.4
RANDOM_SPLIT Decile Table

Decile	Predicted Random_Split	Random_ Split Rate (%)	Cum Random_ Split Rate (%)	Cum Lift	Min Score	Max Score
top	62	50.49	50.41	101	–1.26	1.33
2	62	50.49	50.41	101	–1.27	–1.26
3	61	49.67	50.27	101	–1.38	–1.27
4	62	50.49	50.31	101	–1.54	–1.38
5	61	49.67	50.16	100	–1.60	–1.54
6	60	48.86	49.93	100	–3.07	–1.60
7	62	50.49	50.00	100	–3.19	–3.07
8	61	49.67	50.00	100	–3.28	–3.19
9	62	50.49	50.05	100	–4.71	–3.28
bottom	61	49.67	50.00	100	–13.44	–4.71

OVERFIT is now clean of noise. To test the soundness of the last assertion, I rerun GenIQ with clean OVERFIT data. The resultant decile table, in Table 38.4, displays Cum Lifts within [100, 101]. Hence, OVERFIT is clean of noise and is ready for building a well-fitted model. Note, the decile table in Table 38.4 is a *smart* decile table, which does not require generating a corresponding quasi N-tile analysis.

It is worth of note that model builders, who do not have access to GenIQ, can certainly use logistic or ordinary regression models in place of GenIQ. The results will not be as efficient in finding the most complicated sources of noise, but the linear noise will be identified and eliminated from the original dataset, yielding a nice cleaning of the data.

38.4 Summary

Overfitting, a problem akin to model inaccuracy, is as old as model building itself because it is part of the modeling process. An overfitted model is one that approaches reproducing the training data on which the model is built—by capitalizing on the idiosyncrasies of the training data. I introduce a new solution, based on the data mining feature of the GenIQ Model, to the old problem of overfitting. I illustrate, with a real case study, how the GenIQ Model identifies the complexity of the idiosyncrasies and instructs for deletion of the individuals that contribute to the complexity of the data to produce a dataset clean of noise. The clean dataset is ready for building a well-fitted model. I note that model builders, who do not have access to GenIQ, can use logistic or ordinary regression models in place of GenIQ. The results will assuredly find linear noise for identification and elimination from the original dataset, yielding a nice clean dataset.

39

The Importance of Straight Data: Revisited

39.1 Introduction

The purpose of this chapter is to revisit examples discussed in Chapters 5 and 12. The examples illustrated the importance of straight data—without explanations because the preceding chapters did not cover the material needed to understand the solutions.

At this point, I have provided the required background. Thus, in this chapter, for completeness, I put forward in detail solutions. The solutions use the data mining feature, straightening data, of the GenIQ Model. I start with the example in Chapter 12 and conclude with the example in Chapter 5.

39.2 Restatement of Why It Is Important to Straighten Data

From Section 5.2, there are five reasons why it is important to straighten data:

1. The straight-line (linear) relationship between two continuous variables X and Y *is as simple as it gets*. As X increases (decreases) in its values, Y increases (decreases) in its values. In this case, X and Y are positively correlated. Or, as X increases (decreases) in its values, Y decreases (increases) in its values. In this case, X and Y are negatively correlated. As an example of this setting of simplicity (and everlasting importance), Einstein's E and m have a perfect, positive linear relationship.
2. With linear data, the data analyst without difficulty sees *what is going on within the data*. The class of linear data is the desirable element for good model-building practice.
3. Most marketing models, belonging to the class of innumerable varieties of the linear statistical model, *require linear relationships* between a dependent variable and (a) each predictor variable in a model and (b) *all* predictor variables considered jointly, regarding them as an array of predictor variables that have a multivariate normal distribution.
4. It is well known that *nonlinear models*, attributed with yielding good predictions with nonstraight data, in fact, *do better with straight data*.
5. I have not ignored the feature of symmetry. Not accidentally, there are theoretical reasons for *symmetry and straightness going hand in hand*. Straightening data often makes data symmetric and vice versa. Recall that symmetric data have

values that are in correspondence in size and shape on opposite sides of a dividing line or middle value of the data. The iconic symmetric data profile in statistics is bell-shaped.

39.3 Restatement of Section 12.3.1.1 "Reexpressing INCOME"

I envision an underlying positively sloped straight line running through the 10 points in the PROFIT–INCOME smooth plot in Figure 12.2, even though the smooth trace reveals four severe kinks. Based on the general association test with a test statistic (TS) value of 6, which is *almost* equal to the cutoff score 7, as presented in Chapter 3, I conclude there is an *almost noticeable* straight-line relationship between PROFIT and INCOME. The correlation coefficient for the relationship is a reliable $r_{PROFIT,\ INCOME}$ of 0.763. Notwithstanding these indicators of straightness, the relationship could use some straightening, but clearly, the bulging rule does not apply.

An alternative method for straightening data, especially characterized by nonlinearities, is the GenIQ procedure, a machine-learning, genetic-based data mining method.

39.3.1 Complete Exposition of Reexpressing INCOME

I use the GenIQ Model to reexpress INCOME. The genetic structure, which represents the reexpressed INCOME variable, labeled gINCOME, is defined in Equation 12.3:

$$\text{gINCOME} = \sin(\sin(\sin(\sin(\text{INCOME})))*\text{INCOME}) + \log(\text{INCOME}) \qquad (12.3)$$

The structure uses the nonlinear reexpressions of the trigonometric sine function (four times) and the log (to base 10) function to loosen the "kinky" PROFIT–INCOME relationship. The relationship between PROFIT and INCOME (via gINCOME) is now smooth as the smooth trace reveals no serious kinks in Figure 12.3. Based on TS equals 6, which again is almost equal to the cutoff score of 7, I conclude there is an almost noticeable straight-line PROFIT–gINCOME relationship, a nonrandom scatter about an underlying positively sloped straight line. The correlation coefficient for the reexpressed relationship is a reliable $r_{PROFIT,\ gINCOME}$ of 0.894.

Visually, the effectiveness of the GenIQ procedure in straightening the data is obvious: the sharp peaks and valleys in the original PROFIT smooth plots versus the smooth wave of the reexpressed smooth plot. Quantitatively, the gINCOME-based relationship represents a noticeable improvement of 7.24% (= (0.894 – 0.763)/0.763) increase in correlation coefficient "points" over the INCOME-based relationship.

Two points of note: I previously invoked the statistical factoid that states the log function is typically used to reexpress a dollar-unit variable. Thus, it is not surprising that the genetically evolved structure gINCOME uses the log function. Regarding the use of the log function for the PROFIT variable, I concede that PROFIT could not benefit, no doubt due to the "mini" in the dataset (i.e., the small size of the data). So, I chose to work with PROFIT, not log of PROFIT, for the sake of simplicity (another EDA mandate, even for instructional purposes).

GenIQ model tree

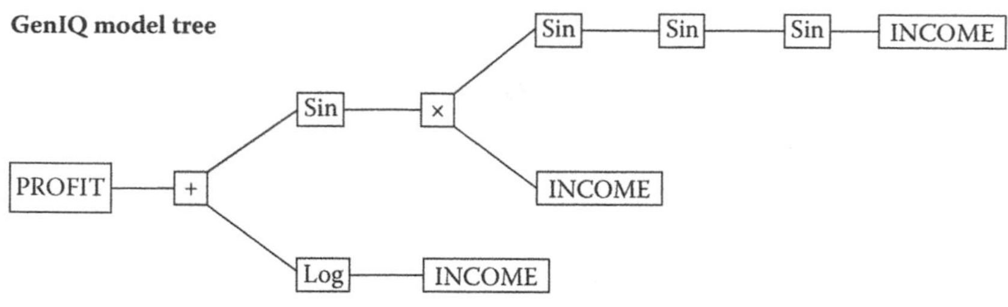

GenIQ model computer code

```
x1 = INCOME;
   x1 = Log (x1);
      x2 = INCOME;
         x2 = sin (x2);
         x2 = sin (x2);
         x2 = sin (x2);
            x3 = INCOME;
      x2 = x2 + x3;
      x2 = sin (x2);
x1 = x1 + x2;
GenIQvar = x1;
gINCOME = GenIQvar;
```

FIGURE 39.1
GenIQ Model for gINCOME.

39.3.1.1 The GenIQ Model Detail of the gINCOME Structure

The GenIQ Model for gINCOME is in Figure 39.1.

39.4 Restatement of Section 5.6 "Data Mining the Relationship of (xx3, yy3)"

Recall, I data mine for the underlying structure of the paired variables (xx3, yy3) using a machine-learning approach to the discipline of evolutionary computation, specifically *genetic programming* (GP). The fruits of my data mining work yield the scatterplot in Figure 5.3. The data mining work is not an expenditure of preoccupied time (i.e., not waiting for time-consuming results) or mental effort as the GP-based data mining (GP-DM) is a machine-learning adaptive intelligent process, which is quite effective for straightening data. The data mining software used is the GenIQ Model, which renames the data-mined variable with the prefix GenIQvar. Data-mined (xx3, yy3) is relabeled (xx3, GenIQvar(yy3)).

39.4.1 The GenIQ Model Detail of the GenIQvar(yy3) Structure

The GenIQ Model for GenIQvar(yy3) is in Figure 39.2.

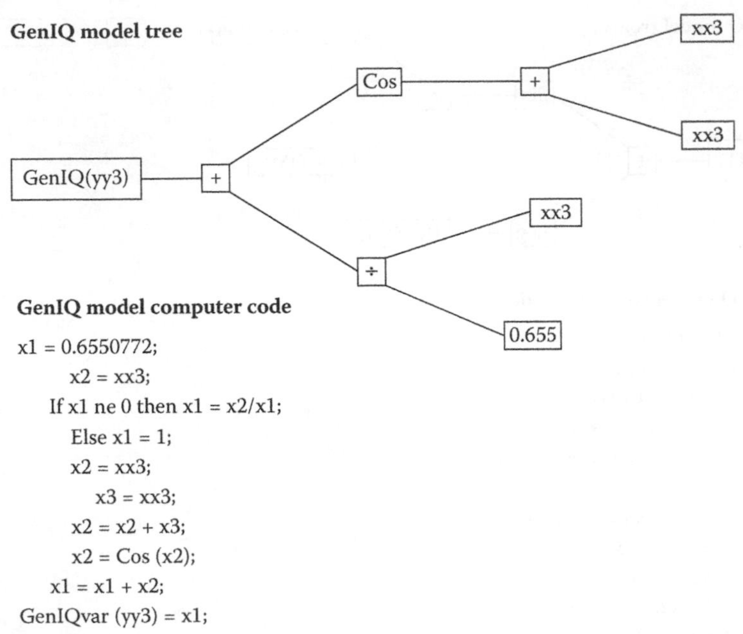
GenIQ model tree
GenIQ(yy3)
+
Cos
+
xx3
xx3
÷
xx3
0.655
GenIQ model computer code
x1 = 0.6550772;
x2 = xx3;
If x1 ne 0 then x1 = x2/x1;
Else x1 = 1;
x2 = xx3;
x3 = xx3;
x2 = x2 + x3;
x2 = Cos (x2);
x1 = x1 + x2;
GenIQvar (yy3) = x1;

FIGURE 39.2
GenIQ Model for GenIQvar(yy3).

39.5 Summary

I revisit examples discussed in Chapters 5 and 12. The examples illustrate the importance of straight data—without explanations because the preceding chapters did not cover the material needed to understand the solutions.

At this point, I have provided the background required. Thus, for completeness, I put forward in detail solutions in this chapter. The solutions use the data mining feature, straightening data, of the GenIQ Model. I start with the example in Chapter 12 and conclude with the example in Chapter 5. Now, I have completed the illustrations of why it is important to straighten data.

40

*The GenIQ Model: Its Definition and an Application**

40.1 Introduction

Using a variety of techniques, regression modelers build the everyday models that maximize expected response and profit based on the results of marketing programs, solicitations, and the like. Standard techniques include the statistical methods of classical discriminant analysis (DA) as well as the logistic regression model (LRM) and the ordinary least squares (OLS) regression model. A recent addition to the regression modelers' arsenal is the machine-learning (ML) method of artificial neural networks (ANNs). Another newcomer is the GenIQ Model, an ML alternative to OLS regression model and LRM. It is the focus of this chapter and is presented in full detail.

First, I provide background on the concept of optimization as optimization techniques provide the estimation of all models. Then, I introduce *genetic modeling*, the ML optimization approach that serves as the engine for the GenIQ Model. As the ubiquitous marketing objectives are to maximize expected response and profit for developing marketing strategies, I demonstrate how the GenIQ Model serves to meet those objectives. Actual case studies explicate further the potential of the GenIQ Model.

40.2 What Is Optimization?

Whether in business or model building, optimization is central to the decision-making process. In both theory and practice, an optimization technique involves selecting the best (or most favorable) condition within a given environment. To distinguish between available choices, an objective function (also known as a fitness function) must be predetermined. The choice, which corresponds to the extreme value† of the objective function, is the best outcome that constitutes the details of the solution to the problem.

Modeling techniques are developed to find a specific solution to a problem. For example, in marketing, one such problem is to predict sales. The OLS regression technique is a model formulated to address sales prediction. The framing of the regression problem is finding the regression equation such that the prediction errors (the difference between actual and

* This chapter is based on an article in *Journal of Targeting, Measurement and Analysis for Marketing*, 9, 3, 2001. Used with permission.

† If the optimization problem seeks to minimize the objective function, then the extreme value is the smallest; if it seeks to maximize, then the extreme value is the largest.

predicted sales) are small.* The objective function is the prediction error, making the best equation the one that minimizes that prediction error. Calculus-based methods are used to estimate the best regression equation.

As I discuss further in the chapter, each modeling method addresses its decision problem. The GenIQ Model addresses problems in industries such as direct and database marketing and customer relationship management (CRM). Hence, the GenIQ Model uses genetic modeling as the optimization technique for its solution.

40.3 What Is Genetic Modeling?†

Just as Darwin's principle of the survival of the fittest‡ explains tendencies in human biology, regression modelers can use the same principle to predict the best solution to an optimization problem.§ Each genetic model has an associated fitness function value that indicates how well the model solves, or "fits," the problem. A model with a high fitness value solves the problem better than a model with a lower fitness value and survives and reproduces at a higher rate. Models that are less fit survive and reproduce, if at all, at a lower rate.

If two models are effective in solving a problem, then some of their parts contain some valuable genetic material probabilistically. Recombining the parts of highly fit parent models produces probabilistically offspring models that are better fit at solving the problem than either parent. Offspring models then become the parents of the next generation, repeating the recombination process. After many generations, an evolved model is declared the best-so-far solution of the problem.

Genetic modeling consists of the following steps [1]:

1. Define the fitness function. The fitness function allows for identifying good or bad models, after which refinements advance the goal of producing the best model.
2. Select the set of functions (e.g., the set of arithmetic operators [addition, subtraction, multiplication, division]; log and exponential) and variables (predictors X_1, X_2, ..., X_n

* The definition of *small* (technically called mean squared error) is the average of the squared differences between actual and predicted values.

† Genetic modeling as described in this chapter is formally known as genetic programming. I choose the term *modeling* instead of *programming* because the latter term, which has its roots in computer sciences, does not connote the activity of model building to data analysts with statistics or quantitative backgrounds.

‡ When people hear the phrase "survival of the fittest," most think of Charles Darwin. Well, interestingly, he did not coin the term but did use it 10 years later in the fifth edition of his still-controversial *On the Origin of Species*, published in 1869. British philosopher Herbert Spence first used the phrase "survival of the fittest"—after reading Charles Darwin's *On the Origin of Species* (1859)—in his *Principles of Biology* (1864), in which he drew parallels between his own economic theories and Darwin's biological ones, writing, "This survival of the fittest is that which Mr. Darwin has called 'natural selection.'" Darwin first used Spencer's new phrase "survival of the fittest" as a synonym for "natural selection" in the fifth edition of *On the Origin of Species*, published in 1869.

§ The focus of this chapter is optimization, but genetic modeling has been applied to a variety of problems: optimal control, planning, sequence induction, empirical discovery and forecasting, symbolic integration, and discovering mathematical identities.

and numerical values) that are believed to be related to the problem at hand (the dependent variable Y).* An initial population of random models is generated using the preselected set of functions and variables.

3. Calculate the fitness of each model in the population by applying the model to a training set, a sample of individuals along with their values on the predictor variables $X_1, X_2, \ldots, X_n$ and the dependent variable Y. Thus, every model has a fitness value reflecting how well it solves the problem.
4. Create a new population of models by mimicking the natural genetic operators: The genetic operators applied to models in the current population selected with a probability based on fitness (i.e., the fitter the model, the more likely the model is to be selected).
   a. *Reproduction*: Copy models from the current population into the new population.
   b. *Crossover*: Create two offspring models for the new population by genetically recombining randomly chosen parts of two parent models from the current population.
   c. *Mutation*: Introduce random changes to some models from the current population into the new population.

The model with the highest fitness value produced in a generation is the *best-of-generation* model, which is the solution, or an approximate solution, to the problem.

40.4 Genetic Modeling: An Illustration

Consider the process of building a response model, for which the dependent variable RESPONSE assumes two values: yes and no. I designate the best model as one with the highest R-squared† value. Thus, the fitness function is the formula for R-squared. (Analytical note: I am using the R-squared measure only for illustrative purposes. R-squared is not the fitness function of the GenIQ Model. The GenIQ Model fitness function is in Section 40.8.)

I have to select functions and variables that are related to the problem at hand (e.g., predicting RESPONSE). Function and variable selection engages theoretical rationale or empirical expertise. Function selection sometimes takes on a rapid pace of trial and error.

I have two variables, X_1 and X_2, to use as predictors of RESPONSE. Thus, the variable set contains X_1 and X_2. I add the numerical value "b" to the variable set based on prior experience. I define the function set to contain the four arithmetic operations and the exponential function (exp), also based on prior experience.

* Effectively, I have chosen a genetic alphabet.

† I know that R-squared is not an appropriate fitness function for a 0–1 dependent variable model. Perhaps I should use the likelihood function of the logistic regression model as the fitness measure for the response model or an example with a continuous (profit) variable and the R-squared fitness measure. There is more about the appropriate choice of fitness function for the problem at hand in a further section.

20%
20%
exp
"+"
%
"−"
20%
20%
"*"
20%

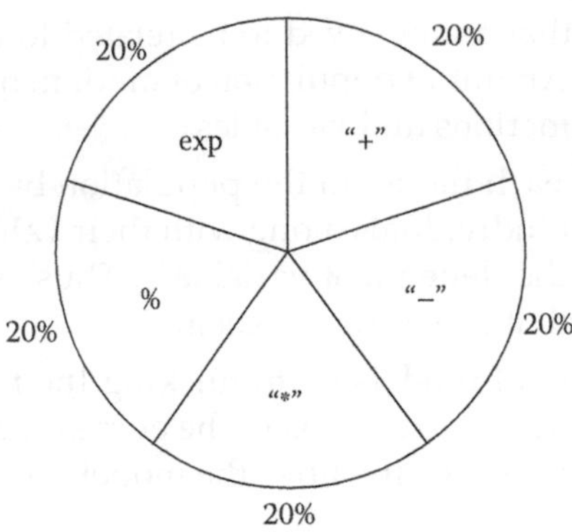

FIGURE 40.1
Unbiased function roulette wheel.

12.5%
12.5%
b
"+"
12.5%
12.5%
X2
"−"
X1
"*"
12.5%
12.5%
exp
%
12.5%
12.5%

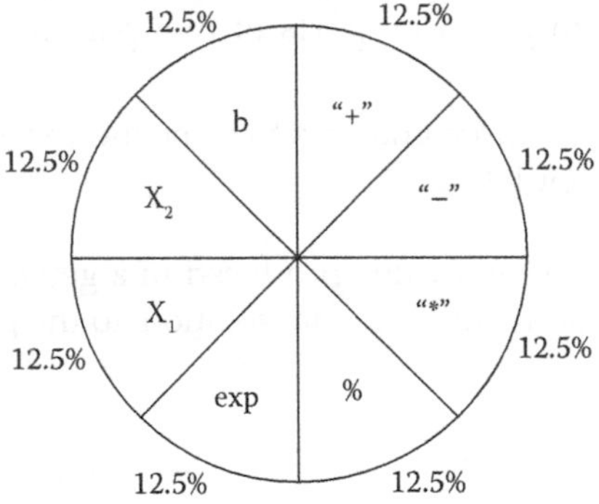

FIGURE 40.2
Unbiased function-variable roulette wheel.

Generating the initial population of random models is done with an unbiased function roulette wheel (Figure 40.1) and an unbiased function-variable roulette wheel (Figure 40.2). The slices of the function wheel are of equal size, namely 20%. The slices of the function-variable wheel are of equal size, namely 12.5%. Note, the division symbol "%" is used to denote the protected division. The protected division means that division by zero, which is undefined, is set to the value 1.

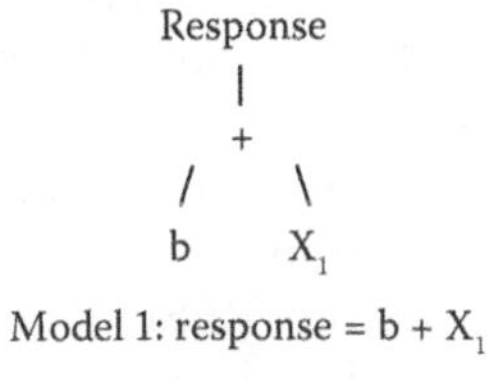

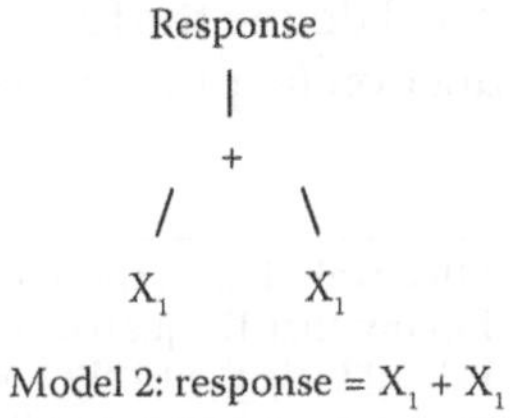

To generate the first random model, I spin the function wheel. The pointer of the wheel falls on slice "+." Next, I spin the function-variable wheel, and the pointer lands on slice X_1. With two following spins of the function-variable wheel, the pointer lands on slices "X_1" and "b," successively. I decide to stop evolving the model at this point. The resultant random model (Model 1) is in Figure 40.3 as a rooted point-label tree.

I generate the second random model, in Figure 40.4, by spinning the function wheel once, then by spinning the function-variable wheel twice. The pointer lands on slices "+," X_1, and X_1, successively. Similarly, I generate three additional random models, Models 3, 4, and 5 in Figures 40.5, 40.6, and 40.7, respectively.

Response
+
b
X1
Model 1: response = b + X1

FIGURE 40.3
Random Model 1.

Response
+
X1
X1
Model 2: response = X1 + X1

FIGURE 40.4
Random Model 2.

Response
|
*
/ \
X_1 X_1

Model 3: response = $X_1 * X_1$

FIGURE 40.5
Random Model 3.

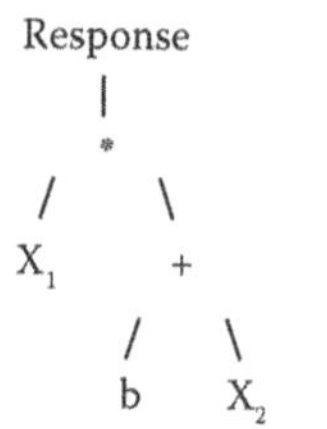

Model 4: response = $X_1 * (b + X_2)$

FIGURE 40.6
Random Model 4.

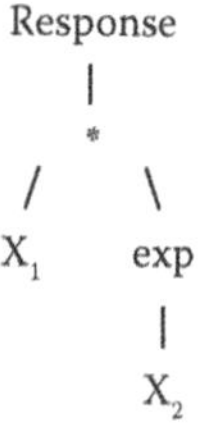

Model 5: response = $X_1 * \exp(X_2)$

FIGURE 40.7
Random Model 5.

Thus, I have generated the initial population of five random models (genetic population size is five).

Each of the five models in the population is assigned a fitness value in Table 40.1 to indicate how well it solves the problem of predicting RESPONSE. Because I am using R-squared as the fitness function, I apply each model to a training dataset to calculate its R-squared value. Model 1 produces the highest R-squared value, 0.52, and Model 5 produces the lowest R-squared value, 0.05.

Fitness for the population itself is calculable. The *total fitness of the population* is the sum of the fitness values among all models in the population. Here, the total population fitness is 1.53 (Table 40.1).

40.4.1 Reproduction

After the creation of an initial population of random models, all subsequent populations of models evolve with adaptive intelligence via the implementation of the genetic operators and the mechanism of selection *proportional to fitness* (PTF). Reproduction is the process by which models are duplicated or copied based on selection PTF. Selection PTF is model fitness value divided by total population fitness (see Table 40.1). For example, Model 1 has a PTF value of 0.34 (= 0.52/1.53).

Reproduction PTF means that a model with a high PTF value has a high probability of being selected for inclusion in the next generation. The reproduction operator implementation is with a biased model roulette wheel (Figure 40.8), where the sizes of slices are according to PTF values.

The operation of reproduction proceeds as follows. The spin of the biased model roulette wheel determines the essential qualities of copying models—which models to copy and how many times to copy. The pointer determines the model to copy without alteration and put into the next generation. Spinning the wheel in Figure 40.8, say 100 times, produces on average the following selection: 34 copies of Model 1, 27 copies of Model 2, 25 copies of Model 3, 11 copies of Model 4, and 3 copies of Model 5.

TABLE 40.1
Initial Population

	Fitness Value (R-Squared)	PTF (Fitness/Total)
Model 1	0.52	0.34
Model 2	0.41	0.27
Model 3	0.38	0.25
Model 4	0.17	0.11
Model 5	0.05	0.03
Population total fitness	1.53	

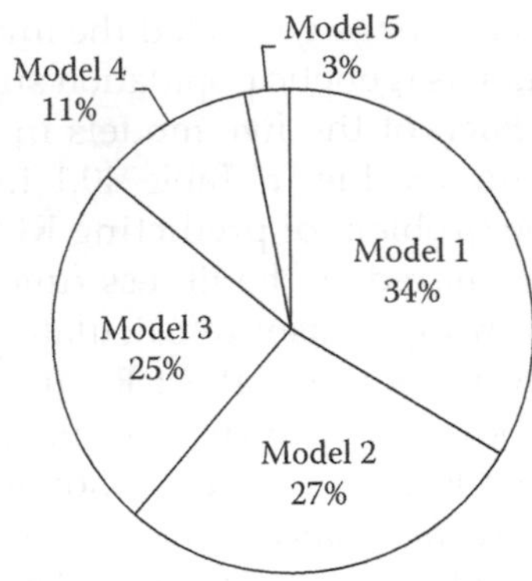

FIGURE 40.8
Biased model roulette wheel.

40.4.2 Crossover

The crossover (sexual recombination) operation is on two parent models by recombining randomly chosen parts of the two parent models. The expectation is that the offspring models are fitter than either parent model.

The crossover operation works with selection PTF. An illustration makes this operation easy to understand. Consider the parent models in Figures 40.9 and 40.10. The operation begins by randomly selecting an internal point (a function) in the tree for the crossover site.

Say, for instance, that the crossover sites are the lower "+" and "*" for Parents 1 and 2, respectively.

The crossover fragment for a parent is the subtree that has at its root the crossover site function. Crossover fragments for Parents 1 and 2 are in Figures 40.11 and 40.12, respectively.

Offspring 1, in Figure 40.13, from Parent 1 is produced by deleting the crossover fragment of Parent 1 and then inserting the crossover fragment of Parent 2 at the crossover point of Parent 1. Offspring 2, Figure 40.14, from Parent 2 produced by deleting the crossover fragment of Parent 2 and then inserting the crossover fragment of Parent 1 at the crossover point of Parent 2.

Parent 1

```
    +
   / \
 X2   +
     / \
    b   X1
```

FIGURE 40.9
Parent 1.

Parent 2

```
      –
     / \
    *   X4
   / \
  c   X3
```

FIGURE 40.10
Parent 2.

Crossover fragment 1

```
   +
  / \
 b   X1
```

FIGURE 40.11
Crossover fragment 1.

Crossover fragment 2

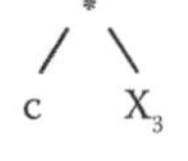

FIGURE 40.12
Crossover fragment 2.

40.4.3 Mutation

The mutation operation begins by selecting a point at random within a tree. This mutation point can be an internal point (a function) or an external or terminal point (a variable or numerical value). The mutation operation either *replaces* a randomly generated function with another function (from the function set previously defined) or *inverts*[*] the terminals of the subtree whose root is the randomly selected internal point.

For example, Model I, Figure 40.15, undergoes mutation by replacing the function "–" with "+," resulting in mutated Model I.1 in

* When a subtree has more than two terminals, the terminals are randomly permutated.

Figure 40.16. Model I is also mutated by *inverting* the terminal points c and X3, resulting in mutated Model I.2 in Figure 40.17.

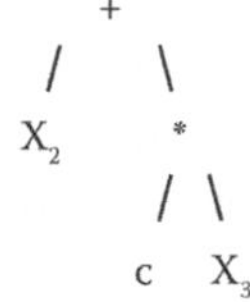

FIGURE 40.13
Offspring 1.

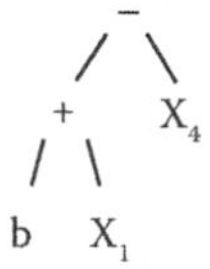

FIGURE 40.14
Offspring 2.

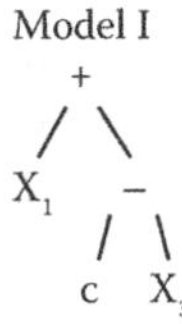

FIGURE 40.15
Model I for mutation.

Model I.1
+
X_1 +
c X_3

FIGURE 40.16
Mutated Model I.1.

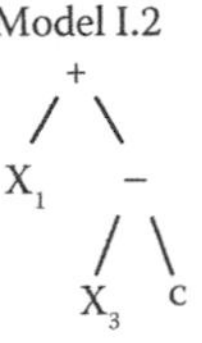

FIGURE 40.17
Mutated Model I.2.

40.5 Parameters for Controlling a Genetic Model Run

There are several control parameters requiring preset values before evolving a genetic model.

1. Genetic population size, the number of models randomly generated and subsequently evolved.
2. The maximum number of generations to run until there is no improvement in the fitness function values.
3. Reproduction probability, the percentage of the population that is copied. If population size is 100 and reproduction probability is 10%, then 10 models from each generation are selected (with reselection allowed) for reproduction. Selection is based on PTF.
4. Crossover probability, the percentage of the population that is used for crossover. If population size is 100 and crossover probability is 80%, then 80 models from each generation are selected (with reselection allowed) for crossover. Selection is based on PTF. Models are paired at random.
5. Mutation probability, the percentage of the population that is used for mutation. If population size is 100, and mutation 10%, then 10 models from each generation are selected (with reselection allowed) for mutation. Selection is based on PTF.
6. Termination criterion, the single model with the largest fitness value over all generations, the so-called best-so-far model, is declared the result of a run.

40.6 Genetic Modeling: Strengths and Limitations

Genetic modeling has strengths and limitations like any methodology. Perhaps the most important strength of genetic modeling is that it is a workable alternative to statistical models, which are highly parametric with sample size restrictions. Statistical models require, for their estimated coefficients, algorithms depending on smooth, unconstrained functions with the existence of derivatives (well-defined slope values). In practice, the functions (response surfaces) are noisy, multimodal, and frequently discontinuous. In contrast, genetic models are robust, assumption-free, nonparametric models and perform well on large and small samples. The only requirement is a fitness function, which can be designed to ensure that the genetic model does not perform worse than any other statistical model.

Genetic modeling has shown itself to be effective for solving large optimization problems as it can efficiently search through response surfaces of very large datasets. Also, genetic modeling can be used to learn complex relationships, making it a viable data mining tool for rooting out valuable pieces of information.

A potential limitation of genetic modeling is in the setting of the genetic modeling parameters: genetic population size and reproduction, crossover, and mutation probabilities. The parameter settings are, in part, data and problem dependent; thus, proper settings require experimentation. Fortunately, new theories and empirical studies are continually providing rules of thumb for these settings as application areas broaden. These guidelines* make genetic modeling an accessible approach for regression modelers not formally trained in genetic modeling. Even with the "correct" parameter settings, genetic models do not guarantee the optimal (best) solution. Further, genetic models are only as good as the definition of the fitness function. Precisely defining the fitness function sometimes requires expert experimentation.

40.7 Goals of Marketing Modeling

Marketers typically attempt to improve the effectiveness of their marketing strategies by targeting their best customers or prospects. They use a model to identify individuals who are likely to respond to or generate profit† from a campaign, solicitation, and the like. The model provides, for each individual, estimates of the probability of response or estimates of the contribution of profit. Although the precision of these estimates is important, the performance of the model is measured at an aggregated level as reported in a decile analysis.

Marketers have defined the *Cum Lift*, found in the decile analysis, as the relevant measure of model performance. Based on the selection of individuals by the model, marketers create a "push-up" list of individuals likely to respond or contribute to profit to obtain an advantage over a random selection of individuals.

The Cum Response Lift is an index of the expected incremental responses with a selection based on a response model over the expected responses based on a random selection (the chance model). Similarly, the Cum Profit Lift is an index of the expected incremental profit with a selection based on a profit model over the expected profit with a random selection (the chance model). The concept of Cum Lift and the steps of the construction in a decile analysis are discussed in Chapter 26.

It should be clear that a model that produces a decile analysis with more responses or profit in the upper (top, second, third, or fourth) deciles is a better model than a model with fewer responses or less profit in the upper deciles. This concept is the motivation for the GenIQ Model.

40.8 The GenIQ Response Model

The GenIQ approach to modeling is to address the ubiquitous objective concerning regression modelers across a multitude of industry sectors (e.g., direct, database, or telemarketing; business analytics; risk analytics; consumer credit; customer life cycle; financial services

* See http://www.geniq.net/GenIQModelFAQs.html from my website.

† I use the term *profit* as a stand-in for any measure of an individual's worth, such as sales per order, lifetime sales, revenue, number of visits, or number of purchases.

marketing; and the like), namely, maximizing response and profit. The GenIQ Model uses the genetic methodology to optimize *explicitly* the desired criterion: *maximize the upper deciles*. Consequently, the GenIQ Model allows regression modelers to build response and profit models in ways that are not possible with current statistical models.

The GenIQ Response Model is theoretically superior—with respect to maximizing the upper deciles—to a response model built with alternative response techniques because of the explicit nature of the fitness function. The formulation of the fitness function is beyond the scope of this chapter. But, suffice it to say, the fitness function seeks to fill the upper deciles with as many responses as possible, equivalently, to maximize the Cum Response Lift.

Alternative response techniques such as DA, LRM, and ANN only maximize the desired criterion implicitly. Their optimization criterion (fitness function) serves as a surrogate for the desired criterion. DA, with the assumption of bell-shaped data, is defined to maximize the ratio of between-group sum of squares to the within-group sum of squares explicitly.

LRM, with the two assumptions of independence of responses and an S-shape relationship between predictor variables and response, is defined to maximize the logistic likelihood (LL) function.

The ANN, a highly parametric method, is typically defined to minimize explicitly mean squared error (MSE).

40.9 The GenIQ Profit Model

The GenIQ Profit Model is theoretically superior—on maximizing the upper deciles—to the OLS regression and ANN. The GenIQ Profit Model uses the genetic methodology with a fitness function that explicitly addresses the desired modeling criterion. The fitness function is defined *to fill the upper deciles with as much profit as possible that, equivalently, maximizes the Cum Profit Lift*.

The fitness function for OLS and ANN models minimizes MSE, which serves as a surrogate for the desired criterion.

OLS regression has another weakness in a marketing application. A key assumption of the regression technique is that the dependent variable data must follow a bell-shaped curve. If a violation of this assumption is severe, the resultant model may not be valid. Unfortunately, profit data are not bell-shaped. For example, a 2% response rate yields 98% nonresponders with profit values of zero dollars or some nominal cost associated with nonresponse. Data with a concentration of 98% of a single value cannot be spread out to form a bell-shaped distribution.

There is still another data issue when using OLS with marketing data. Lifetime value (LTV) is an important marketing performance measure. LTV typically has a distribution with positive skewness. The log is the appropriate transformation to reshape positively skewed data into a bell-shaped curve. However, using the log of LTV as the dependent variable in OLS regression does not guarantee that other OLS assumptions are not subject to violation.* Accordingly, attempts at modeling profit with OLS regression are questionable or difficult.

The GenIQ Response and Profit Models have no restriction on the dependent variable. The GenIQ Models produce accurate and precise predictions with a dependent variable of

* The error structure of the OLS equation may not necessarily be normally distributed with zero mean and constant variance, in which case the modeling results are questionable and additional transformations may be needed.

any shape.* GenIQ Models are insensitive to the shape of the dependent variable because the GenIQ estimation uses the genetic methodology, which is inherently nonparametric and assumption-free.

In fact, due to its nonparametric and assumption-free estimation, the GenIQ Models place no restriction on the interrelationship among the predictor variables. The GenIQ Models are unaffected by any degree of correlation among the predictor variables. In contrast, OLS and ANN, as well as DA and LRM, can tolerate only a "moderate" degree of intercorrelation among the predictor variables to ensure a stable calculation of their models. Severe degrees of intercorrelation among the predictor variables often lead to inestimable models.

Moreover, the GenIQ Models have no restriction on sample size. The GenIQ Models, whether built on small samples or large samples, are equally predictive to the extent the data permit. OLS, DA, and, somewhat less, ANN and LRM† models require at least a "moderate" size sample.‡

40.10 Case Study: Response Model

Cataloguer ABC requires a response model based on a recent direct mail campaign, which produces a 0.83% response rate (dependent variable is RESPONSE). ABC's consultant built an LRM using three variables based on the techniques discussed in Chapter 10:

1. RENT_1 is a composite variable measuring the ranges of rental cost.§
2. ACCTS_1 is a composite variable measuring the activity of various financial accounts.¶
3. APP_TOTL is the number of inquiries.

The logistic response model is defined in Equation 40.1 as

$$\text{Logit of RESPONSE} = -1.9 + 0.19\text{*APP\_TOTL} - 0.24\text{*RENT\_1} - 0.25\text{*ACCTS\_1} \quad (40.1)$$

The LRM response validation decile analysis in Table 40.2 shows the performance of the model over chance (i.e., no model). The decile analysis shows a model with good performance in the upper deciles: Cum Lifts for top, second, third, and fourth deciles are 264, 174, 157, and 139, respectively. Note that the model may not be as good as initially believed. There is some degree of unstable performance through the deciles. That is, the number of responses does not decrease steadily through the deciles. This unstable performance, which is characterized by jumps in deciles 3, 5, 6, and 8, is probably due to (1) an unknown relationship between the predictor variables and RESPONSE or (2) an important predictor variable not included in this model. Note that only perfect models have perfect performance throughout the deciles. Good models have some jumps, albeit minor ones.

I build a GenIQ Response Model based on the same three variables used in the LRM. The GenIQ response tree is in Figure 40.18. The validation decile analysis, in Table 40.3,

* The dependent variable can be bell-shaped or skewed, bimodal or multimodal, and continuous or discontinuous.

† There are specialty algorithms for logistic regression with small sample size.

‡ Statisticians do not agree on how "moderate" a moderate sample size is. I drew a sample of statisticians to determine the size of a moderate sample; the average was 5,000.

§ Four categories of rental cost: less than $200 per month, $200–$300 per month, $300–$500 per month, and greater than $500 per month.

¶ Financial accounts include bank cards, department store cards, installment loans, and so on.

TABLE 40.2
LRM Response Decile Analysis

Decile	Number of Individuals	Number of Responses	Decile Response Rate (%)	Cumulative Response Rate (%)	Cum Lift Response
top	1,740	38	2.20	2.18	264
2	1,740	12	0.70	1.44	174
3	1,740	18	1.00	1.30	157
4	1,740	12	0.70	1.15	139
5	1,740	16	0.90	1.10	133
6	1,740	20	1.10	1.11	134
7	1,740	8	0.50	1.02	123
8	1,740	10	0.60	0.96	116
9	1,740	6	0.30	0.89	108
bottom	1,740	4	0.20	0.83	100
Total	**17,400**	**144**	**0.83**		

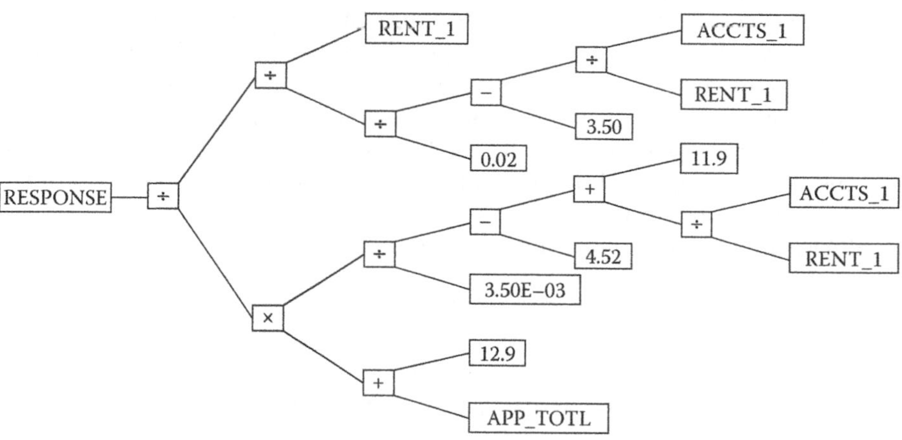

FIGURE 40.18
GenIQ Response tree.

TABLE 40.3
GenIQ Response Decile Analysis

Decile	Number of Individuals	Number of Responses	Decile Response Rate (%)	Cumulative Response Rate (%)	Cum Lift Response
top	1,740	44	2.50	2.53	306
2	1,740	18	1.00	1.78	215
3	1,740	10	0.60	1.38	167
4	1,740	10	0.60	1.18	142
5	1,740	14	0.80	1.10	133
6	1,740	10	0.60	1.02	123
7	1,740	12	0.70	0.97	117
8	1,740	10	0.60	0.92	111
9	1,740	8	0.50	0.87	105
bottom	1,740	8	0.50	0.83	100
Total	**17,400**	**144**	**0.83**		

TABLE 40.4

Comparison: LRM and GenIQ Response

Decile	LRM	GenIQ	GenIQ Improvement over LRM (%)
top	264	306	16.0
2	174	215	23.8
3	157	167	6.1
4	139	142	2.2
5	133	133	–0.2
6	134	123	–8.2
7	123	117	–4.9
8	116	111	–4.6
9	108	105	–2.8
bottom	100	100	—

shows a model with very good performance in the upper deciles: Cum Lifts for top, second, third, and fourth deciles are 306, 215, 167, and 142, respectively. In contrast with the LRM, the GenIQ Model has only two minor jumps, in deciles 5 and 7. The implication is the genetic methodology has evolved a better model because it has uncovered a nonlinear relationship among the predictor variables with RESPONSE. This comparison between LRM and GenIQ is conservative as GenIQ used the same three predictor variables used in LRM. As I discuss in Chapter 41, the strength of GenIQ is finding its best set of variables for the prediction task at hand.

The GenIQ Response Model is defined in Equation 40.2* as:

$$\text{GenIQvar\_RESPONSE} = \frac{7.0E-5*\text{RENT\_1}**3}{(\text{ACCTS\_1}-3.50*\text{RENT\_1})*(12.9+\text{APP\_TOTL})*(\text{ACCTS\_1}-7.38*\text{RENT\_1})} \quad (40.2)$$

GenIQ does not outperform LRM across all the deciles in Table 40.4. However, GenIQ yields noticeable Cum Lift improvements for the important top three deciles: 16.0%, 23.8%, and 6.1%, respectively.

40.11 Case Study: Profit Model

Telecommunications Company ATMC seeks to build a zip-code-level model to predict usage, dependent variable TTLDIAL1. Based on the techniques discussed in Chapter 12, the variables used in building an OLS regression model are as follows:

1. AASSIS_1 is a composite of public assistance-related census variables.
2. ANNTS_2 is a composite of ancestry census variables.

* The GenIQ Response Model is written conveniently in a standard algebraic form as all the divisions indicated in the GenIQ response tree (e.g., ACCTS_1/RENT_1) are not undefined (division by zero). If at least one indicated division is undefined, then I would have to report the GenIQ Model computer code, as reported in previous chapters.

3. FEMMAL_2 is a composite of gender-related variables.
4. FAMINC_1 is a composite variable measuring the ranges of home value.*

The OLS profit (usage) model is defined in Equation 40.3 as

$$\text{TTLDIAL1} = 1.5 + -0.35{*}\text{AASSIS\_1} + 1.1{*}\text{ANNTS\_2} + 1.4{*}\text{FEMMAL\_2} + 2.8{*}\text{FAMINC\_1} \quad (40.3)$$

The OLS profit validation decile analysis in Table 40.5 shows the performance of the model over chance (i.e., no model). The decile analysis shows a model with good performance in the upper deciles: Cum Lifts for the top, second, third, and fourth deciles are 158, 139, 131, and 123, respectively.

I build a GenIQ Profit Model based on the same four variables used in the OLS model. The GenIQ profit tree is in Figure 40.19. The validation decile analysis in Table 40.6 shows a model with very good performance in the upper deciles: Cum Lifts for the top, second, third, and fourth deciles are 198, 167, 152, and 140, respectively. This comparison between OLS and GenIQ is conservative as GenIQ was assigned the same four predictor variables used in OLS. (Curiously, GenIQ only used three of the four variables.) As mentioned in the previous section, which discussed the comparison between LRM and GenIQ, I discuss in the next chapter why the OLS–GenIQ comparison here is conservative.

The GenIQ Profit (Usage) Model is defined in Equation 40.4 as

$$\text{GenIQvar\_TTLDIAL1} = 5.95 + \text{FAMINC\_1} + (\text{FAMINC\_1} + \text{AASSIS\_1}){*}((0.68{*}\text{FEMMAL\_2}){*}(\text{AASSIS\_1}\ 3.485)) \quad (40.4)$$

GenIQ does outperform OLS across all the deciles in Table 40.7. GenIQ yields noticeable Cum Lift improvements down to the seventh decile; improvements range from 25.5% in the top decile to 6.9% in the seventh decile.

TABLE 40.5

Decile Analysis OLS Profit (Usage) Model

Decile	Number of Customers	Total Dollar Usage	Average Usage	Cumulative Average Usage	Cum Lift Usage
top	1,800	$38,379	$21.32	$21.32	158
2	1,800	$28,787	$15.99	$18.66	139
3	1,800	$27,852	$15.47	$17.60	131
4	1,800	$24,199	$13.44	$16.56	123
5	1,800	$26,115	$14.51	$16.15	120
6	1,800	$18,347	$10.19	$15.16	113
7	1,800	$20,145	$11.19	$14.59	108
8	1,800	$23,627	$13.13	$14.41	107
9	1,800	$19,525	$10.85	$14.01	104
bottom	1,800	$15,428	$8.57	$13.47	100
Total	18,000	$242,404	$13.47		

* Five categories of home value: less than $100,000; $100,000–$200,000; $200,000–$500,000; $500,000–$750,000; and greater than $750,000.

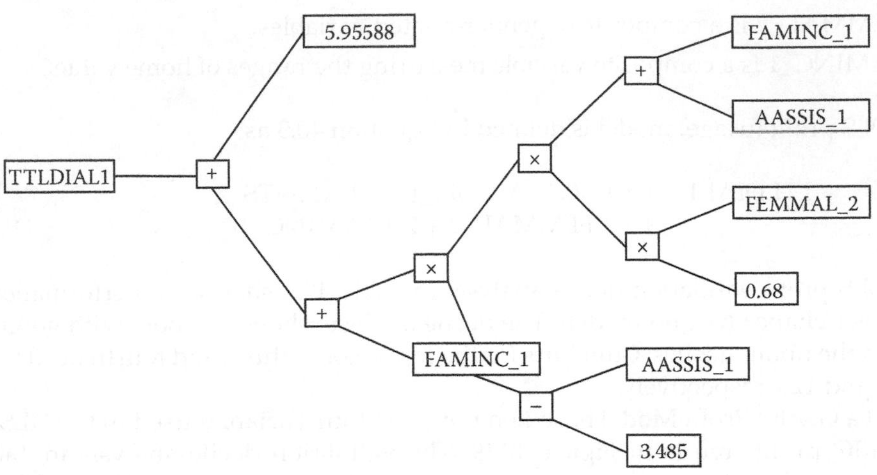

FIGURE 40.19
GenIQ Profit tree.

TABLE 40.6

Decile Analysis GenIQ Profit (Usage) Model

Decile	Number of Customers	Total Dollar Usage	Average Usage	Cumulative Average Usage	Cum Lift Usage
top	1,800	$48,079	$26.71	$26.71	198
2	1,800	$32,787	$18.22	$22.46	167
3	1,800	$29,852	$16.58	$20.50	152
4	1,800	$25,399	$14.11	$18.91	140
5	1,800	$25,115	$13.95	$17.91	133
6	1,800	$18,447	$10.25	$16.64	124
7	1,800	$16,145	$8.97	$15.54	115
8	1,800	$17,227	$9.57	$14.80	110
9	1,800	$15,125	$8.40	$14.08	105
bottom	1,800	$14,228	$7.90	$13.47	100
Total	**18,000**	**$242,404**	**$13.47**		

TABLE 40.7

Comparison: OLS and GenIQ Profit (Usage)

Decile	OLS	GenIQ	GenIQ Improvement Over OLS (%)
top	158	198	25.5
2	139	167	20.0
3	131	152	16.2
4	123	140	14.1
5	120	133	10.9
6	113	124	9.3
7	108	115	6.9
8	107	110	2.7
9	104	105	0.6
bottom	100	100	—

40.12 Summary

All standard statistical modeling techniques involve optimizing a fitness function to find a specific solution to a problem. The popular ordinary and logistic regression techniques, which seek accurate prediction and classification, respectively, optimize the fitness functions of MSE and the LL, respectively. Optimization computations employ calculus-based methods.

I present a new modeling technique, the GenIQ Model, which seeks maximum performance (response or profit) from diversified marketing programs, solicitations, and the like. The GenIQ Model optimizes the fitness function Cum Lift. The optimization computations of GenIQ use the genetic methodology not the usual calculus. I provided a compendious proem to genetic methodology with an illustration and inspection of its strengths and limitations.

The GenIQ Model is theoretically superior—on maximizing Cum Lift—to the ordinary regression model and LRM because of its clearly and fully formulated fitness function. The GenIQ fitness function explicitly seeks to fill the upper deciles with as many responses or as much profit as possible, equivalently, to maximize the Cum Lift. Standard statistical methods only implicitly maximize the Cum Lift as their fitness functions (MSE and LL) serve as a surrogate maximizing Cum Lift.

Last, I demonstrated the potential of the new technique with response and profit model illustrations. The GenIQ Response Model illustration yields noticeable Cum Lift improvements over logistic regression for the important first three decile ranges: 16.0%, 23.8%, and 6.1%, respectively. The GenIQ Profit Model illustration yields noticeable Cum Lift improvements in ordinary regression down through the seventh decile; improvements range from 25.5% in the top decile to 6.9% in the seventh decile.

Reference

1. Koza, J., *Genetic Programming: On the Programming of Computers by Means of Natural Selection*, MIT Press, Cambridge, MA, 1992.

41

*Finding the Best Variables for Marketing Models**

41.1 Introduction

Finding the best possible subset of variables to put in a model has been a frustrating exercise. Many methods of variable selection exist, but none of them is perfect. Moreover, they do not create new variables, which would enhance the predictive power of the original variables themselves. Furthermore, none uses a criterion that addresses the specific needs of marketing models. I present the GenIQ Model as a methodology that uses genetic modeling to find the best variables for marketing models. Most significant, the GenIQ Model addresses uniquely the specific requirement of marketing models, namely, to maximize the Cum Lift.

41.2 Background

The literature is replete with extensive research on the problem of finding the best subset of variables to define the best model. Existing methods, based on theory, search heuristics, and rules of thumb, each use a unique criterion to build the best model. Selection criteria break into two groups: one based on criteria involving classical hypothesis testing and the other involving residual error sum of squares.† Different criteria typically produce different subsets. The number of variables in common with the different subsets is not necessarily large, and the sizes of the subsets can vary considerably.

Essentially, the problem of variable selection is to examine certain subsets and select the subset that either maximizes or minimizes an appropriate criterion. Two subsets are obvious: the best single variable and the complete set of variables. The problem lies in selecting an intermediate subset that is better than both of these extremes. Therefore, the issue is how to find the *necessary variables* among the complete set of variables by deleting both irrelevant variables (variables not affecting the dependent variable) and *redundant variables* (variables not adding anything to the dependent variable) [1].

Reviewed next are five widely used variable selection methods. The first four methods are in all statistical software packages,‡ and the last is the favored rule-of-thumb approach used by many statistical modelers. The test statistic (TS) for the first three methods uses

* This chapter is based on an article in *Journal of Targeting, Measurement and Analysis for Marketing*, 9, 3, 2001. Used with permission.

† Other criteria are based on information theory and Bayesian rules.

‡ SAS/STAT Manual. See PROC REG and PROC LOGISTIC, www.support.asa.com, 2011.

either the F statistic for a continuous dependent variable or the G statistic for a binary dependent variable (e.g., response that assumes only two values, yes/no). The TS for the fourth method is either R-squared for a continuous dependent variable or the Score statistic for a binary dependent variable. The fifth method uses the popular correlation coefficient r.

1. *Forward selection (FS)*: This method adds variables to the model until no remaining variable (outside the model) can add anything significant to the dependent variable. FS begins with no variable in the model. For each variable, the TS, a measure of the contribution of the variable to the model, is calculated. The variable with the largest TS value, which is greater than a preset value C, is added to the model. Then, each remaining variable goes through the process of calculating and evaluating its TS value to determine whether the variable enters the model. Thus, variables are added to the model one by one until no remaining variable produces a TS value greater than C. Once a variable is in the model, it remains there.
2. *Backward elimination (BE)*: This method deletes variables one by one from the model until all remaining variables contribute something significant to the dependent variable. BE begins with a model that includes all variables. Variables are then deleted from the model one by one until all the variables remaining in the model have TS values greater than C. At each step, a variable is subject to deletion if it has the smallest contribution to the model (i.e., with the smallest TS value that is less than C).
3. *Stepwise (SW)*: This method is a modification of the FS approach and differs in that variables already in the model do not necessarily stay. As in FS, SW adds variables to the model one at a time. Variables that have TS values greater than C enter the model. After a variable enters, however, SW looks at all the variables already included to delete any variable that does not have a TS value greater than C.
4. *R-squared (R-sq)*: This method finds several subsets of different sizes that best predict the dependent variable. R-sq finds subsets of variables that best predict the dependent variable based on the appropriate TS. The best subset of size k has the largest TS value. For a continuous dependent variable, TS is the popular measure R-sq, the coefficient of (multiple) determination, which measures the proportion of explained variance of the dependent variable by a multiple regression model. For a binary dependent variable, TS is the theoretically correct but less-known Score statistic.* R-sq finds the best one-variable model, the best two-variable model, and so forth. However, it is unlikely that one subset will stand out as clearly the best as TS values often bunch together. For example, they are equal in value when rounding at the, say, third place after the decimal point.† R-sq generates some subsets of each size, which allows the modeler to select a subset, possibly using nonstatistical measures.
5. *Rule-of-thumb top-k variables (top-k)*: This method selects the top ranked variables regarding their association with the dependent variable. The correlation coefficient r measures the association of each variable with the dependent variable.

* R-squared theoretically is not the appropriate measure for a binary dependent variable. However, many analysts use it with varying degrees of success.

† For example, consider two TS values: 1.934056 and 1.934069. These values are equal when rounding occurs at the third place after the decimal point: 1.934.

The variables are ranked by their absolute r values,* from largest to smallest. The top-k ranked variables are considered the best subset. If the statistical model with the top-k variables indicates that each variable is statistically significant, then the set of k variables is declared the best subset. If any variable is not statistically significant, then the variable is removed and replaced by the next ranked variable. Each variable in the resultant set of variables is subject to significance testing and removal if not significant. The process of testing and removal of variables repeats until only significant variables define the statistical model.

41.3 Weakness in the Variable Selection Methods

While the mentioned methods produce reasonably good models, each method has a drawback specific to its selection criterion. A detailed discussion of the weaknesses is beyond the scope of this chapter. However, there are two common weaknesses, which do merit attention [2,3]. *First, the selection criteria of these methods do not explicitly address the specific needs of marketing models, namely, to maximize the Cum Lift.*

Second, these methods cannot identify structure in the data. They find the best subset of variables without digging into the data, a feature that is necessary for finding important variables or structures. Therefore, variable selection methods without data mining capability cannot generate the *enhanced best subset.* The following illustration clarifies this weakness. Consider the complete set of variables, X_1, X_2, ..., X_{10}. Any variable selection method will only find the best combination of the original variables (say, X_1, X_3, X_7, X_{10}) but can never automatically transform a variable (say, transform X_1 to log X_1) if it were needed to increase the information content (predictive power) of that variable. Furthermore, none of these methods can generate a reexpression of the original variables (perhaps X_3/X_7) if the constructed variable were to offer more predictive power than the original component variables combined. In other words, current variable selection methods cannot find the enhanced best subset that needs to include transformed and reexpressed variables (possibly X_1, X_3, X_7, X_{10}, log X_1, X_3/X_7). A subset of variables without the potential of new variables offering enhanced predictive power limits clearly the modeler in building the best model.

Specifically, these methods fail to identify structure of the types discussed next.

Transformed variables with a preferred shape: A variable selection procedure should have the ability to transform an individual variable, if necessary, to induce a symmetric distribution. Symmetry is the preferred shape of an individual variable. For example, the workhorses of statistical measures—the mean and variance—are based on the symmetric distribution. A skewed distribution produces inaccurate estimates for means, variances, and related statistics, such as the correlation coefficient. Analyses based on a skewed distribution provide typically questionable findings. Symmetry facilitates the interpretation of the effect of the variable in an analysis. A skewed distribution is difficult to examine because most of the observations are bunched together at either end of the distribution.

A variable selection method also should have the ability to straighten nonlinear relationships. A linear or straight-line relationship is the preferred shape when considering two variables. A straight-line relationship between independent and dependent variables is an

* Absolute r value means that the sign is ignored. For example, if r = −0.23, then absolute r = +0.23.

assumption of the popular statistical linear model. (Remember, a linear model is defined as a sum of weighted variables, such as $Y = b_0 + b_1*X_1 + b_2*X_2 + b_3*X_3$.)* Moreover, a straight-line relationship among all the independent variables considered jointly is also a desirable property [4]. Straight-line relationships are easy to interpret: A unit of increase in one variable produces an expected constant increase in a second variable.

Constructed variables from the original variables using simple arithmetic functions: A variable selection method should have the ability to construct simple reexpressions of the original variables. Sum, difference, ratio, or product variables potentially offer more information than the original variables themselves. For example, when analyzing the efficiency of an automobile engine, two important variables are miles traveled and fuel used (gallons). However, it is well known that the ratio variable of miles per gallon is the best variable for assessing the performance of the engine.

Constructed variables from the original variables using a set of functions (e.g., arithmetic, trigonometric, or Boolean functions): A variable selection method should have the ability to construct complex reexpressions with mathematical functions to capture the complex relationships in the data and offer potentially more information than the original variables themselves. In an era of data warehouses and the Internet, big data consisting of hundreds of thousands to millions of individual records and hundreds to thousands of variables are commonplace. Relationships among many variables produced by so many individuals are sure to be complex, beyond the simple straight-line pattern. Discovering the mathematical expressions of these relationships, although difficult without theoretical guidance, should be the hallmark of a high-performance variable selection method. For example, consider the well-known relationship among three variables: the lengths of the three sides of a right triangle. A powerful variable selection procedure would identify the relationship among the sides, even in the presence of measurement error: The longer side (diagonal) is the square root of the sum of squares of the two shorter sides.

In sum, these two weaknesses suggest that a high-performance variable selection method for marketing models should find the best subset of variables that maximizes the Cum Lift criterion. In the sections that follow, I reintroduce the GenIQ Model of Chapter 40, this time as a high-performance variable selection technique for marketing models.

41.4 Goals of Modeling in Marketing

Marketers typically attempt to improve the effectiveness of their campaigns, solicitations, and the like by targeting their best customers or prospects. They use a model to identify individuals who are likely to respond (or generate profit†) from their marketing efforts. The model provides an individual's estimate of the probability of response (or contribution of profit). Although the precision of these estimates is important, the performance of the model is measured at an aggregated level as reported in a decile analysis.

* The weights or coefficients (b_0, b_1, b_2, and b_3) are derived to satisfy some criteria, such as to minimize the mean-squared error used in ordinary least squares regression or to minimize the joint probability function used in logistic regression.

† I use the term *profit* as a stand-in for any measure of an individual's worth, such as sales per order, lifetime sales, revenue, number of visits, or number of purchases.

Marketers have defined the *Cum Lift*, found in the decile analysis, as the relevant measure of model performance. Based on the selection of individuals by the model, marketers create a push-up list of individuals likely to respond (or contribute to profit) to obtain an advantage over a random selection of individuals.

The Cum Response Lift is an index of the expected incremental responses with a selection based on a response model over the expected responses based on a random selection (no model, chance). Similarly, the Cum Profit Lift is an index of the expected incremental profit with a selection based on a profit model over the expected profit with a random selection (no model, chance). The concept of Cum Lift and the steps of the construction in a decile analysis are in Chapter 26.

It should be clear at this point that a model that produces a decile analysis with more responses (or profit) in the upper deciles (top, second, third, or fourth) is a better model than a model with fewer responses (or less profit) in the upper deciles. This concept is the motivation for the GenIQ Model. The GenIQ approach to modeling addresses specifically the objectives concerning marketers, namely, maximizing response (or profit) from their marketing efforts. The GenIQ Model uses the genetic methodology to optimize explicitly the desired criterion: maximize the upper deciles. Consequently, the GenIQ Model allows a regression modeler to build response and profit models in ways that are not possible with current methods.

The GenIQ Response and Profit Models are theoretically superior—on maximizing the upper deciles—to response and profit models built with alternative techniques because of the explicit nature of the fitness function. The actual formulation of the fitness function is beyond the scope of this chapter: It suffices to say the fitness function seeks to fill the upper deciles with as many responses (or as much profit as possible), equivalently, to maximize response (or profit) Cum Lift.

Due to the explicit nature of its fitness criterion and the way it evolves models, the GenIQ Model offers a high-performance variable selection for marketing models. The high-performance variable selection is apparent once I illustrate the GenIQ variable selection process in the next section.

41.5 Variable Selection with GenIQ

The best way of explaining variable selection with the GenIQ Model is to illustrate how GenIQ identifies structure in data. In this illustration, I demonstrate finding structure for a response model. GenIQ works equally well for a profit model, with a nominally defined profit continuous dependent variable.

Cataloguer ABC requires a response model built on a recent mail campaign that produces a 3.54% response rate. In addition to the RESPONSE dependent variable, there is a set of nine candidate predictor variables, whose measurements were taken prior to the mail campaign.

1. AGE_Y: Knowledge of customer's age (1 = if known, 0 = if not known)
2. OWN_TEL: Presence of a telephone in the household (1 = yes, 0 = no)
3. AVG_ORDE: Average dollar order
4. DOLLAR_2: Dollars spent within last 2 years

5. PROD_TYP: Number of different products purchased
6. LSTORD_M: Number of months since last order
7. FSTORD_M: Number of months since first order
8. RFM_CELL: Recency/frequency/money cells (1 = best to 5 = worst)*
9. PROMOTION: Number of promotions customer has received

To get an initial read on the information content (predictive power) of the variables, I perform a correlation analysis, which provides the correlation coefficient† for each candidate predictor variable with RESPONSE in Table 41.1. The top four variables in descending order of the magnitude‡ of the strength of association are DOLLAR_2, RFM_CELL, PROD_TYP, and LSTORD_M.

I perform five logistic regression analyses (with RESPONSE) corresponding to the five variable selection methods. The resulting best subsets among the nine original variables are in Table 41.2. Surprisingly, the forward, backward, and SW methods produced the identical subset (DOLLAR_2, RFM_CELL, LSTORD_M, AGE_Y). Because these methods produced a subset size of four, I set the subset size to four for the R-sq and top-k methods.

TABLE 41.1

Correlation Analysis: Nine Original Variables with RESPONSE

Rank	Variable	Correlation Coefficient (r)
top	DOLLAR_2	0.11
2	RFM_CELL	–0.10
3	PROD_TYP	0.08
4	LSTORD_M	–0.07
5	AGE_Y	0.04
6	PROMOTION	0.03
7	AVG_ORDE	0.02
8	OWN_TEL	0.10
9	FSTORD_M	0.01

TABLE 41.2

Best Subsets among the Nine Original Variables

	DOLLAR_2	RFM_ CELL	LSTORD_M	AGE_Y	AVG_ ORDE
FS	x	x	x	x	
BE	x	x	x	x	
SW	x	x	x	x	
R-sq	x		x	x	x
Top-4	x	x	x		x
Frequency	5	4	5	4	2

* RFM_CELL will be treated as a scalar variable.

† I know that the correlation coefficient with or without scatterplots is a crude gauge of predictive power.

‡ The direction of the association is not relevant. That is, the sign of the coefficient is ignored.

A common subset size, namely four, allows for a fair comparison across all methods. R-sq and top-k produced different best subsets, which include DOLLAR_2, LSTORD_M, and AVG_ORDE. It is interesting to note that the most frequently used variables are DOLLAR_2 and LSTORD_M in the "Frequency" row in Table 41.2. The validation performance of the five logistic models regarding Cum Lift is in Table 41.3. Assessment of model performance at the decile level is as follows:

1. At the top decile, R-sq produces the worst-performing model: Cum Lift 239 versus Cum Lifts 252–256 for the other models.
2. At the second decile, R-sq produces the worst-performing model: Cum Lift 198 versus Cum Lifts 202–204 for the other models.
3. At the third decile, R-sq produces the best-performing model: Cum Lift 178 versus Cum Lifts 172–174 for the other models.

Similar findings are at the other depths-of-file.

To facilitate the comparison of the five statistics-based variable selection methods and the GenIQ Model, I use a single measure of model performance for the five methods, AVG—the average performance of the five models, the average of the Cum Lifts across the five methods for each decile in Table 41.3.

41.5.1 GenIQ Modeling

This section requires an understanding of the genetic methodology and the parameters for controlling a genetic model run (as discussed in Chapter 40).

I set the parameters for controlling the GenIQ Model to run as follows:

1. Population size: 3,000 (models)
2. Number of generations: 250
3. Percentage of the population copied: 10%
4. Percentage of the population used for crossover: 80%
5. Percentage of the population used for mutation: 10%

TABLE 41.3

LRM Performance Comparison by Variable Selection Methods: Cum Lifts

Decile	FS	BE	SW	R-sq	Top-4	AVG
top	256	256	256	239	252	252
2	204	204	204	198	202	202
3	174	174	174	178	172	174
4	156	156	156	157	154	156
5	144	144	144	145	142	144
6	132	132	132	131	130	131
7	124	124	124	123	121	123
8	115	115	115	114	113	114
9	107	107	107	107	107	107
bottom	100	100	100	100	100	100

The GenIQ-variable set consists of the nine candidate predictor variables. For the GenIQ-function set, I select the arithmetic functions (addition, subtraction, multiplication, and division); some Boolean operators (and, or, xor, greater/less than); and the log function (Ln). The log function* is helpful in symmetrizing typically skewed dollar amount variables, such as DOLLAR_2. I anticipate that DOLLAR_2 would be part of a genetically evolved structure defined with the log function. Of course, RESPONSE is the dependent variable.

At the end of the run, 250 generations of copying/crossover/mutation have evolved 750,000 (=250×3,000) models according to selection proportional to fitness (PTF). Each model undergoes evaluation regarding how well it solves the problem of filling the upper deciles with responders. Good models having more responders in the upper deciles are more likely to contribute to the next generation of models; poor models having fewer responders in the upper deciles are less likely to contribute to the next generation of models. Consequently, the last generation consists of 3,000 high-performance models, each with a fitness value indicating how well the model solves the problem. The top fitness values, typically the 18 largest values,† define a *set of 18 best models* with equivalent performance (filling the upper deciles with a virtually identical large number of responders).

The set of variables defining one of the best models has variables in common with the set of variables defining another best model. The common variables comprise the regarded best subset. The mean incidence of a variable across the set of best models provides a measure for determining the best subset. The GenIQ-selected best subset of original variables consists of variables with mean incidence greater than 0.75.‡ The variables that meet this cutoff score reflect an honest determination of necessary variables on the criterion of maximizing the deciles.

Returning to the illustration, GenIQ provides the mean incidence of the nine variables across the set of 18 best models in Table 41.4. Thus, the GenIQ-selected best subset consists of five variables: DOLLAR_2, RFM_CELL, PROD_TYP, AGE_Y, and LSTORD_M.

This genetic-based best subset has four variables in common with the statistics-based best subsets (DOLLAR_2, RFM_CELL, LSTORD_M, and AGE_Y). Unlike the statistics-based

TABLE 41.4

Mean Incidence of Original Variables across the Set of 18 Best Models

Variable	Mean Incidence
DOLLAR_2	1.43
RFM_CELL	1.37
PROD_TYP	1.22
AGE_Y	1.11
LSTORD_M	0.84
PROMOTION	0.67
AVG_ORDE	0.37
OWN_TEL	0.11

* The log to the base 10 also symmetrizes dollar amount variables.

† Top fitness values typically bunch together with equivalent values. The top fitness values are considered equivalent in that their values are equal when rounded at, say, the third place after the decimal point. Consider two fitness values: 1.934056 and 1.934069. These values are equal when rounding occurs at, say, the third place after the decimal point: 1.934.

‡ The mean incidence cutoff score of 0.75 has been empirically predetermined.

methods, GenIQ finds value in PROD_TYP and includes it in its best subset in Table 41.5. It is interesting to note that the most frequently used variables are DOLLAR_2 and LSTORD_M, in the "Frequency" row in Table 41.5.

At this point, I can assess the predictive power of the genetic-based and statistics-based best subsets by comparing logistic regression models (LRMs) with each subset. However, after identifying GenIQ-evolved structure, I choose to make a more fruitful comparison.

41.5.2 GenIQ Structure Identification

Just as in nature, where structure is the consequence of natural selection as well as sexual recombination and mutation, the GenIQ Model evolves structure via selection PTF (natural selection), crossover (sexual recombination), and mutation. *The GenIQ fitness leads to structure evolved on the criterion of maximizing the deciles.* An important structure is in the best models—typically, the models with the four largest fitness values.

Continuing with the illustration, GenIQ has evolved several structures or *GenIQ-constructed variables.* The GenIQ Model (Figure 41.1) has the largest fitness value and reveals five new variables, NEW_VAR1 through NEW_VAR5. Additional structures are in the remaining three best models: NEW_VAR6 to NEW_VAR8 (in Figure 41.2), NEW_VAR9 (in Figure 41.3), and NEW_VAR10 (in Figure 41.4).

TABLE 41.5

Best Subsets among Original Variables: Statistics- and Genetic-Based Variable Selection Methods

Method	DOLLAR_2	RFM_ CELL	LSTORD_M	AGE_Y	AVG_ ORDE	PROD_ TYP
FS	x	x	x	x		
BE	x	x	x	x		
SW	x	x	x	x		
R-sq	x		x	x	x	
Top-4	x	x	x		x	
GenIQ	x	x	x	x		x
Frequency	6	5	6	5	2	1

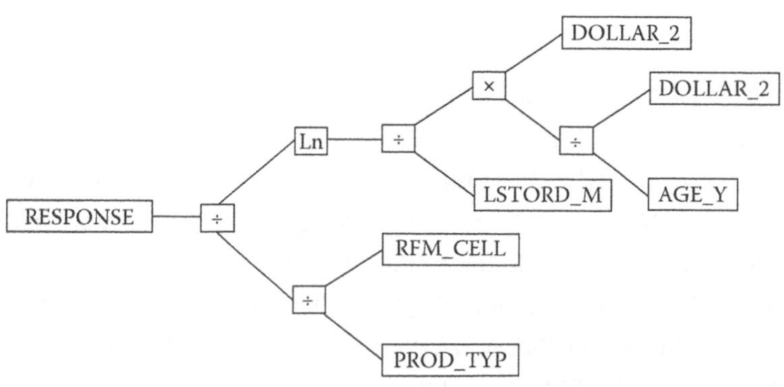

FIGURE 41.1
GenIQ Model, best 1 (top of top of four models).

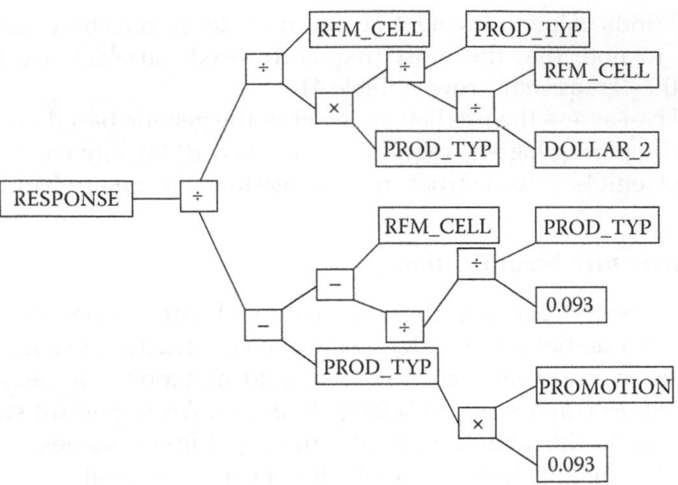

FIGURE 41.2
GenIQ Model, best 2.

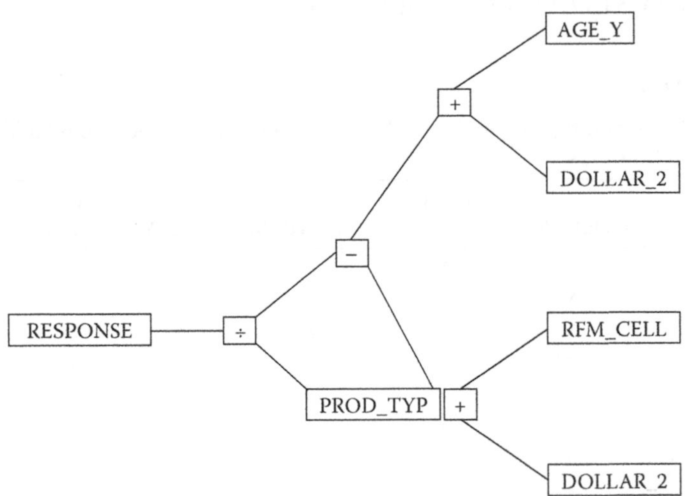

FIGURE 41.3
GenIQ Model, best 3.

1. NEW_VAR1 = DOLLAR_2/AGE_Y; if Age_Y = 0, then NEW_VAR1 = 1
2. NEW_VAR2 = (DOLLAR_2)*NEW_VAR1
3. NEW_VAR3 = NEW_VAR2/LSTORD_M; if LSTORD_M = 0, then NEW_VAR3 = 1
4. NEW_VAR4 = Ln(NEW_VAR3); if NEW_VAR3 greater than 0, then NEW_VAR4 = 1
5. NEW_VAR5 = RFM_CELL/PROD_TYP; if PROD_TYP = 0, then NEW_VAR5 = 1
6. NEW_VAR6 = RFM_CELL/DOLLAR_2; if DOLLAR_2 = 0, then NEW_VAR6 = 1
7. NEW_VAR7 = PROD_TYP/NEW_VAR6; if NEW_VAR6 = 0, then NEW_VAR7 = 1
8. NEW_VAR8 = NEW_VAR7*PROD_TYP

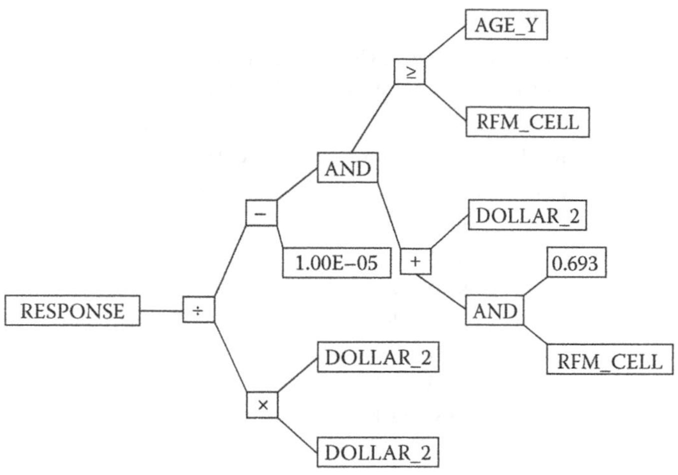

FIGURE 41.4
GenIQ Model, best 4.

9. NEW_VAR9 = (AGE_Y/DOLLAR_2) − (RFM_CELL/DOLLAR_2); if DOLLAR_2 = 0, then NEW_VAR9 = 1
10. NEW_VAR10 = 1 if AGE_Y ≥ RFM_CELL; otherwise = 0

To get a read on the predictive power of the new GenIQ-constructed variables, I perform a correlation analysis for each of the nine original variables and the 10 new variables with RESPONSE. Some new variables have a stronger association with RESPONSE than the original variables. Specifically, the following associations (larger correlation coefficient, ignoring the sign) are observed in Table 41.6.

1. NEW_VAR7, NEW_VAR5, NEW_VAR8, and NEW_VAR1 have a stronger association with RESPONSE than the best original variable, DOLLAR_2.
2. NEW_VAR10 and NEW_VAR4 fall between the second- and third-best original variables, RFM_CELL and PROD_TYP.
3. NEW_VAR2 and NEW_VAR3 are ranked eleventh and twelfth in importance before the last two original predictor variables, AGE_Y and PROMOTION.

41.5.3 GenIQ Variable Selection

The GenIQ-constructed variables plus the GenIQ-selected variables comprise the *enhanced best subset* that reflects an honest determination of necessary variables on the criterion of maximizing the deciles. For the illustration data, the enhanced set consists of 15 variables: DOLLAR_2, RFM_CELL, PROD_TYP, AGE_Y, LSTORD_M, and NEW_VAR1 through NEW_VAR10. The assessment of the predictive power of the enhanced best set is the comparison between LRMs with the genetic-based best subset and with the statistics-based best subset.

TABLE 41.6

Correlation Analysis: 9 Original and 10 GenIQ Variables with RESPONSE

Rank	Variable	Correlation Coefficient (*r*)
top	NEW_VAR7	0.16
2	NEW_VAR5	0.15
3	NEW_VAR8	0.12
4	NEW_VAR1	0.12
5	DOLLAR_2	0.11
6	RFM_CELL	–0.10
7	NEW_VAR10	0.10
8	NEW_VAR4	0.10
9	PROD_TYP	0.08
10	LSTORD_M	–0.07
11	NEW_VAR2	0.07
12	NEW_VAR3	0.06
13	NEW_VAR9	0.05
14	AGE_Y	0.04
15	PROMOTION	0.03
16	NEW_VAR6	–0.02
17	AVG_ORDE	0.02
18	OWN_TEL	0.01
19	FSTORD_M	0.01

Using the enhanced best set, I perform five logistic regression analyses corresponding to the five variable selection methods. The resultant genetic-based best subsets are in Table 41.7. The forward, backward, and SW methods produced different subsets (of size 4). R-sq(4) and top-4 also produced different subsets. It appears that New_VAR5 is the most important variable (i.e., most frequently used) as all five methods select it (row "Frequency" in Table 41.7 equals 5). LSTORD_M is of second importance, as four of the five methods select it (row "Frequency" equals 4). DOLLAR_2, RFM_CELL and AGE_Y are the least important, as only one method selects them (row "Frequency" equals 1).

To assess the gains in predictive power of the genetic-based best subset over the statistics-based best subset, I define AVG-g as the average measure of model validation performance for the five methods for each decile.

Comparison of AVG-g and AVG (average model performance based on the statistics-based set) indicates noticeable gains in predictive power obtained by the GenIQ variable selection technique in Table 41.8. The percentage gains range from an impressive 6.4% (at the fourth decile) to a slight 0.7% (at the ninth decile). The mean percentage gain for the most actionable depth-of-file, the top four deciles, is 3.9%.

This illustration demonstrates the power of the GenIQ variable selection technique over the current statistics-based variable selection methods. GenIQ variable selection is a high-performance method for marketing models with data mining capability. This method is significant in that it finds the best subset of variables to maximize the Cum Lift criterion.

TABLE 41.7

Best Subsets among the Enhanced Best Subset Variables

Method	DOLLAR_2	RFM_CELL	PROD_TYP	AGE_Y	LSTORD_M	NEW_VAR1	NEW_VAR4	NEW_VAR5
FS			x		x		x	x
BE	x			x	x			x
SW			x		x		x	x
R-sq			x		x	x		x
Top-4		x				x	x	x
Frequency	1	1	3	1	4	2	3	5

TABLE 41.8

Model Performance Comparison Based on the Genetic-Based Best Subsets: Cum Lifts

Decile	FS	BE	SW	R-sq	Top-4	AVG-g	AVG	Gain (%)
top	265	260	262	265	267	264	252	4.8
2	206	204	204	206	204	205	202	1.2
3	180	180	180	178	180	180	174	3.0
4	166	167	167	163	166	166	156	6.4
5	148	149	149	146	149	148	144	3.1
6	135	137	137	134	136	136	131	3.3
7	124	125	125	123	125	124	123	1.0
8	116	117	117	116	117	117	114	1.9
9	108	108	108	107	108	108	107	0.7
bottom	100	100	100	100	100	100	100	0.0

41.6 Nonlinear Alternative to Logistic Regression Model

The GenIQ Model offers a nonlinear alternative to the inherently linear LRM. Accordingly, an LRM is a linear approximation of a potentially nonlinear response function, which is typically noisy, multimodal, and discontinuous. The LRM together with the GenIQ-enhanced best subset of variables provides an unbeatable combination of traditional statistics improved by the genetic-based machine learning of the GenIQ Model. However, this *hybrid GenIQ–LRM model* is still a linear approximation of a potentially nonlinear response function. The GenIQ–Model itself—as defined by the entire tree with all its structure—is a nonlinear superstructure with a strong possibility for further improvement over the hybrid GenIQ–LRM and, of course, over LRM. Because the degree of nonlinearity in the response function is never known, the best approach is to compare the GenIQ Model with the hybrid GenIQ–LRM model. If the improvement is determined to be stable and noticeable, then the GenIQ Model should be used.

Continuing with the illustration, the GenIQ Model Cum Lifts are in Table 41.9. The GenIQ Model offers noticeable improvements over the performance of the hybrid GenIQ–LRM model (AVG-g). The percentage gains (fifth column from the left) range from an impressive 7.1% (at the top decile) to a respectable 1.2% (at the ninth decile). The mean percentage gain for the most actionable depth of file, the top four deciles, is 4.6%.

I determine the improvements of the GenIQ Model over the performance of the LRM (AVG). The mean percentage gain (rightmost column) for the most actionable depth of file, the top four deciles, is 8.6%, which includes a huge 12.2% in the top decile.

Note, a set of four separate GenIQ Models is needed to obtain the reported decile performance levels because no single GenIQ Model could be evolved to provide gains for all upper deciles. The GenIQ Models that produced the top, second, third, and fourth deciles are in Figures 41.1, 41.5, 41.6, and 41.7, respectively. The GenIQ Model that produced the fifth through bottom deciles is in Figure 41.8.

TABLE 41.9

Model Performance Comparison of LRM and GenIQ Model: Cum Lifts

				GenIQ Gain Over	
Decile	**AVG-g (Hybrid)**	**AVG (LRM)**	**GenIQ**	**Hybrid (%)**	**LRM (%)**
top	264	252	283	7.1	12.2
2	205	202	214	4.4	5.6
3	180	174	187	3.9	7.0
4	166	156	171	2.9	9.5
5	148	144	152	2.8	5.9
6	136	131	139	2.5	5.9
7	124	123	127	2.2	3.2
8	117	114	118	1.3	3.3
9	108	107	109	1.2	1.9
bottom	100	100	100	0.0	0.0

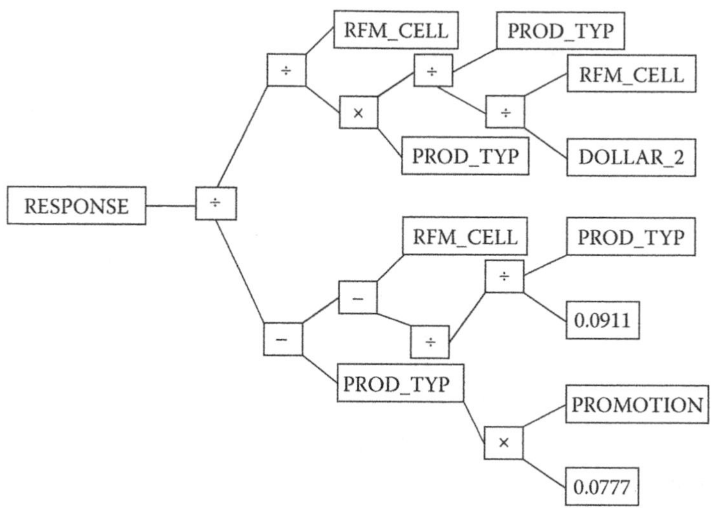

FIGURE 41.5
Best GenIQ Model, second of top four models.

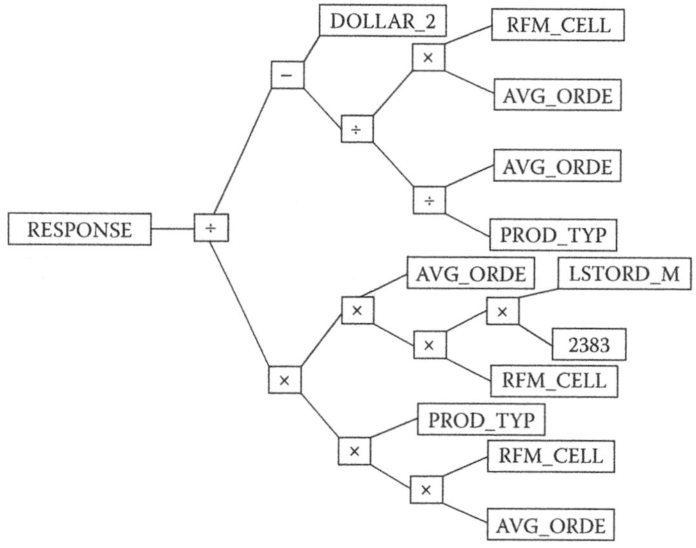

FIGURE 41.6
Best GenIQ Model, third of top four models.

A set of GenIQ Models is required when the response function is nonlinear with noise, multipeaks, and discontinuities. The capability of GenIQ to generate many models with desired performance gains reflects the flexibility of the GenIQ paradigm. It allows for intelligent adaptive modeling of the data to account for the variation of an apparent nonlinear response function.

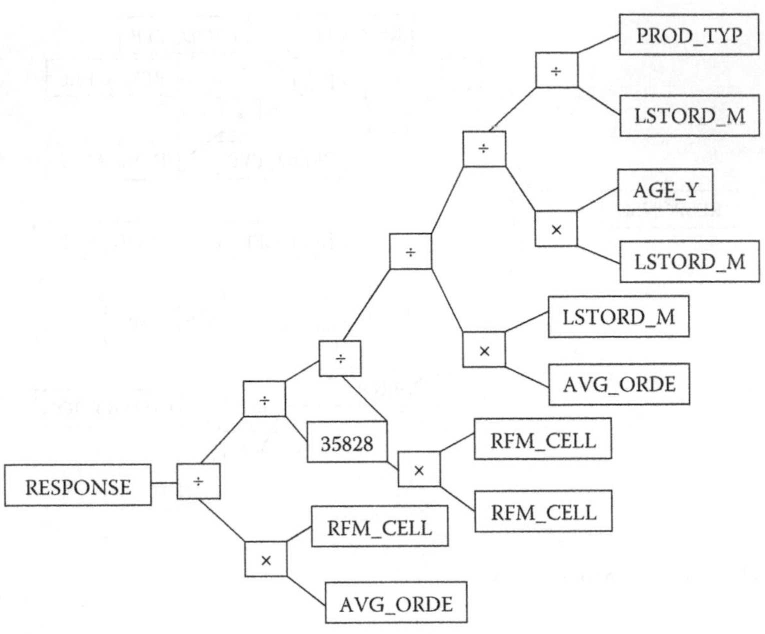

FIGURE 41.7
Best GenIQ Model, fourth of top four models.

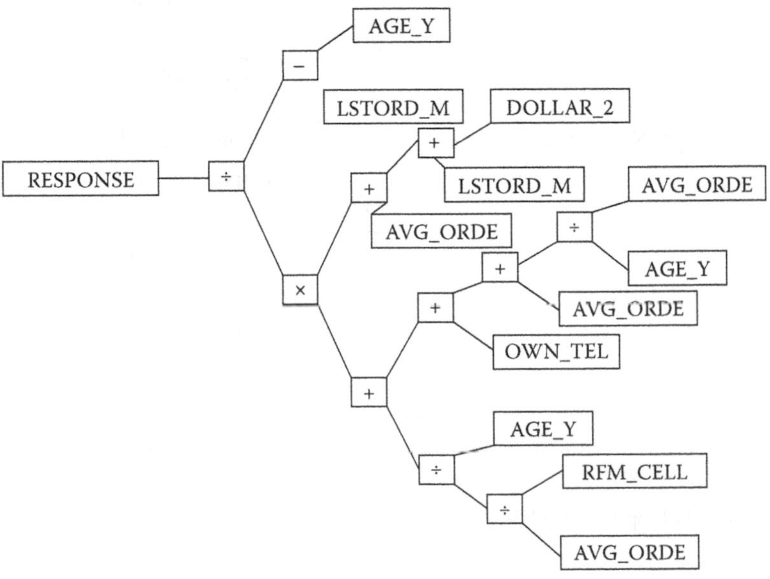

FIGURE 41.8
Best GenIQ Model for fifth through bottom deciles.

This illustration shows the power of the GenIQ Model as a nonlinear alternative to the LRM. GenIQ provides a two-step procedure for response modeling. First, build the best hybrid GenIQ–LRM model. Second, select the best GenIQ Model. If the GenIQ Model offers a stable and noticeable improvement over the hybrid model, then the GenIQ Model is the preferred response model.

As previously mentioned, the GenIQ Model works equally well for finding structure in a profit model. Accordingly, the GenIQ Model is a nonlinear alternative to the OLS regression model. The GenIQ Model offers potentially stable and noticeable improvement over OLS and the hybrid GenIQ–OLS model.

41.7 Summary

After framing the problem of variable selection with the five popular statistics-based methods, I pointed out two common weaknesses of the methods. Each hinders its capacity to achieve the desired requirement of a marketing model: neither identifying structure nor explicitly maximizing the Cum Lift criterion.

I present the GenIQ Model as a genetic-based approach for variable selection for marketing models. The GenIQ Response and Profit Models are theoretically superior—on maximizing the upper deciles—to response and profit models built with logistic and ordinary regression models, respectively, because of the nature of their fitness function. The GenIQ fitness function seeks explicitly to fill the upper deciles with as many responses or as much profit as possible. Standard statistical methods maximize the Cum Lift implicitly because their fitness functions serve as a surrogate for maximizing Cum Lift.

Using a response model illustration, I demonstrate the GenIQ Model as a high-performance variable selection method with data mining capability for finding an important structure to maximize the Cum Lift criterion. Starting with nine candidate predictor variables, the statistics-based variable selection methods identified five predictor variables in defining its best subsets. GenIQ also identified five predictor variables, of which four were in common with the statistics-based best subsets. Also, GenIQ evolved 10 structures (new variables), of which four had a stronger association with response than the best original predictor variable. Two new variables fell between the second- and third-best original predictor variables. As a result, GenIQ created the enhanced best subset of 15 variables.

The GenIQ variable selection method outperformed the statistics-based variable selection methods. I built LRMs for the five statistics-based variable selection methods using the enhanced best subset and compared its AVG-g with the AVG of the LRM for the five statistics-based methods using the original nine variables. Comparison of AVG-g and AVG indicated noticeable gains in predictive power: The percentage gains range from an impressive 6.4% to a slight 0.7%. The mean percentage gain for the most actionable depth of file, the top four deciles, is 3.9%.

Last, I advance the GenIQ Model itself as a nonlinear alternative to the standard regression models. LRM together with the GenIQ-enhanced best subset of variables provides an unbeatable combination of traditional statistics improved by machine learning. However, this hybrid GenIQ–LRM model is still a linear approximation of a potentially nonlinear response function. The GenIQ Model itself—as defined by the entire tree with all its structure—is a nonlinear superstructure with a strong possibility for further improvement over the hybrid GenIQ–LRM model. For the response illustration, the set of GenIQ

Models produces noticeable improvements in the performance of the hybrid GenIQ–LRM model. The percentage gains range from an impressive 7.1% to a respectable 1.2%. The mean percentage gain for the most actionable depth of file, the top four deciles, is 4.6%.

References

1. Dash, M., and Liu, H., Feature selection for classification, in *Intelligent Data Analysis,* Elsevier Science, New York, 1997, pp. 131–156.
2. Ryan, T.P., *Modern Regression Methods,* Wiley, New York, 1997.
3. Miller, A.J., *Subset Selection in Regression,* Chapman and Hall, London, 1990.
4. Fox, J., *Applied Regression Analysis, Linear Models, and Related Methods,* Sage, Thousand Oaks, CA, 1997.

42

Interpretation of Coefficient-Free Models

42.1 Introduction

The statistical ordinary least squares (OLS) regression model is the reference thought of when marketers hear the words "new kind of model." Model builders use the regression concept and its prominent characteristics when judiciously evaluating an alternative modeling technique. This reference of thought is because the ordinary least squares regression paradigm is the underpinning for the solution to the ubiquitous prediction problem. Marketers with a limited statistical background undoubtedly draw on their educated notions of the regression model before accepting a new technique. New modeling techniques are evaluated by the new coefficients they produce. If the new coefficients impart comparable information to the prominent characteristic of the regression model—the regression coefficient—then the new technique passes the first line of acceptance. If the new coefficients do not channel similar information, then the technique is subject to certain rejection. A quandary arises when a new modeling technique, like some machine-learning methods, produces models with no coefficients. The primary purpose of this chapter is to present a method for calculating a quasi-regression coefficient, which provides a frame of reference for evaluating and using coefficient-free models. Secondarily, the quasi-regression coefficient serves as a trusty assumption-free alternative to the regression coefficient, based on an implicit and almost never tested assumption necessary for reliable interpretation.

42.2 The Linear Regression Coefficient

The redoubtable regression coefficient, formally known as the OLS linear regression coefficient, linear-RC(ord), enjoys everyday use in marketing analysis and modeling. The common definition of the *linear-RC(ord) for predictor variable* X is *the predicted (expected) constant change in the dependent variable Y associated with a unit change in* X. The usual mathematical expression for the coefficient is in Equation 42.1. Although correctly stated, the definition requires a commentary to have a thorough understanding for expert use. I parse the definition of linear-RC(ord) and then provide two illustrations of the statistical measure of the open work of the linear regression paradigm, which serves as a backdrop for presenting the quasi-regression coefficient. Note, hereafter I use the short form ordinary regression for ordinary least squares regression because least squares is implied.

$$\text{Linear-RC(ord)} = \frac{\text{Predicted change in Y}}{\text{Unit change in X}} \tag{42.1}$$

Consider the simple ordinary linear regression model of Y on X based on a sample of (X_i, Y_i) points: pred_Y = a + b*X.

1. *Simple* means one predictor variable X is used.
2. *Ordinary* connotes that the dependent variable Y is continuous.
3. *Linear* has a dual meaning. Explicitly, linear denotes that the model is the sum of the weighted predictor variable b*X and the constant *a*. It connotes implicitly the linearity assumption that the true relationship between Y and X is straight line.
4. *Unit change in X* means the difference of value 1 for two X values ranked in ascending order, X_r and X_{r+1}, that is, $X_{r+1} - X_r = 1$.
5. *Change in Y* means the difference in predicted Y, pred_Y, values corresponding to (X_r, pred_Y_r) and (X_{r+1}, pred_Y_{r+1}): pred_Y_{r+1} − pred_Y_r.
6. *Linear-RC(ord)* indicates that the expected change in Y is constant, namely, b.

42.2.1 Illustration for the Simple Ordinary Regression Model

Consider the simple linear regression of Y on X based on the 10 observations in Dataset A in Table 42.1. The satisfaction of the linearity assumption lies in the plot of Y versus X. There is an observed positive straight-line relationship between the two variables. The X–Y plot and the plot of residual versus predicted Y both suggest that the resultant regression model in Equation 42.2 is reliable. (These types of plots are discussed in Chapter 10. The plots for these data are not shown.) Accordingly, the model provides a level of assurance that the estimated linear-RC(ord) value of 0.7967 is a reliable point estimator of the true linear regression coefficient of X. Thus, for each unit change in X, between observed X values of 19 and 78, the expected constant change in Y is 0.7967.

$$\text{pred\_Y} = 22.2256 + 0.7967 * X \quad (42.2)$$

42.2.2 Illustration for the Simple Logistic Regression Model

Consider the simple logistic regression of response Y on X based on the 10 observations in Dataset B in Table 42.2. Recall, the logistic regression model (LRM) predicts the logit Y and is *linear* in the same ways as the OLS regression model. It is a linear model, as it is defined by the sum of weighted predictor variables and a constant, and has the linearity assumption

TABLE 42.1

Dataset A

Y	X	pred_Y
86	78	84.3688
74	62	71.6214
66	58	68.4346
65	53	64.4511
64	51	62.8576
62	49	61.2642
61	48	60.4675
53	47	59.6708
52	38	52.5004
40	19	37.3630

TABLE 42.2

Dataset B

Y	X	pred_lgt Y	pred_prb Y
1	45	1.4163	0.8048
1	35	0.2811	0.5698
1	31	–0.1729	0.4569
1	32	–0.0594	0.4851
1	60	3.1191	0.9577
0	46	1.5298	0.8220
0	30	–0.2865	0.4289
0	23	–1.0811	0.2533
0	16	–1.8757	0.1329
0	12	–2.3298	0.0887

that the underlying relationship between the *logit* Y and X is straight line. Accordingly, the definition of the simple logistic linear regression coefficient, *linear-RC(logit) for predictor variable X,* is *the expected constant change in the logit Y associated with a unit change in X.*

The smooth plot of logit Y versus X is inconclusive regarding the linearity between the logit Y and X, undoubtedly due to only 10 observations. However, the plot of residual versus predicted logit Y suggests that the resultant regression model in Equation 42.3 is reliable. (The smooth plot is discussed in Chapter 10. Plots for these data are not shown.) Accordingly, the model provides a level of assurance about the reliability of the regression coefficient. Specifically, the estimated linear-RC(logit) value of 0.1135 is a reliable point estimator of the true linear logistic regression coefficient of X. Thus, for each unit change in X, between observed X values of 12 and 60, the expected constant change in the logit Y is 0.1135.

$$\text{pred\_lgt Y} = -3.6920 + 0.1135{*}X \tag{42.3}$$

42.3 The Quasi-Regression Coefficient for Simple Regression Models

I present the quasi-regression coefficient, quasi-RC, for the simple regression model of Y on X. The *quasi-RC for predictor variable X* is *the expected change—not necessarily constant—in the dependent variable Y per unit change in* X. The distinctness of the quasi-RC is that it offers a generalization of the linear-RC. It has the flexibility to measure nonlinear relationships between the dependent and predictor variables. I outline the method for calculating the quasi-RC and motivate its utility by applying the quasi-RC to the ordinary regression illustration. Then, I continue the use of the quasi-RC method for the logistic regression illustration, showing how it works for linear and nonlinear predictions.

42.3.1 Illustration of Quasi-RC for the Simple Ordinary Regression Model

Continuing with the simple ordinary regression illustration, I outline the steps, referencing the columns in Table 42.3, for deriving the quasi-RC(ord):

1. Score the data to obtain the predicted Y, pred_Y (Column 4).
2. Rank the data in ascending order by X and form the pair (X_r, X_{r+1}) (Columns 1 and 2).

TABLE 42.3
Calculations for Quasi-RC(ord)

X_r	X_r+1	change_X	pred_Y_r	pred_Y_r+1	change_Y	quasi-RC(ord)
–	19	–	–	37.3630	–	–
19	38	19	37.3630	52.5005	15.1374	0.7967
38	47	9	52.5005	59.6709	7.1704	0.7967
47	48	1	59.6709	60.4676	0.7967	0.7967
48	49	1	60.4676	61.2643	0.7967	0.7967
49	51	2	61.2643	62.8577	1.5934	0.7967
51	53	2	62.8577	64.4511	1.5934	0.7967
53	58	5	64.4511	68.4346	3.9835	0.7967
58	62	4	68.4346	71.6215	3.1868	0.7967
62	78	16	71.6215	84.3688	12.7473	0.7967

3. Calculate the change in X: $X_{r+1} - X_r$ (Column 3 = Column 2 – Column 1).
4. Calculate the change in predicted Y: $pred\_Y_{r+1} - pred\_Y_r$ (Column 6 = Column 5 – Column 4).
5. Calculate the quasi-RC(ord) for X: change in predicted Y divided by change in X (Column 7 = Column 6/Column 3).

The quasi-RC(ord) is constant across the nine (X_r, X_{r+1}) intervals and equals the estimated linear RC(ord) value of 0.7967. In superfluity, it is constant for each and every unit change in X within each of the X intervals: between 19 and 38, 38 and 47, …, and 62 and 78. The constant change is no surprise as the predictions are from a linear model. Also, the plot of pred_Y versus X (not shown) indicates a perfectly positively sloped straight line whose slope is 0.7967. (Why?)

42.3.2 Illustration of Quasi-RC for the Simple Logistic Regression Model

By way of further motivation for the quasi-RC methodology, I apply the five steps to the logistic regression illustration with the appropriate changes to accommodate working with logit units for deriving the quasi-RC(logit). I outline the steps, referencing the columns in Table 42.4, for deriving the quasi-RC(logit):

1. Score the data to obtain the predicted logit Y, pred_lgt Y (Column 4).
2. Rank the data in ascending order by X and form the pair (X_r, X_{r+1}) (Columns 1 and 2).
3. Calculate the change in X: $X_{r+1} - X_r$ (Column 3 = Column 2 – Column 1).
4. Calculate the change in predicted logit Y: pred_lgt Y_{r+1} – pred_lgt Y_r (Column 6 = Column 5 – Column 4).
5. Calculate the quasi-RC for X (logit): change in predicted logit Y divided by change in X (Column 7 = Column 6/Column 3).

The quasi-RC(logit) is constant across the nine (X_r, X_{r+1}) intervals and equals the estimated linear-RC(logit) value of 0.1135. Again, in superfluity, it is constant for each and every unit change within the X intervals: between 12 and 16, 16 and 23, …, and 46 and 60. Again, this is no surprise as the predictions are from a linear model. Also, the plot of predicted logit Y versus X (not shown) shows a perfectly positively sloped straight line with a slope of 0.1135.

TABLE 42.4
Calculations for Quasi-RC(logit)

X_r	X_r+1	change_X	pred_lgt_r	pred_ lgt_r+1	change_lgt	quasi-RC(logit)
–	12	–	–	−2.3298	–	–
12	16	4	−2.3298	−1.8757	0.4541	0.1135
16	23	7	−1.8757	−1.0811	0.7946	0.1135
23	30	7	−1.0811	−0.2865	0.7946	0.1135
30	31	1	−0.2865	−0.1729	0.1135	0.1135
31	32	1	−0.1729	−0.0594	0.1135	0.1135
32	35	3	−0.0594	0.2811	0.3406	0.1135
35	45	10	0.2811	1.4163	1.1352	0.1135
45	46	1	1.4163	1.5298	0.1135	0.1135
46	60	14	1.5298	3.1191	1.5893	0.1135

Thus far, the two illustrations show how the quasi-RC method works and holds up for simple linear predictions, that is, predictions produced by linear models with one predictor variable.

In the next section, I show how the quasi-RC method works with a simple nonlinear model in its efforts to provide an honest attempt at imparting regression coefficient-like information. A nonlinear model is defined, in earnest, as a model that is not a linear model, that is, it is not a sum of weighted predictor variables. The simplest nonlinear model—the probability of response—is a restatement of the simple logistic regression of response Y on X, defined in Equation 42.4.

$$\text{Probability of response Y} = \exp(\text{logit Y})/(1 + \exp(\text{logit Y})) \quad (42.4)$$

Clearly, this model is nonlinear. It is said to be nonlinear in its predictor variable X, which means that the expected change in the probability of response *varies* as the unit change in X varies through the range of observed X values. Accordingly, the quasi-RC(prob) for predictor variable X is the expected change—not necessarily constant—in the probability Y per unit change in X. In the next section, I conveniently use the logistic regression illustration to show how the quasi-RC method works with nonlinear predictions.

42.3.3 Illustration of Quasi-RC for Nonlinear Predictions

Continuing with the logistic regression illustration and modifying the steps for the quasi-RC(logit) account for working in probability units. The first two steps reference Columns 3 and 4 in Table 42.2 and the remaining steps (3 through 6) reference the columns in Table 42.5. The six steps for deriving the quasi-RC(prob) are:

1. Score the data to obtain the predicted logit of Y, pred_lgt Y (Column 3, Table 42.2).
2. Convert the pred_lgt Y to predicted probability Y, pred_prb Y (Column 4, Table 42.2). The conversion formula is as follows: probability Y equals exp(logit Y) divided by the sum of 1 plus exp(logit Y).
3. Rank the data in ascending order by X and form the pair (X_r, X_{r+1}) (Columns 1 and 2, Table 42.5).
4. Calculate change in X: $X_{r+1} - X_r$ (Column 3 = Column 2 – Column 1, Table 42.5).
5. Calculate change in probability Y: pred_prb Y_{r+1} – pred_prb Y_r (Column 6 = Column 5 – Column 4, Table 42.5).

TABLE 42.5
Calculations for Quasi-RC(prob)

X_r	X_r+1	change_X	prob_Y_r	prob_Y_ r+1	change_ prob	quasi- RC(prob)
–	12	–	–	0.0887	–	–
12	16	4	0.0887	0.1329	0.0442	0.0110
16	23	7	0.1329	0.2533	0.1204	0.0172
23	30	7	0.2533	0.4289	0.1756	0.0251
30	31	1	0.4289	0.4569	0.0280	0.0280
31	32	1	0.4569	0.4851	0.0283	0.0283
32	35	3	0.4851	0.5698	0.0847	0.0282
35	45	10	0.5698	0.8048	0.2349	0.0235
45	46	1	0.8048	0.8220	0.0172	0.0172
46	60	14	0.8220	0.9577	0.1357	0.0097

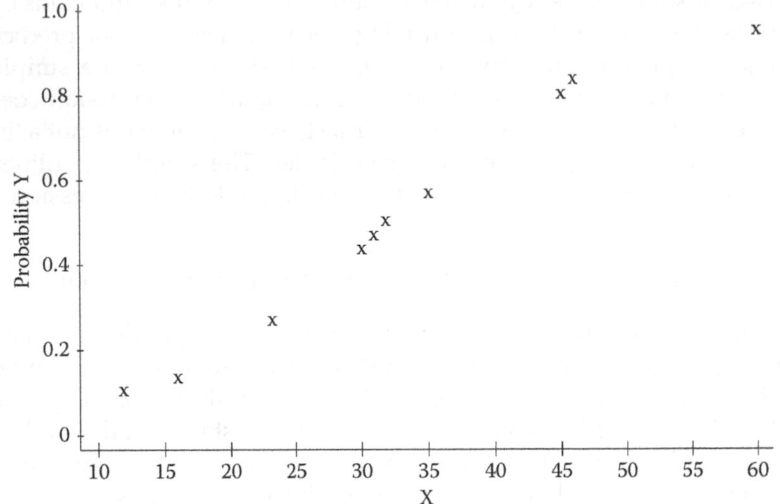

FIGURE 42.1
Plot of Probability Y versus X.

6. Calculate the quasi-RC(prob) for X: Change in probability Y divided by the change in X (Column 7 = Column 6/Column 3, Table 42.5).

Quasi-RC(prob) varies as X—in a nonlinear manner—goes through its range between 12 and 60. The quasi-RC values for the nine intervals are 0.0110, 0.0172, …, 0.0097, respectively, in Table 42.5. The nonlinearity is no surprise as the general relationship between probability of response and a given predictor variable has a theoretical prescribed nonlinear S-shape (known as an ogive curve). The plot of the probability of Y versus X in Figure 42.1 reveals this nonlinearity, although the limiting 10 points may make it too difficult to see.

Thus far, the three illustrations show how the quasi-RC method works and holds up for linear and nonlinear predictions based on the simple one predictor variable regression model. In the next section, I extend the method beyond the simple one predictor variable regression model to *the everymodel,* any multiple linear or nonlinear regression model, or any coefficient-free model.

42.4 Partial Quasi-RC for the Everymodel

The interpretation of the regression coefficient in the multiple (two or more predictor variables) regression model essentially remains the same as its meaning in the simple regression model. The regression coefficient is formally called the *partial* linear regression coefficient, partial linear-RC, which connotes that the model has other variables whose effects are partialled out of the relationship between the dependent variable and the predictor variable under consideration. The *partial linear-RC for predictor variable X* is *the expected constant change in the dependent variable Y associated with a unit change in X when the other variables are held constant.* This interpretation of the partial linear regression coefficient is universally accepted (as discussed in Section 15.7).

The reading of the partial linear-RC for a given predictor variable is based on an implicit assumption that the statistical adjustment—which removes the effects of the other variables from the dependent variable and the predictor variable—produces a linear relationship between the dependent variable and the predictor variable. Although the workings of statistical adjustment are theoretically sound, it does not guarantee linearity between the adjusted dependent and adjusted predictor variables. In general, an assumption based on the property of linearity is tenable. In the present case of statistical adjustment, the likelihood of the linearity assumption holding tends to decrease as the number of other variables increases. Interestingly, it is not a customary effort to check the validity of the linearity assumption, which could render the partial linear-RC questionable.

The quasi-RC method provides the partial quasi-RC as a trusty assumption-free alternative measure of the "expected change in the dependent variable" without reliance on statistical adjustment and restriction of a linear relationship between the dependent variable and the predictor variable. Formally, the partial quasi-RC for predictor variable X is the expected change—not necessarily constant—in the dependent variable Y associated with a unit change in X when the other variables are held constant. The quasi-RC method provides flexibility, which enables the data analyst to:

1. Validate an overall linear trend in the dependent variable versus the predictor variable for given values within the *other variables* region (i.e., given the other variables are held constant). For linear regression models, the method serves as a diagnostic to test the linearity assumption of the partial linear-RC. If the test result is positive (i.e., a nonlinear pattern emerges), which is a symptom of an incorrect structural form of the predictor variable, then a remedy can be inferred (i.e., a choice of reexpression of the predictor variable to induce linearity with the dependent variable).
2. Consider the liberal view of a nonlinear pattern in the dependent variable versus the predictor variable for given values within the other variables region. For nonlinear regression models, the quasi-RC method provides an exploratory data analysis (EDA) procedure to uncover the underlying structure of the expected change in the dependent variable.
3. Obtain coefficient-like information from coefficient-free models. This information encourages the use of black box machine-learning methods, characterized by the absence of regression-like coefficients.

In the next section, I outline the steps for calculating the partial quasi-RC for the everymodel. I provide an illustration using a multiple LRM to show how the method works and how to interpret the results. In the last section of this chapter, I apply the partial quasi-RC method to the coefficient-free GenIQ Model presented in Chapter 40.

42.4.1 Calculating the Partial Quasi-RC for the Everymodel

Consider the everymodel for predicting Y based on four predictor variables X_1, X_2, X_3, and X_4. The calculations and guidelines for the partial quasi-RC for X_1 are as follows:

1. To affect the "holding constant" of the other variables $\{X_2, X_3, X_4\}$, consider the typical values of the *M-spread common region*. For example, the M20-spread common region consists of the individuals whose $\{X_2, X_3, X_4\}$ values are common to the individual M20-spreads (the middle 20% of the values) for each of the other variables, that is, common to M20-spread for X_2, X_3, and X_4. Similarly, the M25-spread common region consists of the individuals whose $\{X_2, X_3, X_4\}$ values are common to the individual M25-spreads (the middle 25% of the values) for each of the other variables.
2. The size of the common region is clearly based on the number and measurement of the other variables. A rule of thumb for sizing the region for reliable results is as follows: The initial M-spread common region is M20. If partial quasi-RC values seem suspect, then increase the common region by 5%, resulting in an M25-spread region. Increase the common region by 5% increments until the partial quasi-RC results are trustworthy. Note, a 5% increase is a nominal 5% because increasing each of the other variable's M-spread by 5% does not necessarily increase the common region by 5%.
3. For any of the other variables, whose measurement is coarse (includes a handful of distinct values), its individual M-spread may need to be decreased by 5% intervals until the partial quasi-RC values are trustworthy.
4. Score the data to obtain the predicted Y for all individuals in the common M-spread region.
5. Rank the scored data in ascending order by X_1.
6. Divide the data into equal-sized slices by X_1. In general, if the expected relationship is linear, as when working with a linear model and testing the partial linear-RC linearity assumption, then start with five slices. Increase the number of slices as required to obtain a trusty relationship. If the expected relationship is nonlinear, as when working with a nonlinear regression model, then start with 10 slices. Increase the number of slices as required to obtain a trusty relationship.
7. The number of slices, indexed by *i*, depends on two matters of consideration: the size of the M-spread common region and the measurement of the predictor variable for which partial quasi-RC is being derived. If the common region is small, then a large number of slices tends to produce unreliable quasi-RC values. If the region is large, then a large number of slices, which otherwise does not pose a reliability concern, may produce untenable results, such as offering too liberal a view of the overall pattern. If the measurement of the predictor variable is coarse, the number of slices equals the number of distinct values.
8. Calculate the minimum, maximum, and median of X_1 within each slice and form the pair (median $X_{\text{slice } i}$, median $X_{\text{slice } i+1}$).
9. Calculate the change in X_1: median $X_{\text{slice } i+1}$ − median $X_{\text{slice } i}$.
10. Calculate the median of the predicted Y within each slice and form the pair (median pred_$Y_{\text{slice } i}$, median $Y_{\text{slice } i+1}$).

11. Calculate the change in predicted Y: median $pred\_Y_{slice\ i+1}$ – median $pred\_Y_{slice\ i}$.
12. Calculate the partial quasi-RC for X_1: change in predicted Y divided by change in X_1.

42.4.2 Illustration for the Multiple Logistic Regression Model

Consider the illustration in Chapter 41 of Cataloger ABC, who requires a response model built on a recent mail campaign. I build an LRM for predicting RESPONSE based on four predictor variables:

1. DOLLAR_2: Dollars spent within last 2 years
2. LSTORD_M: Number of months since last order
3. RFM_CELL: Recency/frequency/money cells (1 = best to 5 = worst)
4. AGE_Y – knowledge of customer's age (1 = if known; 0 = if not known)

The RESPONSE model is shown in Equation 42.5

$$\begin{aligned} pred\_lgt\ RESPONSE &= -3.004 + 0.00210*DOLLAR\_2 \\ &\quad -0.1995*RFM\_CELL - 0.0798*LSTORD\_M + 0.5337*AGE\_Y \end{aligned} \tag{42.5}$$

I detail the calculation for deriving the LRM partial quasi-RC(logit) for DOLLAR_2 in Table 42.6.

1. Score the data to obtain pred_lgt RESPONSE for all individuals in the M-spread common region.
2. Rank the scored data in ascending order by DOLLAR_2 and divide the data into five slices, in Column 1, by DOLLAR_2.
3. Calculate the minimum, maximum, and median of DOLLAR_2, in Columns 2, 3, and 4, respectively, for each slice and form the pair (median $DOLLAR\_2_{slice\ i}$, median $DOLLAR\_2_{slice\ i+1}$) in Columns 4 and 5, respectively.
4. Calculate the change in DOLLAR_2: median $DOLLAR\_2_{slice\ i+1}$ – median $DOLLAR\_2_{slice\ i}$ (Column 6 = Column 5 – Column 4).
5. Calculate the median of the predicted logit RESPONSE within each slice and form the pair (median pred_lgt $RESPONSE_{slice\ i}$, median pred_lgt $RESPONSE_{slice\ i+1}$) in Columns 7 and 8.
6. Calculate the change in pred_lgt RESPONSE: median pred_lgt $RESPONSE_{slice\ i+1}$ – median pred_lgt $RESPONSE_{slice\ i}$ (Column 9 = Column 8 – Column 7).
7. Calculate the partial quasi-RC(logit) for DOLLAR_2: the change in the pred_lgt RESPONSE divided by the change in DOLLAR_2 for each slice (Column 10 = Column 9/Column 6).

The interpretation of the LRM partial quasi-RC(logit) for DOLLAR_2 follows:

1. For Slice 2, which has minimum and maximum DOLLAR_2 values of 43 and 66, respectively, the partial quasi-RC(logit) is 0.0012. The value 0.0012 means that for each unit change in DOLLAR_2 between 43 and 66, the expected constant change in the logit RESPONSE is 0.0012.

TABLE 42.6

Calculations for LRM Partial Quasi-RC(logit): DOLLAR_2

Slice	min_ DOLLAR_2	max_ DOLLAR_2	med_ $DOLLAR\_2_{\_slice\ i}$	med_ $DOLLAR\_2_{\_slice\ i+1}$	change_ DOLLAR_2	med_lgt_r	med_lgt_r+1	change_lgt	quasi-RC (logit)
1	0	43	–	40	–	–	–3.5396	–	–
2	43	66	40	50	10	–3.5396	–3.5276	0.0120	0.0012
3	66	99	50	80	30	–3.5276	–3.4810	0.0467	0.0016
4	99	165	80	126	46	–3.4810	–3.3960	0.0850	0.0018
5	165	1,293	126	242	116	–3.3960	–3.2219	0.1740	0.0015

2. Similarly, for Slices 3, 4, and 5, the expected constant changes in the logit RESPONSE within the corresponding intervals are 0.0016, 0.0018, and 0.0015, respectively. Note, for Slice 5, the maximum DOLLAR_2 value, in Column 3, is 1,293.
3. At this point, the pending implication is that there are four levels of expected change in the logit RESPONSE associated by DOLLAR_2 across its range from 43 to 1,293.
4. However, the partial quasi-RC plot for DOLLAR_2 of the relationship of the smooth predicted logit RESPONSE (Column 8) versus the smooth DOLLAR_2 (Column 5), in Figure 42.2, indicates there is a single expected constant change across the DOLLAR_2 range, as the variation among slice-level changes is reasonably due to sample variation. This last examination supports the decided implication that the linearity assumption of the partial linear-RC for DOLLAR_2 is valid. Thus, I accept the expected constant change of the partial linear-RC for DOLLAR_2, 0.00210 (from Equation 42.5).
5. Alternatively, the quasi-RC method provides a trusty assumption-free estimate of the partial linear-RC for DOLLAR_2, *partial quasi-RC(linear)*, defined as the regression coefficient of the simple ordinary regression of the smooth logit predicted RESPONSE on the smooth DOLLAR_2, Columns 8 and 5, respectively. The partial quasi-RC(linear) for DOLLAR_2 is 0.00159 (details not shown).

In sum, the quasi-RC methodology provides alternatives that only the data analyst, who is intimate with the data, can assess. They are threefold: (1) Accept the partial quasi-RC after evaluating the variation among slice-level changes in the partial quasi-RC plot as nonrandom. (2) Accept the partial linear-RC (0.00210) after the partial quasi-RC plot validates the linearity assumption. (3) Accept the trusty partial quasi-RC(linear) estimate (0.00159) after the partial quasi-RC plot validates the linearity assumption. Of course, the default alternative is to accept outright the partial linear-RC without testing the linearity assumption. Note, the small difference in magnitude between the trusty and the "true" estimates of the partial linear-RC for DOLLAR_2 is not typical, as the next discussion shows.

I calculate the LRM partial quasi-RC(logit) for LSTORD_M, using six slices corresponding to the distinct values of LSTORD_M, in Table 42.7.

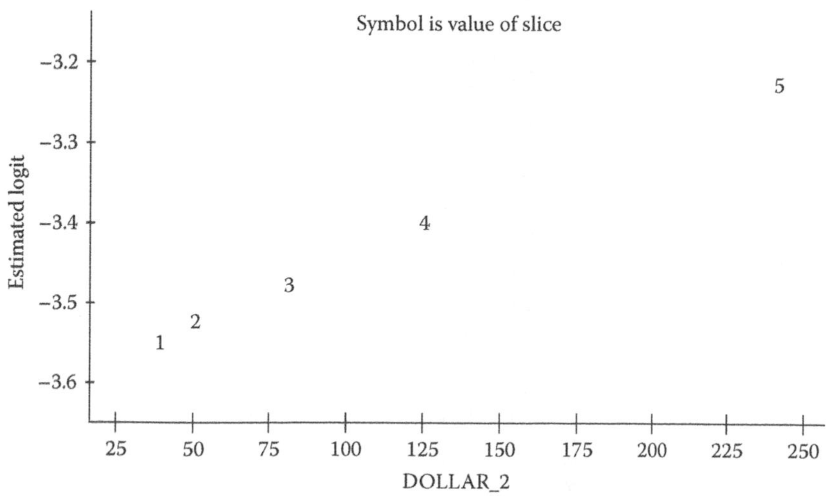

FIGURE 42.2
Visual display of LRM partial quasi-RC(logit) for DOLLAR_2.

TABLE 42.7

Calculations for LRM Partial Quasi-RC(logit): LSTORD_M

Slice	min_ LSTORD_M	max_ LSTORD_M	med_ LSTORD_M_r	med_ LSTORD_M_r+1	change_ LSTORD_M	med_lgt_r	med_lgt_r+1	change_lgt	quasi-RC (logit)
1	1	1	–	1	–	–	–3.2332	–	–
2	1	3	1	2	1	–3.2332	–3.2364	–0.0032	–0.0032
3	3	3	2	3	1	–3.2364	–3.3982	–0.1618	–0.1618
4	3	4	3	4	1	–3.3982	–3.5049	–0.1067	–0.1067
5	4	5	4	5	1	–3.5049	–3.5727	–0.0678	–0.0678
6	5	12	5	6	1	–3.5727	–3.5552	0.0175	0.0175

1. The partial quasi-RC plot for LSTORD_M of the relationship between the smooth predicted logit RESPONSE and the smooth LSTORD_M, in Figure 42.3, is clearly nonlinear with expected changes in the logit RESPONSE: −0.0032, −0.1618, −0.1067, −0.0678, and 0.0175.
2. The implication is that the linearity assumption for the LSTORD_M does not hold. There is not an expected constant change in the logit RESPONSE as implied by the prescribed interpretation of the partial linear-RC for LSTORD_M, −0.0798 (in Equation 42.5).
3. The secondary implication is that the structural form of LSTORD_M is not correct. The S-shaped nonlinear pattern suggests creating quadratic and cubic reexpressions of LSTORD_M and testing both versions of LSTORD_M for model inclusion.
4. Satisfyingly, the partial quasi-RC(linear) value of −0.0799 (from the simple ordinary regression of the smooth predicted logit RESPONSE on the smooth LSTORD_M) equals the partial linear-RC value of −0.0798. The implications are: (1) The partial linear-RC provides the *average* constant change in the logit RESPONSE across the LSTORD_M range of values from 1 to 66 and (2) the partial quasi-RC provides a more accurate reading of the changes on the six presliced intervals across the LSTORD_M range.

Forgoing the details, the LRM partial quasi-RC plots for both RFM_CELL and AGE_Y support the linearity assumption of the partial linear-RC. Thus, the partial quasi-RC(linear) and the partial linear-RC values should be equivalent. In fact, they are: For RFM_CELL, the partial quasi-RC(linear) and the partial-linear RC are −0.2007 and −0.1995, respectively; for AGE_Y, the partial quasi-RC(linear) and the partial-linear RC are 0.5409 and 0.5337, respectively.

In sum, this illustration shows that the workings of the quasi-RC methodology perform quite well on the linear predictions based on multiple predictor variables. It suffices to say, by converting the logits into probabilities—as was done in the simple logistic regression illustration in Section 42.3.3—the quasi-RC approach performs equally well with nonlinear predictions based on multiple predictor variables.

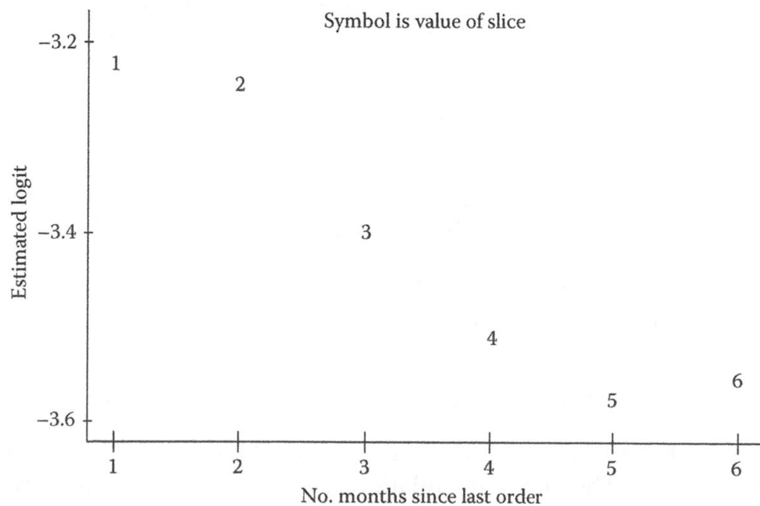

FIGURE 42.3
Visual display of LRM partial quasi-RC(logit) for LSTORD_M.

42.5 Quasi-RC for a Coefficient-Free Model

The linear regression paradigm, with over two centuries of theoretical development and practical use, has made the equation form—the sum of weighted predictor variables ($Y = b_0 + b_1X_1 + b_2X_2 + \ldots + b_nX_n$)—the icon of predictive models. Using the aged coefficients is the reason why the new machine-learning techniques of the last half-century are evaluated by the coefficients they produce. If the new coefficients impart comparable information to the regression coefficient, then the new technique passes the first line of acceptance. If the new coefficients do not channel similar information, then the technique is subject to certain rejection. Ironically, some machine-learning methods offer better predictions without the use of coefficients. The burden of acceptance of the coefficient-free model lies with the extraction of something familiar and trusting. The quasi-RC procedure provides data analysts and marketers with the comfort and security of coefficient-like information for evaluating and using the coefficient-free machine-learning models.

Machine-learning models without coefficients can assuredly enjoy the quasi-RC method. One of the most popular coefficient-free models is the regression tree, for example, chi-squared automatic interaction detection (CHAID). The regression tree has a unique equation form of "if–then–else" rules, which has rendered its interpretation virtually self-explanatory and has freed it from a burden of acceptance and the unnoticed absence of the regression coefficient. In contrast, most machine-learning methods, like artificial neural networks (ANNs), have not enjoyed an easy first line of acceptance. Even their proponents have called ANNs as black boxes. Ironically, ANNs do have coefficients (actually, interconnecting weights between input and output layers), but no formal effort has been made to translate them into coefficient-like information. The genetic GenIQ Model has no outright coefficients. Numerical values are sometimes part of the genetic model, but they are not coefficient-like in any way—just genetic material that evolved as necessary for accurate prediction.

The quasi-RC method as discussed so far works nicely on the linear and nonlinear regression model. In the next section, I illustrate how the quasi-RC technique works and how to interpret its results for a nonregression, nonlinear coefficient-free model, such as GenIQ Model as presented in Chapter 40. As expected, the quasi-RC technique works with ANN models and CHAID or classification and regression trees (CART) models.

42.5.1 Illustration of Quasi-RC for a Coefficient-Free Model

Again, consider the illustration in Chapter 41 of Cataloger ABC, who requires a response model built on a recent mail campaign. I select the best number 3 GenIQ Model (in Figure 41.3) for predicting RESPONSE based on four predictor variables:

1. DOLLAR_2: Dollars spent within last 2 years
2. PROD_TYP: Number of different products
3. RFM_CELL: Recency/frequency/money cells (1 = best to 5 = worst)
4. AGE_Y: Knowledge of customer's age (1 = if known, 0 = if not known)

The GenIQ partial quasi-RC(prob) table and plot for DOLLAR_2 are in Table 42.8 and Figure 42.4, respectively. The plot of the relationship between the smooth predicted probability RESPONSE (GenIQ-converted probability score) and the smooth DOLLAR_2 is clearly nonlinear, which is considered reasonable, due to the inherently nonlinear nature of the GenIQ Model. The implication is that partial quasi-RC(prob) for DOLLAR_2 reliably

TABLE 42.8

Calculations for GenIQ Partial Quasi-RC(prob): DOLLAR_2

Slice	min_ DOLLAR_2	max_ DOLLAR_2	med_ DOLLAR _2_r	med_ DOLLAR_2_r+1	change_ DOLLAR_2	med_prb_r	med_prb_r+1	change_prb	quasi-RC (prob)
1	0	50	–	40	–	–	0.031114713	–	–
2	50	59	40	50	10	0.031114713	0.031117817	0.000003103	0.000000310
3	59	73	50	67	17	0.031117817	0.031142469	0.000024652	0.000001450
4	73	83	67	79	12	0.031142469	0.031154883	0.000012414	0.000001034
5	83	94	79	89	10	0.031154883	0.031187925	0.000033043	0.000003304
6	94	110	89	102	13	0.031187925	0.031219393	0.000031468	0.000002421
7	110	131	102	119	17	0.031219393	0.031286803	0.000067410	0.000003965
8	131	159	119	144	25	0.031286803	0.031383536	0.000096733	0.000003869
9	159	209	144	182	38	0.031383536	0.031605964	0.000222428	0.000005853
10	209	480	182	253	71	0.031605964	0.032085916	0.000479952	0.000006760

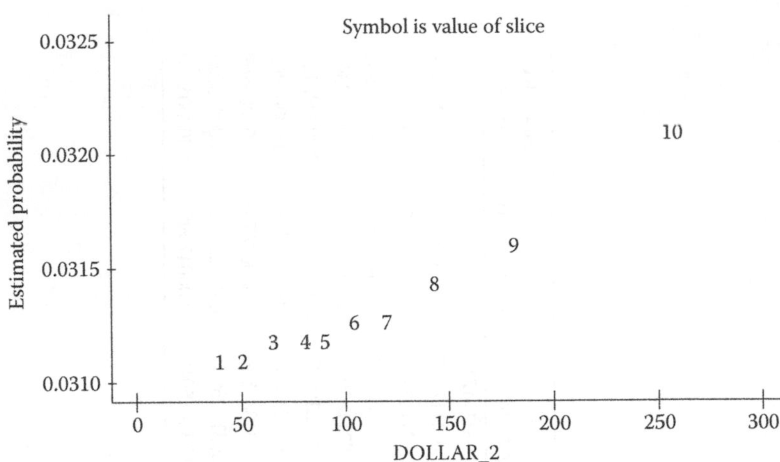

FIGURE 42.4
Visual display of GenIQ partial quasi-RC(prob) for DOLLAR_2.

reflects the expected changes in probability RESPONSE. The interpretation of the partial quasi-RC(prob) for DOLLAR_2 is as follows: For Slice 2, which has minimum and maximum DOLLAR_2 values of 50 and 59, respectively, the partial quasi-RC(prob) is 0.000000310. The value 0.000000310 means that, for each unit change in DOLLAR_2 between 50 and 59, the expected constant change in the probability RESPONSE is 0.000000310. Similarly, for Slices 3, 4, … , 10, the expected constant changes in the probability RESPONSE are 0.000001450, 0.000001034, … , 0.000006760, respectively.*

The GenIQ partial quasi-RC(prob) table and plot for PROD_TYP are in Table 42.9 and Figure 42.5, respectively. Because PROD_TYP assumes distinct values between 3 and 47, albeit more than a handful, I use 20 slices to take advantage of the granularity of the quasi-RC plotting. The interpretation of the partial quasi-RC(prob) for PROD_TYP can follow the literal rendition of "for each and every unit change" in PROD_TYP as done for DOLLAR_2. However, as the quasi-RC technique provides alternatives, the following interpretations are also available:

1. The partial quasi-RC plot of the relationship between the smooth predicted probability RESPONSE and the smooth PROD_TYP suggests two patterns. For Pattern 1, for PROD_TYP values between 6 and 15, the unit changes in probability RESPONSE can be viewed as sample variation masking an expected constant change in probability RESPONSE. The masked expected constant change can be determined by the average of the unit changes in probability RESPONSE corresponding to PROD_TYP values between 6 and 15. For Pattern 2, for PROD_TYP values greater than 15, the expected change in probability RESPONSE is increasing in a nonlinear manner, which follows the literal rendition for each and every unit change in PROD_TYP.
2. If the data analyst comes to judge the details in the partial quasi-RC(prob) table or plot for PROD_TYP as much ado about sample variation, then the partial quasi-RC(linear) estimate can be used. Its value, 0.00002495, is from the regression coefficient from the simple ordinary regression of the smooth predicted RESPONSE on the smooth PROD_TYP (Columns 8 and 5, respectively, Table 42.9).

* Note that the maximum values for DOLLAR_2 in Tables 42.6 and 42.8 are not equal. This is because they are based on different M-spread common regions as the GenIQ Model and LRM use different variables.

TABLE 42.9

Calculations for GenIQ Partial Quasi-RC(prob): PROD_TYP

Slice	min_ PROD_TYP	max_ PROD_TYP	med_ PROD_TYP_r	med_ PROD_TYP_r+1	change_ PROD_TYP	med_ prb_r	med_ prb_r+1	change_ prob	quasi_RC (prob)
1	3	6	–	6	–	–	0.031103	–	–
2	6	7	6	7	1	0.031103	0.031108	0.000004696	0.000004696
3	7	8	7	7	0	0.031108	0.031111	0.000003381	–
4	8	8	7	8	1	0.031111	0.031113	0.000001986	0.000001986
5	8	8	8	8	0	0.031113	0.031113	0.000000000	–
6	8	9	8	8	0	0.031113	0.031128	0.000014497	–
7	9	9	8	9	1	0.031128	0.031121	–0.000006585	–0.000006585
8	9	9	9	9	0	0.031121	0.031136	0.000014440	–
9	9	10	9	10	1	0.031136	0.031142	0.000006514	0.000006514
10	10	11	10	10	0	0.031142	0.031150	0.000007227	–
11	11	11	10	11	1	0.031150	0.031165	0.000015078	0.000015078
12	11	12	11	12	1	0.031165	0.031196	0.000031065	0.000031065
13	12	13	12	12	0	0.031196	0.031194	–0.000001614	–
14	13	14	12	13	1	0.031194	0.031221	0.000026683	0.000026683
15	14	15	13	14	1	0.031221	0.031226	0.000005420	0.000005420
16	15	16	14	15	1	0.031226	0.031246	0.000019601	0.000019601
17	16	19	15	17	2	0.031246	0.031305	0.000059454	0.000029727
18	19	22	17	20	3	0.031305	0.031341	0.000036032	0.000012011
19	22	26	20	24	4	0.031341	0.031486	0.000144726	0.000036181
20	26	47	24	30	6	0.031486	0.031749	0.000262804	0.000043801

TABLE 42.10

Calculations for GenIQ Partial Quasi-RC(prob): RFM_CELL

Slice	min_RFM_ CELL	max_RFM_ CELL	med_RFM_ CELL_r	med_RFM_ CELL_r+1	change_ RFM_CELL	med_prb_r	med_prb_r+1	change_prb	quasi-RC (prob)
1	1	3	–	2	–	–	0.031773	–	–
2	3	4	2	3	1	0.031773	0.031252	–0.000521290	–0.000521290
3	4	4	3	4	1	0.031252	0.031137	–0.000114949	–0.000114949
4	4	4	4	4	0	0.031137	0.031270	0.000133176	–
5	4	5	4	5	1	0.031270	0.031138	–0.000131994	–0.000131994
6	5	5	5	5	0	0.031138	0.031278	0.000140346	–

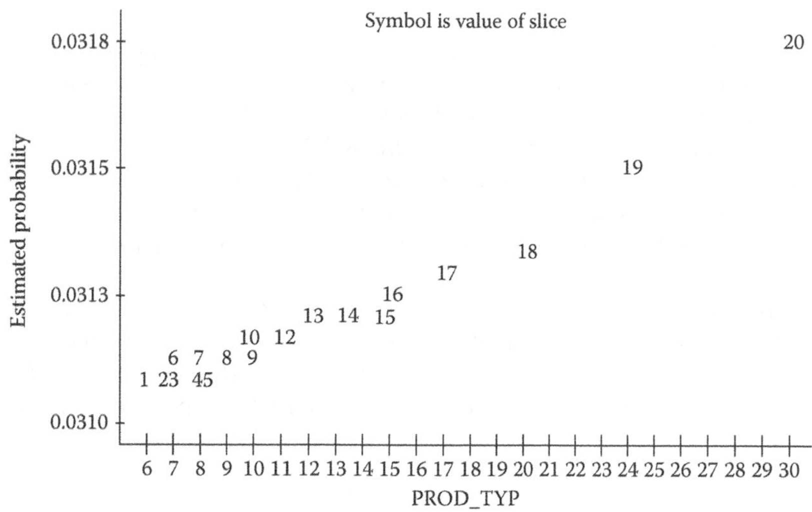

FIGURE 42.5
Visual display of GenIQ partial quasi-RC(prob) for PROD_TYP.

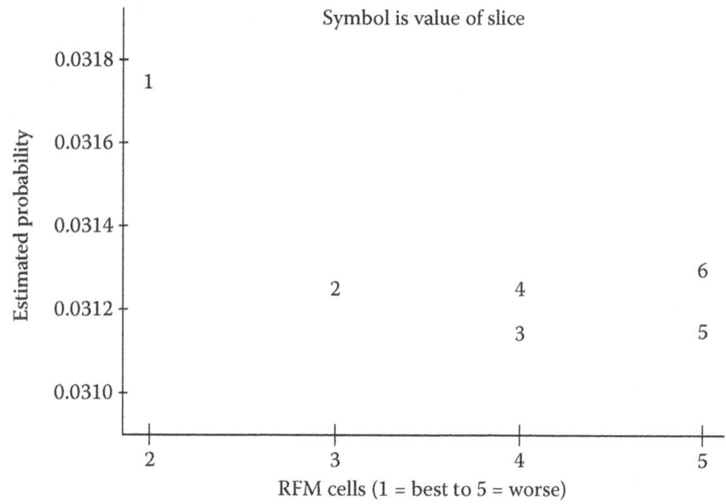

FIGURE 42.6
Visual display of GenIQ partial quasi-RC(prob) for RFM_CELL.

The GenIQ partial quasi-RC(prob) table and plot for RFM_CELL are in Table 42.10 and Figure 42.6, respectively. The partial quasi-RC of the relationship between the smooth predicted probability RESPONSE and the smooth RFM_CELL suggests an increasing expected change in probability. Recall, RFM_CELL is an interval-level variable with a reverse scale: 1 = best to 5 = worst. Thus, RFM_CELL has clearly expected a nonconstant change in probability. The plot has *double* smooth points at both RFM_CELL = 4 and RFM_CELL = 5, for which the *double-smoothed* predicted probability RESPONSE is the average of the reported probabilities. For RFM_CELL = 4, the twin points are 0.031252 and 0.031137. Thus, the double-smoothed predicted probability RESPONSE is 0.311945. Similarly, for RFM_CELL = 5, the double-smoothed predicted probability RESPONSE is 0.31204. The interpretation of the partial quasi-RC(prob) for RFM_CELL can follow the literal rendition for each and every unit change in RFM_CELL.

The GenIQ partial quasi-RC(prob) table and plot for AGE_Y are in Table 42.11 and Figure 42.7, respectively. The partial quasi-RC plot of the relationship between the smooth predicted probability RESPONSE and the smooth RFM_CELL is an uninteresting expected linear change in probability. The plot has double-smoothed points at both AGE_Y = 1, for which the double-smoothed predicted probability RESPONSE is the average of the reported probabilities. For AGE_Y = 1, the twin points are 0.031234 and 0.031192. Thus, the double-smoothed predicted probability RESPONSE is 0.31213. The interpretation of the partial quasi-RC(prob) for AGE_Y can follow the literal rendition for each and every unit change in AGE_Y.

In sum, this illustration shows how the quasi-RC methodology works on a nonregression, nonlinear coefficient-free model. The quasi-RC procedure provides data analysts and marketers with the sought-after comfort and security of coefficient-like information for evaluating and using coefficient-free machine-learning models like GenIQ.

TABLE 42.11

Calculations for GenIQ Partial Quasi-RC(prob): AGE_Y

Slice	min_ AGE_Y	max_ AGE_Y	med_ AGE_Y_r	med_AGE_ Y_r+1	change_ AGE_Y	med_ prb_r	med_ prb_r+1	change_ prb	quasi-RC (prob)
1	0	1	–	1	–	–	0.031177	–	–
2	1	1	1	1	0	0.031177	0.031192	0.000014687	–
3	1	1	1	1	0	0.031192	0.031234	0.000041677	–

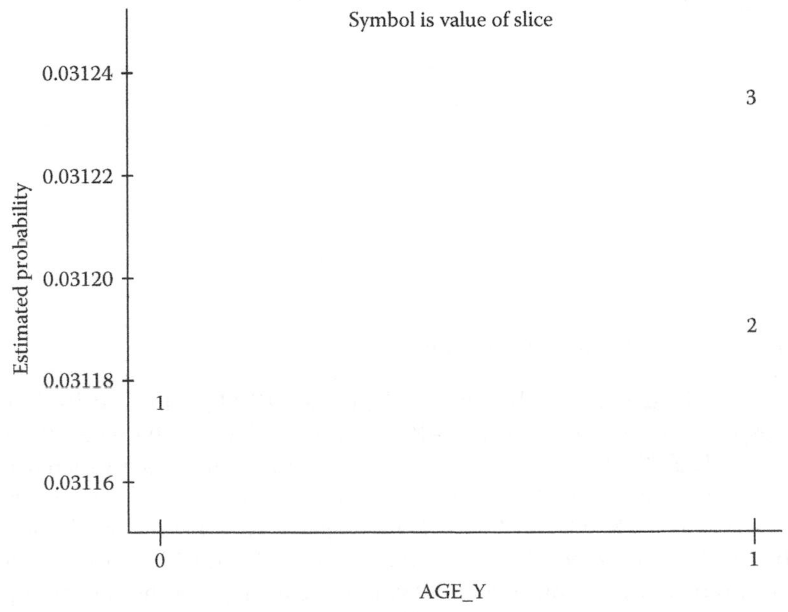

FIGURE 42.7
Visual Display of GenIQ Partial Quasi RC(prob) for AGE_Y.

42.6 Summary

The redoubtable regression coefficient enjoys everyday use in marketing analysis and modeling. Model builders and marketers use the regression coefficient when interpreting the tried-and-true regression model. I restate that the reliability of the regression coefficient is based on the workings of the linear statistical adjustment. The adjustment removes the effects of the other variables from the dependent variable and the predictor variable, producing a linear relationship between the dependent variable and the predictor variable.

In the absence of another measure, model builders and marketers use the regression coefficient to evaluate new modeling methods. The analytical dependency of the regression coefficient leads to a quandary as some of the newer methods have no coefficients. As a counter step, I present the quasi-regression coefficient (quasi-RC), which provides information similar to the regression coefficient for evaluating and using coefficient-free models. Moreover, the quasi-RC serves as a trusty assumption-free alternative to the regression coefficient when the linearity assumption is not met.

I provide illustrations with the simple one predictor variable linear regression models to highlight the importance of the satisfaction of linearity assumption for accurate reading of the regression coefficient itself as well as its effect on the predictions of the model. With these illustrations, I outline the method for calculating the quasi-RC. A comparison between the actual regression coefficient and the quasi-RC shows perfect agreement, which advances the trustworthiness of the new measure.

Then, I extend the quasi-RC for the everymodel, which is any linear, nonlinear regression, nonregression or any coefficient-free model. Formally, the partial quasi-RC for predictor variable X is the expected change—not necessarily constant—in the dependent variable Y associated with a unit change in X when the other variables are held constant. With a multiple logistic regression illustration, I compare and contrast the logistic partial linear-RC with the partial quasi-RC. The quasi-RC methodology provides alternatives that only the data analyst, who is intimate with the data, assess. They are threefold: (1) Accept the partial quasi-RC after evaluating the variation among slice-level changes in the partial quasi-RC plot as nonrandom. (2) Accept the partial linear-RC after the partial quasi-RC plot validates the linearity assumption. (3) Accept the trusty partial quasi-RC(linear) estimate after the partial quasi-RC plot validates the linearity assumption. Of course, the default alternative is to accept outright the partial linear-RC without testing the linearity assumption.

Last, I illustrate the quasi-RC methodology for the coefficient-free GenIQ Model. The quasi-RC procedure provides me with the sought-after comfort and security of coefficient like information for evaluating and using the coefficient-free GenIQ Model.

I provide the SAS© routine for constructing the *M-spread common region in Chapter 44.*

43

Text Mining: Primer, Illustration, and TXTDM Software

43.1 Introduction

Text mining is data mining of textual data. Accordingly, all the data mining techniques, tips, and tricks discussed in this book apply to performing text mining of words, phrases, sentences, documents, and even a body of documents. Before text miners start in earnest, they first take one small step of converting text into numbers and one giant leap of drawing out hidden information within an accumulation of text gathered for study.

The purpose of this chapter is to improve the user-friendliness and usability of text mining. As such, the chapter has three objectives: First, to serve as a primer, readable, brief though detailed, about what text mining encompasses and how to conduct basic text mining; second, to illustrate text mining with a small body of text, yet interesting in its content; and third, to make text mining available to the reader by providing my SAS© subroutines, named *TXTDM*. The subroutines are also available for downloading from my website: http://www.geniq.net/articles.html#section9.

43.2 Background

Text mining is data mining of textual data. After turning text into numbers, all the data mining techniques, tips, and tricks discussed in this book can be used to perform text mining of words, phrases, sentences, documents, and even a body of documents, in short, a "bag of words" [1]. The framework of text mining comes with several terms and concepts not found in ordinary data mining. Before text miners start in earnest, they first take one small step of converting text into numbers, and one giant leap of drawing out hidden information within the bag gathered for study.

Curiously, the origin of the word *text* gives us the definition of what is being text-mined. Text dates back to the late 14th century when its meaning is "wording of anything written."* Thus, text mining is finding patterns and their meanings in the wordings of anything written. This centuries-old definitional phrase reaffirms text mining is text data mining. In the 1990s, text mining applications have exploded due to the Internet's burgeoning

* From Old French *texte*, Old North French *tixte*,"text, book; Gospels" (12th c.).

textual data arrival of website pages, social media, buyers' reviews, bloggers' commentaries, insurance claims, diagnostic interviews, and so on.

Text mining is one part of four other overlapping disciplines referred to as text processing. The disciplines include:

- *Natural language processing* (NLP) is a field of language processing by computers. Computer programs are developed to recognize human speech in various languages but mainly English.*
- *Computational linguistics* (CL) originated with efforts in the United States in the 1950s to use computers to automatically translate texts from foreign languages, particularly Russian scientific journals, into English [2]. Today, computational linguistics is advancing computer interaction with the written and spoken language to further the process of producing language in various settings of conversation between two or more people as a feature of a book, play, or movie.
- *Information retrieval* (IR) is the process of locating information resources relevant to an information request from other information assets that can be drawn on by a person or organization to function effectively. Google is perhaps the best IR search engine.
- *Machine learning* (ML) was coined by Samuel in 1959 and is the field of study that assigns computers the ability to learn without being explicitly programmed [3]. In other words, ML investigates ways in which the computer can acquire knowledge directly from data and thus learn to solve problems.

43.2.1 Text Mining Software: Free versus Commercial versus TXTDM

Perhaps, the difficult part of starting a text mining project is choosing which software to execute the text mining. There are many choices, but they fall into two categories: free and commercial. As for TXTDM, I will discuss in which group it belongs.

Free software is free, and commercial software is expensive. If free is too good to be true, it is—most of the time. Free software, also known of open source, can be easily downloaded from the Internet. However, for the unskilled and even the information technology (IT) professional, installing free software is invariably fraught with problems. The set of problems encountered upon downloading and installing free software is a mixture of error messages: "Download and install were not successful," "path not found," "file not found," "format not supported," "specified driver(s) not found," and so on.

The IT professional, knowing there is no direct technical support, goes online to the community of users of the freeware to submit a "problem/issue/other" ticket. At this point, the ticket response is usually quick and plentiful. But, the real problem is the quality of the answers. The answers come from experts, self-acclaimed experts, and like-minded know-it-all users who just want to be "helpful." Filtering out the correct answer is fastest if the IT inquirer is proficient in the areas of systems and architecture development. If not, the IT person with the problem will not only have the problem of finding the correct answer but also will need further help to act on the answer. In contrast, for commercial software, this scenario is an anomaly. If such an installation issue does occur, there is technical support via phone, e-mail, and online remote access.

* I worked with a blind computer programmer. He would talk to his computer to script his statistical programs. Then, his computer would read back the code for him to check for bugs. My colleague is one of the most unforgettable persons I ever met in my life.

An extensive account of difficulties met when utilizing free and commercial text mining software is provided by Francis [4].

When the open source installation is complete, the text miner goes to the documentation, which is a serious issue for most users because it is considered to be "simply a mess" (https://opensource.com/life/16/2/book-review-how-make-sense-any-mess; http://www.catb.org/esr/writings/taoup/html/documentationchapter.html). The salient value of the open source is its abundance of features, especially the latest routines that provide richer output. Unfortunately, these features, poorly documented, are not utilized, at least by users who are not well-versed in the subject. To a lesser extent, commercial software documentation has a greater emphasis on conveying parameter syntax for many function operators. When the user requires clarification, direct technical support is available.

Last but not least, the core difference between free and commercial software is the quality of the product. Free software originally was started by theoreticians of the subject matter. Once this initial phase of product developmental was complete, the open source was truly open to view, modify, extend, or transform by all users of varying levels of knowledge of the application domain—from true expert to enthusiastic novice. All such alterations are now available without rigorous or any meaningful testing for conceptual soundness and numerical accuracy. In other words, there are no expressed warranty and liability. Users employ the software at their risk. Thus, the quality of free software ranges between slightly dubitable. In contrast, commercial software is developed by specialists, highly educated at the top PhD universities, who work in a team environment that ensures the closed-source propriety software is of the highest quality.

The proposed TXTDM is neither free nor expensive. TXTDM is a simple, inexpensive solution to acquire a text mining application using only the Base SAS product. The price of Base SAS as the necessary module of the SAS platform is strategically inexpensive. SAS has earned its long-standing worldwide customer base, which is impressive given its humble beginnings (founded by two newly minted statisticians in 1976) of being among the first handful of statistical analysis mainframe packages. Data miners, seeking a utile text mining application, can avail themselves of TXTDM as their employers likely have SAS.

The basic text mining application has a core functionality. The front-end of the text mining process is to convert the unstructured text dataset into a structured spreadsheet-like dataset in which the columns represent the individuals' written words and the rows represent the individuals' presence (ones) or absence (zeros) of mentioned words. After the text-to-number transformation is complete, the resultant dataset is ready for any data mining.

When developers of text mining applications start putting their signatures on their applications, they either meaningfully or unwittingly implant their biases of any combination of NLP, CL, IR, and ML as features in their text mining algorithms. The back-end of the basic text mining application now has items that are more decorative than useful but that are not essential. To the latter distinction, TXTDM is the essential text mining analytics.

43.3 Primer of Text Mining

As text mining is data mining of textual data, data miners have to convert text into numbers. Plausibly, the data miner may not realize that they have been converting textual data, namely categorical data, into numbers by *dummifying* categorical variables.

Consider, GENDER that assumes "male," "female," and "not recorded." Dummifying GENDER involves creating three corresponding dummy variables, as follows:

MALE = 1 if GENDER = male; otherwise MALE = 0
FEMALE = 1 if GENDER = female; otherwise FEMALE = 0
NOT_RECORDED = 1 if GENDER = not recorded; otherwise NOT_RECORDED = 0

Note, all statistical techniques use all but one dummy variable; the left-out dummy variable is called the reference variable.

The first step in the series of text mining tasks is to convert *unstructured* text (e.g., words, phrases, sentences, and documents) into a scaled-up version of dummifying categorical variables, yielding a *structured* table of numbers. This one small step of text-to-numbers conversion is the input for the one giant leap of the text-mining discovery of finding hidden meaning within a document. The structured table of numbers is like a spreadsheet of columns (i.e., words) and rows (i.e., individuals' wordings of anything written). The columns represent words. The rows represent an individual's pool of words on the given topic of study. The row can be keywords, phrases, one sentence or two, a paragraph, or even a full document. The text dataset consisting of all rows represents a body of documents, called a *corpus*.

The necessary preprocessing steps with word-level properties before the text-to-numbers conversion are as follows:

1. *Tokenization*: The process of breaking up a group (e.g., phrases, sentences) of text into words, called tokens.
2. *Stop Words*: Words for removal before or after processing. Small words (e.g., a, an, the, and, I, on, it, in, with, of, am, are, and so on) are removed because they carry little if any content or information.
3. *Stemming*: The process of reducing words to their word stems (root forms) by crudely chopping off the ends (e.g., ly, ed, ing). The results of stemming remove the grammatical distinction between the stemmed words. For example, for the words modeling and modeled, *ing* and *ed* are cut off, yielding the stem word: model.
   a. Lemmatization is a smart form of stemming by the use of vocabulary, analysis of word formation, and part of speech to remove inflectional endings, so the stemmed word retains its context. Lemmatization is part of NLP and not used in the basic text mining process.
4. *Normalization*: Grooming the documents to ensure the preprocessed corpus is successfully text-mined in the first run. If no grooming is carried out, manual fixing of the corpus requires at least a second run of the program. Regardless of first and second passes of grooming, all text mining programs, free, commercial, and TXTDM, require a final stage of manual fixing (as well as manual preprocessing).
   a. Grammatical syntax and characters are main issues of preprocessing.
      i. The period and the comma. Remove it virtually wherever it is in the corpus. Possible exceptions are, for example, Ph.D. and PhD.
      ii. Remove nonalphanumeric characters, such as =, [ ], ?, @, &, *, !, and ( ).

iii. Missing spellings always cause some repairs (e.g., wrods, publisht).
iv. Initial capitalization of words. Ensure all words are in lowercase.
v. A prized-grooming hint, which I discovered, to maximize the corpus quality is the *hyphenated connection* between one seemingly small word with a corresponding domain-based related word. This approach ensures an expected word phrase has a proper chance of showing its value. An illustration of hyphenated connections follows.

Consider a model characterized by having coefficients versus not having coefficients. If one includes the words no, yes, and coefficient (stemmed of coefficients), the low frequencies of no and yes will surely result in either no and yes being eliminated from the corpus or remaining in the corpus but contributing noise to the analysis.

The prized tip is to create and include in the corpus the words no-coefficient and yes-coefficient; I have experienced excellent results from the hyphenated connection.

43.4 Statistics of the Words

There are four basic statistics of the words in the corpus.

1. *Presence/absence*: After the initial execution of the text mining program, each word is a column, and the documents* are rows in the structured numerical table. The words are dummy variables such that each row consists of values of 1 and 0 depending on the presence and absence, respectively, of the word in the document.
2. *Term frequency* (TF): The number of times a given term (word) in a corpus (all rows, all documents) is present is the statistic—the *term frequency* for that word.
   a. If a word appears often in a document, TF assumes a large value.
   b. If a word has a large TF value in a small number of documents, then the word's *importance* is very high.
   c. If a word is present in virtually all documents, then the word's importance is very low.
   d. If a word has a small TF value in a single document, then its importance is low.
   e. If a word has a large TF value in a single document, then its importance is high.
   f. If a word occurs in virtually all documents, then its importance is very low.
3. *Document frequency* (DF): The number of documents in which a given word is present is the statistic—the *document frequency* for that word.
   a. If a word appears in many documents, then the word has a high value of DF.
   b. If a word appears in many documents, then the word has low importance.

* Documents can be phrases, sentences, documents themselves—anything written.

4. *Term frequency–inverse document frequency* (TF–IDF): This statistic computes a weight for each term that reflects its importance. TF–IDF uses the statistics TF, DF, and IDF. TF–IDF adjusts TF by the scalar IDF. When a word appears in many documents, it is considered unimportant, and the IDF value is lowered, perhaps near zero. When a word is relatively unique and appears in few documents, the IDF value grows large because the word is important. The definition of TF–IDF involves the following equations:

$$\text{IDF(i)} = \log(\text{N}/(\text{DF(i)})) \tag{43.1}$$

where i is the index for the ith word, word(i), DF(i) is the frequency of word(i) in all documents, and N is the number of words in all documents.

$$\text{TF–IDF} = \text{TF(i)*IDF(i)} \tag{43.2}$$

Applying the definition of IDF(i) in Equation 43.1 yields

$$\text{TF–IDF} = \text{TF(i)*}\log(\text{N}/(\text{DF(i)}). \tag{43.3}$$

$$\text{If DF(i)} = 0 \text{ then IDF(i)} = \log(\text{N}), \text{ and TF–IDF} = \text{TF(i)*}\log(\text{N}). \tag{43.4}$$

The TF–IDF statistic offers the text miner two types of analyses: an unweighted text mining analysis with raw frequency TFs or a weighted text mining analysis using TF–IDF to account for the importance of words as discussed earlier.

43.5 The Binary Dataset of Words in Documents

The text-to-numbers conversion yields a binary dataset, where the columns are word variables (that assume the values of zeros and ones), and the rows are documents (that are wordings of anything written). The binary dataset can work well with virtually all statistical techniques. Lest we rush to our favorite technique (assuming the problem goal is known), it is prudent to discuss the characteristics of this apparently unimpressive dataset.

Depending on the sizes of the documents and the number of documents, the binary dataset might be safely called big data. Regardless of the size of the dataset, the statistical method would implement without the corpus knowing (negative bias) the meaning of its text-converted zeros and ones. However, the text mining procedure would not be working against this seemingly negative bias because text mining has the advantage of computer-intensive processing in finding patterns in words that have varying levels of frequency and predictiveness.

Another interesting feature of the zero-one dataset is that almost all the values are zeros. That is, the zero-one dataset is sparse. Here again, the text mining procedure is not at a disadvantage, unlike most statistical techniques, because it expects sparse data* and uses the sparseness upon text-mining implementation. Lastly, the zero-one dataset has no missing

* By way of the prescribed preprocessing, and the statistics TF, DF, and TF–IDF.

values. Missing data are the biggest nemesis of virtually all statistical methodologies but are not an issue in text mining.

Given the uniqueness and features of the zero-one dataset, can the text miner rush to analyze the data like the statistical analyst? The answer is yes. However, the two most popular problem-solving tasks are clustering and prediction (or classification) of a continuous (or categorial) dependent variable. In the next section, I illustrate a novel example of clustering and prediction using the same data.

43.6 Illustration of TXTDM Text Mining

I illustrate text mining based on the topic of Chapter 40, The GenIQ Model. I conducted a small survey of 21 individuals who, after reading the chapter, responded to the question: Which model, the GenIQ Model or the pair of the traditional ordinary least squares (OLS) and logistic regression (LR) models, do you prefer based on the advantages and disadvantages of each?

The 21-comment survey is the corpus, called TEXT, on which I text mine. The objective of the text mining project is to build:

1. A classification model to estimate the probability of being favorably disposed to the GenIQ Model. This type of classification model is typically called *text categorization* in the text mining literature.
2. A clustering model to group individuals on their preferences for the GenIQ Model over the pair OLS–LR models.

Note that I use a small dataset for ease of discussing TXTDM, ensuring that every detail of the TXTDM process is tractable and understood. Needless to say, TXTDM handily accommodates big data as well as small.

I discuss the implementation of TXTDM in step-by-step detail:

1. I review the 21-document corpus to observe there are about 10 words in each document (num_vars =10), and no word in the corpus has a length greater than 25 letters (max_varlen = $25). The two parameters, num_vars and max_varlen, are necessary for loading TEXT into TXTDM.
   a. GenIQ Model is a machine-learning model that uses genetic programming.
   b. GenIQ Model has no assumptions, is nonparametric, and black box.
   c. GenIQ Model has no coefficients; therefore, it is uninterpretable.
   d. GenIQ Model is data defining; it does not fit the data.
   e. GenIQ Model is a machine-learning alternative to basic regression.
   f. GenIQ Model automatically does data mining for new variables.
   g. GenIQ Model automatically findings new variables.
   h. GenIQ Model has automatic variable selection for creating new variables.
   i. GenIQ Model requires no data prep.
   j. GenIQ Model optimizes cumlift and the decile table.

k. GenIQ Model is uninterpretable as it has no coefficients.
l. OLS and logistic regressions are benchmarks for newer prediction models.
m. OLS and logistic equations have coefficients.
n. OLS and logistic are interpretable because they have coefficients.
o. OLS and logistic are reliable, accurate, and interpretable.
p. OLS and logistic are not black box and have equations.
q. OLS and logistic have many variable selection methods.
r. OLS and logistic require data prep, which is time-consuming.
s. OLS and logistic require data prep, have coefficients, not black box.
t. OLS and logistic have coefficients; therefore, they are not black box.
u. OLS and logistic require data prep, have equations and are not black box.

2. I preprocess the corpus according to Section 43.3. I modify the following words:
   a. I change GenIQ Model to GenIQModel and OLS and logistic to OLS-Logistic, because I want the paired words as a proper noun.
   b. Either of these two connected words can serve as the dependent variable in the building of the classification model. I choose GenIQModel as the dependent variable and rename it for clarity. Thus, the dependent variable is defined as:

      GenIQ_FAVORED=1 if survey respondent favors GenIQ over OLS-Logistic based are the provided description of GenIQ in Chapter 40.

      GenIQ_FAVORED=0 (equivalent to respondent favors OLS_Logistic based on the description of OLS-Logistic).
   c. I use the hyphenated connection for terms "no assumptions" because "no" is considered a stop-word, and if no-assumption is not in the corpus, this key feature of GenIQ (i.e., not requiring any assumption) would not be captured.
   d. Additional terms, which are hyphenated include: genetic-programming, no-coefficients, data-defining, not-fitting-data, alter-regression, new-variables, variable-selection, no-data-prep, decile-table, newer-prediction, yes-equations, not-black-box, and no-equations.
3. The preprocessed TEXT is now ready for the text-to-numbers conversion.
   a. GenIQModel machine-learning genetic-programming
   b. GenIQModel no-assumptions nonparametric black-box
   c. GenIQModel no-coefficients uninterpretable
   d. GenIQModel data-defining no-fitting-the-data
   e. GenIQModel machine-learning alt-regression
   f. GenIQModel data-mining new-variables
   g. GenIQModel new-variables
   h. GenIQModel variable-selection new-variables
   i. GenIQModel no-data-prep
   j. GenIQModel optimizes cumlift decile-table
   k. GenIQModel uninterpretable no-coefficients
   l. OLS-Logistic benchmarks newer-prediction

m. OLS-Logistic equations yes-coefficients
n. OLS-Logistic interpretable yes-coefficients
o. OLS-Logistic reliable accurate interpretable
p. OLS-Logistic not-black-box yes-equations
q. OLS-Logistic variable-selection
r. OLS-Logistic data-prep time-consuming
s. OLS-Logistic data-prep yes-coefficients not-black-box
t. OLS-Logistic yes-coefficients not-black-box
u. OLS-Logistic data-prep yes-equations not-black-box

4. I load preprocessed TEXT into TXTDM by executing the subroutine in Appendix 43.A. The print out of TEXT is in Table 43.1.
5. The next step creates *intermediary* binary words via running the subroutine in Appendix 43.B. The output is the *log*, in Table 43.2. The log shows all the words in the corpus with a corresponding word with the prefix _COL1 (e.g., _COL1GenIQModel=GenIQModel).
6. The log, in Table 43.2, is copied as is, except for the exclusion of the last two variables _COL1 and ID, and pasted (after &varlist=) into the next subroutine in Appendix 43.C.
7. The next step is creating the *final* binary words via re-running the subroutine in Appendix 43.C. The output is the log, in Table 43.3, which consists of the *final* set of words. Two points of note: (1) variables ID and _COL1 are back, and (2) this subroutine serves to replace the originally hyphenated connected words with underscored connected words (e.g., yes_coefficients replaces yes-coefficients).

   When copying all the words from the log, from GenIQModel through yes_equations, make sure to exclude ID and _COL1. Thus, the set of final words is in Table 43.4. Note, GenIQ_FAVORED, the dependent variable, is defined by GenIQModel and OLS_Logistic. Thus, the latter two variables are not in the final words.
8. The next step calculates the basic statistics of text mining by running the subroutine in Appendix 43.D). The values of TF, DF, Num_Docs, and N are in Tables 43.5 through 43.7. Note, this subroutine creates a small dataset WORDS from dataset TEXT.
9. The next step is appending the dependent variable GenIQ_FAVORED for the classification model by running the subroutine in Appendix 43.E. Recall that the GenIQ_FAVORED values are obtained from Step #2.

The subroutine in Appendix 43.F, executes logistic regression of dependent variable GenIQ_FAVORED with words selected by GenIQ's unique variable-selection feature (not shown, but remarked upon by respondent ID #8). Text miners without access to GenIQ obviously use their preferred variable selection approach. The logistic regression output, which defines the GenIQ_FAVORED Model, is in Table 43.8.

Glaringly, the maximum likelihood estimates indicate the model is amiss because all the p-values (last column Pr > ChiSq) are equal or close to 1.0. The model is problematic as all the words are nonsignificant. A mathematical *warning* accompanies the model output of this seemingly useless model. The warning states there is a "complete separation of data." This warning, not an error message, means (1) the model perfectly predicts all

TABLE 43.1
TEXT Dataset

ID	c01	c02	c03	c04	c05	c06	c07	c08	c09	c10
1	GenIQModel	machine-learning	genetic-programming							
2	GenIQModel	no-assumptions	nonparametric	black-box						
3	GenIQModel	no-coefficients	uninterpretable							
4	GenIQModel	data-defining	no-fitting-the-data							
5	GenIQModel	machine-learning	alt-regression							
6	GenIQModel	data-mining	new-variables							
7	GenIQModel	new-variables								
8	GenIQModel	variable-selection	new-variables							
9	GenIQModel	no-data-prep								
10	GenIQModel	optimizes	cumlift	decile-table						
11	GenIQModel	uninterpretable	no-coefficients							
12	OLS-Logistic	benchmarks	newer-prediction							
13	OLS-Logistic	equations	yes-coefficients							
14	OLS-Logistic	interpretable	yes-coefficients							
15	OLS-Logistic	reliable	accurate	interpretable						
16	OLS-Logistic	not-black-box	yes-equations							
17	OLS-Logistic	variable-selection								
18	OLS-Loglstlc	data-prep	time-consuming							
19	OLS-Logistic	data-prep	yes-coefficients	not-black-box						
20	OLS-Logistic	yes-coefficients	not-black-box							
21	OLS-Logistic	data-prep	yes-equations	not-black-box						

TABLE 43.2

Log of Intermediary Binary Words

```
224  %put &varlist;
_COL1GenIQModel=GenIQModel_COL1OLS_Logistic=OLS_Logistic_COL1accurate=accurate
_COL1alt_regression=alt_regression_COL1benchmarks=benchmarks_COL1black_box=black_box
_COL1cumlift=cumlift_COL1data_defining=data_defining_COL1data_mining=data_mining
_COL1data_prep=data_prep_COL1decile_table=decile_tab1e_COL1equations=equations
_COL1genetic_programming=genetic_programming_COL1interpretable=interpretable
_COL1machine_learning=machine_learning_COL1new_variables=new_variables
_COL1newer_prediction=newer_prediction_COL1no_assumptions=no_assumptions
_COL1no_coefficients=no_coefficients_COL1no_data_prep=no_data_prep
_COL1no_fitting_the_data=no_fitting_the_data_COL1nonparametric=nonparametric
_COL1not_black_box=not_black_box_COL1optimizes=optimizes_COL1reliable=reliable
_COL1time_consuming=time_consuming_COL1uninterpretable=uninterpretable
_COL1variable_selection=variable_selection_COL1yes_coefficients=yes_coefficients
_COL1yes_equations=yes_equations_COL1=ID=
```

TABLE 43.3

Log of Final Binary Words

```
GLOBAL VARLIST GenIQModel ID OLS_Logistic_COL1 accurate alt_regression benchmarks black_box
cumlift data_defining data_mining data_prep/decile_table equations genetic_programming
interpretable machine_learning new_variables newer_prediction no_assumptions no_coefficients
no_data_prep no_fitting_the_data nonparametric not_black_box optimizes reliable time_consuming
uninterpretable variable_selection yes_coefficients yes_equations
```

TABLE 43.4

Final Words

GenIQModel	cumlift	genetic_programming	no_coefficients	reliable
OLS_Logistic	data_defining	interpretable	no_data_prep	time_consuming
accurate	data_mining	machine_learning	no_fitting_the_data	uninterpretable
alt_regression	data_prep	new_variables	nonparametric	variable_selection
benchmarks	decile_table	newer_prediction	not_black_box	yes_coefficients
black_box	equations	no_assumptions	optimizes	yes_equations

observations to their respective groups and (2) the maximum likelihood estimates are not unique, and the model fit is questionable. The condition for this situation, which is common with logistic regression, is usually a small dataset with binary predictor variables, as is the case under consideration.

The complete separation warning does not preclude the model from having a quality of practical use. Large datasets can also have complete separation. Unlike with small datasets, where the separation is sometimes apparent, the size of the large dataset makes it virtually impossible to detect the separation. In this case, the perfect model has usefulness, in that it informs the model builder of the existence of separation. The model builder can use this information to improve the quality of the set of candidate predictor variables.

TABLE 43.5

TF—WORD FREQUENCY

ID	GenIQModel	OLS_Logistic	accurate	alt_regression	benchmarks	black_box	cumlift	data_defining	data_mining	data_prep	decile_table	equations	genetic_programming	interpretable	machine_learning	new_variables	newer_prediction	no_assumptions	no_coefficients	no_data_prep	no_fitting_the_data	non_parametric	not_black_box	optimizes	reliable	time_consuming	uninterpretable	variable_selection	yes_coefficients	yes_equations
1	1	0	0	0	0	0	0	0	0	0	0	0	1	0	1	0	0	0	0	0	0	0	0	0	0	0	0	0	0	0
2	1	0	0	0	0	1	0	0	0	0	0	0	0	0	0	0	0	1	0	0	0	1	0	0	0	0	0	0	0	0
3	1	0	0	0	0	0	0	0	0	0	0	0	0	0	0	0	0	0	1	0	0	0	0	0	0	0	1	0	0	0
4	1	0	0	0	0	0	0	1	0	0	0	0	0	0	0	0	0	0	0	0	1	0	0	0	0	0	0	0	0	0
5	1	0	0	1	0	0	0	0	0	0	0	0	0	0	1	0	0	0	0	0	0	0	0	0	0	0	0	0	0	0
6	1	0	0	0	0	0	0	0	1	0	0	0	0	0	0	1	0	0	0	0	0	0	0	0	0	0	0	0	0	0
7	1	0	0	0	0	0	0	0	0	0	0	0	0	0	0	1	0	0	0	0	0	0	0	0	0	0	0	0	0	0
8	1	0	0	0	0	0	0	0	0	0	0	0	0	0	0	1	0	0	0	0	0	0	0	0	0	0	0	1	0	0
9	1	0	0	0	0	0	0	0	0	0	0	0	0	0	0	0	0	0	0	1	0	0	0	0	0	0	0	0	0	0
10	1	0	0	0	0	0	1	0	0	0	1	0	0	0	0	0	0	0	0	0	0	0	0	1	0	0	0	0	0	0
11	1	0	0	0	0	0	0	0	0	0	0	0	0	0	0	0	0	0	1	0	0	0	0	0	0	0	1	0	0	0
12	0	1	0	0	1	0	0	0	0	0	0	0	0	0	0	0	1	0	0	0	0	0	0	0	0	0	0	0	0	0
13	0	1	0	0	0	0	0	0	0	0	0	1	0	0	0	0	0	0	0	0	0	0	0	0	0	0	0	0	1	0
14	0	1	0	0	0	0	0	0	0	0	0	0	0	1	0	0	0	0	0	0	0	0	0	0	0	0	0	0	1	0
15	0	1	1	0	0	0	0	0	0	0	0	0	0	1	0	0	0	0	0	0	0	0	0	0	1	0	0	0	0	0
16	0	1	0	0	0	0	0	0	0	0	0	0	0	0	0	0	0	0	0	0	0	0	1	0	0	0	0	0	0	1
17	0	1	0	0	0	0	0	0	0	0	0	0	0	0	0	0	0	0	0	0	0	0	0	0	0	0	0	1	0	0
18	0	1	0	0	0	0	0	0	0	1	0	0	0	0	0	0	0	0	0	0	0	0	0	0	0	1	0	0	0	0
19	0	1	0	0	0	0	0	0	0	1	0	0	0	0	0	0	0	0	0	0	0	0	1	0	0	0	0	0	1	0
20	0	1	0	0	0	0	0	0	0	0	0	0	0	0	0	0	0	0	0	0	0	0	1	0	0	0	0	0	1	0
21	0	1	0	0	0	0	0	0	0	1	0	0	0	0	0	0	0	0	0	0	0	0	1	0	0	0	0	0	0	1

TABLE 43.6

DF—DOCUMENT FREQUENCY

GenIQModel	OLS_Logistic	accurate	alt_regression	benchmarks	black_box	cumlift	data_defining	data_mining	data_prep	decile_table	equations	genetic_programming	interpretable	machine_learning	new_variables	newer_prediction	no_assumptions	no_coefficients	no_data_prep	no_fitting_the_data	non_parametric	not_black_box	optimizes	reliable	time_consuming	uninterpretable	variable_selection	yes_coefficients	yes_equations
11	10	1	1	1	1	1	1	1	3	1	1	1	2	2	3	1	1	2	1	1	1	4	1	1	1	2	2	4	2

TABLE 43.7

Number of Documents and Number of Words

Number of Documents	Number of Words
21	30

TABLE 43.8

Maximum Likelihood Estimates of the GenIQ_FAVORED Model

Parameter	DF	Estimate	Standard Error	Wald Chi-Square	Pr > ChiSq
Intercept	1	−9.2065	172.9	0.0028	0.9575
data_prep	1	7.3E-17	99.8194	0.0000	1.0000
time_consuming	1	−442E-16	223.2	0.0000	1.0000
accurate	1	8.32E-16	172.9	0.0000	1.0000
yes_coefficients	1	3.22E-16	99.8194	0.0000	1.0000
not_black_box	1	−435E-16	193.3	0.0000	1.0000
interpretable	1	−42E-15	223.2	0.0000	1.0000
equations	1	−42E-15	223.2	0.0000	1.0000
benchmarks	1	−455E-16	199.6	0.0000	1.0000
machine_learning	1	18.4128	199.6	0.0085	0.9265
no_assumptions	1	18.4128	199.6	0.0085	0.9265
no_data_prep	1	18.4128	199.6	0.0085	0.9265
no_coefficients	1	18.4128	186.7	0.0097	0.9215
data_defining	1	18.4128	199.6	0.0085	0.9265
data_mining	1	−114E-18	141.1	0.0000	1.0000
variable_selection	1	−269E-16	141.1	0.0000	1.0000
new_variables	1	18.4128	141.2	0.0170	0.8962
alt_regression	1	3.9E-16	141.1	0.0000	1.0000
cumlift	1	18.4128	199.6	0.0085	0.9265

When separation is detected, there are several strategies to eliminate the problem. Often predictor variables with large standard errors are the likely culprits. The modeling process should exclude these predictor variables. Another strategy is to combine some binary predictor variable, where it is likely that it assumes a value zero (or one), where every observation has the value zero (or one), or no observation has the value zero (or one) (5). Additionally, a warning of complete separation might signal there is at least one predictor variable that is a duplicate or a close kin of the dependent variable under a different name. The act of identifying and removing the duplicate predictor yields a reliable set of candidate predictor variables.

With small datasets, such as WORDS, complete separation may still not be obvious. Producing a perfectly separated model does not invalidate the information within the data. The viability of the GenIQ_FAVORED Model continues after presenting the scored WORDS dataset.

The scored WORDS dataset, in Table 43.9, is created by applying the GenIQ_FAVORED Model via the subroutine in Appendix 43.E. The scored file shows the words defining the model and the respondent ID, which is the third from the last column in the Table 43.9. The actual and predicted GenIQ_FAVORED values, namely, GenIQ_FAVORED and prob_GenIQ_FAVORED, respectively, are in the last two columns.

TABLE 43.9
WORDS Dataset Scored by GenIQ_FAVORED Model

data_prep	time_consuming	accurate	yes_coefficients	not_black_box	interpretable	equations	benchmarks	machine_learning	no_assumptions	no_data_prep	no_coefficients	data_defining	data_mining	variable_selection	new_variables	alt_regression	cumlift	ID	GenIQ_FAVORED	Prob_GenIQ_FAVORED
0	0	0	0	0	0	0	0	1	0	0	0	0	0	0	0	0	0	1	1	0.99990
0	0	0	0	0	0	0	0	0	1	0	0	0	0	0	0	0	0	2	1	0.99990
0	0	0	0	0	0	0	0	0	0	0	1	0	0	0	0	0	0	3	1	0.99990
0	0	0	0	0	0	0	0	0	0	0	0	1	0	0	0	0	0	4	1	0.99990
0	0	0	0	0	0	0	0	1	0	0	0	0	0	0	0	1	0	5	1	0.99990
0	0	0	0	0	0	0	0	0	0	0	0	0	1	0	1	0	0	6	1	0.99990
0	0	0	0	0	0	0	0	0	0	0	0	0	0	0	1	0	0	7	1	0.99990
0	0	0	0	0	0	0	0	0	0	0	0	0	0	1	1	0	0	8	1	0.99990
0	0	0	0	0	0	0	0	0	0	1	0	0	0	0	0	0	0	9	1	0.99990
0	0	0	0	0	0	0	0	0	0	0	0	0	0	0	0	0	1	10	1	0.99990
0	0	0	0	0	0	0	0	0	0	0	1	0	0	0	0	0	0	11	1	0.99990
0	0	0	0	0	0	0	1	0	0	0	0	0	0	0	0	0	0	12	0	0.00010
0	0	0	1	0	0	1	0	0	0	0	0	0	0	0	0	0	0	13	0	0.00010
0	0	0	1	0	1	0	0	0	0	0	0	0	0	0	0	0	0	14	0	0.00010
0	0	1	0	0	1	0	0	0	0	0	0	0	0	0	0	0	0	15	0	0.00010
0	0	0	0	1	0	0	0	0	0	0	0	0	0	0	0	0	0	16	0	0.00010
0	0	0	0	0	0	0	0	0	0	0	0	0	0	1	0	0	0	17	0	0.00010
1	1	0	0	0	0	0	0	0	0	0	0	0	0	0	0	0	0	18	0	0.00010
1	0	0	1	1	0	0	0	0	0	0	0	0	0	0	0	0	0	19	0	0.00010
0	0	0	1	1	0	0	0	0	0	0	0	0	0	0	0	0	0	20	0	0.00010
1	0	0	0	1	0	0	0	0	0	0	0	0	0	0	0	0	0	21	0	0.00010

Empirically, the scored WORDS dataset confirms the complete separation of data as prob_GenIQ_FAVORED values, 0.09990 and 0.00010, corresponding to perfect classification of respondents who favor and do not favor GenIQ over OLS_Logistic, respectively. In short, holding a ruler below the row of ID #11 clearly shows a complete separation of the data. The complete separation of the GenIQ_FAVORED Model does not affect the usability for its intended purpose of illustrating the text mining analytic process. The analysis going forward does not suffer from the perfect model, as will be observed.

In support of the rule of thumb: Too many predictor variables in a model cause multicollinearity, a condition of highly correlated predictor variables in a regression model,

I heuristically determine whether the binary words cause collinearity in the model. I calculate the average correlation of the words in the model by running the subroutine in Appendix 43.G. The average correlation is quite low, 0.10722, which indicates the model does not suffer from multicollinearity.

43.7 Analysis of the Text-Mined GenIQ_FAVORED Model

As previously discussed, the text-mined GenIQ_FAVORED Model perfectly predicts the likelihood of respondents who favor GenIQ over OLS_Logistic with a probability of 0.99990. Similarly, the model perfectly predicts the likelihood of respondents who do not favor GenIQ over OLS_Logistic with a probability of 0.00010. Next is the text mining analysis of the GenIQ_FAVORED Model from the patterns the word profiles in the scored WORDS dataset in Table 43.9.

43.7.1 Text-Based Profiling of Respondents Who Prefer GenIQ

First, I consider the respondents with IDs from #1 to #11 who favor the GenIQ Model. I take a quick tally from the sparse table to determine (1) the number of words used by each respondent and (2) the number of respondents who used each word. All respondents who favor GenIQ have only one word, which points to the feature they feel is the distinctive characteristic element of GenIQ. For the number of words used, for example, Respondent ID #1 views GenIQ as a machine-learning model. Respondent ID #2 views GenIQ as a model with no assumptions, which is consistent with Respondent ID #1 because machine-learning GenIQ does not have any assumptions.

For the number of respondents, three respondents use one word, new-variables; two respondents (different but overlap) use two words, machine-learning and no-coefficients; six respondents (different but overlap) use one word among no-assumptions, no-data-prep, data-defining, data-mining, variable-selection, alt-regression (alternative-to-regression), and cumlift. There are no responders who favor GenIQ that use the remaining eight words.

To build the text-based profile of GenIQ, I consider all the words (without the hyphenated connection to give a grammatical style profile narrative) used by the top 11 respondents. The collective profile of those who favor GenIQ Model over OLS_Logistic is:

> The GenIQ Model is a machine-learning alternative regression model to OLS and LR models. GenIQ optimizes Cum Lift, whereas OLS and LR optimize mean squared error and the log-likelihood function, respectively. Thus, for applications where model performance uses Cum Lift, GenIQ is preferred. As a machine-learning method (using genetic programming), GenIQ has no assumptions, no coefficients. GenIQ automatically, with no data preparation, lets the data define (data-defining) the model as opposed to OLS and LR, where they fit the data to the model. Additionally, as a data-defining method, GenIQ has a unique data-mining feature, which automatically performs the variable selection, which includes the creation of new variables.

In sum, the profiling of respondents who prefer GenIQ reveals that text mining quite accurately captures the essential features that make GenIQ a unique machine-learning alternative to the traditional OLS and LR models. The profiling only uses 10 (starting from machine learning, first row, ninth column: first occurrence of Value 1, Table 43.9) of the 30 words in the corpus.

43.7.2 Text-Based Profiling of Respondents Who Prefer OLS-Logistic

Next, I consider the respondents with IDs from #12 to #21 who favor OLS-Logistic regression models. I take a quick tally from the sparse table to determine (1) the number of words used by each respondent and (2) the number of respondents who used each word. There are three respondents, ID #12, ID #16, and ID #17, who favor OLS-Logistic with one word, that is, benchmarks, not-a-black-box, and variable-selection, respectively. Each word identifies the feature the three respondents feel is the salient element of OLS-Logistic. There is one respondent, ID #19, who uses three words (i.e., data-prep, yes-coefficients, and not-black-box). The remaining six respondents uses two words among yes-coefficients, equations, interpretable, accurate, data-prep, and time-consuming.

For the number of respondents, there are four (different with some overlap) who use two words: yes-coefficients and not-black-box. Three respondents use one word: data-prep. Two respondents use one word: interpretable. Five respondents (different with some overlap) use one word from among time-consuming, accurate, equations, benchmarks, and variable-selection.

To build the text-based profile of OLS-Logistic, I consider all the words used by the lower 10 respondents, ID #12 to ID #21. The aggregated profile of those who favor OLS-Logistic over the GenIQ Model is:

> The OLS and LR models provide benchmarks for an alternative technique. They are traditional methods, which are defined by coefficients in ordinary equations. The latter characteristics make the models highly interpretable, unlike the black-box GenIQ defined with no coefficients. Thus, OLS and LR models are not black-box. Their prominent features are accurate point estimates and many variable selection approaches (e.g., forward, backward, stepwise, and R-square). A well-known weakness of the models is mandatory data preparation, which is quite time-consuming.

The profiling of respondents who prefer OLS-Logistic renders a succinct narrative of the two traditional models. The profiling uses 8 (the first 7 words across 10 lower positioned respondents and variable-selection in Column 16, Table 43.9) of the 30 words, not surprisingly different from those used in the profiling of respondents who prefer GenIQ.

A concluding remark to assure the newbie text miner, likely engrossed with the new methodology, does not forget to exercise proper due diligence: As with ordinary data mining, text mining requires proper training of the model, validating with in-sample holdout datasets, and out-of-sample datasets.

43.8 Weighted TXTDM

I provide the subroutines for a weighted TXTDM in Appendices 43.8 through 43.13.

The effects of weighting, which reflects the importance of the words by using the formula of Equations 43.1 through 43.4, on the small TEXT survey produced equivalent results as the unweighted text mining due to the small size of WORDS dataset.

43.9 Clustering Documents

For the text categorization model of GenIQ_FAVORED, the objective is to estimate the probability of a respondent belonging to either category GenIQ_FAVORED(=1) or OLS_Logistic(GenIQ_FAVORED=0). Within text categorization, the category is known a priori as there is, in the WORDS dataset, an additional column that indicates the known category of the document/row (respondent in the case of the GenIQ_FAVORED model).

Another type of text-mining analysis, perhaps more popular than text categorization is *clustering documents*. The objective of clustering documents is to determine the number of classes (categories) to which the documents belong. Unlike text categorization, clustering document seeks to add a column to the WORDS dataset, the values of which are nominal, representing classes. Thus, clustering documents assign each document to one of the classes uncovered.

43.9.1 Clustering GenIQ Survey Documents

I illustrate the clustering documents process with the GenIQ survey used in the text categorization model. There are many clustering methods. A well-known and often-used clustering technique is k-means. The objective of the k-means algorithm is to divide the sample of p dimensions into k clusters so that the within-cluster sum of squares is minimized (6). K-means is a classical clustering method adapted to documents. It is widely used for document clustering and is relatively efficient (7). K-means is not the author's go-to clustering technique for data mining.

I prefer the clustering method of oblique principal components for directly clustering *variables*, which in turn, with a short subroutine, easily *clusters the documents* and assigns them to one of the cluster classes (segments). I use SAS Proc VARCLUS for this approach and refer to the application as *TXTCLUS*. I illustrate the TXTCLUS implementation for clustering documents in step-by-step detail for building the Clustering GenIQ Survey Documents Model. For clarity, I reiterate the objective at hand: to group respondents (documents) on their preferences of the GenIQ Model over OLS-Logistic.

1. I start with the WORDS dataset in Step 8 from the text categorization model of GenIQ_FAVORED.
2. I briefly explain the VARCLUS algorithm (8), then run TXTCLUS subroutines to yield output, TXTCLUS model, and analysis.

   The VARCLUS procedure divides a set of numeric variables (words) into disjoint clusters. Each cluster is a linear combination, actual a principle component, of the variables in the cluster. The cluster components are oblique (i.e., the components are correlated). In contrast, an ordinary principal component analysis generates components, which are computed from all the variables under consideration, and the principal components are uncorrelated.

   Hence, the VARCLUS algorithm is an oblique principle component analysis. The VARCLUS procedure creates an output dataset ("coef" from code outstat=coef, found in the subroutine in Appendix 43.N). Dataset coef is used with the SAS SCORE procedure to compute component scores for each cluster. Scoring, in the next step, is used to assign each document to its class, effectively adding another column, SEGMENT, to dataset WORD_NLBS. WORD_NLBS is identical to dataset WORDS without the long labels due to dummifying the words. Displaying VARLCUS output with long labels would result in hard to read tables.

3. I set VARCLUS with WORDS dataset as the input file and set number of classes equal to two (i.e., seeking a two-group solution) based on the obvious content of the WORDS dataset. I run subroutine VARCLUS for Two-Cluster Solution with seven words. The set of these seven words is a result of performing several iterative runs deleting those words that have low R-squared values, a statistic that I explain in the output of the final VARCLUS solution in Table 43.10. As noted below, I continue the analysis with the dataset WORD_NLBS, not WORDS.

In Table 43.10, the VARCLUS output R-squared values are read as follows:

1. In the first column, "Cluster" indicates the clusters, Cluster 1 and Cluster 2.
2. In the second column, "Variables" (words) indicates which words comprise the cluster. Cluster 1 consists of data-prep, yes-coefficients, not-black-box, data-mining, and new-variables. Cluster 2 consists of machine-learning and alt-regression (alternative to regression).
3. The third column, "R-squared with Own Cluster," is the squared correlation of the word with its cluster component.
4. The fourth column, "R-squared with Next Closest," is the next highest squared correlation of the word with a cluster component.
5. "Own Cluster" values should be higher than the R-squared with any other cluster. "Next Closest" should be a low value if the clusters are well-separated.
6. The last column, "1–R**2 Ratio," is the ratio of 1 minus the value in the "Own Cluster" column to 1 minus the value in the "Next Closest" column. For example, for data-prep in Cluster 1:
   a. 1–R**2 for Own Cluster = 0.6274 (=1 – 0.3726)
   b. 1–R**2 for Next Closest = 0.9852 (=1 – 0.0148).
   c. 1–R**2 Ratio = 0.6368 (=0.6274/0.9852).

TABLE 43.10

VARCLUS 2-Class Solution of WORD_NLBS Dataset

		R-Squared with		
Cluster	Variable	Own Cluster	Next Closest	1–R**2 Ratio
Cluster 1	data_prep	0.3726	0.0148	0.6368
	yes_coefficients	0.3207	0.0209	0.6938
	not_black_box	0.5188	0.0209	0.4914
	data_mining	0.3244	0.0044	0.6786
	new_variables	0.4548	0.0148	0.5534
Cluster 2	machine_learning	0.8446	0.0059	0.1563
	alt_regression	0.8446	0.0028	0.1558

7. Low 1–R**2 Ratios indicate well-separated clusters, and the clustering solution is declared good.
8. For Cluster 1, the 1–R**2 Ratio values (0.6368, 0.6938, 0.4914, 0.6786, and 0.5534) indicate Cluster 1 construction, based on the five words that defined it, is moderately good because its ratio values are moderate.
9. For Cluster 2, the 1–R**2 Ratio values (0.1563 and 0.1558) indicate Cluster 2 construction, based on the two words that defined it, is good because its ratio values are low.
10. Lest one forgets, the dataset is small. Thus, the clustering of GenIQ survey documents is good.

The next step is to assess the content of the clusters. Cluster 1 consists of data_prep, yes_coefficients, not_black_box, data_mining, and new_variables. Cluster 1 does not appear to represent respondents who favor OLS-Logistic as data_mining and new_variables are clearly not features of the OLS and LR models. As for Cluster 2, which consists of machine_learning and alt_regression (alternative to regression), it obviously reflects respondents who favor GenIQ. What explains the quality of the two-cluster solution given the content issue of Cluster 1?

The standard output of reporting R-squared values prevents a thorough assessment of Cluster 1 because the values do not indicate the direction of the correlation between the word and its cluster. Thus, the correlation coefficient between a word and its cluster is needed. The calculation of correlation coefficient must follow the scoring of the clusters. I run the subroutine in Appendix 43.O, but do not discuss its output until discussing directions of data_mining and new_variables with Cluster 1.

I run the subroutine in Appendix 43.P to determine whether the signs of the correlations of data_mining and new_variables are correct (i.e., negative). The output in Table 43.11 indicates that data_mining and new_variables have negative correlations with its Cluster 1. Thus, the five words correctly correlated with Cluster 1 define Cluster 1 as representing respondents who favor OLS-Logistic. Cluster 2 represents respondents who favor GenIQ. In the final analysis, Cluster 1 is about OLS-Logistic, and Cluster 2 is about GenIQ.

Now, I review the outputs of scoring the WORD_NLBS dataset. The first piece of output, Table 43.12, adds two columns to the dataset, Clus1 and Clus2, which are the scores of the two clusters. The respondent assignment corresponds to its larger score. For example, Respondent ID #1 has Clus1 = –0.23 and Clus2 = 1.52. Therefore,

TABLE 43.11

Correlation Coefficient of Words with Its Cluster 1

Rank	Word	Corr_Clus1
1	not_black_box	0.7203
2	data_prep	0.6104
3	yes_coefficients	0.5663
4	data_mining	–0.5696
5	new_variables	–0.6744

TABLE 43.12

Scoring VARCLUS for Two-Cluster Solution on WORD_NLBS Dataset

data_prep	yes_ coefficients	not_ black_ box	data_ mining	new_ variables	machine_ learning	alt_ regression	Clus1	Clus2	ID	GenIQ Model
0	0	0	0	0	1	0	−0.23	1.52	1	1
0	0	0	0	0	0	0	−0.23	−0.29	2	1
0	0	0	0	0	0	0	−0.23	−0.29	3	1
0	0	0	0	0	0	0	−0.23	−0.29	4	1
0	0	0	0	0	1	1	−0.23	4.01	5	1
0	0	0	1	1	0	0	−2.49	−0.29	6	1
0	0	0	0	1	0	0	−1.18	−0.29	7	1
0	0	0	0	1	0	0	−1.18	−0.29	8	1
0	0	0	0	0	0	0	−0.23	−0.29	9	1
0	0	0	0	0	0	0	−0.23	−0.29	10	1
0	0	0	0	0	0	0	−0.23	−0.29	11	1
0	0	0	0	0	0	0	−0.23	−0.29	12	0
0	1	0	0	0	0	0	0.48	−0.29	13	0
0	1	0	0	0	0	0	0.48	−0.29	14	0
0	0	0	0	0	0	0	−0.23	−0.29	15	0
0	0	1	0	0	0	0	0.67	−0.29	16	0
0	0	0	0	0	0	0	−0.23	−0.29	17	0
1	0	0	0	0	0	0	0.62	−0.29	18	0
1	1	1	0	0	0	0	2.23	−0.29	19	0
0	1	1	0	0	0	0	1.38	−0.29	20	0
1	0	1	0	0	0	0	1.52	−0.29	21	0

Respondent ID #1 is assigned to Cluster 2 (favoring GenIQ). Because Respondent #1 is correctly classified. In contrast for Respondent ID #2 whose cluster scores are Clus1 = −0.23 and Clus2 = −0.29, Respondent ID #2 is assigned to Cluster 1 (favoring OLS-Logistic), yet is actually a GenIQ believer (GenIQModel = 1). Thus, Respondent #2 is misclassified. I discuss the misclassification of Respondent ID #2 and others misclassified respondents.

The second output in Table 43.13 simply replaces columns Clus1 and Clus2 by the column SEGMENT in the WORD_NLBS dataset.

It is instructive to analyze how well the clustering performs. Respondents with IDs #11–#21 belong to Clus1, which is the OLS-Logistic segment. These assignments are all correct as their values of GenQIModel = 0 (equivalent to OLS_Logistic = 1). Thus, the accuracy rate in Clus1 segment is 100%.

As for the 11 respondents with IDs #1–#11, the accuracy rate is not good. The Clus2 accuracy rate is 45.5% (= 5/11). The five correct assignments are for respondents with IDs #1, and #5–#8, in Table 43.14. The 11 respondents are actual GenIQ believers. However, the boxed-in respondents with IDs #2, #3, #4, # 9, #10, and #11, are incorrectly assigned as OLS-Logistic because their Clus2 values are larger than their Clus1 values. Recall, Clus1 is about OLS-Logistic, and Clus2 is about GenIQ. Why did this happen?

For the first 11 respondents of concern, their cluster paired values (−0.23, −0.29) are reasonably equivalent. I maintain the six paired values are *equivalent* due to the small dataset. I would like to take a *liberal approach* of declaring the paired values statistically nonsignificant (i.e., the values are equal within a margin of error). In this case, I randomly assign three of the six respondents in question to GenIQModel, and the three remaining respondents stay as were originally assigned. This yields a Clus2 accuracy rate of 72.7% (= (5+3)/11).

Next, I evaluate the performance of the clustering of documents. For ease of discussion, I generate a cross-tabulation of GenIQModel and SEGMENT in Table 43.15.

The total correct classification rate (TCCR) is equal to 71.43% (= (10 + 5)/21). In the absence of a baseline (chance model), no reliable declaration can be made on the performance level of the GenIQModel-TCCR value 71.43%. The chance model, ChanceModel-TCCR, is calculated by running the subroutine in Appendix 43.Q.

The formula of the ChanceModel-TCCR is the sum of the Total columns' squared percentages of GenIQModel = 0, namely, 47.62%*47.62%, and GenIQModel = 1, namely, 52.38%*52.38%. The calculation yields ChanceModel-TCCR = 50.11%. Thus, the incremental gain in classification of GenIQ-Clustering Model over chance is 42.53% = ((71.43% − 50.11%)/50.11%) in the lower panel of Table 43.15. Moreover, the relationship between GenIQ Model and the SEGMENT is significant with a Chi-Squared p-value of 0.0146, which is surprising for a small sample size (21). The calculations for these findings are in Appendix 43.Q.

Finally, I test the liberal approach mentioned earlier to compare the liberal GenIQ-Clustering Model versus the Chance Model. I run the subroutine in Appendix 43.R. The output is in Table 43.16.

The liberal GenIQModel-TCCR = 85.71% (=18/21). Thus, the incremental gain of the liberal GenIQModel-TCCR over ChanceModel–TCCR is 71.04% (85.71% − 50.11%)/50.11%) in Table 43.16. Also, the relationship between liberal GenIQModel and the SEGMENT is very significant with a Chi-Squared p-value of 0.0006. The calculations for these findings are in Appendix 43.R.

TABLE 43.13

SEGMENT Appended to Scored WORD_NLBS Dataset

data_prep	yes_ coefficients	not black_ box	data_ mining	new_ variables	machine_ learning	alt_ regression	SEGMENT	ID	GenIQ Model
0	0	0	0	0	1	0	Clus2	1	1
0	0	0	0	0	0	0	Clus1	2	1
0	0	0	0	0	0	0	Clus1	3	1
0	0	0	0	0	0	0	Clus1	4	1
0	0	0	0	0	1	1	Clus2	5	1
0	0	0	1	1	0	0	Clus2	6	1
0	0	0	0	1	0	0	Clus2	7	1
0	0	0	0	1	0	0	Clus2	8	1
0	0	0	0	0	0	0	Clus1	9	1
0	0	0	0	0	0	0	Clus1	10	1
0	0	0	0	0	0	0	Clus1	11	1
0	0	0	0	0	0	0	Clus1	12	0
0	1	0	0	0	0	0	Clus1	13	0
0	1	0	0	0	0	0	Clus1	14	0
0	0	0	0	0	0	0	Clus1	15	0
0	0	1	0	0	0	0	Clus1	16	0
0	0	0	0	0	0	0	Clus1	17	0
1	0	0	0	0	0	0	Clus1	18	0
1	1	1	0	0	0	0	Clus1	19	0
0	1	1	0	0	0	0	Clus1	20	0
1	0	1	0	0	0	0	Clus1	21	0

TABLE 43.14

Cluster Values by GenIQModel

Clus1	Clus2	ID	GenIQ Model
−0.23	1.52	1	1
−0.23	−0.29	2	1
−0.23	−0.29	3	1
−0.23	−0.29	4	1
−0.23	4.01	5	1
−2.49	−0.29	6	1
−1.18	−0.29	7	1
−1.18	−0.29	8	1
−0.23	−0.29	9	1
−0.23	−0.29	10	1
−0.23	−0.29	11	1

TABLE 43.15

GenIQModel versus Chance Model

GenIQModel Frequency Percent Row Pct Col Pct	SEGMENT Clus1	Clus2	Total
0	10 47.62 100.00 62.50	0 0.00 0.00 0.00	10 - - 47.62
1	6 28.57 54.55 37.50	5 23.81 45.45 100.00	11 - - 52.38
Total	16 76.19	5 23.81	21 100.00

Statistics for Table of GenIQModel by SEGMENT

Statistic	DF	Value	Prob
Chi-Square	1	5.9659	0.0146

Gen IQModel_ TCCR	Chance_ TCCR	Gain_ Over_ Chance
71.43%	50.11%	42.53%

43.9.1.1 Conclusion of Clustering GenIQ Survey Documents

Lest one forgets, the purpose of clustering documents is to create a SEGMENT variable in the WORD_NLBS dataset to segment documents. Whereas, for text categorization, there must be a class variable to build the text mining predictive model. The presentments of the two versions of assessment of the clustering documents are instructive. Also, the

TABLE 43.16

Liberal GenIQModel versus Model

GenIQModel	SEGMENT		
Frequency Percent Row Pct Col Pct	1	2	Total
0	10 47.62 100.00 76.92	0 0.00 0.00 0.00	10 – – 47.62
1	3 14.29 27.27 23.08	8 38.10 72.73 100.00	11 – – 52.38
Total	13 61.90	8 38.10	21 100.00

Statistics for Table of GenIQModel by SEGMENT

Statistic	DF	Value	Prob
Chi-Square	1	11.7483	0.0006

Gen IQModel_ TCCR	Chance_ TCCR	Gain_ Over_ Chance
85.71%	50.11%	71.04%

discussions are included to suggest what the text miner should think about when finalizing a clustering document model.

43.10 Summary

This chapter serves as a resource for data miners interested in text mining. It includes a primer—a tractable illustration to embolden a newbie text miner to undertake one of the popular techniques addressing the bursting of textual data due to the Internet and its ancillaries, such as social media, blogs, and so on. Lastly, the chapter provides the affordable TXTDM text mining software, written in high-level SAS Base and SAS/STAT, which converts easily into virtually any language. TXTDM is a central text mining program, fully supported by direct contact with SAS technical support. TXTDM offers an attractive alternative to free software, with its infamous lack of technical support, or to expensive commercial text mining products.

Appendix

This appendix is for TXTDM, which consists of the following subroutines coded in SAS Base and SAS/STAT.

Appendix 43.A Loading Corpus TEXT Dataset

```
%let num_vars=10;
%let max_varlen=$25.;

data TEXT;
infile datalines dlm = ' ' missover;
input ID (c01-c&num_vars) (:&max_varlen);
datalines;
1 GenIQModel machine-learning genetic-programming
2 GenIQModel no-assumptions nonparametric black-box
3 GenIQModel no-coefficients uninterpretable
4 GenIQModel data-defining no-fitting-the-data
5 GenIQModel machine-learning alt-regression
6 GenIQModel data-mining new-variables
7 GenIQModel new-variables
8 GenIQModel variable-selection new-variables
9 GenIQModel no-data-prep
10 GenIQModel optimizes cumlift decile-table
11 GenIQModel uninterpretable no-coefficients
12 OLS-Logistic benchmarks newer-prediction
13 OLS-Logistic equations yes-coefficients
14 OLS-Logistic interpretable yes-coefficients
15 OLS-Logistic reliable accurate interpretable
16 OLS-Logistic not-black-box yes-equations
17 OLS-Logistic variable-selection
18 OLS-Logistic data-prep time-consuming
19 OLS-Logistic data-prep yes-coefficients not-black-box
20 OLS-Logistic yes-coefficients not-black-box
21 OLS-Logistic data-prep yes-equations not-black-box
;
PROC PRINT data = TEXT;
title2' TEXT Dataset ';
run;
```

Appendix 43.B Intermediate Step Creating Binary Words

```
PROC TRANSPOSE data=TEXT out=TEXT_transp;
var c01-c&num_vars;
by ID;
run;

data TEXT;
set TEXT_transp;
_COL1 = COL1;
run;
```

```
PROC TRANSREG data=TEXT DESIGN;
model class (_COL1 / ZERO='x');
output out = TEXT (drop = Intercept _NAME_ _TYPE_ );
id ID;
run;

PROC SQL noprint;
select trim(name)||'='||substr(trim(name),6)
into :varlist separated by ' '
from dictionary.columns
where libname eq "WORK" and memname eq "TEXT";
quit;
%put &varlist;
```

Appendix 43.C Creating the Final Binary Words

```
%let varlist=
_COL1GenIQModel=GenIQModel_COL1OLS_Logistic=OLS_Logistic
   _COL1accurate=accurate
_COL1alt_regression=alt_regression_COL1benchmarks=benchmarks
   _COL1black_box=black_box
_COL1cumlift=cumlift_COL1data_defining=data_defining
   _COL1data_mining=data_mining
_COL1data_prep=data_prep _COL1decile_table=decile_table _COL1equations=equations
_COL1genetic_programming=genetic_programming _COL1interpretable=interpretable
_COL1machine_learning=machine_learning _COL1new_variables=new_variables
_COL1newer_prediction=newer_prediction _COL1no_assumptions=no_assumptions
_COL1no_coefficients=no_coefficients _COL1no_data_prep=no_data_prep
_COL1no_fitting_the_data=no_fitting_the_data _COL1nonparametric=nonparametric
_COL1not_black_box=not_black_box _COL1optimizes=optimizes _COL1reliable=reliable
_COL1time_consuming=time_consuming _COL1uninterpretable=uninterpretable
_COL1variable_selection=variable_selection _COL1yes_coefficients=yes_coefficients
_COL1yes_equations=yes_equations;

PROC DATASETS library=work nolist;
modify TEXT;
rename &varlist;
quit;

PROC CONTENTS data=TEXT
out = vars (keep = name type)
noprint;
run;

PROC SQL noprint;
select name into : varlist separated by ' '
```

```
from vars;
quit;
%put _global_;
```

Appendix 43.D Calculate Statistics TF, DF, NUM_DOCS, and N (=Num of Words)

```
libname tm 'c://0-tm';

PROC SORT data=TEXT; by ID;
run;

PROC SUMMARY data=TEXT;
var _numeric_;
output out=sums (drop=ID) sum=;
by ID;
run;

data tm.WORDS;
retain ID;
set sums;
ID+1;
drop _TYPE_ _FREQ_;
run;

PROC SUMMARY data=tm.WORDS;
var _numeric_;
output out=sums (drop=ID) sum=;
by ID;
run;

data tm.WORDS;
retain ID;
set sums;
ID+1;
drop _TYPE_ _FREQ_;
run;

PROC PRINT data=tm.WORDS noobs;
title2' WORD FREQUENCY - TF (Zero-One WORD dataset) ';
run;

PROC SUMMARY data=tm.WORDS;
var _numeric_;
output out=tm.DF (drop= ID _TYPE_ _FREQ_) sum=;
run;
```

```
PROC PRINT data=tm.df noobs;
title2' DOCUMENT FREQUENCY for given Word - DF ';
run;

PROC SUMMARY data=tm.words;
var ID;
output out=tm.NUM_DOCS (drop= _TYPE_ _FREQ_) max=NUM_DOCS;
run;

PROC PRINT data=tm.NUM_DOCS;
title2' NUMBER of DOCUMENTS ';
run;

data tm.TOT_WORDS;
set tm.WORDS;
drop ID;
if _n_=1;
array nums(*) _numeric_;
TOT_WORDS=dim(nums)-1;
keep TOT_WORDS;
run;

PROC PRINT data=tm.TOT_WORDS;
title2 ' N - NUMBER of WORDS in ALL DOCUMENTS ';
run;
```

Appendix 43.E Append GenIQ_FAVORED to WORDS Dataset

```
libname tm 'c://0-tm';
title ' ';
title2 ' ';
data GenIQ_FAVORED;
infile datalines dlm = ' ' missover;
input ID GenIQ_FAVORED;
datalines;
1 1
2 1
3 1
4 1
5 1
6 1
7 1
8 1
9 1
10 1
11 1
```

```
12 0
13 0
14 0
15 0
16 0
17 0
18 0
19 0
20 0
21 0
;
run;

PROC SORT data=GenIQ_FAVORED; by ID;
PROC SORT data=tm.WORDS; by ID;
run;

data tm.WORD_GenIQ_FAVORED;
merge tm.WORDS GenIQ_FAVORED;
by ID;
wt=1;
run;

PROC PRINT data=tm.WORD_RESP;
run;
```

Appendix 43.F Logistic GenIQ_FAVORED Model

```
%let varlist=
  data_prep time_consuming accurate yes_coefficients not_black_box interpretable equations
  benchmarks machine_learning no_assumptions no_data_prep no_coefficients data_defining
  data_mining variable_selection new_variables alt_regression cumlift;

PROC LOGISTIC data= tm.WORD_GenIQ_FAVORED nosimple des outest=coef;
model GenIQ_FAVORED = &varlist;
run;

PROC SCORE data=tm.WORD_GenIQ_FAVORED predict type=parms
       score=coef out=score;
var &varlist;
run;

data score;
set score;
logit=GenIQ_FAVORED2;
prob_GenIQ_FAVORED=exp(logit)/(1+ exp(logit));
```

```
PROC SORT data=score; by descending prob_GenIQ_FAVORED;
run;

PROC PRINT data=score noobs;
var &varlist ID GenIQ_FAVORED prob_GenIQ_FAVORED;
format prob_GenIQ_FAVORED 5.4;
run;
```

Appendix 43.G Average Correlation among Words

```
libname tm "c://0-tm";
title " AVG_CORR of WORDS ";

%let varlist=
  data_prep time_consuming accurate yes_coefficients not_black_box interpretable equations
  benchmarks machine_learning no_assumptions no_data_prep no_coefficients
  data_defining data_mining variable_selection new_variables alt_regression cumlift;

data num_vars;
set tm.WORDS;
keep &varlist;
data num_vars;
set num_vars;
if _n_=1;
array nums(*) _numeric_;
num_vars=dim(nums);
keep num_vars;
call symputx ('num_vars',num_vars);
run;

%put &num_vars;
PROC PRINT data=num_vars;
title2 ' num_vars ';
run;

PROC CORR data=tm.WORDS out=out noprint;
var &varlist;
run;

data out1;
set out;
if _type_='MEAN' or _type_='STD' or _type_='N' then delete;
drop _type_;
array vars (&num_vars) &varlist;
```

```
array pos (&num_vars) x1 - x&num_vars;
do i= 1 to &num_vars;
pos(i)=abs(vars(i));
end;
drop &varlist i;
run;

data out2;
set out1;
array poss (&num_vars) x1- x&num_vars;
do i= 1 to &num_vars;
if poss(i) =1 then poss(i)=.;
drop i;
end;
run;

PROC MEANS data=out2 sum noprint;
output out=out3 sum=;
run;

data out4;
set out3;
sum_=sum(of x1-x&num_vars);
sum_div2= sum_/2;
bot= ((_freq_*_freq_)-_freq_)/2;
avg_corr= sum_div2/bot;
run;

data AVG_CORR;
set out4;
keep AVG_CORR;
proc print data=AVG_CORR;
run;
```

Appendix 43.H Creating TF–IDF

```
title2' creating tf_idf ';
libname tm 'c://0-tm';
options mprint symbolgen;

data tm.TOT_WORDS;
set tm.WORDS;
drop ID;
if _n_=1;
array nums(*) _numeric_;
TOT_WORDS=dim(nums)-1;
```

```
* minus 1 because of TOT_WORDS;
keep TOT_WORDS;
call symputx ('TOT_WORDS',TOT_WORDS);
run;
%put &TOT_WORDS;

%let varlist=
GenIQModel OLS accurate classification computer_program cumlift
decile_table interpretable logistic machine_learning no_coefficient no_equation
prediction regression reliable specifies statistical workhorses;

PROC PRINT data=tm.NUM_DOCS;
title' NUM_DOCS ';
run;

PROC SUMMARY data=tm.WORDS;
var _numeric_;
output out=tm.df (drop= ID _TYPE_ _FREQ_) sum=;
run;

data tm.df;
set tm.df;
array words(&tot_words) &varlist;
array df(&tot_words) df1-df&tot_words;
do i=1 to &tot_words;
df(i)=words(i);
drop i &varlist;
end;
run;

PROC PRINT data=tm.df noobs;
title' words_inrows - df';
run;

data tm.tf;
set tm.WORDS;
array words(&tot_words) &varlist;
array tf(&tot_words) tf1-tf&tot_words;
do i=1 to &tot_words;
tf(i)=words(i);
drop i &varlist;
end;
run;

PROC PRINT data=tm.tf;
title' tf ';
run;

data tm.tf;
set tm.tf; m=1;
```

```
data tm.df;
set tm.df; m=1;

data tm.num_docs;
set tm.num_docs; m=1;
run;

data tm.tf_idf;
merge tm.tf tm.df tm.num_docs;
by m;
drop m;
run;

PROC PRINT data=tm.tf_idf;
title' tf_idf ';
run;

data tm.tf_idf;
set tm.tf_idf;
array tf(&tot_words) tf1-tf&tot_words ;
array df(&tot_words) df1-df&tot_words;
array idf(&tot_words) idf1 - idf&tot_words;
array tf_idf(&tot_words) tf_idf1 - tf_idf&tot_words;

do i=1 to dim(df);
if df(i)=0 then
idf(i)= log(num_docs);else
idf(i)= log(num_docs/(df(i)));
tf_idf(i)=tf(i)*idf(i);

keep ID tf_idf1 - tf_idf&tot_words;
end;
run;

PROC PRINT data=tm.tf_idf;
title' tf_idf ';
run;
```

Appendix 43.I WORD_TF–IDF Weights by Concat of WORDS and TF–IDF

```
libname tm "c://0-tm";
options symbolgen mprint;

data tm.tot_words;
set tm.WORDS;
```

```
drop ID;
if _n_=1;
array nums(*) _numeric_;
TOT_WORDS=dim(nums)-1;
keep TOT_WORDS;
call symputx ('TOT_WORDS',TOT_WORDS);
run;

%put &TOT_WORDS;
%let varlist=
  GenIQModel OLS accurate classification computer_program cumlift
  decile_table interpretable logistic machine_learning no_coefficient no_equation prediction
  regression reliable specifies statistical workhorses;

data _null_;
set tm.tf_idf;
array word(&tot_words) &varlist;
array tf_idf(&tot_words) tf_idf1- tf_idf&tot_words;
do i=1 to &tot_words;
 call symputx('word'|| left(put(i,2.)),vname(word(i)));
 call symputx('tf_idf' || left(put(i,2.)),vname(tf_idf(i)));
end;
run;

%macro concat_word_tf_idf;
data tm.concat_word_tf_idf;
set tm.tf_idf;
rename %do i=1 %to &tot_words;
          &&tf_idf&i=&&word&i.._&&tf_idf&i
       %end;;
run;

%mend concat_word_tf_idf;
%concat_word_tf_idf

PROC CONTENTS data=tm.concat_word_tf_idf
run;

PROC CONTENTS data=tm.concat_word_tf_idf
out = vars (keep = name type);
run;

PROC SQL noprint;
select name into : varlist separated by ' '
from vars;
quit;
%put _global_;
```

Appendix 43.J WORD_RESP WORD_TF–IDF RESP

```
libname tm 'c://0-tm';
options pageno=1;
title ' ';
title2 ' ';

PROC SORT data=tm.concat_word_tf_idf; by ID;
PROC SORT data=tm.word_resp; by ID;
run;

data tm.word_word_tf_idf_resp;
retain ID;
merge
tm.concat_word_tf_idf
tm.word_resp; by ID;
run;

PROC PRINT data=tm.word_word_tf_idf_resp;
run;
```

Appendix 43.K Stemming

```
data tm.word_resp;
set tm.word_resp;
interpretable=sum(of interpretable:);
uninterpretable=sum(of uninterpretable:);
machine_learning=sum(of machine_learning:);
not_black_box=sum(of not_black_box:);
new_variables=sum(of new_variables:);
no_coefficient=sum(of no_coefficient:);
yes_coefficient=sum(of yes_coefficient:);
yes_equation=sum(of yes_equation:);
wt=1;
run;

PROC CONTENTS data=tm.word_resp;
run;
```

Appendix 43.L WORD Times TF–IDF

```
%let varlist=
  GenIQModel OLS accurate classification computer_program cumlift
```

```
  decile_table interpretable logistic machine_learning no_coefficient no_equation prediction
  regression reliable specifies statistical workhorses;

PROC SORT data=tm.tf_idf; by ID;
PROC SORT data=tm.word_resp; by ID;
run;

data tm.word_tf_idf_resp;
retain ID;
merge
tm.tf_idf tm.word_resp; by ID;
run;

data tm.word_tf_idf_resp;
set tm.word_tf_idf_resp;
array words(*) &varlist;
array tf_idf(*) tf_idf1-tf_idf&tot_words;
array word_wted(*) wted1- wted&tot_words;
do i = 1 to dim(words);
word_wted(i)=words(i)*tf_idf(i);
drop i;
end;
run;

PROC PRINT;
run;
```

Appendix 43.M Dataset Weighted with Words for Profile

```
%let varlist=
  GenIQModel OLS accurate classification computer_program cumlift
  decile_table interpretable logistic machine_learning no_coefficient no_equation prediction
  regression reliable specifies statistical workhorses;

data _null_;
set tm.word_tf_idf_resp;
array word(&tot_words) &varlist;
array word_wted(*) wted1 - wted&tot_words;
do i=1 to &tot_words;
call symputx('word'|| left(put(i,2.)),vname(word(i)));
call symputx('word_wted' || left(put(i,2.)),vname(word_wted(i)));
end;
run;

%macro word_wted;
data tm.word_wted;
```

```
set tm.word_tf_idf_resp;
rename %do i=1 %to &tot_words;
       &&word_wted&i=&&word&i.._&&word_wted&i
   %end;;
drop tf_idf1-tf_idf&tot_words;
run;

%mend word_wted;
%word_wted

PROC PRINT data=tm.WORD_WTED;
run;

PROC CONTENTS data= tm.WORD_WTED
out = vars (keep = name type);
run;
PROC SQL noprint;
select name into : varlist separated by ' '
from vars;
quit;
%put _global_ ;
```

Appendix 43.N VARCLUS for Two-Class Solution

```
libname tm 'c://0-tm';
options pageno=1;
title' ';
title ' VARCLUS - 2-Cluster Solution ';

ods listing;
ods html;
%let varlist=
data_prep yes_coefficients not_black_box data_mining new_variables
machine_learning alt_regression;

data tm.WORDS_NLBS;
set tm.WORDS;
attrib _ALL_ label=' ';
run;

PROC VARCLUS data= tm.WORDS_NLBS MINC=2 MAXC=2 simple outstat=coef;
ods select Rsquare;
var &varlist;
run;
ods html close;
```

Appendix 43.O Scoring VARCLUS for Two-Cluster Solution

```
%let clusoltn=2;
title2 "The &clusoltn Cluster Solution";

%let varlist=
   data_prep yes_coefficients not_black_box data_mining new_variables
   machine_learning alt_regression;

data Coef&clusoltn;
set Coef;
if _ncl_ = . or _ncl_ = &clusoltn;
drop _ncl_;
run;

PROC SCORE data=tm.words_nlbs score=Coef&clusoltn out=scored;
var &varlist;
run;

PROC SORT data=scored;by descending GenIQModel;

PROC PRINT data=scored noobs;var &varlist clus1-clus2 ID GenIQModel;
format clus1 clus2 5.2;
run;

* Assigning the Individual to the Classified Cluster-Segment;
data scored_classified;
set scored ;
temp=max(clus1, clus2);
     if clus1 = temp then predictd = clus1;
else if clus2 = temp then predictd = clus2;
run;

data tm.scored_classified (drop=temp);
set scored_classified;
temp=max(clus1, clus2);
     if clus1 = temp then SEGMENT = 'clus1';
else if clus2 = temp then SEGMENT = 'clus2';
run;

PROC PRINT data=tm.scored_classified noobs;
var &varlist SEGMENT ID GenIQModel;
run;
```

Appendix 43.P Direction of Words with Its Cluster 1

```
%let Clus=Clus1;
title " r of &Clus with Correlates ";
```

```
%let varlistclus1=
data_prep yes_coefficients not_black_box
data_mining new_variables;

PROC CORR data=tm.scored_classified rank
outp=out noprint;
var &varlistclus1;
with &Clus;
run;

PROC PRINT data=out;
title' out ';
run;

data out1;
set out;
if _TYPE_='MEAN' then delete;
if _TYPE_='STD' then delete;
drop _NAME_;
run;

PROC PRINT data=out1;
title' out1 ';
run;

PROC TRANSPOSE data=out1
out=out2 (rename=(_1=n _2=Corr_&Clus ) ) prefix=_;
run;

data out2;
set out2;
word=_NAME_;
run;

data out3;
set out2;

PROC SORT data=out3; by descending Corr_&clus;
run;

data words;
set out3;
Rank+1;
keep Rank word Corr_&clus;
run;

PROC PRINT data=words noobs;var Rank word Corr_&clus ;
format corr_&clus 6.4;
run;
```

Appendix 43.Q Performance of GenIQ Model versus Chance Model

```
libname tm 'c://0-tm';
options pageno=1;
title' ';

PROC FREQ data=tm.scored_classified;
table GenIQModel*SEGMENT /chisq sparse out=D;
run;

PROC TRANSPOSE data=D out=transp;
run;

data IMPROV;
retain GenIQMODEL_TCCR;
set transp;
drop _LABEL_;
if _NAME_="GenIQModel" then delete;
if _NAME_="SEGMENT" then delete;
if _NAME_="PERCENT" then CHANCE_TCCR=(((col1+col2)**2)+((col3+col4)**2))/10000;
if _NAME_="COUNT" then GenIQMODEL_TCCR=((col1+col4)/sum(of col1-col4))/1;
GAIN_OVER_CHANCE= ((GenIQMODEL_TCCR- CHANCE_TCCR)/CHANCE_TCCR);
if GAIN_OVER_CHANCE=. then delete;
run;

PROC PRINT data=IMPROV;
var GenIQMODEL_TCCR CHANCE_TCCR GAIN_OVER_CHANCE;
format CHANCE_TCCR GenIQMODEL_TCCR GAIN_OVER_CHANCE percent8.2;
run;
```

Appendix 43.R Performance of Liberal-Cluster Model versus Chance Model

```
data Liberal_Cluster;
input GenIQModel SEGMENT Count @@;
datalines;
0 1 10 0 2 0
1 1 3 1 2 8
;
PROC FREQ data=Liberal_Cluster;
table GenIQModel*SEGMENT /chisq sparse out=D;
weight count;
run;
```

```
PROC TRANSPOSE data=D out=transp;
run;

data tccr;
retain GenIQMODEL_TCCR;
set transp;
drop _LABEL_;
if _NAME_="GenIQModel" then delete;
if _NAME_="SEGMENT" then delete;
if _NAME_="PERCENT" then CHANCE_TCCR=(((col1+col2)**2)+((col3+col4)**2))/10000;
if _NAME_="COUNT" then GenIQMODEL_TCCR=((col1+col4)/sum(of col1-col4))/1;
GAIN_OVER_CHANCE= ((GenIQMODEL_TCCR- CHANCE_TCCR)/CHANCE_TCCR);
if GAIN_OVER_CHANCE=. then delete;
run;

PROC PRINT data=tccr;
var GenIQMODEL_TCCR CHANCE_TCCR GAIN_OVER_CHANCE;
format CHANCE_TCCR GenIQMODEL_TCCR GAIN_OVER_CHANCE percent8.2;
run;
```

References

1. Harris, Z.S., Distributional structure, *Word*, 10, 146–162, 1954.
2. Hutchins, J., Retrospect and prospect in computer-based translation, in *Proceedings of MT Summit VII*, pp. 30–44, 1999.
3. Samuel, A.L., Some studies in machine learning using the game of checkers, *IBM Journal of Research and Development*, 3(3), 210–229, 1959.
4. Francis, L.A., *Taming Text: An Introduction to Text Mining, Casualty Actuarial Society Forum*, Winter, 2006, p. 2.
5. Allison, P.D., *Convergence Failures in Logistic Regression*, Paper 360-2008, SAS Global Forum, 2008.
6. Hartigan, J.A., *Clustering Algorithms*, Wiley, New York, 1975.
7. Weiss, S.M., Indurkhya, N., Zhang, T., and Damerau, F.J., *Predictive Methods for Analyzing Unstructured Information*, Springer, New York, 2004.
8. SAS/STAT 14.1; *User's Guide*. http://support.sas.com/documentation/cdl/en.

44

Some of My Favorite Statistical Subroutines

This chapter includes specific subroutines referenced throughout the book and generic subroutines for some second edition chapters for which the data no longer exist. Lastly, I provide some of my favorite statistical subroutines that are helpful in almost all analyses. The subroutines are also available for downloading from my website: http://www.geniq.net/articles.html#section9

44.1 List of Subroutines

- Smoothplots (Mean and Median) of Chapter 5—X1 versus X2
- Smoothplots of Chapter 10—Logit and Probability
- Average Correlation of Chapter 16—Among Var1 Var2 Var3
- Bootstrapped Decile Analysis of Chapter 40
- H-Spread Common Region of Chapter 42
- Favorite—Proc Corr with Option Rank, Vertical Output
- Favorite—Decile Analysis—Response
- Favorite—Decile Analysis—Profit
- Favorite—Smoothing Time-Series Data (Running Medians of Three)
- Favorite—First Cut Is the Deepest—Among Variables with Large Skew Values

44.2 Smoothplots (Mean and Median) of Chapter 5—X1 versus X2

```
%let Y=X1;
%let X=X2;
%let slice_X=10;

title ' ';
data IN;
input X1 X2;
cards;
13 14
17 19
54 43
23 88
11 77
```

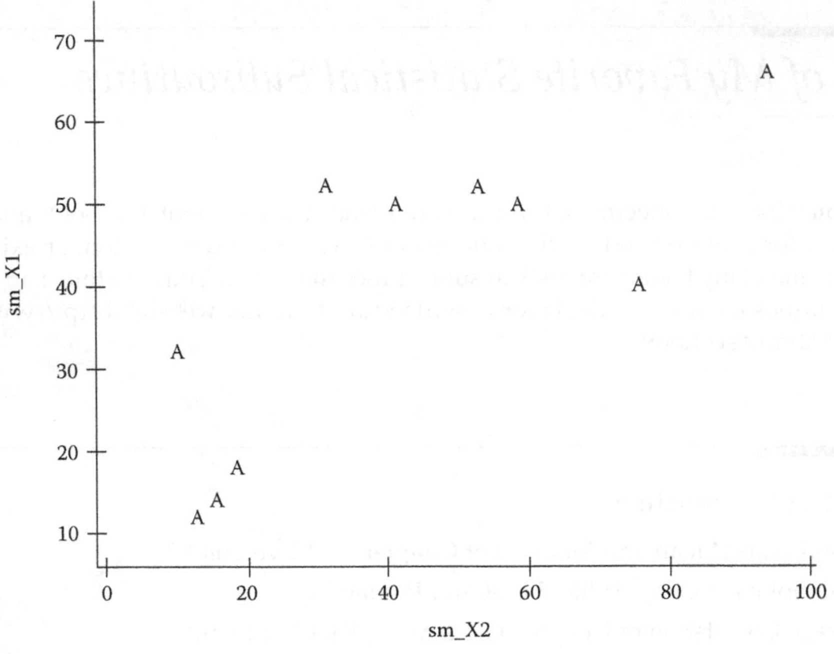

Smooth Mean plot for X1 and X2.

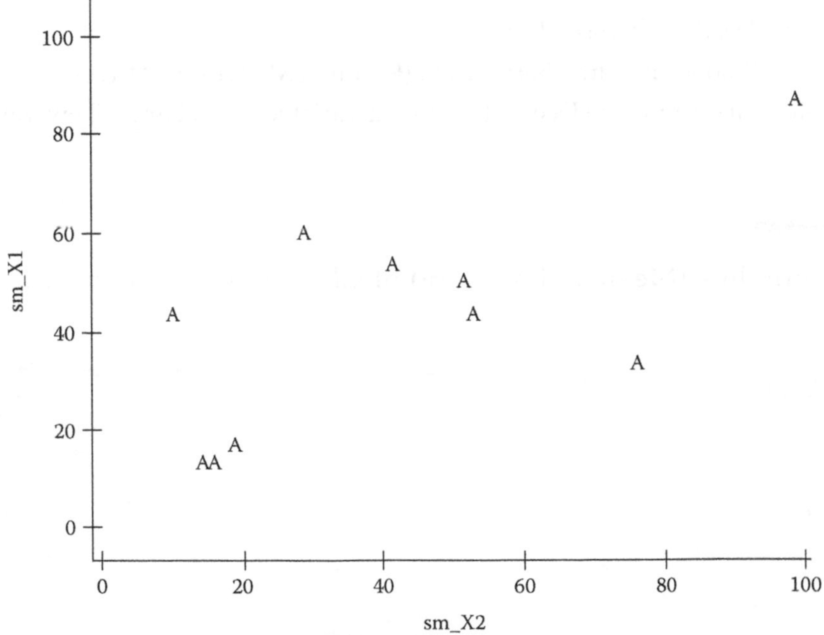

Smooth Median plot for X1 and X2.

```
09 33
32 53
10 12
51 52
13 14
17 19
43 10
88 98
77 25
33 76
53 41
12 15
52 53
83 76
43 41
13 14
17 19
32 53
10 12
51 52
13 14
17 19
43 10
88 98
77 25
33 76
53 41
12 15
52 53
83 76
43 41
;
run;

data smooth;
set IN;
wt=1;
run;

PROC PRINT;
run;

data score;
set smooth;
keep wt &Y &X;
run;

data notdot;
set score;
if &X ne .;
```

```
PROC MEANS data=notdot sum noprint; var wt;
output out=samsize (keep=samsize) sum=samsize;
run;

data scoresam (drop=samsize);
set samsize score;
retain n;
if _n_=1 then n=samsize;
if _n_=1 then delete;
run;

PROC SORT data=scoresam; by descending &X;
run;

data score_X;
set scoresam;
if &X ne . then cum_n+wt;
if &X = . then slice_X =.;
else slice_X=floor(cum_n*&slice_X/(n+1));
drop cum_n n;
run;

PROC SUMMARY data=score_X nway;
class slice_X;
var &X &Y;
output out=smout_&X mean = sm_&X sm_&Y/noinherit;
run;

title 'Mean Smoothplot - X1 vs. X2';
PROC PRINT data=smout_&X;
run;

PROC PLOT data=smout_&X HPCT=80 VPCT=80;
plot sm_&Y*sm_&X;
run;

title ' ';
PROC SUMMARY data=score_X nway;
class slice_X ;
var &X &Y;
output out=smout_&X median = sm_&X sm_&Y/noinherit;
run;

PROC PRINT data=smout_&X;
title 'Median Smoothplot - X1 vs. X2';
run;

PROC PLOT data=smout_&X HPCT=80 VPCT=80;
plot sm_&Y*sm_&X;
run;
quit;
```

44.3 Smoothplots of Chapter 10—Logit and Probability

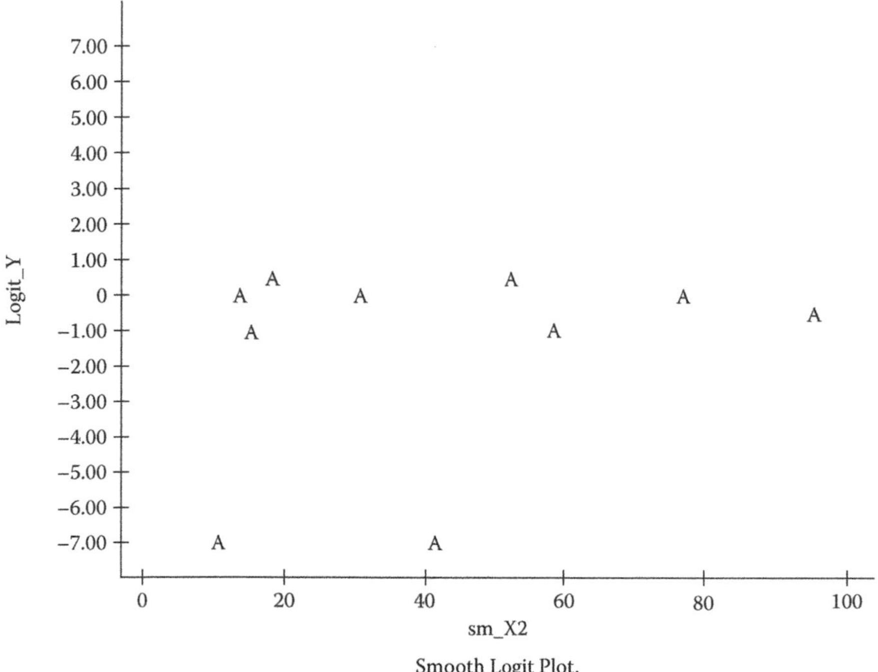

Smooth Logit Plot.

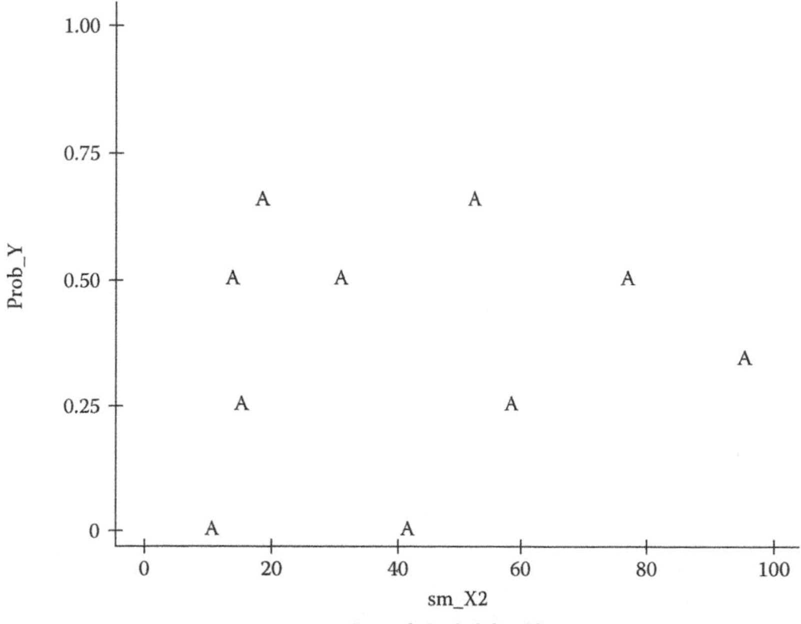

Smooth Probability Plot.

```
%let Y=Y;
%let X=X2;
%let slice_X=10;

data IN;
input X1 X2;
cards;
13 14
17 19
54 43
23 88
11 77
09 33
32 53
10 12
51 52
13 14
17 19
43 10
88 98
77 25
33 76
53 41
12 15
52 53
83 76
43 41
13 14
17 19
32 53
10 12
51 52
13 14
17 19
43 10
88 98
77 25
33 76
53 41
12 15
52 53
83 76
43 41
;
run;

data IN;
set IN;
Y=RAND('BERNOULLI',1/3);
```

```
data smooth;
set IN;
wt=1;
run;

PROC PRINT;
run;

data score;
set smooth;
keep wt &Y &X;
run;

data notdot;
set score;
if &X ne.;

PROC MEANS data=notdot sum noprint; var wt;
output out=samsize (keep=samsize) sum=samsize;
run;

data scoresam (drop=samsize);
set samsize score;
retain n;
if _n_=1 then n=samsize;
if _n_=1 then delete;
run;

PROC SORT data=scoresam; by descending &X;
run;

data score_X;
set scoresam;
if &X ne . then cum_n+wt;
if &X = . then slice_X =.;
else slice_X=floor(cum_n*&slice_X/(n+1));
drop cum_n n;
run;

PROC SUMMARY data=score_X nway;
class slice_X;
var &X &Y;
output out=smout_&X mean= sm_&X sm_&Y/noinherit;
run;

PROC PRINT data=smout_&X;
run;

data sliced_X;
set smout_&X;
```

```
Logit_&Y=log( sm_Y/(1-sm_Y));
if sm_&Y=1 then Logit_Y= 7;
if sm_&Y=0 then Logit_Y=-7;
Prob_&Y= exp(Logit_Y)/((1+exp(Logit_Y)));
run;

PROC PRINT data=sliced_X;
run;

PROC PLOT data=sliced_X HPCT=80 VPCT=80;
plot Logit_&Y*sm_&X /vaxis=-7 to +7 by 1;
format logit_&Y 6.2;
title 'Smooth Logit Plot';
run;

PROC PLOT data=sliced_X HPCT=80 VPCT=80;
plot Prob_Y*sm_&X /vaxis=0 to 1 by 0.25 ;
format Prob_&Y 6.2;
title 'Smooth Probability Plot';
run;
quit;
```

44.4 Average Correlation of Chapter 16—Among Var1 Var2 Var3 (Table 44.1)

TABLE 44.1
AVG_CORR

AVG_CORR
0.91308

```
%let varlist =
Var1 Var2 Var3;
title2 " AVG_CORR of &varlist ";

%let numvars=3;

data dat1;
input Var1 Var2 Var3 :4.0;
cards;
1234 2345 3456
5678 4567 8798
1256 0978 4567
;
run;
```

```
PROC CORR data=dat1 out=out;
var &varlist;
run;

data out1;
set out;
if _type_='MEAN' or _type_='STD' or _type_='N' then delete;
drop _type_;
array vars (&numvars)
&varlist;

array pos (&numvars) x1 - x&numvars;
do i= 1 to &numvars;
pos(i)=abs(vars(i));
end;
drop
&varlist i;
run;

data out2;
set out1;
array poss (&numvars) x1- x&numvars;
do i= 1 to &numvars;
if poss(i) =1 then poss(i)=.;
drop i;
end;
run;

PROC PRINT;
run;

PROC MEANS data=out2 sum;
output out=out3 sum=;

PROC PRINT;
run;

data out4;
set out3;
sum_=sum(of x1-x&numvars);
sum_div2= sum_/2;
bot= ((_freq_*_freq_)-_freq_)/2;
AVG_CORR= sum_div2/bot;
run;

data avg_corr;
set out4;
keep avg_corr;
PROC PRINT;
run;
```

44.5 Bootstrapped Decile Analysis of Chapter 29—Using Data from Table 23.4 (Table 44.2)

TABLE 44.2

Bootstrapped Decile Analysis

samsize_bs = 16003, n_sampl_bs = 50								
Decile	**Number of Individuals**	**Number of Responders**	**Response Rate (%)**	**Cum Response Rate (%)**	**Cum Single-Sample Lift (%)**	**Cum Bootstrap Lift (%)**	**Bootstrap Margin of Error (80%)**
top	1,600	1,118	69.88	69.88	314	315	7.3
2	1,600	637	39.81	54.84	247	247	4.3
3	1,601	332	20.74	43.47	195	195	3.7
4	1,600	318	19.88	37.57	169	171	3.5
5	1,600	165	10.31	32.12	144	145	2.0
6	1,601	165	10.31	28.48	128	128	1.5
7	1,600	158	9.88	25.83	116	116	1.5
8	1,601	256	15.99	24.60	111	111	0.8
9	1,600	211	13.19	23.33	105	105	0.6
bottom	1,600	199	12.44	22.24	100	100	0.0
	16,003	**3,559**					

```
options source nonotes;
options nomprint nomlogic nosymbolgen;

Y=RESPONSE;
%let data_in=IN; /* add wt=1 to dataset IN */
%let depvar=Y;
%let indvars=_X11 - _X13 _X19 _X21;

%let samsize_bs=16003;
%let n_sampl_bs=50;

PROC SURVEYSELECT data=&data_in method=urs out=sample
  n=&samsize_bs rep=&n_sampl_bs outhits;
run;

%macro loop;
%do rep=1 %to &n_sampl_bs;

data Replicate&Rep;
set sample;
if Replicate=&Rep;
run;

%let dsn=Replicate&Rep;
ods exclude ODDSRATIOS;
```

```
PROC LOGISTIC data=&dsn nosimple noprint des outest=coef;
model &depvar = &indvars;
run;

PROC SCORE data=&dsn predict type=parms score=coef
out=score;
var &indvars;
run;

data score;
set score;
estimate=&depvar.2;

data notdot;
set score;
if estimate ne .;

PROC MEANS data=notdot sum noprint; var wt;
output out=samsize (keep=samsize) sum=samsize;
run;

data scoresam (drop=samsize);
set samsize score;
retain n;
if _n_=1 then n=samsize;
if _n_=1 then delete;
run;

PROC SORT data=scoresam; by descending estimate;
run;

data score;
set scoresam;
if estimate ne . then cum_n+wt;
if estimate = . then dec=.;
else dec=floor(cum_n*10/(n+1));
run;

PROC SUMMARY data=score missing;
class dec;
var &depvar wt;
output out=sum_dec sum=sum_can sum_wt;

data sum_dec;
set sum_dec;
avg_can=sum_can/sum_wt;
run;

data avg_rr;
set sum_dec;
```

```
if dec=.;
keep avg_can;
run;

data sum_dec1;
set sum_dec;
if dec=. or dec=10 then delete;
cum_n +sum_wt;
r =sum_can;
cum_r +sum_can;
cum_rr=(cum_r/cum_n)*100;
avg_cann=avg_can*100;
run;

data avg_rr;
set sum_dec1;
if dec=9;
keep avg_can;
avg_can=cum_rr/100;
run;

%let scoresam=&Rep;
data scoresam&Rep;
set avg_rr sum_dec1;
retain n;
if _n_=1 then n=avg_can;
if _n_=1 then delete;
lift&Rep = (cum_rr/n);
if dec ne .;
keep dec lift&Rep;
run;

PROC SORT data=scoresam&Rep; by dec;
run;
%end;

data combine;
merge %do i=1 %to &n_sampl_bs;
scoresam&i
%end;;
by dec;
run;

data bs_lift_SE;
set combine;
bs_est=mean(of lift:);
bs_std=std(of lift:);
bs_SE=1.28*bs_std;
keep dec bs_est bs_SE;
run;
```

```
ods exclude ODDSRATIOS;
PROC LOGISTIC data=&data_in nosimple noprint des outest=coef;
model &depvar = &indvars;
run;

PROC SCORE data=&data_in predict type=parms score=coef
out=score;
var &indvars;
run;

data score;
set score;
estimate=&depvar.2;

data notdot;
set score;
if estimate ne .;

PROC MEANS data=notdot sum noprint; var wt;
output out=samsize (keep=samsize) sum=samsize;
run;

data scoresam (drop=samsize);
set samsize score;
retain n;
if _n_=1 then n=samsize;
if _n_=1 then delete;
run;

PROC SORT data=scoresam; by descending estimate;
run;

data score;
set scoresam;
if estimate ne . then cum_n+wt;
if estimate = . then dec=.;
else dec=floor(cum_n*10/(n+1));
run;

PROC SUMMARY data=score missing;
class dec;
var &depvar wt;
output out=sum_dec sum=sum_can sum_wt;
run;

data sum_dec;
set sum_dec;
avg_can=sum_can/sum_wt;
run;

data avg_rr;
set sum_dec;
```

```
if dec=.;
keep avg_can;
run;

data sum_dec1;
set sum_dec;
if dec=. or dec=10 then delete;
cum_n +sum_wt;
r =sum_can;
cum_r +sum_can;
cum_rr=(cum_r/cum_n)*100;
avg_cann=avg_can*100;
run;

data avg_rr;
set sum_dec1;
if dec=9;
keep avg_can;
avg_can=cum_rr/100;
run;

data scoresam;
set avg_rr sum_dec1;
retain n;
if _n_=1 then n=avg_can;
if _n_=1 then delete;
lift=(cum_rr/n);
if dec ne .;
_2SAM_EST=2*lift;
keep dec _2SAM_EST lift;
run;

data boot;
merge
bs_lift_SE scoresam;
lift_bs=_2SAM_EST-bs_est;
keep dec _2SAM_EST bs_est lift_bs bs_SE;
run;

%end;
%mend;

dm 'clear log';
%loop

ods exclude ODDSRATIOS;
PROC LOGISTIC data=&data_in nosimple noprint des outest=coef;
model &depvar = &indvars;
freq wt;
run;
```

```
PROC SCORE data=&data_in predict type=parms score=coef
out=score;
var &indvars;
run;

data score;
set score;
estimate=&depvar.2;
label
estimate='estimate';
run;

data notdot;
set score;
if estimate ne .;

PROC MEANS data=notdot sum noprint; var wt;
output out=samsize (keep=samsize) sum=samsize;
run;

data scoresam (drop=samsize);
set samsize score;
retain n;
if _n_=1 then n=samsize;
if _n_=1 then delete;
run;

PROC SORT data=scoresam; by descending estimate;
run;

data score;
set scoresam;
if estimate ne . then cum_n+wt;
if estimate = . then dec=.;
else dec=floor(cum_n*10/(n+1));
run;

PROC SUMMARY data=score missing;
class dec;
var &depvar wt;
output out=sum_dec sum=sum_can sum_wt;

data sum_dec;
set sum_dec;
avg_can=sum_can/sum_wt;
run;

data avg_rr;
set sum_dec;
if dec=.;
keep avg_can;
run;
```

```
data sum_dec1;
set sum_dec;
if dec=. or dec=10 then delete;
cum_n +sum_wt;
r =sum_can;
cum_r +sum_can;
cum_rr=(cum_r/cum_n)*100;
avg_cann=avg_can*100;
run;

data avg_rr;
set sum_dec1;
if dec=9;
keep avg_can;
avg_can=cum_rr/100;
run;

data scoresam;
set avg_rr sum_dec1;
retain n;
if _n_=1 then n=avg_can;
if _n_=1 then delete;
lift=(cum_rr/n);
if dec=0 then decc=' top ';
if dec=1 then decc=' 2 ';
if dec=2 then decc=' 3 ';
if dec=3 then decc=' 4 ';
if dec=4 then decc=' 5 ';
if dec=5 then decc=' 6 ';
if dec=6 then decc=' 7 ';
if dec=7 then decc=' 8 ';
if dec=8 then decc=' 9 ';
if dec=9 then decc='bottom';
if dec ne .;
run;

PROC SORT data= scoresam; by dec;
PROC SORT data= boot; by dec;
run;

data scoresam_bs;
merge
scoresam boot; by dec;
run;

options label;
title1' ';
title2" samsize_bs=&samsize_bs, n_sampl_bs=&n_sampl_bs ";
```

```
PROC PRINT data=scoresam_bs d split='*' noobs;
var decc sum_wt r avg_cann cum_rr lift lift_bs bs_SE;
label decc='DECILE'
sum_wt ='NUMBER OF*INDIVIDUALS'
r ='NUMBER OF*RESPONDERS'
cum_r ='CUM No. CUSTOMERS w/* RESPONDERS'
avg_cann ='RESPONSE *RATE (%)'
cum_rr ='CUM RESPONSE * RATE (%)'
lift ='C U M*Single-Sample*LIFT (%)'
lift_bs ='C U M*BOOTSTRAP*LIFT (%)'
bs_SE='BOOTSTRAP*MARGIN of*ERROR (80%)';
sum sum_wt r;
format sum_wt r cum_n cum_r comma8.0;
format avg_cann cum_rr 6.2;
format lift lift_bs 3.0;
format bs_SE 5.1;
run;
```

44.6 H-Spread Common Region of Chapter 42

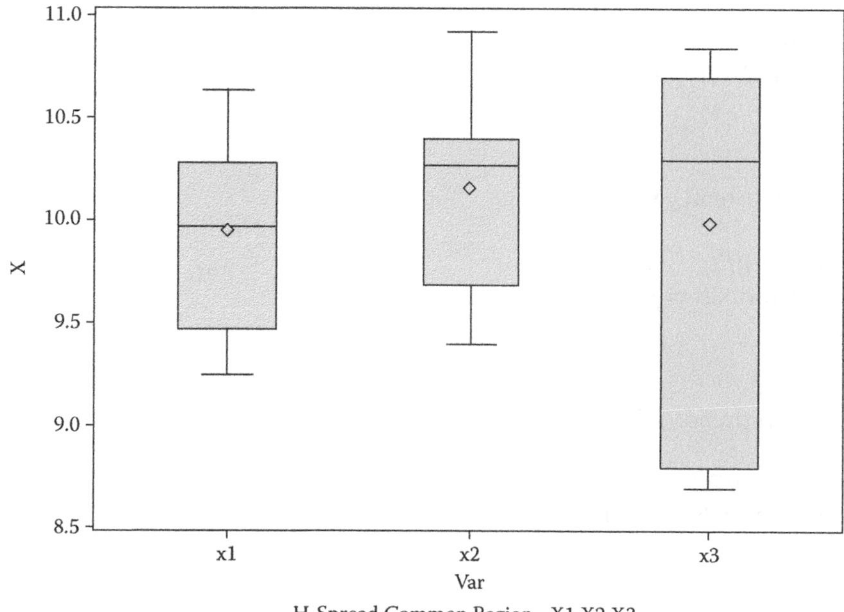

H-Spread Common Region - X1 X2 X3.

```
%let spread=50;
title "H-spread&spread";
```

```
data IN;
call streaminit(12345);

do id=1 to 120;
X1 = RAND('NORMAL',10, 1);
X2 = RAND('NORMAL',10, 1.5);
X3 = RAND('NORMAL',10, 2);
output;
end;
run;

PROC RANK data=IN groups=100 out=OUT;
var X1 X2 X3;
ranks X1r X2r X3r;
run;

PROC PRINT data=OUT;
run;

PROC RANK data=IN groups=100 out=OUT;
var X1-X3;
ranks X1r X2r X3r;
run;

data H_spread&spread._X1;
set out;
rhp=(100-&spread)/2;
if x1r=>(rhp-1) and x1r<=(99-rhp);
keep id x1 x1r;
run;

data H_spread&spread._X2;
set out;
rhp=(100-&spread)/2;
if x2r=>(rhp-1) and x2r<=(99-rhp);
keep id x2 x2r;
run;

data H_spread&spread._X3;
set out;
rhp=(100-&spread)/2;
if x3r=>(rhp-1) and x3r<=(99-rhp);
keep id x3 x3r;
run;

PROC SORT data=H_spread&spread._X1; by id;
PROC SORT data=H_spread&spread._X2; by id;
PROC SORT data=H_spread&spread._X3; by id;
run;
```

```
data H_spread&spread._X1X2X3;
merge
H_spread&spread._X1 (in=var_x1)
H_spread&spread._X2 (in=var_x2)
H_spread&spread._X3 (in=var_x3);
by id;
if var_x1=1 and var_x2=1 and var_x3=1;
run;

PROC MEANS data=H_spread&spread._X1X2X3 mean n;
var X1-X3;
run;

data H_spread&spread._X1;
set H_spread&spread._X1X2X3;
var='x1';
x=x1;
keep id x var;

data H_spread&spread._X2;
set H_spread&spread._X1X2X3;
var='x2';
x=x2;
keep id x var;

data H_spread&spread._X3;
set H_spread&spread._X1X2X3;
var='x3';
x=x3;
keep id x var;

data H_spread&spread._X1X2X3;
set H_spread&spread._X1 H_spread&spread._X2 H_spread&spread._X3;

PROC PRINT;
run;

PROC SORT; by var;
proc print data= H_spread&spread._X1X2X3;
run;

PROC BOXPLOT data=H_spread&spread._X1X2X3;
plot x*var;
run;
```

44.7 Favorite—Proc Corr with Option Rank, Vertical Output (Table 44.3)

TABLE 44.3
Proc Corr with Option Rank, Vertical Output

Rank	Predictor	CorrCoef_ with_TARGET	N	p-value
1	X4	0.27713	9	0.4858
2	X1	0.06562	8	0.8832
3	X2	−0.01183	9	0.9769
4	X3	0.00614	8	0.9891

```
data dat1;
input X1 - X4: 3.0 TARGET 1.0;
cards;
123 234 345 456 1
. 756 . 654 0
234 843 654 867 1
123 234 345 456 0
654 856 534 654 1
234 543 854 867 1
123 834 845 456 0
654 756 534 654 0
234 543 654 867 0
;
run;

PROC PRINT;
run;

data dat1;
set dat1;
wt=1;
run;

PROC CORR data=dat1 rank fisher;
ods output fisherpearsoncorr=out;
var x1-x4;
with target;
freq wt;
run;

ods listing;
PROC PRINT data=out;
run;

data out1;
set out;
abs_corr = abs(corr);
```

```
CorrCoef_with_TARGET=corr;
Predictor=var;
N=nobs;
keep var corr n pvalue abs_corr CorrCoef_with_TARGET Predictor;
run;

PROC SORT data=out1; by descending abs_corr;
run;

data out2;
set out1;
if abs_corr ge .0;
if CorrCoef_with_TARGET = . then delete;

data out2;
set out2;
Rank=_n_;
run;

PROC PRINT data=out2 noobs;
var Rank Predictor CorrCoef_with_TARGET n pvalue;
run;
```

44.8 Favorite—Decile Analysis—Response (Table 44.4)

TABLE 44.4

Decile Analysis—Response

Y (RESPONSE) Regressed on X1 X2 X3					
Decile	**Number of Individuals**	**Number of Responders**	**Response Rate (%)**	**Cum Response Rate (%)**	**Cum Lift (%)**
top	4	4	100.0	100.0	250
2	4	2	50.00	75.00	188
3	4	2	50.00	66.67	167
4	4	1	25.00	56.25	141
5	4	2	50.00	55.00	138
6	4	1	25.00	50.00	125
7	4	1	25.00	46.43	116
8	4	1	25.00	43.75	109
9	4	1	25.00	41.67	104
bottom	4	1	25.00	40.00	100
	40	**16**			

```
%let data_in=IN;
%let depvar=Y;
%let indvars=X1 X2 X3;
```

```
data &data_in;
input &depvar &indvars wt;
cards;
1    63.28405135    -62.89590924    0.31725    1
1    -7.965165127    9.077917498    0.29397    1
1    -40.8721149    41.85990786    0.40705    1
1    108.8084024    -107.6672824    0.25316    1
1    3.071713061    -2.215322147    0.40705    1
1    44.96645653    -44.18664467    0.25316    1
1    2.328170141    -1.89973146    0.24562    1
1    89.08870743    -88.21705972    0.42732    1
1    30.1080088    -29.0253107    0.24562    1
1    -11.14966201    11.97082199    0.25316    1
1    24.6912264    -23.85538734    0.25316    1
1    33.46889223    -32.68556731    0.40705    1
1    51.82377813    -51.4138173    0.40705    1
1    70.28970224    -69.42221865    0.24562    1
1    -95.85890655    97.00002655    0.40705    1
1    77.53692092    -77.19292134    0.26126    1
0    3.309578275    -3.261180349    0.24562    1
0    10.12748375    -9.549172853    0.25316    1
0    -12.88207239    13.97592671    0.29397    1
0    -17.32877567    18.18516658    0.31111    1
0    -70.59773747    71.24695425    0.31111    1
0    43.27915239    -42.13803238    0.24562    1
0    -7.880514668    8.995154718    0.25316    1
0    40.93399103    -40.09173673    0.25316    1
0    81.07550795    -80.35859121    0.24562    1
0    -7.965165127    9.063100546    0.24562    1
0    36.93492473    -35.95553062    0.28211    1
0    23.23610469    -22.80766601    0.25339    1
0    0    1.141120008    0.24562    1
0    0    0.939629385    0.25316    1
0    81.17218438    -80.76633346    0.24562    1
0    21.67949378    -20.97110166    0.24562    1
0    61.36545177    -60.91557128    0.25316    1
0    61.36545177    -60.95549093    0.31725    1
0    77.90838509    -77.58149603    0.28481    1
0    77.90838509    -77.60466917    0.24562    1
0    77.90838509    -77.08023514    0.32738    1
0    16.48495995    -15.88724129    0.40705    1
0    39.74610442    -38.99089853    0.24562    1
0    30.7499237    -29.94045894    0.24562    1
;
run;

PROC LOGISTIC data=&data_in nosimple des outest=coef;
model &depvar = &indvars;
freq wt;
run;
```

```
PROC SCORE data=&data_in predict type=parms score=coef
out=score;
var &indvars;
run;

data score;
set score;
estimate=&depvar.2;
run;

data notdot;
set score;
if estimate ne.;

PROC MEANS data=notdot sum noprint; var wt;
output out=samsize (keep=samsize) sum=samsize;
run;

data scoresam (drop=samsize);
set samsize score;
retain n;
if _n_=1 then n=samsize;
if _n_=1 then delete;
run;

PROC SORT data=scoresam; by descending estimate;
run;

data score;
set scoresam;
if estimate ne . then cum_n+wt;
if estimate = . then dec=.;
else dec=floor(cum_n*10/(n+1));
run;

PROC SUMMARY data=score missing;
class dec;
var &depvar wt;
output out=sum_dec sum=sum_can sum_wt;

data sum_dec;
set sum_dec;
avg_can=sum_can/sum_wt;
run;

data avg_rr;
set sum_dec;
if dec=.;
keep avg_can;
run;
```

```
data sum_dec1;
set sum_dec;
if dec=. or dec=10 then delete;
cum_n +sum_wt;
r =sum_can;
cum_r +sum_can;
cum_rr=(cum_r/cum_n)*100;
avg_cann=avg_can*100;
run;

data avg_rr;
set sum_dec1;
if dec=9;
keep avg_can;
avg_can=cum_rr/100;
run;

data scoresam;
set avg_rr sum_dec1;
retain n;
if _n_=1 then n=avg_can;
if _n_=1 then delete;
lift=(cum_rr/n);
if dec=0 then decc=' top ';
if dec=1 then decc=' 2 ';
if dec=2 then decc=' 3 ';
if dec=3 then decc=' 4 ';
if dec=4 then decc=' 5 ';
if dec=5 then decc=' 6 ';
if dec=6 then decc=' 7 ';
if dec=7 then decc=' 8 ';
if dec=8 then decc=' 9 ';
if dec=9 then decc='bottom';
if dec ne .;
run;

title1 ' ';
title2' Decile Analysis based on ';
title3" &depvar (RESPONSE) regressed on &indvars ";

PROC PRINT data=scoresam d split='*' noobs;
var decc sum_wt r avg_cann cum_rr lift;
label decc='DECILE'
   sum_wt ='NUMBER OF*INDIVIDUALS'
   r ='NUMBER OF*RESPONDERS'
   cum_r ='CUM No. CUSTOMERS w/* RESPONSES'
   avg_cann ='RESPONSE *RATE (%)'
   cum_rr ='CUM RESPONSE * RATE (%)'
   lift =' C U M *LIFT (%)';
sum sum_wt r;
```

```
format sum_wt r cum_n cum_r comma10.;
format avg_cann cum_rr 5.2;
format lift 3.0;
run;
```

44.9 Favorite—Decile Analysis—Profit (Table 44.5)

TABLE 44.5
Decile Analysis—Profit

Y (PROFIT) Regressed on X1 X2					
Decile	**Number of Customers**	**Total Profit**	**Decile Mean Profit**	**Decile Cum Profit**	**Cum Lift**
top	2	$26,114.11	$13,057.06	$13,057.06	401
2	2	$24,014.11	$12,007.06	$12,532.06	385
3	2	$2,896.50	$1,448.25	$8,837.45	272
4	2	$3,265.30	$1,632.65	$7,036.25	216
5	2	$3,170.38	$1,585.19	$5,946.04	183
6	2	$1,716.88	$858.44	$5,098.11	157
7	2	$1,237.14	$618.57	$4,458.17	137
8	2	$1,526.48	$763.24	$3,996.31	123
9	2	$595.96	$297.98	$3,585.38	110
bottom	2	$541.96	$270.98	$3,253.94	100
	20	**$65,078.82**			

```
%let data_in=IN;
%let depvar=Y;
%let indvars=X1 X2;

data &data_in;
input &indvars &depvar wt;
cards;
1 0.64417 14212.99 1
0 0.05839 908.11 1
0 0.06754 538.77 1
1 0.21690 1548.25 1
1 0.50600 11701.12 1
0 0.02847 13.70 1
0 0.26161 1575.19 1
0 0.29051 1602.65 1
0 0.04119 528.26 1
0 0.05310 618.37 1
1 0.44417 12312.99 1
```

```
0 0.06839 978.11 1
0 0.07754 738.77 1
1 0.31690 1348.25 1
1 0.51600 11901.12 1
0 0.04847 17.70 1
0 0.28161 1595.19 1
0 0.31051 1662.65 1
0 0.05119 578.26 1
0 0.06310 698.37 1
;
run;

PROC REG data=&data_in outest=coeff;
estimate:model Y = &indvars;
run;

PROC SCORE data=&data_in
out=score (keep= wt estimate Y )
predict SCORE=coeff TYPE=PARMS;
var&indvars;
run;

data notdot;
set score;
if estimate ne.;

PROC MEANS data=notdot sum noprint; var wt;
output out=samsize (keep=samsize) sum=samsize;
run;

data scoresam (drop=samsize);
set samsize score;
retain n;
if _n_=1 then n=samsize;
if _n_=1 then delete;
run;

PROC SORT data=scoresam; by descending estimate;
run;

data score;
set scoresam;
if estimate ne . then cum_n+wt;
if estimate = . then dec=.;
else dec=floor(cum_n*10/(n+1));
run;
```

```
PROC SUMMARY Data=Score missing;
class dec;
var Y wt;
output out=sum_dec sum=sum_Y sum_wt;

data sum_dec;
set sum_dec;
avg_Y=sum_Y/sum_wt;
run;

data avg_fix;
set sum_dec;
if dec ne .;
keep sum_Y sum_wt;

PROC SUMMARY data=avg_fix;
var sum_Y sum_wt;
output out=fix_dec sum=num_Y tot_cus;
run;

data avg_ss;
set fix_dec;
avg_Y=num_Y/tot_cus;
keep avg_Y;
run;

data sum_dec1;
set sum_dec;
if dec=. or dec=10 then delete;
cum_n +sum_wt;
s =sum_Y;
cum_s +sum_Y;
cum_ss=(cum_s/cum_n);
avg_Ys=avg_Y;
run;

data scoresam;
set avg_ss sum_dec1;
retain n;
if _n_=1 then n=avg_Y;
if _n_=1 then delete;
lift=(cum_ss/n)*100;
if dec=0 then decc=' top';
if dec=1 then decc=' 2 ';
if dec=2 then decc=' 3 ';
if dec=3 then decc=' 4 ';
```

```
if dec=4 then decc=' 5 ';
if dec=5 then decc=' 6 ';
if dec=6 then decc=' 7 ';
if dec=7 then decc=' 8 ';
if dec=8 then decc=' 9 ';
if dec=9 then decc='bottom';
if dec ne .;
run;

title1 ' ';
title2' Decile Analysis based on ';
title3" &depvar (PROFIT) regressed on &indvars ";

PROC PRINT data=scoresam d split='*' noobs;
var decc sum_wt s avg_Ys cum_ss lift;
label decc='DECILE'
sum_wt ='NUMBER OF*CUSTOMERS'
s ='TOTAL*PROFIT'
cum_s ='CUM No. CUSTOMERS w/* PROFIT'
avg_Ys =' DECILE* MEAN PROFIT'
cum_ss ='DECILE* CUM PROFIT'
lift =' C U M * LIFT ';
sum sum_wt s;
format s dollar14.2;
format sum_wt cum_n cum_s comma10.;
format avg_Ys cum_ss dollar10.2;
format lift 3.0;
run;
```

44.10 Favorite—Smoothing Time-Series Data (Running Medians of Three) (Table 44.6)

TABLE 44.6

Smoothing Time-Series Data

				Dataset IN					
TS X1	X2	X3	X4	X5	X6	X7	X8	X9	X10
TS 23	45	36	57	65	19	29	44	33	56
				Smooth Sequence of Xs					
smmed1	smmed2	smmed3	smmed4	smmed5	smmed6	smmed7	smmed8	smmed9	smmed10
45	36	45	57	57	29	29	33	44	33

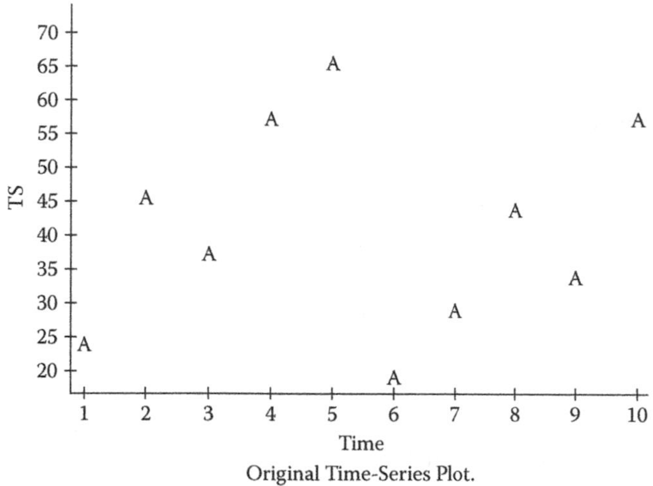

Original Time-Series Plot.

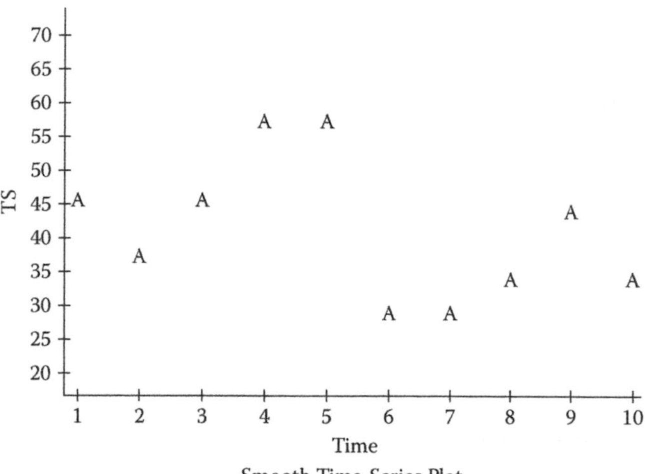

Smooth Time-Series Plot.

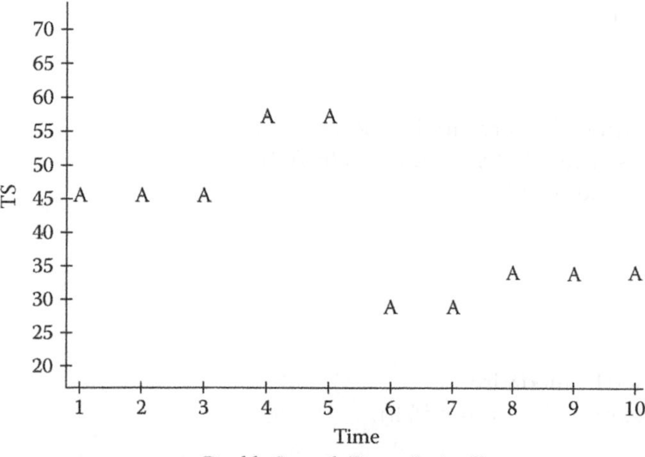

Double Smooth Times-Series Plot.

```
%let ln=10;

data IN;
input TS $2. X1 – X&ln;
cards;
TS 23 45 36 57 65 19 29 44 33 56
;
run;

PROC PRINT;
title2' dataset IN ';
run;

PROC TRANSPOSE data=IN out=tposed;
id TS;
var X1-X&ln;
run;

PROC PRINT data=tposed;
title2 ' dataset tposed ';
run;

data tposed;
set tposed;
Time+1;
run;

PROC PRINT;
title2 ' dataset tposed with Time';
run;

PROC PRINT data=IN;
run;

PROC PLOT data=tposed vpercent=90 hpercent=90;
plot TS *time/ haxis=1 to &ln by 1 vaxis=20 to 70 by 5;
title3' Original Time-Series Plot';
run;

data Medians_of_3;
set IN;
array d{*} X1-X&ln;
array med [&ln] med1 - med&ln;
array smmed[&ln] smmed1 - smmed&ln;
do i=1 to dim(d);
med{i}=.;
if i <=dim(d)-2 then do;
```

```
med{i} = median(d{i},d{i+1},d{i+2});
smmed(i+1) = med(i);
end;
end;
smmed(1) = median( d(2), d(3), (3*d(2) -2*d(3)) );
smmed(dim(d))= median( d(dim(d)), d(dim(d)-1), (3*d(dim(d)-1) -2*d(dim(d))));
drop i;
run;


PROC PRINT;
var smmed1-smmed&ln;
title2 ' Smooth Sequence of Xs ';
run;


PROC TRANSPOSE data=Medians_of_3 out=tposed;
id TS;
var smmed1 - smmed&ln;
run;


PROC PRINT data=tposed;
title2 ' dataset sm tposed ';
run;


data tposed;
set tposed;
Time+1;
run;


PROC PRINT;
title2 ' dataset sm tposed with Time';
run;


PROC PLOT data=tposed vpercent=90 hpercent=90;
plot TS *time/ haxis=1 to &ln by 1 vaxis=20 to 70 by 5;
title3' Smooth Time-Series Plot ';
run;

%let ln=10;

data IN;
input  TS $2. X1 - X&ln;
cards;
TS  36 45 45 57 57 29 29 33 33 44
;
run;
```

```
PROC PRINT;
title2' dataset IN ';
run;

data Medians_of_3;
set IN;
array d{*} X1-X&ln;
array med [&ln] med1 - med&ln;
array smmed[&ln] smmed1 - smmed&ln;
do i=1 to dim(d);
med{i}=.;
if i <=dim(d)-2 then do;
med{i} = median(d{i},d{i+1},d{i+2});
smmed(i+1) = med(i);
end;
end;
smmed(1) = median( d(2), d(3), (3*d(2) -2*d(3)) );
smmed(dim(d))= median( d(dim(d)), d(dim(d)-1), (3*d(dim(d)-1) -2*d(dim(d))));
drop i j;
run;

PROC PRINT;
var
smmed1-smmed&ln;
title2 ' Smooth Sequence of Xs ';
run;

PROC TRANSPOSE data=Medians_of_3 out=tposed;
id TS;
var smmed1 - smmed&ln;
run;

PROC PRINT data=tposed;
title2 ' dataset  sm tposed ';
run;

data tposed;
set tposed;
Time+1;
run;

PROC PRINT;
title2 ' dataset sm tposed with Time';
run;

PROC PLOT data=tposed vpercent=90 hpercent=90;
plot TS *time/ haxis=1 to &ln by 1 vaxis=20 to 70 by 5;
title3' Double Smooth Time-Series Plot ';
run;
quit;
```

44.11 Favorite—First Cut Is the Deepest—Among Variables with Large Skew Values

Whether using big data, small data, statistical methods, or machine-learning techniques, good data practice is cutting, from a list of variables, those whose skewness values are relatively extreme. Adhering to strict statistics, "extreme" for skewness means outside the open interval (–2, +2). In practical statistics, extreme is relative to the observed skewness values. In Figure 44.10, I would first delete the top five largest variables, X19, X20, X18, X21, and X22. Then, proceed with the analysis. If results are not acceptable, then delete variables further down the list. Repeat deletion until results are good (Table 44.7).

TABLE 44.7
Variables Ranked by Skew

Obs	Variable	Skew
1	X19	34.0829
2	X20	20.6845
3	X18	16.6941
4	X21	14.4500
5	X22	11.2846
6	X23	9.5358
7	X5	3.3244
8	X8	2.9486
9	X7	2.8639
10	X6	2.8225
11	X4	2.7528
12	X3	2.7430
13	X1	1.3352
14	X16	1.0474
15	X15	1.0149
16	X9	1.0103
17	X10	0.9850
18	X17	0.9355
19	X14	0.8487
20	X13	0.7779
21	X2	0.7630
22	X12	0.7158
23	X11	–0.0225
24	**X24**	**–0.4209**

```
PROC MEANS data=IN skew;
var X1-X24;
output out=skews skew=;
run;

PROC PRINT data=skews;
run;

PROC TRANSPOSE data=skews
out=Tskews (rename=(Col1=Skew _NAME_=Variable));
var X1-X24;
run;

PROC SORT data=Tskews; by descending Skew;
run;

PROC PRINT data=Tskews;
run;
```

Index

N

O

P

Q

R

T

U

V

W

Z